AQA A-level

CHISLEH

KU-395-523

Mathematics

For A-level Year 1 and AS

1

Authors

Sophie Goldie

Val Hanrahan

Cath Moore

Jean-Paul Muscat

Susan Whitehouse

Series editors

Roger Porkess

Catherine Berry

Consultant editor

Heather Davis

HODDER
EDUCATION

AN HACHETTE UK COMPANY

Hachette UK's policy is to use papers that are natural, renewable and recyclable products and made from wood grown in sustainable forests. The logging and manufacturing processes are expected to conform to the environmental regulations of the country of origin.

Orders: please contact Bookpoint Ltd, 130 Park Drive, Milton Park, Abingdon, Oxon OX14 4SE. Telephone: (44) 01235 827720. Fax: (44) 01235 400454. Email education@bookpoint.co.uk Lines are open from 9 a.m. to 5 p.m., Monday to Saturday, with a 24-hour message answering service. You can also order through our website: www.hoddereducation.co.uk

ISBN: 978 1 4718 5286 2

© Sophie Goldie, Val Hanrahan, Jean-Paul Muscat, Roger Porkess, Susan Whitehouse and MEI 2017

First published in 2017 by
Hodder Education,
An Hachette UK Company
Carmelite House
50 Victoria Embankment
London EC4Y 0DZ
www.hoddereducation.co.uk

Impression number   10 9 8 7 6 5 4 3 2 1
Year   2021   2020   2019   2018   2017

Cover photo © Tim Gainey/Alamy Stock Photo

Typeset in Bembo Std, 11/13 pts. by Aptara®, Inc.

Printed in Italy

A catalogue record for this title is available from the British Library.

Contents

*Please note that the marks stated on the example questions are to be used as a guideline only, AQA have not reviewed and approved the marks.

Contents

Mathematics is not only a beautiful and exciting subject in its own right but also one that underpins many other branches of learning. It is consequently fundamental to our national wellbeing.

This book covers the content of AS Mathematics and so provides a complete course for the first of the two years of Advanced Level study. The requirements of the second year are met in a second book.

Between 2014 and 2016 A-level Mathematics and Further Mathematics were very substantially revised, for first teaching in 2017. Major changes include increased emphasis on

- Problem solving
- Proof
- Use of ICT
- Modelling
- Working with large data sets in statistics.

This book embraces these ideas. Chapter 1 is on **problem solving** and this theme is continued throughout the book with several spreads based on the problem solving cycle. In addition a large number of exercise questions involve elements of problem solving; these are identified by the **PS** icon beside them. The ideas of **mathematical proof** and rigorous logical argument are also introduced in Chapter 1 and are then involved in suitable exercise questions throughout the book. The same is true of **modelling**; the modelling cycle is introduced in the first chapter and the ideas are reinforced through the rest of the book. Questions which involve an element of modelling are identified by the **M** icon.

The use of **technology**, including graphing software, spreadsheets and high specification calculators, is encouraged wherever possible, for example in the Activities used to introduce some of the topics in Pure mathematics, and particularly in the analysis and processing of **large data sets** in Statistics. A large data set is provided at the end of the book but this is essentially only for reference. It is also available online as a spreadsheet (www.hoddereducation.co.uk/AQAMathsYear1) and it is in this form that readers are expected to store and work on this data set, including answering the exercise questions that are based on it. Places where ICT can be used are highlighted by a **T** icon.

Throughout the book the emphasis is on understanding and interpretation rather than mere routine calculations, but the various exercises do nonetheless provide plenty of scope for practising basic techniques. The exercise questions are split into three bands. Band 1 questions (indicated by a green bar) are designed to reinforce basic understanding, while most exercises precede these with one or two questions designed to help students bridge the gap between GCSE and AS Mathematics; these questions are signposted by a 🌉 icon. These include a 'thinking' question which addresses a key stumbling block in the topic and a multiple choice question to test key misconceptions. Band 2 questions (yellow bar) are broadly typical of what might be expected in an examination: some of them cover routine techniques; others are design to provide some stretch and challenge for readers. Band 3 questions (red bar) explore round the topic and some of them are rather more demanding. In addition, extensive online support, including further questions, is available by subscription to MEI's Integral website, http://integralmaths.org. (Please note that these external links are not being entered in an AQA approval process.)

In addition to the exercise questions, there are five sets of questions, called Practice questions, covering groups of chapters. All of these sets include identified questions requiring **problem solving** **PS**, **mathematical proof** **MP**, **use of ICT** **T** and **modelling** **M**.

The book is written on the assumption that readers have been successful in GCSE Mathematics, or its equivalent, and are reasonably confident and competent with that level of mathematics. There are places where the work depends on knowledge from earlier in the book and this is flagged up in the margin in Prior knowledge boxes. This should be seen as an invitation to those who have problems with the particular topic to revisit it earlier in book. At the end of each chapter there is a summary of the new knowledge that readers should have gained.

Two common features of the book are Activities and Discussion points. These serve rather different purposes. The Activities are designed to help readers get into the thought processes of the new work that they are about to meet; having done an Activity, what follows will seem much easier. The Discussion points invite readers to talk about particular points with their fellow students and their teacher and so enhance their understanding. Another feature is a Caution icon ❗, highlighting points where it is easy to go wrong.

The authors have taken considerable care to ensure that the mathematical vocabulary and notation are used correctly in this book, including those for variance and standard deviation, as defined in the AQA specification for AS Level in Mathematics. In the paragraph on notation for sample variance and sample standard deviation (page 344), it explains that the meanings of 'sample variance', denoted by s^2, and 'sample standard deviation', denoted by s, are defined to be calculated with divisor $(n-1)$. In early work in statistics it is common practice to introduce these concepts with divisor n rather than $(n-1)$. However there is no recognised notation to denote the quantities so derived. Students should be aware of the variations in notation used by manufacturers on calculators and know what the symbols on their particular models represent.

Answers to all exercise questions and practice questions are provided at the back of the book, and also online at www.hoddereducation.co.uk/AQAMathsYear1. Full step-by-step worked solutions to all of the practice questions are available online at www.hoddereducation.co.uk/AQAMathsYear1. All answers are also available on Hodder Education's Dynamic Learning platform. (Please note that these additional links have not been entered into the AQA approval process.)

Finally a word of caution. This book covers the content of AS Level Mathematics and is designed to help provide readers with the skills and knowledge for the examination. However, it is not the same as the specification, which is where the detailed examination requirements are set out. So, for example, the book uses a data set about cycling accidents to give readers experience of working with a large data set. Examination questions will test similar ideas but they will be based on different data sets; for more information about these sets readers should consult the specification. Similarly, in the book cumulative binomial tables are used in the explanation of the output from a calculator, but such tables will not be available in examinations. Individual specifications will also make it clear how standard deviation is expected to be calculated. So, when preparing for the examination, it is essential to check the specification.

Catherine Berry

Roger Porkess

Prior knowledge

This book builds on GCSE work, much of which is assumed knowledge.

The order of the chapters has been designed to allow later ones to use and build on work in earlier chapters. The list below identifies cases where the dependency is particularly strong.

The Statistics and Mechanics chapters are placed in separate sections of the book for easy reference, but it is expected that these will be studied alongside the Pure mathematics work rather than after it.

- The work in **Chapter 1: Problem solving** pervades the whole book

- **Chapter 3: Quadratic equations** and graphs requires some manipulation of surds (chapter 2)

- **Chapter 4: Equations and inequalities** uses work on solving quadratic equations (chapter 3)

- **Chapter 5: Coordinate geometry** requires the use of quadratic equations (chapter 3) and simultaneous equations (chapter 4)

- **Chapter 6: Trigonometry** requires some use of surds (chapter 2) and quadratic equations (chapter 3)

- **Chapter 7: Polynomials** builds on the work on quadratic equations (chapter 3)

- **Chapter 8: Graphs and transformations** brings together work on quadratic graphs (chapter 3), trigonometric graphs (chapter 6) and polynomial graphs (chapter 7)

- **Chapter 9: The binomial expansion** builds on polynomials (chapter 7)

- **Chapter 10: Differentiation** draws on a number of techniques, including work on indices (chapter 2), quadratic equations (chapter 3), coordinate geometry (chapter 5) and polynomial graphs (chapter 8)

- **Chapter 11: Integration** follows on from differentiation (chapter 10)

- **Chapter 12: Vectors** builds on coordinate geometry (chapter 5)

- **Chapter 13: Logarithms and exponentials** builds on work on indices (chapter 2)

- **Chapter 15: Data processing, presentation and interpretation** follows on from data collection (chapter 14)

- **Chapter 17: The binomial distribution** draws on ideas from probability (chapter 16) and the binomial expansion (chapter 9)

- **Chapter 18: Hypothesis testing** uses ideas from probability (chapter 16) and the binomial distribution (chapter 17)

- **Chapter 19: Kinematics** requires fluency with quadratic equations (chapter 3) and simultaneous equations (chapter 4)

- **Chapter 20: Forces** ties in with work on vectors (chapter 12), although these two chapters could be covered in either order

- **Chapter 21: Variable acceleration** uses differentiation (chapter 10) and integration (chapter 11).

Acknowledgements

The Publishers would like to thank the following for permission to reproduce copyright material.

Questions from past AS and A Level Mathematics papers are reproduced by permission of MEI and OCR. Question 5 on page 322 is taken from OCR, Core Mathematics Specimen Paper H867/02, 2015. The answer on page 541 is also reproduced by permission of OCR.

Practice questions have been provided by Chris Little (p288–290), Neil Sheldon (p399–402), Rose Jewell (p487–489), and MEI (p96–98, p186–189).

p.309 The smoking epidemic-counting the cost, HEA, 1991: Health Education Authority, reproduced under the NICE Open Content Licence: www.nice.org.uk/Media/Default/About/Reusing-our-content/Open-content-licence/NICE-UK-Open-Content-Licence-.pdf; **p.310** Young People Not in Education, Employment or Training (NEET): February 2016, reproduced under the Open Government Licence www.nationalarchives.gov.uk/doc/open-government-licence/version/3/; **p.334** The World Bank: Mobile cellular subscriptions (per 100 people): http://data.worldbank.org/indicator/IT.CEL.SETS.P2; **p.340** Historical monthly data for meteorological stations: https://data.gov.uk/dataset/historic-monthly-meteorological-station-data, reproduced under the Open Government Licence www.nationalarchives.gov.uk/doc/open-government-licence/version/3/; **p.341** Table 15.26 (no.s of homicides in England & Wales at the start and end of C20th): https://www.gov.uk/government/statistics/historical-crime-data, reproduced under the Open Government Licence www.nationalarchives.gov.uk/doc/open-government-licence/version/3/; **p.362** Environment Agency: Risk of flooding from rivers and the sea, https://flood-warning-information.service.gov.uk/long-term-flood-risk/map?map=RiversOrSea, reproduced under the Open Government Licence www.nationalarchives.gov.uk/doc/open-government-licence/version/3/

Photo credits

p.1 © Kittipong Faengsrikum/Demotix/Press association Images; **p.19** © Randy Duchaine/Alamy Stock Photo; **p.32** © Gaby Kooijman/123RF.com; **p.53** © StockbrokerXtra/Alamy Stock Photo; **p.65** © polifoto/123RF.com; **p.99** (top) © Sakarin Sawasdinaka/123RF.com; **p.99** (lower) © Rico Koedder/123RF.com; **p.130** © Edward R. Pressman Film/The Kobal Collection; **p.148** © ianwool/123RF.com; **p.174** © ullsteinbild/TopFoto; **p.180** © Emma Lee/Alamy Stock Photo; **p.190** © Bastos/Fotolia; **p.229** © NASA; **p.247** © Graham Moore/123RF.com; **p.264** © BioPhoto Associates/Science Photo Library; **p.291** (top) © Jack Sullivan/Alamy Stock Photo; **p.291** (lower) © Ludmila Smite/Fotolia; **p.297** © arekmalang/Fotolia; **p.350** (top) © molekuul/123.com; **p.350** (lower) © Wavebreak Media Ltd/123RF.com; **p.372** (top) ©Tom Grundy/123RF.com; **p.372** (lower) © George Dolgikh/Fotolia; **p.383** © Fotoatelie/Shutterstock; **p.403** ©Tan Kian Khoon/Fotolia; **p.406** ©Volodymyr Vytiahlovskyi/123RF.com; **p.434** © Stocktrek Images, Inc./Alamy Stock Photo; **p.437** © Dr Jeremy Burgess/Science Photo Library; **p.438** © Herbert Kratky/123RF.com; **p.440** © Matthew Ashmore/Stockimo/Alamy Stock Photo; **p.445** © NASA; **p.450** ©V Kilian/Mauritius/Superstock; **p.452** © scanrail/123RF.com; **p.472** © FABRICE COFFRINI/AFP/GettyImages; **p.484** © Peter Bernik/123RF.com.

Every effort has been made to trace all copyright holders, but if any have been inadvertently overlooked, the Publishers will be pleased to make the necessary arrangements at the first opportunity.

1

Problem solving

The authorities of a team sport are planning to hold a World Cup competition. They need to decide how many teams will come to the host country to compete in the 'World Cup finals'.

The World Cup finals will start with a number of groups. In a group, each team plays every other team once. One or more from each group will qualify for the next stage.

The next stage is a knock-out competition from which one team will emerge as the world champions.

Every team must be guaranteed at least three matches to make it financially viable for them to take part.

To ensure a suitable length of competition, the winners must play exactly seven matches.

→   How many teams can take part in the World Cup finals?

→   What general rules apply to the number of teams in a competition like this, involving groups and then a knock-out stage?

1

1 Solving problems

Mathematics is all about solving problems. Sometimes the problems are 'real-life' situations, such as the 'World Cup finals' problem on the previous page. In other cases, the problems are purely mathematical.

The problem solving cycle

One common approach to solving problems is shown in Figure 1.1. It is called the **problem solving cycle**.

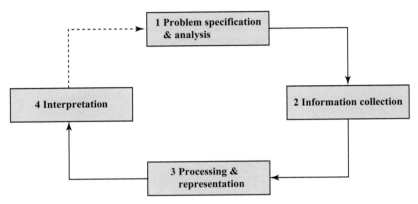

Figure 1.1 The problem solving cycle

In the **Problem specification and analysis** stage, you need to formulate the problem in a way which allows mathematical methods to be used. You then need to analyse the problem and plan how to go about solving it. Often, the plan will involve the collection of information in some form. The information may already be available or it may be necessary to carry out some form of experimental or investigational work to gather it.

In the World Cup finals problem, the specification is given.

> • All teams play at least 3 matches
> • The winner plays exactly 7 matches

Figure 1.2 World Cup finals specification

You need to analyse the problem by deciding on the important variables which you need to investigate as shown in Figure 1.3.

> • Number of teams
> • Number of groups
> • Number to qualify from each group

Figure 1.3 Variables to investigate

You also need to plan what to do next. In this case, you would probably decide to try out some examples and see what works.

In the **Information collection** stage, you might need to carry out an experiment or collect some data. In the World Cup finals problem, you do not need to collect data, but you will try out some possibilities and then draw on your experience from the experimentation, as shown in Figure 1.4.

> Try 6 groups with 2 teams qualifying from each group
> So there are 12 teams at the start of the knock-out stage
> 6 teams in the following round
> 3 teams in the following round
> – doesn't work!
>
> Try 2 groups with 2 qualifying from each group
> So there are 4 teams at the start of the knock-out stage
> 2 teams in the following round, so that must be the final.
> – works

Figure 1.4

In the **Processing and representation** stage, you will use suitable mathematical techniques, such as calculations, graphs or diagrams, in order to make sense of the information collected in the previous stage.

In the World Cup finals problem, you might draw a diagram showing how the tournament progresses in each case.

It is often best to start with a simple case. In this problem, Figure 1.5 shows just having semi-finals and a final in the knock-out stage.

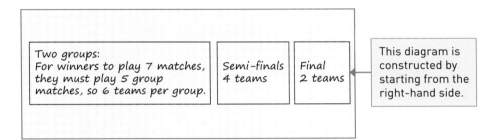

Figure 1.5

Discussion points

➜ Are there any other patterns?

➜ Find and illustrate some more solutions that work.

One possible solution to the World Cup finals problem has now been found: this solution is that there are twelve teams, divided into two groups of six each.

You may have noticed that for the solutions that do work, the number of groups has to be 2, 4, 8, 16, . . . (so that you have a suitable number of teams in the knock-out stage), and that the number of teams in each group must be at least four (so that all teams play at least three matches).

In the **Interpretation** stage, you should report on the solutions to the problem in a way which relates to the original situation. You should also reflect on your solutions to decide whether they are satisfactory. For many sports, twelve teams would not be considered enough to take part in World Cup finals, but this solution could be appropriate for a sport which is not played to a high level in many countries. At this stage of the problem solving process, you might need to return to the problem specification stage and gather further information. In this case, you would need to know more about the sport and the number of teams who might wish to enter.

Discussion points

- ➜ What other possible solutions to the World Cup finals problem are there?
- ➜ Are any of them unsuitable?
- ➜ How do you know if you have found all the possible solutions?
- ➜ What do you think would be the best solution if the sport were football?
- ➜ What about rugby, cricket, baseball and netball?
- ➜ What criteria have you used to decide?

Using algebra

In many problems, using algebra is helpful in formulating and solving a problem.

Example 1.1

OAB is a 60° sector of a circle of radius 12 cm. A complete circle, centre Q, touches OA, OB and the arc AB.

Find the radius of the circle with centre Q.

Solution

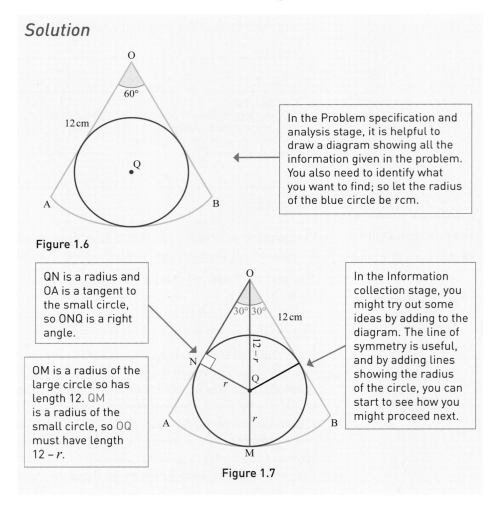

In the Problem specification and analysis stage, it is helpful to draw a diagram showing all the information given in the problem. You also need to identify what you want to find; so let the radius of the blue circle be r cm.

Figure 1.6

QN is a radius and OA is a tangent to the small circle, so ONQ is a right angle.

OM is a radius of the large circle so has length 12. QM is a radius of the small circle, so OQ must have length $12 - r$.

In the Information collection stage, you might try out some ideas by adding to the diagram. The line of symmetry is useful, and by adding lines showing the radius of the circle, you can start to see how you might proceed next.

Figure 1.7

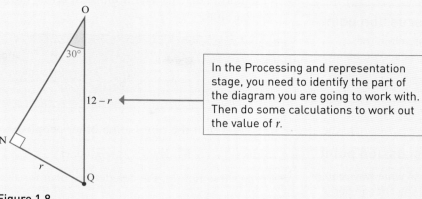

In the Processing and representation stage, you need to identify the part of the diagram you are going to work with. Then do some calculations to work out the value of r.

Figure 1.8

ONQ is a right-angled triangle. The angle at O is 30°, the opposite side has length r, and the hypotenuse has length $12 - r$.

$$\sin 30° = \frac{r}{12 - r}$$

$$0.5(12 - r) = r$$

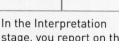

$\sin 30° = 0.5$

$$12 - r = 2r$$

$$12 = 3r$$

$$r = 4$$

The radius of the circle centre Q is 4 cm.

In the Interpretation stage, you report on the solution in terms of the original problem.

Make sure you consider whether your answer is sensible. You can see from Figure 1.7 that the diameter of the circle centre Q must be less than 12 cm, so a radius of 4 cm (giving diameter 8 cm) is sensible. To check this, try drawing the diagram for yourself.

Discussion points

➜ What would the value of r be if the sector angle was 120° instead of 60°?

➜ What about other angles?

The modelling cycle

Another approach to problem-solving is described as modelling. You use it when you are trying to get at the mathematics underlying real-life situations. Here is a simple example to show you the process (often, the problem is more complicated).

Katie is planning to walk from Land's End to John O'Groats to raise money for charity. She wants to know how long it will take her. The distance is 874 miles.

This is the first stage of her planning calculation.

> I can walk at 4 mph.
>
> Distance = speed × time
>
> 874 = 4 × time
>
> Time = 219 hours
>
> $\frac{219}{24}$ = 9.1 . . .
>
> So it will take 10 days.

Katie knows that 10 days is unrealistic. She changes the last part of her calculation to

$$\frac{219}{8} = 27.375...$$

So it will take 28 days.

Katie finds out the times some other people have taken on the same walk:

45 days   51 days   70 days   84 days

Seeing these figures, she decides her plan is still unrealistic.

Look at the flow chart showing the modelling process (Figure 1.9). Identify on the chart the different stages of Katie's work.

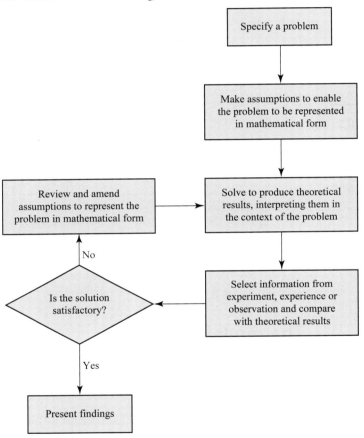

Figure 1.9 Modelling flow chart

Exercise 1.1

 ① Match the description with the equivalent expression.

multiple of 3	$x + 7$
odd number	$3n$
7 more than x	$2y$
twice as big as y	$2n + 1$

 ② A rectangle has length x and width 3 less than x. Which expression represents its area?

A $2x + 3$ B $x(x - 3)$
C $4x - 6$ D $x^2 + 3x$

 ③ Beth and Polly are twins and their sister Louise is 2 years older than them. The total of their ages is 32 years.

What are the ages of the three girls?

④ Anya has 400 metres of fencing to create a paddock for some ponies. She wants the paddock to be twice as long as it is wide.

What will the area of the paddock be?

⑤ A train has 8 coaches, some of which are first class coaches and the rest are standard class coaches. A first class coach seats 48 passengers, a standard class 64.

The train has a seating capacity of 480.

How many standard class coaches does the train have?

⑥ Karim buys 18 kg of potatoes. Some of these are old potatoes at 22p per kilogram, the rest are new ones at 36p per kilogram.

Karim pays with a £5 note and receives 20p change.

What weight of new potatoes did he buy?

⑦ Decode this message received by Space Defence Control.

MKW PWCCFQ USFGWJOWWM LU FSSARFGKLQC SOWFUW UWQH KWOS XW

QWWH PRAW CFOFGMLG ABQFDRBMU XLMK KLCK USWWH OFUWA CBQU

⑧ In a multiple choice test of 25 questions, four marks are given for each correct answer and two marks are deducted for each wrong answer. One mark is deducted for any question which is not attempted.

James scores 55 marks and wants to know how many questions he got right. He can't remember how many questions he did not attempt, but he doesn't think it was very many.

How many questions did James get right?

⑨ In Figure 1.10, there are 6 vertices and 15 different lines. A diagram like this is called a mystic rose.

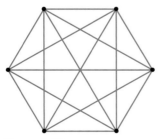

Figure 1.10

Another mystic rose has 153 different lines. How many vertices does it have?

⑩ Priya works in the scheduling department of a railway. She wants to find a model for the time t minutes taken for a journey of k km involving s stops along the way. She finds that each stop adds about 5 minutes to the journey (allowing for slowing down before the station and speeding up afterwards, as well as the time spent at the station).

(i) Priya writes down this model for the journey time:

$$t = 5s + \tfrac{1}{2}k$$

According to Priya's model, what would be the time for a journey of 200 km with two stops on the way?

(ii) Do you think that Priya's model is realistic? Give reasons for your answer.

(iii) After obtaining more data, Priya finds that a better model is

$$t = 3 + 5s + \tfrac{1}{2}k$$

A train sets out at 13:45 and arrives at its destination 240 km away at 16:25. Using Priya's second model, how many stops do you think it made?

How confident can you be about your answer?

⑪ A printer quotes the cost £C of printing n business cards as

$$C = 25 + 0.05n$$

(i) Find the cost of printing
   (a) 100 cards
   (b) 1000 cards

(ii) The printer offers a different formula for large numbers of cards:

$$C = 50 + 0.03n$$

How many cards need to be ordered for the second formula to be cheaper than the first?

(iii) The printer wants to offer a third price structure that would work out cheaper than the second if over 2000 cards are ordered.

Suggest a possible formula for this price structure.

⑫ When the fraction $\frac{2}{11}$ is written as a decimal the answer is 0.1818 recurring. The length of the recurring pattern is 2. What can you say about the length of the recurring pattern for $\frac{q}{p}$ where q is less than p?

⑬ At a supermarket till you have the choice of joining a queue with a small number of people (e.g. 2) each with a lot of shopping, or a large number of people (e.g. 5) each with a small amount of shopping.

Construct a model to help you decide which you should choose.

2 Writing mathematics

The symbol $\Rightarrow$ means **leads to** or **implies** and is very helpful when you want to present an argument logically, step by step.

Look at Figure 1.11.

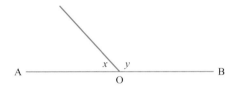

Figure 1.11

You can say

AOB is a straight line $\Rightarrow \angle x + \angle y = 180°$.

Another way of expressing the same idea is to use the symbol $\therefore$ which means **therefore.**

AOB is a straight line $\therefore \angle x + \angle y = 180°$.

A third way of writing the same thing is to use the words *if . . . then . . .*

If AOB is a straight line, *then* $\angle x + \angle y = 180°$.

You can write the symbol $\Rightarrow$ the other way round, as $\Leftarrow$. In that case it means 'is implied by' or 'follows from'.

In the example above you can write

AOB is a straight line $\Leftarrow \angle x + \angle y = 180°$.

and in this case it is still true.

> ### Discussion point
> → Is the statement
>
> AOB is a straight line $\Leftarrow \angle x + \angle y = 180°$
>
> logically the same as
>
> $\angle x + \angle y = 180°$ $\Rightarrow$ AOB is a straight line?

 Note

In situations like this where both the symbols $\Rightarrow$ and $\Leftarrow$ give true statements, the two symbols are written together as $\Leftrightarrow$. You can read this as 'implies and is implied by' or 'is equivalent to'.

Thus

AOB is a straight line $\Leftrightarrow \angle x + \angle y = 180°$.

In the next two examples, the implication is true in one direction but not in the other.

Example 1.2

Write one of the symbols ⇒, ⇐ and ⇔ between the two statements A and B.

A: Nimrod is a cat.

B: Nimrod has whiskers.

Solution

You can write A ⇒ B. Since all cats have whiskers and Nimrod is a cat, it must be true that Nimrod has whiskers.

However, you cannot write A ⇐ B. That is saying 'Nimrod has whiskers', implies 'Nimrod is a cat'. However, there are other sorts of animals that have whiskers, so you cannot conclude that Nimrod is a cat.

Nor can you write A ⇔ B because that is saying that *both* A ⇒ B and A ⇐ B are true; in fact only the first is true.

Example 1.3

Write one of the symbols ⇒, ⇐ or ⇔ between the two statements A and B.

A: The hands of the clock are at right angles.

B: The clock is showing 3 o'clock.

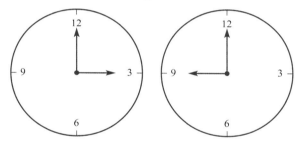

Figure 1.12

Solution

You can write A ⇐ B because it is true that at 3 o'clock the hands are at right angles.

However you cannot write A ⇒ B because there are other times, for example 9 o'clock, when the hands are at right angles.

Nor can you write A ⇔ B because that is saying that *both* A ⇒ B and A ⇐ B are true; in fact only the second is true.

The words 'necessary' and 'sufficient'

Two other words that you will sometimes find useful are **necessary** and **sufficient**.

- **Necessary** is often used (or implied) when giving the conditions under which a statement is true. Thus a necessary condition for a living being to be a spider is that it has eight legs. However, this is not a **sufficient** condition because there are other creatures with eight legs, for example scorpions.

- The word sufficient is also used when giving conditions under which you can be certain that a statement is true. Thus being a spider is a sufficient condition for a creature to have eight legs (but not a necessary one).

With a little thought you will see that if A is a sufficient condition for B, then $A \Rightarrow B$. Similarly if A is a necessary condition for B, then $A \Leftarrow B$.

> A living being is a spider $\Rightarrow$ it has eight legs.
>
> A living being has eight legs $\Leftarrow$ it is a spider.

Discussion point

→ If A is both a necessary and sufficient condition for B, what symbol should connect A and B?

The converse of a theorem

When theorems, or general results, are involved it is quite common to use the word **converse** to express the idea behind the $\Leftarrow$ symbol.

Thus, for the triangle ABC (Figure 1.13), Pythagoras' theorem states

$$\text{Angle BCA} = 90° \Rightarrow AB^2 = BC^2 + CA^2$$

and its converse is

$$\text{Angle BCA} = 90° \Leftarrow AB^2 = BC^2 + CA^2.$$

Figure 1.13

Note

It would be more usual to write the converse the other way round, as

$AB^2 = BC^2 + CA^2 \Rightarrow$ Angle BCA = 90°.

Since in this case both the theorem and its converse are true, you would be correct to use the $\Leftrightarrow$ symbol.

$$\text{Angle BCA} = 90° \Leftrightarrow AB^2 = BC^2 + CA^2.$$

You may find it easier to think in terms of *If . . . then . . .*, when deciding just what the converse of a result or theorem says. For example, suppose you had to write down the converse of the theorem that the angle in a semicircle is 90°. This can be stated as:

> *If* AB is a diameter of the circle through points A, B and C, *then* $\angle ACB = 90°$.

To find the converse, change over the *If* and the *then*, and write the statement so that it reads sensibly. In this case it becomes:

> *If* $\angle ACB = 90°$, *then* AB is a diameter of the circle through points A, B and C.

Discussion points

→ Is the converse of this theorem true?

→ State a theorem for which the converse is not true.

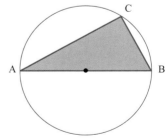

Figure 1.14

① Is the following statement true or false?

n is a multiple of 4 $\Leftarrow n^2$ is a multiple of 4

Justify your answer.

 ② A = being a giraffe, B = having a long neck

Which statement is equivalent to 'Giraffes have long necks'?

A $A \Rightarrow B$ B $A \Leftarrow B$

C $A \Leftrightarrow B$ D if B then A

③ In each case, write one of the symbols $\Rightarrow, \Leftarrow$ or $\Leftrightarrow$ between the two statements A and B.

(i)   A: The object is a cube
      B: The object has six faces

(ii) A: Jasmine has spots
      B: Jasmine is a leopard

(iii) A: The polygon has four sides
      B: The polygon is a quadrilateral

(iv) A: Today is 1 January
      B: Today is New Year's Day

(v)   A: $x = 29$
      B: $x > 10$

(vi) A: This month has exactly 28 days
      B: This month is February and it is not a leap year

(vii) A: PQRS is a rectangle
      B: PQRS is a parallelogram

(viii) A: The three sides of triangle T are equal
      B: The three angles of triangle T are equal

④ For each of the following statements, state the converse, and state whether the converse is true.

(i)   If a triangle has two sides equal, then it has two angles equal.

(ii) If Fred murdered Alf, then Alf is dead.

(iii) ABCD is a square $\Rightarrow$ Each of the angles of ABCD is 90°.

(iv) A triangle with three equal sides has three equal angles.

(v)   If it is sunny, then Struan goes swimming.

⑤ In each case, write one of the symbols $\Rightarrow, \Leftarrow$ or $\Leftrightarrow$ between the two statements P and Q.

(i)   P: $x^3 = x$
       Q: $x = -1$

(ii)   P: $xy = 0$
       Q: $x = 0$ and $y = 0$

(iii) P: $\dfrac{p}{q} = \dfrac{q}{p}$
       Q: $p = q$

(iv)   P: $x = 3$
       Q: $x^2 = 9$

(v)   P: $6 < x < 8$
       Q: $x = 7$

(vi)   The number n is a positive integer greater than 1.
       P: n has exactly two factors
       Q: n is a prime number

(vii) P: m and n are odd integers
       Q: $m + n$ is an even integer

(viii) O and P are points.
       P: $0\,\text{cm} \leqslant OP < 5\,\text{cm}$
       Q: P lies inside the sphere with centre O and radius 5 cm.

⑥ For each of the following statements, state the converse, and state whether the converse is true.

(i)   If x is an integer, then x^2 is an integer.

(ii) For a quadrilateral PQRS: if P, Q, R and S all lie on a circle, then $\angle PQR + \angle PSR = 180°$.

(iii) If $x = y$, then $x^2 = y^2$.

(iv) In Figure 1.15, lines l and m are parallel $\Rightarrow \angle x = \angle y$

(v) n is an odd prime number $\Rightarrow n > 2$

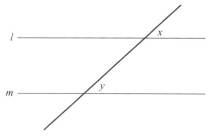

Figure 1.15

⑦ To show that an integer n is divisible by 5, is it

   (i)   necessary

   (ii)   sufficient

to show that it ends in zero?

⑧ For both of the statements below state

   (a)   whether it is true or false

   (b)   the converse

   (c)   whether the converse is true or false.

In the case of a 'false' answer, explain how you know that it is false.

   (i)   ABCDEF is a regular hexagon $\Rightarrow$ The six internal angles are all equal.

   (ii)   ABCDEF is a regular hexagon $\Rightarrow$ All the six sides are the same length.

⑨ ABC and XYZ are two triangles. Louie makes this statement.
'Together AB = XY, BC = YZ and angle ABC = angle XYZ $\Rightarrow$ Triangles ABC and XYZ are congruent.'

   (i)   Say whether Louie's statement is true or false. If your answer is 'false', explain how you know that it is false.

   (ii)   State the converse of the statement.

   (iii)   Say whether the converse is true or false. If your answer is 'false', explain how you know that it is false.

⑩ A, B, C and D are statements, each of which is either true or false. We also know that:

$$A \Rightarrow B, \quad B \Leftrightarrow C,$$
$$C \Rightarrow A \quad \text{and} \quad D \Leftarrow B$$

How many of the statements can be true? (There is more than one answer.)

ACTIVITY 1.1

Choose any three consecutive integers and add them up. What number is always a factor of the sum of three consecutive integers? Can you be certain that this is true for all possible sets of three consecutive integers?

3 Proof

In Activity 1.1 on the left, you formed a **conjecture** – an idea, or theory, supported by evidence from the cases you tested. However, there are an infinite number of possible sets of three consecutive integers. You may be able to test your conjecture for a thousand, or a million, or a billion, sets of consecutive integers, but you have still not proved it. You could program a computer to check your conjecture for even more cases, but this will still not prove it. There will always be other possibilities that you have not checked.

Proof by deduction

Proof by deduction consists of a logical argument as to why the conjecture must be true. This will often require you to use algebra.

Example 1.4 shows how proof by deduction can be used to prove the result from Activity 1.1 above.

Example 1.4

Proof by deduction

Prove that the sum of three consecutive integers is always a multiple of 3.

Solution

Let the first integer be x

$\Rightarrow$ the next integer is $x + 1$ and the third integer is $x + 2$

$\Rightarrow$ the sum of the three integers $= x + (x + 1) + (x + 2)$

$$= 3x + 3$$

$$= 3(x + 1)$$

The sum of the three integers has a factor of 3, whatever the value of x.

$\Rightarrow$ the sum of three consecutive integers is always a multiple of 3.

Proof by exhaustion

With some conjectures you can test all the possible cases. An example is

97 is a prime number.

If you show that none of the whole numbers between 2 and 96 is a factor of 97, then it must indeed be a prime number. However a moment's thought shows that you do not need to try all of them, only those that are less than $\sqrt{97}$, i.e. from 2 up to 9.

Discussion point

➜ It was not strictly necessary to test all the numbers from 2 up to 9. Which numbers were necessary and which were not?

$97 \div 2 = 48\frac{1}{2}$ No $\qquad$ $97 \div 3 = 32\frac{1}{3}$ No $\qquad$ $97 \div 4 = 24\frac{1}{4}$ No

$97 \div 5 = 19\frac{2}{5}$ No $\qquad$ $97 \div 6 = 16\frac{1}{6}$ No $\qquad$ $97 \div 7 = 13\frac{6}{7}$ No

$97 \div 8 = 12\frac{1}{8}$ No $\qquad$ $97 \div 9 = 10\frac{7}{9}$ No

Since none of the numbers divides into 97 exactly, 97 must be a prime.

This method of proof, by trying out all the possibilities, is called **proof by exhaustion**. In theory, it is the possibilities that get exhausted, not you!

Disproving a conjecture

Of course, not all conjectures are true! Sometimes you may need to disprove a conjecture.

To disprove a conjecture all you need is a single **counter-example**. Look at this conjecture:

$n^2 + n + 1$ is a prime number

$n = 1 \Rightarrow n^2 + n + 1 = 3$ $\qquad$ and 3 is prime

$n = 2 \Rightarrow n^2 + n + 1 = 7$ $\qquad$ and 7 is prime

$n = 3 \Rightarrow n^2 + n + 1 = 13$ $\qquad$ and 13 is prime

So far so good, but when you come to $n = 4$

$n = 4 \Rightarrow n^2 + n + 1 = 21$ $\qquad$ and $21 = 3 \times 7$ is not prime.

Nothing more needs to be said. The case of $n = 4$ provides a counter-example and one counter-example is all that is needed to disprove a conjecture. The conjecture is false.

However, sometimes it is easier to disprove a conjecture by exposing a fault in the argument that led to the proposal of the conjecture.

Exercise 1.3

 ① Prove that the sum of seven consecutive integers is always divisible by seven.

 ② Find a counter-example for the statement '$3n + 6$ is always a multiple of 6'.

③ For each of the statements below, decide whether it is true or false. If it is true, prove it using either proof by deduction or proof by exhaustion, stating which method you are using. If it is false, give a counter-example.

  (i) If n is a positive integer, $n^2 + n$ is always even.

  (ii) An easy way to remember 7 times 8 is that $56 = 7 \times 8$, and the numbers 5, 6, 7 and 8 are consecutive. There is exactly one other multiplication of two single-digit numbers with the same pattern.

  (iii) $x^2 > x \Rightarrow x > 1$.

  (iv) No square number ends in a 2.

  (v) n is prime $\Rightarrow n^2 + n + 11$ is prime.

  (vi) One cannot logically say 'I always tell lies'.

④ For each of the statements below, decide whether it is true or false. If it is true, prove it using either proof by deduction or proof by exhaustion, stating which method you are using. If it is false, give a counter-example.

  (i) The value of $n^2 + n + 41$ is a prime number for all positive integer values of n.

> 🖥 **TECHNOLOGY**
>
> You could use table mode on a scientific calculator to help with this – if there is insufficient memory to look at all the values needed, try different start values.

  (ii) The sum of n consecutive integers is divisible by n, where n is a positive integer.

  (iii) Given that $5 \geqslant m > n \geqslant 0$ and that m and n are integers, the equation $m^n = n^m$ has exactly one solution.

  (iv) Given that p and q are odd numbers, $p^2 + q^2$ cannot be divisible by 4.

  (v) It is impossible to draw any conclusion when someone tells you 'I always tell the truth'.

  (vi) The smallest value of $n \ (> 1)$ for which $1^2 + 2^2 + 3^2 + \cdots + n^2$ is a perfect square is 24.

⑤ A rectangular piece of paper is $100\,\text{cm}$ long and $w\,\text{cm}$ wide (Figure 1.16). If it is cut along the dotted line, the two pieces of paper are similar to the original piece.

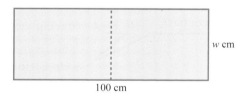

Figure 1.16

Find the value of w.

⑥ Prove that all prime numbers greater than 3 are of the form $6n \pm 1$. State the converse and determine whether it is true or false.

⑦ For each of the statements below, decide whether it is true or false. If it is true, prove it using either proof by deduction or proof by exhaustion, stating which method you are using. If it is false, give a counter-example.

(i) A number is divisible by 9 if the sum of its digits is divisible by 9.

(ii) ABC is a three-digit number.

    (a) ABC is divisible by 3 $\Leftrightarrow$ CBA is divisible by 3.

    (b) ABC is divisible by 5 $\Leftrightarrow$ CBA is divisible by 5.

(iii) An n-sided polygon has exactly $n - 1$ internal right angles $\Rightarrow n = 6$.

⑧ Call the prime numbers, in order, p_1, p_2, p_3 and so on. Thus $p_1 = 2, p_2 = 3, p_3 = 5, \ldots$.

    (i) Show that $p_1 + 1, p_1 p_2 + 1, p_1 p_2 p_3 + 1, p_1 p_2 p_3 p_4 + 1$ are all prime numbers.

    (ii) Explain why $p_1 p_2 p_3 \ldots p_n + 1$ is either prime or has a prime factor greater than p_n for all positive integers of n.

    (iii) How does this allow you to prove that there is an infinite number of prime numbers?

Mountain modelling

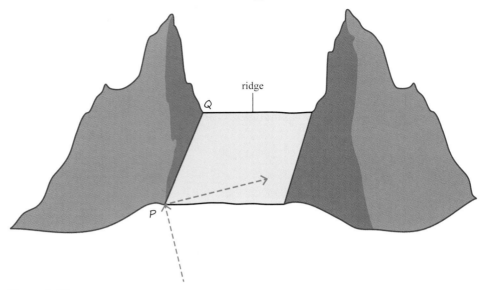

Figure 1.17

Chandra is 75 years old. He is hiking in a mountainous area. He comes to the point P in Figure 1.17. It is at the bottom of a steep slope leading up to a ridge between two mountains. Chandra wants to cross the ridge at the point Q.

The ridge is 1000 metres higher than the point P and a horizontal distance of 1500 metres away. The slope is 800 metres wide. The slope is much too steep for Chandra to walk straight up so he decides to zig-zag across it. You can see the start of the sort of path he might take.

The time is 12 noon. Estimate when Chandra can expect to reach the point Q.

There is no definite answer to this question. That is why you are told to **estimate** the time. You have to follow the problem solving cycle. You will also need to do some **modelling**.

1   **Problem specification and analysis**

    The problem has already been specified but you need to decide how you are going to go about it.

- A key question is how many times Chandra is going to cross the width of the slope. Will it be 2, or 4, or 6 , or . . . ?

- To answer this you have to know the steepest slope Chandra can walk along.

- What modelling assumption are you going to make about the sloping surface?

- To complete the question you will also need to estimate how fast Chandra will walk.

2 Information collection

At this stage you need the answers to two of the questions raised in stage 1: the steepest slope Chandra can walk along and how fast he can walk. Are the two answers related?

3 **Processing and representation**

In this problem the processing and representation stage is where you will draw diagrams to show Chandra's path. It is a three-dimensional problem so you will need to draw true shape diagrams.

Drawing a 3-dimensional object on a sheet of paper, and so in two dimensions, inevitably involves distortion. A true shape diagram shows a 2-dimensional section of the object without any distortion. So a right angle in the object really is 90° in a true shape diagram; similarly parallel lines in the object are parallel in the diagram.

There are a number of possible answers according to how many times Chandra crosses the slope. Always start by taking the easiest case, in this case when he crosses the slope 2 times. This is shown as PR + RQ in Figure 1.18. The point R is half way up the slope ST and the points A and B are directly below R and T.

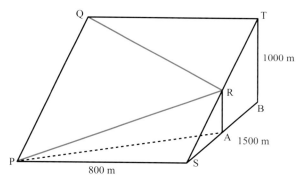

Figure 1.18

Draw the true shape diagrams for triangles SAR, PSR and PAR.

Then use trigonometry and Pythagoras' theorem to work out the angle RPA. If this is not too great, go on to work out how far Chandra walks in crossing the slope and when he arrives at Q.

Now repeat this for routes with more crossings.

4 Interpretation

You now have a number of possible answers.

For this interpretation stage, decide which you think is the most likely time for 75-year-old Chandra to arrive at Q and explain your choice.

LEARNING OUTCOMES

When you have completed this chapter you should be able to:

➤ understand and use the structure of mathematical proof, proceeding from given assumptions through a series of logical steps to a conclusion
➤ use methods of proof, including proof by deduction and proof by exhaustion
➤ disprove by counter-example.

KEY POINTS

1 $\Rightarrow$ means 'implies', 'if...then...' 'therefore'
   $\Leftarrow$ means 'is implied by', 'follows from'
   $\Leftrightarrow$ means 'implies and is implied by', 'is equivalent to'.
2 The converse of $A \Rightarrow B$ is $A \Leftarrow B$.
3 If $A \Leftarrow B$, A is a *necessary* condition for B
   If $A \Rightarrow B$, A is a *sufficient* condition for B.
4 Two methods of proving a conjecture are proof by **deduction** and proof by **exhaustion**.
5 A conjecture can be disproved by finding a **counter-example**.

Surds and indices

The sides of the large square in Figure 2.1 are 8 cm long. Subsequent squares in the tile design are formed by joining the midpoints of the sides of the previous square. The total length of all the lines is $\left(a + b\sqrt{2}\right)$ cm. The numbers a and b are rational. What are the values of a and b?

Figure 2.1

1 Using and manipulating surds

Numbers involving square roots which cannot be written as rational numbers, such as $\sqrt{2}$, are also commonly referred to as **surds**. $\left(\frac{1}{2} + \sqrt{5}\right)$ is another example of a surd. Surds may also involve cube roots, etc., but the focus in this chapter is on square roots. Surds give the *exact* value of a number, whereas a calculator will often only provide an approximation. For example, $\sqrt{2}$ is an exact value, but the calculator value of $1.414213562\ldots$ is only approximate, no matter how many decimal places you have.

Many scientific calculators will simplify expressions containing square roots of numbers for you, but you need to know the rules for manipulating surds in order to handle them in an algebraic setting, such as $(\sqrt{a} + \sqrt{b})(3\sqrt{a} - 4\sqrt{b})$.

The basic fact about the square root of a number is that when multiplied by itself it gives the original number: $\sqrt{5} \times \sqrt{5} = 5$.

Rearranging this gives: $\sqrt{5} = \dfrac{5}{\sqrt{5}}$ and $\dfrac{1}{\sqrt{5}} = \dfrac{\sqrt{5}}{5}$, both of which are useful results.

Surds may be added or subtracted just like other algebraic expressions, keeping the rational numbers and the square roots separate.

Example 2.1

Simplify the following, giving your answers in the simplest surd form.

(i) $\sqrt{32}$

(ii) $4(\sqrt{7} + 3\sqrt{5}) - 3(\sqrt{7} - 2\sqrt{5})$

(iii) $2(\sqrt{x} + \sqrt{y}) - 3(\sqrt{x} - \sqrt{y})$

Solution

(i) $\sqrt{32} = \sqrt{16 \times 2}$

> The largest square number which is a factor of 32 is 16.

$\qquad = \sqrt{16} \times \sqrt{2}$

> Multiply out the brackets.

$\qquad = 4\sqrt{2}$

(ii) $4(\sqrt{7} + 3\sqrt{5}) - 3(\sqrt{7} - 2\sqrt{5})$

> Collect like terms.

$\qquad = 4\sqrt{7} + 12\sqrt{5} - 3\sqrt{7} + 6\sqrt{5}$

$\qquad = \sqrt{7} + 18\sqrt{5}$

(iii) $2(\sqrt{x} + \sqrt{y}) - 3(\sqrt{x} - \sqrt{y})$

$\qquad = 2\sqrt{x} + 2\sqrt{y} - 3\sqrt{x} + 3\sqrt{y}$

$\qquad = -\sqrt{x} + 5\sqrt{y}$

> It's a good idea to write the positive term first since it is easy to lose a minus sign at the start of an expression.

$\qquad = 5\sqrt{y} - \sqrt{x}$

When multiplying two surds, multiply term by term, as shown in the next example.

Example 2.2

Multiply out and simplify:

(i) $(\sqrt{2} + \sqrt{3})(\sqrt{2} + 2\sqrt{3})$

(ii) $(3 + \sqrt{2})(3 - \sqrt{2})$

Solution

(i) $(\sqrt{2} + \sqrt{3})(\sqrt{2} + 2\sqrt{3})$

$$= \sqrt{2}(\sqrt{2} + 2\sqrt{3}) + \sqrt{3}(\sqrt{2} + 2\sqrt{3})$$

$$= (\sqrt{2} \times \sqrt{2}) + (\sqrt{2} \times 2\sqrt{3}) + (\sqrt{3} \times \sqrt{2}) + (\sqrt{3} \times 2\sqrt{3})$$

$$= 2 + 2\sqrt{6} + \sqrt{6} + (2 \times 3)$$

Notice that the first and last terms are rational numbers.

$$= 8 + 3\sqrt{6}$$

(ii) $(3 + \sqrt{2})(3 - \sqrt{2})$

$$= 9 - 3\sqrt{2} + 3\sqrt{2} - (\sqrt{2})^2$$

$$= 9 - 2$$

$$= 7$$

Rationalising the denominator of a surd

In part (ii) of the last example, the terms involving square roots disappeared, leaving an answer that is a rational number. This is because the two numbers to be multiplied together are the factors of the difference of two squares. In this case the squares are of surds and so are not square numbers.

Using the result $(a + b)(a - b) = a^2 - b^2$ with $a = 3$ and $b = \sqrt{2}$ gives

$$(3 + \sqrt{2})(3 - \sqrt{2}) = 3^2 - (\sqrt{2})^2$$

$$= 9 - 2 = 7.$$

This is the basis for a useful technique for simplifying a fraction whose bottom line is a surd. The technique, called **rationalising the denominator**, is illustrated in the next example. It involves multiplying the top line and the bottom line of a fraction by the same expression. This does not change the value of the fraction; it is equivalent to multiplying the fraction by 1.

| **Example 2.3** | Simplify the following, giving your answers in the simplest surd form. |

(i) $\dfrac{4}{\sqrt{2}}$ (ii) $\dfrac{3 + 2\sqrt{6}}{\sqrt{6}}$ (iii) $\dfrac{1}{3 + \sqrt{5}}$

Solution

(i) $\dfrac{4}{\sqrt{2}} = \dfrac{4}{\sqrt{2}} \times \dfrac{\sqrt{2}}{\sqrt{2}}$ ← Multiplying top and bottom by $\sqrt{2}$

$= \dfrac{4\sqrt{2}}{2}$

$= 2\sqrt{2}$

(ii) $\dfrac{3 + 2\sqrt{6}}{\sqrt{6}} = \dfrac{(3 + 2\sqrt{6})}{\sqrt{6}} \times \dfrac{\sqrt{6}}{\sqrt{6}}$ ← Multiplying top and bottom by $\sqrt{6}$

$= \dfrac{3\sqrt{6} + 12}{6}$

$= \dfrac{3(\sqrt{6} + 4)}{6}$

$= \dfrac{\sqrt{6} + 4}{2}$ ← This can also be written as $\dfrac{\sqrt{6}}{2} + 2$

(iii) $\dfrac{1}{3 + \sqrt{5}} = \dfrac{1}{(3 + \sqrt{5})} \times \dfrac{(3 - \sqrt{5})}{(3 - \sqrt{5})}$ ← Rationalising the denominator

$= \dfrac{3 - \sqrt{5}}{3^2 - (\sqrt{5})^2}$

$= \dfrac{3 - \sqrt{5}}{9 - 5}$

$= \dfrac{3 - \sqrt{5}}{4}$

These methods can be generalised to apply to algebraic expressions.

| **Example 2.4** | Simplify the following, giving your answers in the simplest surd form. |

(i) $\dfrac{x\sqrt{y}}{\sqrt{x}}$ (ii) $\dfrac{x + 2\sqrt{y}}{x + \sqrt{y}}$

Discussion points

Try substituting numbers for x and y in the question and answer for (ii). That should help you to appreciate that the result is, in fact, a simpler form.

→ You could start with $x = 3$, $y = 4$.

→ What happens when you substitute $x = 1$, $y = 1$ into the question and answer? Why does this happen?

Solution

(i) $$\frac{x\sqrt{y}}{\sqrt{x}} = \frac{x\sqrt{y}}{\sqrt{x}} \times \frac{\sqrt{x}}{\sqrt{x}}$$

> Multiplying top and bottom by $\sqrt{x}$.

$$= \frac{x\sqrt{yx}}{x}$$

$$= \sqrt{xy}$$

> In algebra it is usual to write letters in alphabetical order.

(ii) $$\frac{x + 2\sqrt{y}}{x + \sqrt{y}} = \frac{(x + 2\sqrt{y})}{(x + \sqrt{y})} \times \frac{(x - \sqrt{y})}{(x - \sqrt{y})}$$

> Rationalising the denominator.

$$= \frac{x^2 - x\sqrt{y} + 2x\sqrt{y} - 2y}{x^2 - (\sqrt{y})^2}$$

$$= \frac{x^2 + x\sqrt{y} - 2y}{x^2 - y}$$

Exercise 2.1

Answer these questions without using your calculator, but do use it to check your answers where possible. Leave all the answers in this exercise in surd form where appropriate.

 ① Explain in each case why the statement is incorrect.

(i) 36 is the highest square factor of 72 so $\sqrt{72} = 2\sqrt{6}$

(ii) 25 is the highest square factor of 75 so $\sqrt{75} = 25\sqrt{3}$

(iii) 4 is the highest square factor of 32 so $\sqrt{32} = 2\sqrt{8}$

 ② Simplify these.

(i) $\sqrt{28}$ (ii) $\sqrt{75}$ (iii) $\sqrt{128}$

 ③ Express as the square root of a single number.

(i) $3\sqrt{6}$ (ii) $5\sqrt{5}$ (iii) $12\sqrt{3}$

④ Simplify by rationalising the denominator.

(i) $\dfrac{4}{\sqrt{2}}$ (ii) $\dfrac{1}{2\sqrt{3}}$ (iii) $\dfrac{6}{\sqrt{8}}$

⑤ Simplify these.

(i) $\dfrac{\sqrt{25}}{\sqrt{49}}$ (ii) $\sqrt{\dfrac{32}{81}}$ (iii) $\sqrt{\dfrac{175}{343}}$

⑥ Simplify these.

(i) $(3 + \sqrt{2}) + (5 + 2\sqrt{2})$

(ii) $(5 + 3\sqrt{2}) - (5 - \sqrt{2})$

(iii) $3(\sqrt{2} - \sqrt{5}) + 5(\sqrt{2} + \sqrt{5})$

(iv) $4(\sqrt{5} - 1) + 4(\sqrt{5} + 1)$

⑦ Simplify these expressions.

(i) $(\sqrt{x} + \sqrt{y}) - (\sqrt{x} - \sqrt{y})$

(ii) $3(\sqrt{a} + 2\sqrt{b}) - (\sqrt{a} - 5\sqrt{b})$

⑧ Simplify these expressions.

(i) $(3 + \sqrt{2})(3 + \sqrt{2})$

(ii) $(3 - \sqrt{2})(3 - \sqrt{2})$

(iii) $(3 + \sqrt{2})(3 - \sqrt{2})$

⑨ Simplify these expressions.

(i) $(\sqrt{7} + \sqrt{2})(\sqrt{7} - 2\sqrt{2})$

(ii) $(\sqrt{7} - \sqrt{2})(\sqrt{7} + 2\sqrt{2})$

(iii) $(\sqrt{p} + \sqrt{q})(\sqrt{p} - 2\sqrt{q})$

(iv) $(\sqrt{p} - \sqrt{q})(\sqrt{p} + 2\sqrt{q})$

In the form
$a + b\sqrt{c}$
means that
it will look
like $5 + \sqrt{3}$
or $\frac{1}{3} - \frac{2}{3}\sqrt{2}$

⑩ Simplify the following by rationalising the denominator.

(i) $\dfrac{1}{\sqrt{3} + 1}$ (ii) $\dfrac{1}{1 - \sqrt{5}}$

(iii) $\dfrac{3}{4 - \sqrt{2}}$ (iv) $\dfrac{3}{\sqrt{5} - 2}$

⑪ Write the following in the form $a + b\sqrt{c}$ where c is an integer and a and b are rational numbers.

(i) $\dfrac{3 + \sqrt{2}}{4 - \sqrt{2}}$ (ii) $\dfrac{2 - \sqrt{3}}{1 + \sqrt{3}}$

(iii) $\dfrac{3 - \sqrt{5}}{\sqrt{5} + 5}$ (iv) $\dfrac{4 + \sqrt{2}}{4 - \sqrt{2}}$

PS ⑫ (i) A square has sides of length $x\,\text{cm}$ and a diagonal of length $8\,\text{cm}$. Use Pythagoras' theorem to find the exact value of x and the area of the square.

(ii) The diagonal of a square is of length $6\,\text{cm}$. Find the exact value of the perimeter of the square.

⑬ A ladder of length $6\,\text{m}$ is placed on horizontal ground with the foot of the ladder $2\,\text{m}$ from the vertical side of a house. How far up the wall does the ladder reach? Give your answer in the simplest possible surd form.

PS ⑭ The cost of building a lighthouse is proportional to the cube of its height, h.

The distance, d, that the top of the lighthouse can be seen from a point at sea level is modelled by $d = \sqrt{2Rh}$, where R is

Figure 2.2

the radius of the Earth and d, R and h are in the same units.

Three possible designs X, Y and Z are considered, in which the top of the lighthouse can be seen at $20\,\text{km}$, $40\,\text{km}$ and $60\,\text{km}$ respectively.

Find the ratios of the costs of designs X, Y and Z.

⑮ Simplify the following by rationalising the denominator.

(i) $\dfrac{\sqrt{3}}{\sqrt{3} - \sqrt{2}}$ (ii) $\dfrac{\sqrt{5} + 2\sqrt{2}}{2\sqrt{5} - \sqrt{2}}$

PS ⑯ (i) On a single sheet of graph paper, or using graphic software, draw the curves with equations $y = \sqrt{x}$ and $y = \sqrt{x + 2}$ for $0 \leqslant x \leqslant 20$.

(ii) Choose two values of x with $0 \leqslant x \leqslant 20$ and **show** that for these values the point

$\left(x, \dfrac{1}{\sqrt{x - 2} - \sqrt{x}} \right)$ lies between

the two curves.

(iii) **Prove** that the point

$\left(x, \dfrac{1}{\sqrt{x - 2} - \sqrt{x}} \right)$ lies between

the two curves.

PS ⑰ (i) Find two numbers a and b with $a \neq b$ so that it is possible to write $a\sqrt{b} + b\sqrt{a}$ in the form $p\sqrt{q}$ where p and q are integers.

(ii) Generalise this result by finding an algebraic connection between a and b such that this is possible.

Prior knowledge

You will have met this at GCSE.

Discussion point

➡ How could you express the equation of the curve defined by Jupiter's moons?

2 Working with indices

Figure 2.3 refers to the moons of the planet Jupiter. For six of its moons, the time, T, that each one takes to orbit Jupiter is plotted against the average radius, r, of its orbit. (The remaining moons of Jupiter would be either far off the scale or bunched together near the origin.)

The curves $T = kr^1$ and $T = kr^2$ (using suitable units and constant k) are also drawn on Figure 2.3. You will see that the curve defined by Jupiter's moons lies somewhere between the two.

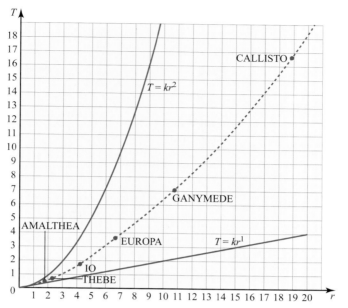

Figure 2.3

You may have suggested that the equation of the curve defined by Jupiter's moons could be written as $T = kr^{\frac{3}{2}}$. If so, well done! This is correct, but what does a power, or index, of $\frac{3}{2}$ mean?

Before answering this question it is helpful to review the language and laws relating to positive whole number indices.

Terminology

In the expression a^m the number represented by a is called the **base** of the expression; m is the **index**, or the **power** to which the base is raised. (The plural of index is indices.)

Multiplication

Multiplying 3^6 by 3^4 gives

$$3^6 \times 3^4 = (3 \times 3 \times 3 \times 3 \times 3 \times 3) \times (3 \times 3 \times 3 \times 3)$$
$$= 3^{10}.$$

Clearly it is not necessary to write down all the 3s like that. All you need to do is to add the powers: $6 + 4 = 10$

and so $\qquad 3^6 \times 3^4 = 3^{6+4} = 3^{10}.$

This can be written in general form as

$$a^m \times a^n = a^{m+n}.$$

Another important multiplication rule arises when a base is successively raised to one power and then another, as for example in $(3^4)^2$:

$$(3^4)^2 = (3^4) \times (3^4)$$
$$= (3 \times 3 \times 3 \times 3) \times (3 \times 3 \times 3 \times 3)$$
$$= 3^8.$$

In this case the powers to which 3 is raised are multiplied: $4 \times 2 = 8$. Written in general form this becomes:

$$(a^m)^n = a^{m \times n}.$$

Division

In the same way, dividing 3^6 by 3^4 gives

$$3^6 \div 3^4 = \frac{(3 \times 3 \times 3 \times 3 \times 3 \times 3)}{(3 \times 3 \times 3 \times 3)}$$
$$= 3^2$$

In this case you subtract the powers: $6 - 4 = 2$

and so $\qquad 3^6 \div 3^4 = 3^{6-4} = 3^2$

This can be written in general form as

$$a^m \div a^n = a^{m-n}.$$

Using these rules you can now go on to give meanings to indices which are not positive whole numbers.

Index zero

If you divide 3^4 by 3^4 the answer is 1, since this is a number divided by itself. However you can also carry out the division using the rules of indices to get

$$3^4 \div 3^4 = 3^{4-4}$$
$$= 3^0$$

and so it follows that $3^0 = 1$.

The same argument applies to 5^0, 2.9^0 or any other (non-zero) number raised to the power zero; they are all equal to 1.

In general $a^0 = 1$.

Negative indices

Dividing 3^4 by 3^6 gives

$$3^4 \div 3^6 = \frac{(3 \times 3 \times 3 \times 3)}{3 \times 3 \times 3 \times 3 \times 3 \times 3}$$
$$3^{4-6} = \frac{1}{3 \times 3}$$

and so $\qquad 3^{-2} = \dfrac{1}{3^2}$

This can be generalised to

$$a^{-m} = \frac{1}{a^m}.$$

Fractional indices

What number multiplied by itself gives the answer 3? The answer, as you will know, is the square root of 3, usually written as $\sqrt{3}$. Suppose instead that the square root of 3 is written 3^p; what then is the value of p?

Since $\qquad 3^p \times 3^p = 3^1$

it follows that $\qquad 3^{2p} = 3^1$

and so $\qquad p = \frac{1}{2}.$

In other words, the square root of a number can be written as that number raised to the power $\frac{1}{2}$, i.e., $\sqrt{a} = a^{\frac{1}{2}}$.

The same argument may be extended to other roots, so that the cube root of a number may be written as that number raised to the power $\frac{1}{3}$, the fourth root corresponds to power $\frac{1}{4}$, and so on.

This can be generalised to

$$\sqrt[n]{a} = a^{\frac{1}{n}}$$

In the example of Jupiter's moons, or indeed the moons or planets of any system, the relationship between T and r is of the form

$$T = kr^{\frac{3}{2}}$$

Squaring both sides gives

$$T \times T = kr^{\frac{3}{2}} \times kr^{\frac{3}{2}}$$

which may be written as $T^2 = cr^3$ (where the constant $c = k^2$).

This is one of Kepler's laws of planetary motion, first stated in 1619.

The use of indices not only allows certain expressions to be written more simply, but also, and this is more important, it makes it possible to carry out arithmetic and algebraic operations (such as multiplication and division) on them. These processes are shown in the following examples.

The following numerical examples show the methods for simplifying expressions using index notation. You can check the answers on your calculator. Once you are confident with these then you can apply the same rules to algebraic examples.

Example 2.5

Write the following numbers as the base 5 raised to a power.

(i) 625   (ii) 1   (iii) $\dfrac{1}{125}$ (iv) $5\sqrt{5}$

Solution

Notice that $5^0 = 1, 5^1 = 5, 5^2 = 25, 5^3 = 125, 5^4 = 625, \ldots$

(i) $625 = 5^4$ (ii) $1 = 5^0$

(iii) $\dfrac{1}{125} = \dfrac{1}{5^3} = 5^{-3}$ (iv) $5\sqrt{5} = 5^1 \times 5^{\frac{1}{2}} = 5^{\frac{3}{2}}$

Example 2.6

Simplify these.

(i) $\left(2^3\right)^4$ (ii) $27^{\frac{1}{3}}$ (iii) $4^{-\frac{5}{2}}$ (iv) $8^{\frac{2}{3}}$

Solution

(i) $\left(2^3\right)^4 = 2^{12} = 4096$

(ii) $27^{\frac{1}{3}} = \sqrt[3]{27} = 3$

(iii) $4^{-\frac{5}{2}} = \left(2^2\right)^{-\frac{5}{2}} = 2^{2 \times \left(-\frac{5}{2}\right)} = 2^{-5} = \dfrac{1}{32}$

(iv) There are two ways to approach this.

(a) $8^{\frac{2}{3}} = \left(8^2\right)^{\frac{1}{3}} = (64)^{\frac{1}{3}} = 4$

(b) $8^{\frac{2}{3}} = \left(8^{\frac{1}{3}}\right)^2 = (2)^2 = 4$

Both give the same answer, and both are correct.

Discussion point

➜ Which of these methods do you prefer? Why?

Example 2.7

Simplify $\left(4\sqrt{2} \times \dfrac{1}{16} \times \sqrt[5]{32}\right)^2$.

Solution

$$\left(4\sqrt{2} \times \frac{1}{16} \times \sqrt[5]{32}\right)^2$$

$$= \left(\left(2^2 \times 2^{\frac{1}{2}}\right) \times \frac{1}{2^4} \times \left(2^5\right)^{\frac{1}{5}}\right)^2$$

> 4, 16 and 32 are all powers of 2.

$$= \left(2^{2+\frac{1}{2}-4+1}\right)^2$$

> You need the same base number to add powers.

$$= \left(2^{-\frac{1}{2}}\right)^2$$

$$= 2^{-1}$$

$$= \frac{1}{2}$$

Mixed bases

If a problem involves a combination of different bases, you may well need to split them, using the rule

$$(a \times b)^n = a^n \times b^n$$

before going on to work with each base in turn.

For example, $\qquad 6^3 = (2 \times 3)^3$

$$= 2^3 \times 3^3$$

which is another way of saying $216 = 8 \times 27$.

Example 2.8

Solve the equation $2^a \times 6^b = 48$

Solution

> Start by writing each number as a product of its prime factors.

$$2^a \times (2 \times 3)^b = 2^4 \times 3$$

$$2^a \times 2^b \times 3^b = 2^4 \times 3$$

$$\Rightarrow \quad 2^{a+b} \times 3^b = 2^4 \times 3$$

Comparing powers this gives the pair of simultaneous equations:

$$a + b = 4$$
$$b = 1$$

So $\quad a = 3, b = 1$

Simplifying sums and differences of fractional powers

The next two examples show a type of simplification which you will often find you need to do.

Example 2.9

Simplify $5^{\frac{1}{2}} - 5^{\frac{3}{2}} + 5^{\frac{5}{2}}$.

Solution

This can be written as

$$\sqrt{5} - 5\sqrt{5} + 5^2\sqrt{5}$$

$$= (1 - 5 + 25)\sqrt{5}$$

$$= 21\sqrt{5}$$

$$5^{\frac{5}{2}} = 5^{2\frac{1}{2}}$$
$$= 5^2 5^{\frac{1}{2}}$$
$$= 5^2\sqrt{5}$$

Example 2.10

Simplify $3x(x + 7)^{\frac{1}{2}} - 2(x + 7)^{\frac{3}{2}}$.

Solution

The expression is $3x(x + 7)^{\frac{1}{2}} - 2(x + 7)(x + 7)^{\frac{1}{2}}$

$(x + 7)^{\frac{1}{2}}$ is a common factor

$$= \left(3x - 2(x + 7)\right)(x + 7)^{\frac{1}{2}}$$

$$= (x - 14)(x + 7)^{\frac{1}{2}}$$

or $(x - 14)\sqrt{(x + 7)}$

Standard form

One useful application of indices is in writing numbers in **standard form**. In standard form, very large or very small numbers are written as a decimal number between 1 and 10 multiplied by a power of 10. This is often used to simplify the writing of large or small numbers and make sensible approximations.

Standard form is closely related to the way our number system works, using words like ten, a hundred, a thousand, a million. This is true also for the metric system. Thus 5.2 kilograms is 5.2×10^3 grams.

Here is a list of common metric prefixes and how they relate to standard form:

deka	$\times 10^1$	deci	$\times 10^{-1}$
hecto	$\times 10^2$	centi	$\times 10^{-2}$
kilo	$\times 10^3$	milli	$\times 10^{-3}$
mega	$\times 10^6$	micro	$\times 10^{-6}$
giga	$\times 10^9$	nano	$\times 10^{-9}$
tera	$\times 10^{12}$	pico	$\times 10^{-12}$
peta	$\times 10^{15}$	femto	$\times 10^{-15}$

Example 2.11

Light travels at a speed of 300 million metres per second. At a certain time Pluto is 5.4 terametres from Earth. How many hours does it take light to travel from Pluto to Earth?

Solution

$$\text{Time taken} = \frac{5.4 \times 10^{12}}{3 \times 10^8}$$

$$= 1.8 \times 10^4 \text{ seconds}$$

$$= \frac{18\,000}{3600} \text{ hours}$$

There are 60×60 seconds in an hour.

$$= 5 \text{ hours}$$

Exercise 2.2

Answer these questions without using your calculator, unless otherwise specified, but do use it to check your answers where possible.

① 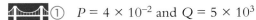 $P = 4 \times 10^{-2}$ and $Q = 5 \times 10^3$

Match each calculation to its correct answer in standard form.

$PQ \quad QP \quad P \div Q \quad Q \div P$

$8 \times 10^{-6} \qquad 1.25 \times 10^{-5}$

$2 \times 10^2 \qquad 0.8 \times 10^{-5}$

$20 \times 10^1 \quad 1.25 \times 10^5 \quad 2 \times 10^{-6}$

② Which answer is equal to $1000 \div$ 1 millionth?

A 10^9 B 10^3 C 10^{-3} D 10^{-9}

③ Write the following numbers as powers of 3.

(i) 81 (ii) 1 (iii) $\dfrac{1}{81}$ (iv) $\dfrac{1}{9}$

④ Write the following as whole numbers or fractions.

(i) 2^{-5} (ii) $27^{\frac{1}{3}}$ (iii) $64^{-\frac{1}{2}}$ (iv) $\left(\dfrac{2}{3}\right)^0$

⑤ Simplify the following, writing your answer in the form x^n.

(i) $x \times x^{\frac{1}{2}}$ (ii) $\sqrt{x^{-6}}$

(iii) $x^{\frac{1}{2}} \times x^{\frac{5}{2}}$ (iv) $\sqrt[5]{x^{\frac{5}{2}}}$

⑥ (i) Find 5.2 million divided by 1.3 thousandths.

(ii) The frequency of a wave is measured in hertz and is the reciprocal of its time period. What is the time period of a wave with frequency 20 kHz?

⑦ Signals sent from landers on Mars travel at the speed of light: $3.0 \times 10^8 \text{ m s}^{-1}$.

The greatest distance of Mars from Earth is 378 million km, and the closest distance is 78 million km.

(i) Write down the greatest and closest distances of Mars from Earth in metres, using standard form.

(ii) Use a calculator to find the greatest and least times that a signal sent from Mars takes to reach Earth.

⑧ Simplify the following, writing your answer in the form ax^n.

(i) $\sqrt{16x^6}$ (ii) $\sqrt{2x^5} \times \sqrt{8x^{-3}}$

(iii) $(x^{-3})^4 \times (2x^{\frac{1}{3}})^6$

⑨ Expand the following.

(i) $x^{\frac{1}{2}}(x^{\frac{3}{2}} - x^{\frac{1}{2}})$ (ii) $a^{-\frac{1}{2}}(a^{\frac{3}{2}} - a^{-\frac{3}{2}})$

(iii) $p^{\frac{2}{3}}(p^{\frac{1}{3}} - p^{-\frac{2}{3}})$

⑩ Simplify the following, writing your answer in the form ax^n.

(i) $(-3x)^3 \div (3x^{-3})$

(ii) $(2\sqrt{x})^2 \div (2x^2)^{-4}$

⑪ Simplify the following.

(i) $2^{\frac{1}{2}} + 2^{\frac{3}{2}}$ (ii) $4^{\frac{1}{3}} - 4^{\frac{4}{3}}$

(iii) $7^{\frac{1}{2}} - 7^{\frac{3}{2}} + 7^{\frac{5}{2}}$

PS ⑫ Solve the equation $4^x \times 3^{2x} = 6$.

PS ⑬ Simplify the following.

(i) $(1 + x)^{\frac{1}{2}} + (1 + x)^{\frac{3}{2}}$

(ii) $6(x + y)^{\frac{1}{2}} - 5(x + y)^{\frac{3}{2}}$

(iii) $(x^2 - 2x + 3)^{\frac{1}{2}} - (x^2 - 2x + 3)^{\frac{3}{2}}$

⑭ Find integers x and y such that $4^x \times 3^y = 24^8$.

⑮ Given that $2^{x+y} = 1$ and $10^{3x-y} = 100$, find x and y.

Hence find 5^{y-x} and x^y.

⑯ A sphere of radius R cm has area A cm^2 and volume V cm^3 (Figure 2.4).

Show that

$A = kV^{\frac{2}{3}}$ where

$k = \sqrt[3]{36\pi}$.

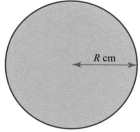

Figure 2.4

⑰ In this question, h is positive and you may use your calculator.

(i) Write down the values of $0.5^1, 0.5^2, 0.5^3, 0.5^4$.

What happens to 0.5^h as $h \to \infty$?

(ii) Write down the values of $0^1, 0^{\frac{1}{2}}, 0^{0.1}, 0^{0.01}$.

What happens to 0^h as h gets closer and closer to zero?

(iii) Write down the values of $1^0, \left(\frac{1}{2}\right)^0, (0.1)^0, (0.01)^0$

What do you think is the value of 0^0?

LEARNING OUTCOMES

When you have completed this chapter you should be able to:
➤ use and manipulate surds
➤ rationalise the denominator
➤ understand and use the laws of indices for all rational exponents.

> **Note**
>
> These are results you need to know.

KEY POINTS

1 Multiplication: $a^m \times a^n = a^{m+n}$
2 Division: $a^m \div a^n = a^{m-n}$
3 Power zero: $a^0 = 1$
4 Negative indices: $a^{-m} = \dfrac{1}{a^m}$

5 Fractional indices: $a^{\frac{1}{n}} = \sqrt[n]{a}$
6 Power of a power: $(a^m)^n = a^{mn}$
7 When simplifying expressions containing square roots you need to
   (i) make the number under the square root as small as possible
   (ii) rationalise the denominator if necessary.

FUTURE USES

■ You need to be able to work with surds when you use the quadratic formula to solve quadratic equations. This is covered in Chapter 3, and you will use quadratic equations in many areas of mathematics.

■ You will need to be able to use indices with confidence, including negative and fractional indices, in the chapters on Differentiation (Chapter 10) and Integration (Chapter 11).

■ The chapter on Exponentials and logarithms (Chapter 13) develops ideas using indices further, including looking at applications such as population growth, compound interest and radioactive decay.

3 Quadratic functions

In this activity you will investigate quadratic graphs using graphing software or a graphical calculator. Figure 3.1 shows the graphs of the quadratic functions $y = x^2$, $y = (x - 2)(x - 5)$, $y = (x + 2)(x - 5)$, and $y = (x + 2)(x + 5)$; they are all related to each other.

Each of the graphs has exactly the same shape, but they are positioned differently. Explain what makes them exactly the same shape. How does each equation relate to the position of the graph?

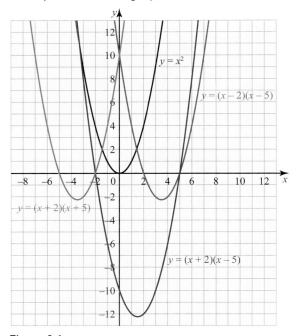

Figure 3.1

> ! You may find that your initial choice of scale (or the default setting) is not suitable for a particular graph. Make sure you know how to edit the axes.

For each of the questions below:

T (a) Use any technology available to you to draw the graphs on the same axes.

📈 **TECHNOLOGY**

You could type $y = (x - a)(x - b)$ and vary the values of a and b.

(b) In what ways are they similar and in what ways are they different? Explain how the equation tells you about the differences.

1 (i) $y = (x - 3)(x + 1)$ (ii) $y = (x - 3)(x - 1)$

2 (i) $y = x(x + 4)$ (ii) $y = (x + 2)(x + 4)$ (iii) $y = (x + 2)(x - 4)$

3 (i) $y = (x - 2)(x - 4)$ (ii) $y = x(x + 4)$ (iii) $y = (x - 4)(x + 4)$

4 (i) $y = (2 + x)(2 - x)$ (ii) $y = (3 + x)(2 - x)$ (iii) $y = (4 + x)(2 - x)$

1 Quadratic graphs and equations

Prior knowledge

You will have met this at GCSE.

The name **quadratic** comes from **'quad'** meaning square, because the variable is squared (for example, x^2). A quadratic expression is any expression that can be rearranged to have the form $ax^2 + bx + c$ where $a \neq 0$ and b and c can be any numbers, positive, negative or zero.

How would you draw the graph of $y = x^2 - 7x + 10$ without using any technology?

You could draw up a table of values and plot the graph like this.

> **Plotting** a graph means marking the points on graph paper and joining them up as accurately as you can.

Table 3.1

x	0	1	2	3	4	5	6	7
x^2	0	1	4	9	16	25	36	49
$-7x$	0	-7	-14	-21	-28	-35	-42	-49
$+10$	10	10	10	10	10	10	10	10
y	10	4	0	-2	-2	0	4	10

📈 **TECHNOLOGY**

Table mode on a scientific calculator will give you the table of values.

> Notice the symmetry in the table. How can you see this on the graph?

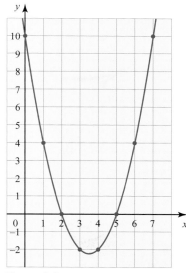

Figure 3.2

Alternatively you could start by solving the equation $x^2 - 7x + 10 = 0$.

The first step is to write it in factorised form $(x - 2)(x - 5) = 0$.

So $x = 2$ or 5.

> when $y = 0$

The curve $y = (x - 2)(x - 5)$ crosses the x axis at $(2, 0)$ and $(5, 0)$ and crosses the y axis at $(0, 10)$.

> when $x = 0$

This gives you three points which you can join up to make a smooth curve like this.

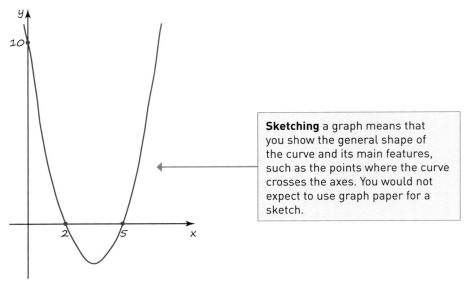

> **Sketching** a graph means that you show the general shape of the curve and its main features, such as the points where the curve crosses the axes. You would not expect to use graph paper for a sketch.

Figure 3.3

Prior knowledge

Quadratic factorisation should be familiar to you. Example 3.1 and Activity 3.1 are included as a reminder.

Factorising quadratic expressions

In the work above, you saw how the factorised form of a quadratic expression helps you to solve the related quadratic equation, and gives you useful information about the graph. Quadratic factorisation is an important skill that you will often need.

Example 3.1

Factorise $x^2 + 7x + 12$.

Solution

Splitting the middle term, $7x$, as $4x + 3x$ you have

$$x^2 + 7x + 12 = x^2 + 4x + 3x + 12$$

$$= x(x + 4) + 3(x + 4)$$

$$= (x + 3)(x + 4)$$

How do you know to split the middle term, $7x$, into $4x + 3x$, rather than say $5x + 2x$ or $9x - 2x$?

The numbers 4 and 3 can be added to give 7 (the middle coefficient) and multiplied to give 12 (the constant term), so these are the numbers chosen.

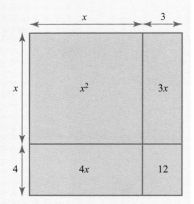

Figure 3.4

The coefficient of x is 7.

$x^2 + 7x + 12$

The constant term is 12.

$4 + 3 = 7$.

$4 \times 3 = 12$.

Note

You can write the x terms the other way around so $3x + 4x$:

$x^2 + 3x + 4x + 12$

$= x(x + 3) + 4(x + 3)$

$= (x + 4)(x + 3)$

There are other methods of quadratic factorisation. If you have already learned another way, and consistently get your answers right, then continue to use it. The method shown in Example 3.1 has one major advantage: it is self-checking. In the last line but one of the solution, you will see that $(x + 4)$ appears twice. If at this point the contents of the two brackets are different, for example $(x + 4)$ and $(x - 4)$, then something is wrong. You may have chosen the wrong numbers, or made a careless mistake, or perhaps the expression cannot be factorised. There is no point in proceeding until you have sorted out why they are different.

You may check your final answer by multiplying it out to get back to the original expression. Here are two common ways of setting this out.

1   Grid method

	x	$+4$
x	x^2	$+4x$
$+3$	$3x$	$+12$

This gives $x^2 + 4x + 3x + 12$

$$= x^2 + 7x + 12 \text{ (as required)}$$

2   Multiplying term by term

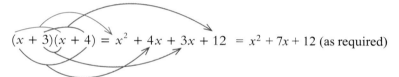

$(x + 3)(x + 4) = x^2 + 4x + 3x + 12 = x^2 + 7x + 12 \text{ (as required)}$

> **!** Before starting the procedure for factorising a quadratic, you should always check that the terms do not have a common factor as in, for example
>
> $$2x^2 - 8x + 6.$$
>
> This can be written as $2(x^2 - 4x + 3)$ and then factorised to give $2(x - 3)(x - 1)$.

> **Note**
> ----------
> You would not expect to draw the lines and arrows in your answers. They have been put in to help you understand where the terms have come from.

ACTIVITY 3.1

Factorise the following, where possible.

(i) $x^2 + 2x + 3x + 6$ (ii) $x^2 + 3x + 2x + 6$ (iii) $x^2 - 3x + 2x - 6$

(iv) $x^2 - 2x - 3x + 6$ (v) $x^2 + 2x - 3x - 6$ (vi) $x^2 - 6x - x + 6$

(vii) $x^2 + x - 6x - 6$ (viii) $x^2 + x + 6x - 6$ (ix) $x^2 - x - 6x + 6$

(x) $x^2 + 6x + x - 6$

Which pairs give the same answers?

The following examples show how you can use quadratic factorisation to solve equations and to sketch curves.

Example 3.2

(i)   Solve $x^2 - 2x - 24 = 0$.

(ii) Sketch the graph of $y = x^2 - 2x - 24$.

Solution

First you look for two numbers that can be added to give -2 and multiplied to give -24:

$$-6 + 4 = -2 \qquad -6 \times (+4) = -24.$$

The numbers are -6 and $+4$ and so the middle term, $-2x$, is split into $-6x + 4x$.

(i)
$$x^2 - 2x - 24 = 0$$
$$\Rightarrow \quad x^2 - 6x + 4x - 24 = 0$$
$$\Rightarrow \quad x(x - 6) + 4(x - 6) = 0$$
$$\Rightarrow \quad (x + 4)(x - 6) = 0$$
$$\Rightarrow \quad x = -4 \text{ or } 6$$

> When two brackets multiply to give zero, one or the other must equal zero.

(ii) From part (i), $y = 0$ when $x = -4$ and when $x = 6$, so these are the points where the curve crosses the x axis.

Also, when $x = 0$, $y = -24$, so you have the point where it crosses the y axis.

The coefficient of x^2 is positive, telling you that the curve is $\cup$-shaped, so you are now in a position to sketch the curve:

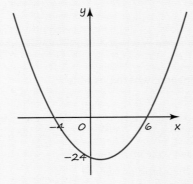

Figure 3.5

> ### Note
>
> When sketching the graph, a way to do this is to
>
> 1. draw a smooth $\cup$-shaped curve.
> 2. draw a horizontal line cutting the curve in two points and label these -4 and 6.
> 3. decide where $x = 0$ will be on this line and draw the y axis through this point.
> 4. finally mark the point $y = -24$ on the y axis.
>
> The **roots** of the equation are the values of x which satisfy the equation. In this case, one root is $x = -4$ and the other root is $x = 6$.

Example 3.3

A perfect square

Solve the equation $x^2 - 20x + 100 = 0$ and hence sketch the graph $y = x^2 - 20x + 100$.

Solution

$$x^2 - 20x + 100 = (x - 10)(x - 10)$$
$$= (x - 10)^2$$

Notice:
$(-10) + (-10) = -20$
$(-10) \times (-10) = +100$

$(x - 10)^2 = 0 \Rightarrow x = 10$ (repeated)

Substituting $x = 0$ gives $y = 100$, which is where the curve crosses the y axis.

Notice that the repeated root at $x = 10$ means that the curve **touches** the x axis.

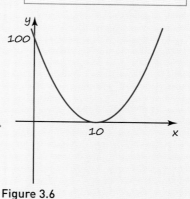

Figure 3.6

> **Note**
>
> The general forms of a perfect square are:
> $$(x + a)^2$$
> $$= x^2 + 2ax + a^2$$
> and
> $$(x - a)^2$$
> $$= x^2 - 2ax + a^2$$
> In Example 3.3, $a = 10$.

Example 3.4

Difference of two squares

Solve the equation $x^2 - 49 = 0$ and hence sketch the graph $y = x^2 - 49$.

Solution

$x^2 - 49$ can be written as $x^2 + 0x - 49$.

Notice that this is $x^2 - 7^2$

$$x^2 + 0x - 49 = (x + 7)(x - 7)$$

$-7 + 7 = 0$ and $(-7) \times 7 = -49$

$$(x + 7)(x - 7) = 0$$
$$\Rightarrow x = -7 \text{ or } x = 7$$

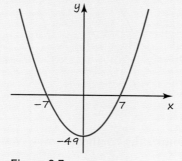

Figure 3.7

> **Note**
>
> Alternatively the equation $x^2 - 49 = 0$ could be rearranged to give $x^2 = 49$, leading to $x \pm 7$.
>
> A difference of two squares may be written in more general form as
> $$a^2 - b^2 = (a + b)(a - b)$$

The previous examples have all started with the term x^2, that is, the coefficient of x^2 has been 1. This is not the case in the next example.

Example 3.5

(i)   Factorise $6x^2 + x - 12$.

(ii) Solve the equation $6x^2 + x - 12 = 0$.

(iii) Sketch the graph $y = 6x^2 + x - 12$.

Solution

(i)   The technique for finding how to split the middle term is now adjusted. Start by multiplying the two outside numbers together:

$$6 \times (-12) = -72.$$

Now look for two numbers which add to give $+1$ (the coefficient of x) and multiply to give -72 (the number found above).

$$(+9) + (-8) = +1 \qquad (+9) \times (-8) = -72$$

Splitting the middle term gives

$$6x^2 + 9x - 8x - 12$$

$$= 3x(2x + 3) - 4(2x + 3)$$

$$= (3x - 4)(2x + 3)$$

> -4 is the largest factor of both $-8x$ and -12.

> $3x$ is the largest factor of both $6x^2$ and $9x$.

(ii) $(3x - 4)(2x + 3) = 0$

$$\Rightarrow 3x - 4 = 0 \quad \text{or} \quad 2x + 3 = 0$$

$$\Rightarrow \qquad x = \frac{4}{3} \quad \text{or} \qquad x = -\frac{3}{2}$$

(iii) The graph is a $\cup$-shaped curve which crosses the x axis at the points where $x = \frac{4}{3}$ and where $x = -\frac{3}{2}$.

When $x = 0$, $y = -12$, so the curve crosses the y axis at $(0, -12)$.

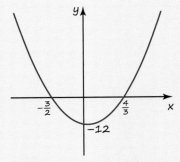

Figure 3.8

> ### Note
>
> The method used in the earlier examples is really the same as this. It is just that in those cases the coefficient of x^2 was 1 and so multiplying the constant term by 1 had no effect.

In the previous examples the coefficient of x^2 was positive, but this is not always the case.

Example 3.6

(i)   Factorise $15 + 2x - x^2$.

(ii)   Hence sketch the graph of $y = 15 + 2x - x^2$.

Solution

(i)   Start by multiplying the two outside numbers together:

$$15 \times -1 = -15$$

> The coefficient of x^2 is -1 so here the outside numbers are 15 and -1.

Now look for two numbers which add to give $+2$ (the coefficient of x) and multiply to give -15 (the number found above).

$$(+5) + (-3) = +2 \qquad (+5) \times (-3) = -15$$

Splitting the middle term gives

$$15 + 5x - 3x - x^2$$
$$= 5(3 + x) - x(3 + x)$$
$$= (5 - x)(3 + x)$$

(ii)   For very large positive or negative values of x, the value of $y = (5 - x)(3 + x)$ will be negative and consequently the curve will be $\cap$-shaped.

It will cross the x axis when $y = 0$, i.e. at $x = +5$ and $x = -3$, and the y axis when $x = 0$, i.e. at $y = 15$.

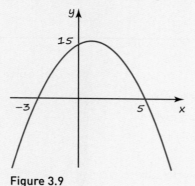

Figure 3.9

In some cases the equation to be solved does not immediately appear to be a quadratic equation.

Example 3.7

Solve the quartic equation $4x^4 - 17x^2 + 4 = 0$.

Solution

Replacing x^2 by z gives the quadratic equation $4z^2 - 17z + 4 = 0$.

$$\Rightarrow (z - 4)(4z - 1) = 0$$

$$\Rightarrow z = 4 \text{ or } 4z = 1$$

> Remember that you replaced x^2 by z earlier.

$$\Rightarrow \text{Either } x^2 = 4 \text{ or } 4x^2 = 1$$

$$\Rightarrow \text{The solution is } x = \pm 2 \text{ or } x = \pm \tfrac{1}{2}.$$

Example 3.8

In a right-angled triangle, the side BC is 1 cm longer than the side AB and the hypotenuse AC is 29 cm long.

Find the lengths of the three sides of the triangle.

Solution

Let the length of AB be x cm. So the length of BC is $x + 1$ cm.
The hypotenuse is 29 cm.

Using Pythagoras' theorem:

$$x^2 + (x + 1)^2 = 29^2$$

$$\Rightarrow x^2 + x^2 + 2x + 1 = 841$$

$$\Rightarrow 2x^2 + 2x - 840 = 0 \longleftarrow$$

Divide by 2 to make the factorisation easier.

$$\Rightarrow x^2 + x - 420 = 0$$

$$\Rightarrow (x + 21)(x - 20) = 0$$

Notice that -21 is rejected since x is a length.

$$\Rightarrow x = -21 \text{ or } x = 20, \text{ giving } x + 1 = 21$$

The lengths of the three sides are therefore 20 cm, 21 cm and 29 cm.

Make sure that you answer the question fully.

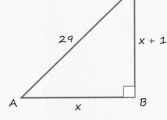

Figure 3.10

Exercise 3.1

 ① Find two numbers that:

(i)　add to give 5 and multiply to give 6

(ii)　add to give 5 and multiply to give -6

(iii)　add to give -14 and multiply to give -15

(iv)　add to give 0 and multiply to give -16

 ② Factorise the following expressions.

(i)　$x^2 + 6x + 8$　　(ii)　$x^2 - 6x + 8$

(iii)　$y^2 + 2y - 8$　　(iv)　$y^2 - 2y - 8$

(v)　$r^2 + 2r - 15$　　(vi)　$r^2 - 2r - 15$

 ③ Factorise the following expressions where possible. If an expression cannot be factorised, say so.

(i)　$s^2 - 4s + 4$　　(ii)　$s^2 + 4s + 4$

(iii)　$p^2 - 4$　　(iv)　$p^2 + 4$

(v)　$a^2 + 2a - 3$　　(vi)　$a^2 + 2a + 3$

④ Solve these equations by factorising.

(i)　$x^2 - 7x + 6 = 0$

(ii)　$x^2 + 7x + 6 = 0$

(iii)　$x^2 - 5x + 6 = 0$

(iv)　$x^2 + 5x + 6 = 0$

(v)　$x^2 - 6x - 7 = 0$

(vi)　$x^2 + 6x - 7 = 0$

⑤ Use the results from Question 4 to draw the graphs of the following functions.

(i)　$y = x^2 - 7x + 6$

(ii)　$y = x^2 + 7x + 6$

(iii)　$y = x^2 - 5x + 6$

(iv)　$y = x^2 + 5x + 6$

(v)　$y = x^2 - 6x - 7$

(vi)　$y = x^2 + 6x - 7$

⑥ Factorise the following expressions.

(i) $4 + 3x - x^2$ (ii) $4 - 3x - x^2$

(iii) $12 + x - x^2$ (iv) $12 - x - x^2$

(v) $35 - 2x - x^2$ (vi) $35 + 2x - x^2$

⑦ Factorise the following expressions.

(i) $2x^2 + 5x + 2$

(ii) $2x^2 - 5x + 2$

(iii) $5x^2 + 11x + 2$

(iv) $5x^2 - 11x + 2$

(v) $2x^2 + 14x + 24$

(vi) $2x^2 - 14x + 24$

⑧ Factorise the following expressions.

(i) $1 + x - 6x^2$ (ii) $1 - x - 6x^2$

(iii) $5 + 3x - 2x^2$ (iv) $5 - 3x - 2x^2$

⌨ TECHNOLOGY

You can use a graphical calculator or graphing software to check your graphs.

⑨ (i) Solve $6x^2 - 7x + 2 = 0$ and hence sketch the graph of $y = 6x^2 - 7x + 2$.

(ii) Solve $9x^2 - 12x + 4 = 0$ and hence sketch the graph of $y = 9x^2 - 12x + 4$.

(iii) Solve $4x^2 - 9 = 0$ and hence sketch the graph of $y = 4x^2 - 9$.

⑩ (i) Solve $12 + 5x - 2x^2 = 0$ and hence sketch the graph of $y = 12 + 5x - 2x^2$.

(ii) Solve $30 - 13x - 3x^2 = 0$ and hence sketch the graph of $y = 30 - 13x - 3x^2$.

(iii) Solve $16 - 25x^2 = 0$ and hence sketch the graph of $y = 16 - 25x^2$.

⑪ Solve the following equations.

(i) $x^4 - 13x^2 + 36 = 0$

(ii) $x^4 - x^2 - 2 = 0$

(iii) $x - 4\sqrt{x} + 3 = 0$

(iv) $x^{\frac{2}{3}} + x^{\frac{1}{3}} - 2 = 0$

(M) ⑫ The length of a rectangular field is 30 m greater than its width, w metres.

(i) Write down an expression for the area $A\,\text{m}^2$ of the field, in terms of w.

(ii) The area of the field is $8800\,\text{m}^2$. Find its width and perimeter.

(M) ⑬ A cylindrical tin of height h cm and radius r cm has surface area, including its top and bottom, of $A\,\text{cm}^2$.

(i) Write down an expression for A in terms of $r, h,$ and π.

(PS) (ii) A tin of height 6 cm has surface area $54\pi\,\text{cm}^2$. What is the volume of the tin?

(PS) (iii) Another tin has the same diameter as height. Its surface area is $150\pi\,\text{cm}^2$. What is its volume?

⑭ A triangle ABC has a right angle at B. The length of AB is $(3x - 2)$ cm and the length of BC is $(x + 1)$ cm. Find the perimeter when the area is $14\,\text{cm}^2$.

⑮ A right-angled triangle has sides of length x, $y^2 - 1$ and $y^2 + 1$, where x and y are integers, and $y > x$.

(i) Find x in terms of y.

(ii) Find the lengths of the sides in the cases where $y = 2, 4,$ and 6.

(PS) (iii) Find the values of y, if they exist, for the right-angled triangles with sides

(a) $20, 99, 101$ (b) $20, 21, 29$.

Note

Sets of numbers connected in this way are called **Pythagorean triples**. There is a wealth of information on the internet about these sets of numbers. Wikipedia is a good place to start.

(M) ⑯ 10 cm squares are cut from the corners of a sheet of cardboard which is then folded to make a box. The sheets of cardboard are twice as long as they are wide.

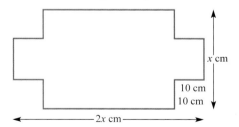

Figure 3.11

(i) For a sheet of cardboard whose width is x cm, find an expression for the volume of the box in terms of x.

(ii) Find the dimensions of the sheet of cardboard needed to make a box with a volume of $8320\,\text{cm}^3$.

2 The completed square form

You will have noticed that all of the quadratic graphs that you have seen have a vertical line of symmetry.

ACTIVITY 3.2

(i) Show, by removing the brackets, that $(x - 3)^2 - 7$ is the same as $x^2 - 6x + 2$.

(ii) Draw the graph of $y = x^2 - 6x + 2$ (you may want to use graphing software to do this). From your graph, write down the equation of the line of symmetry and the coordinates of the turning point. What do you notice?

Repeat (i) and (ii) starting with $(x + 1)^2 + 3$.

(iii) Notice that each time the number in the bracket is half the coefficient of x. This is not a coincidence. Why is that the case?

Look at the curve in Figure 3.12. It is the graph of $y = x^2 - 4x + 5$ and it has the characteristic shape of a quadratic; it is a **parabola**.

Notice that:

- It has a minimum point (the **vertex**) at $(2, 1)$.

- It has a line of symmetry, $x = 2$.

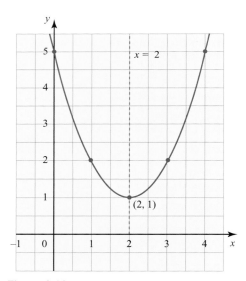

Figure 3.12

In Activity 3.2, you saw that if a quadratic expression is written in the form $(x - p)^2 + q$, you can deduce the vertex and line of symmetry. Writing a quadratic expression in this form is called **completing the square**.

You can write the expression $x^2 - 4x + 5$ in the completed square form like this. Rewrite the expression with the constant term moved to the right

$x^2 - 4x$ $\qquad\qquad +5$

take the coefficient of x $\quad -4$

divide it by 2 $\qquad\qquad -2$

square it $\qquad\qquad\quad +4$.

Add this to the left-hand part and compensate by subtracting it from the constant term on the right

$\qquad x^2 - 4x + 4 \qquad\quad + 5 - 4$.

This can now be written as $(x - 2)^2 + 1$.

This is the completed square form.

The line of symmetry is $x - 2 = 0$ or $x = 2$.

The minimum value is 1. This is when $(x - 2)^2 = 0$, so the vertex is $(2, 1)$.

Example 3.9

Write $x^2 + 5x + 4$ in completed square form.

Hence state the equation of the line of symmetry and the coordinates of the vertex of the curve $y = x^2 + 5x + 4$.

Solution

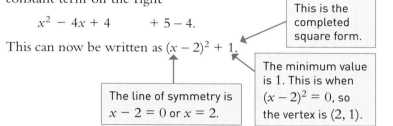

$$x^2 + 5x \qquad\qquad + 4$$
$$= x^2 + 5x + \frac{25}{4} + 4 - \frac{25}{4}$$
$$= \left(x + \frac{5}{2}\right)^2 - \frac{9}{4}$$

$5 \div 2 = \frac{5}{2}$

$\left(\frac{5}{2}\right)^2 = \frac{25}{4}$

This is the completed square form.

The line of symmetry is $x + \frac{5}{2} = 0$, or $x = -\frac{5}{2}$.

The vertex is $\left(-\frac{5}{2}, -\frac{9}{4}\right)$.

Discussion points

The steps in this process ensure that you end up with an expression of the form $(x + p)^2 + q$.

→ How does this work?

→ How is it that the first three terms in the expanded form are always a perfect square?

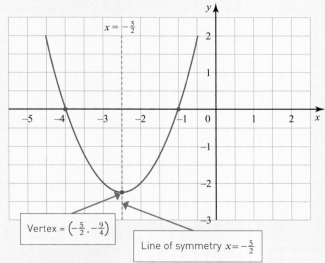

Vertex $= \left(-\frac{5}{2}, -\frac{9}{4}\right)$

Line of symmetry $x = -\frac{5}{2}$

Figure 3.13

For this method to work the coefficient of x^2 must be 1. The next two examples show how the method can be adapted when this is not the case.

Example 3.10

Use the method of completing the square to write down the equation of the line of symmetry and the coordinates of the vertex of the curve $y = 2x^2 + 6x + 5$.

Solution

First take out a factor of 2:

$$2x^2 + 6x + 5 = 2\left[x^2 + 3x + \frac{5}{2}\right]$$

Now proceed as before with the expression in the square brackets:

$$2\left[x^2 + 3x + \frac{5}{2}\right] = 2\left[\left(x + \frac{3}{2}\right)^2 - \left(\frac{3}{2}\right)^2 + \frac{5}{2}\right]$$

$$= 2\left[\left(x + \frac{3}{2}\right)^2 - \frac{9}{4} + \frac{5}{2}\right]$$

$$= 2\left[\left(x + \frac{3}{2}\right)^2 + \frac{1}{4}\right]$$

$$= 2\left(x + \frac{3}{2}\right)^2 + \frac{1}{2}$$

> **Discussion point**
>
> → How can you tell if a quadratic function has a greatest value rather than a least one?

The least value of the expression in the brackets is when $x + \frac{3}{2} = 0$, so the equation of the line of symmetry is $x = -\frac{3}{2}$.

The vertex is $\left(-\frac{3}{2}, \frac{1}{2}\right)$.

Example 3.11

Use the method of completing the square to find the equation of the line of symmetry and the coordinates of the vertex of the curve $y = -x^2 + 6x + 5$.

Solution

Start by taking out a factor of (-1):

$$-x^2 + 6x + 5 = -\left[x^2 - 6x - 5\right]$$

Completing the square for the expression in square brackets gives:

$$-\left[x^2 - 6x - 5\right] = -\left[(x - 3)^2 - 3^2 - 5\right]$$

$$= -\left[(x - 3)^2 - 14\right]$$

$$= -(x - 3)^2 + 14$$

The least value of the expression in the brackets is when $x - 3 = 0$, so the equation of the line of symmetry is $x = 3$.

The vertex is $(3, 14)$. This is a maximum point.

> It is a maximum since the coefficient of x^2 is negative.

ACTIVITY 3.3

This activity is based on the equation $x^2 - 6x + 2 = 0$.

(i)   How do you know that $x^2 - 6x + 2$ cannot be factorised?
(ii) Draw the graph of $y = x^2 - 6x + 2$, either by plotting points or by using a graphical calculator or graphing software, and use your graph to estimate the roots of the equation.
(iii) Substitute each of these values into $x^2 - 6x + 2$.
(iv) How do you know that the roots that you have found are not exact?

Solving a quadratic equation by completing the square

If a quadratic equation has real roots, the method of completing the square can also be used to solve the equation. The steps involved are shown here for the equation used in Activity 3.3. It is a powerful method because it can be used on any quadratic equation.

Example 3.12

(i)   Write the equation $x^2 - 6x + 2$ in the completed square form.

(ii) Hence solve the equation $x^2 - 6x + 2 = 0$.

Solution

(i) $x^2 - 6x + 2 = (x - 3)^2 - 3^2 + 2$
$$= (x - 3)^2 - 7$$

(ii) $(x - 3)^2 - 7 = 0$

$\Rightarrow (x - 3)^2 = 7$

Take the square root of both sides:

$\Rightarrow x - 3 = \pm\sqrt{7}$

$\Rightarrow x \quad = 3 \pm \sqrt{7}$ ⟵ This is an exact answer.

Using your calculator to find the value of $\sqrt{7}$

$\Rightarrow x = 5.646$ or 0.354, to 3 decimal places. ⟵ This is an approximate answer.

The completed square form

① Which of the following are perfect squares?

$x^2 - 4x + 4$ 63

$(x + 1)^2$ $x^2 + 3x + 9$ 49

$(x - 4)^2 - 1$ $x^2 - 2x + 2$

② Choose the point where the function $(x - 3)^2 + 5$ has a minimum point.

A $(-3, 5)$ B $(-3, -5)$

C $(3, 5)$ D $(3, -5)$

③ Multiply out and simplify the following.

(i) $(x + 2)^2 - 3$ (ii) $(x + 4)^2 - 4$

(iii) $(x - 1)^2 + 2$ (iv) $(x - 10)^2 + 12$

(v) $(x - \frac{1}{2})^2 + \frac{3}{4}$ (vi) $(x + 0.1)^2 + 0.99$

④ Solve the following equations without multiplying out, leaving your answers in surd form.

(i) $(x + 1)^2 = 10$ (ii) $(x - 2)^2 = 5$

(iii) $(x - 3)^2 + 3 = 8$

(iv) $(2x + 1)^2 = 3$ (v) $(2x - 3)^2 = 12$

(vi) $(3x - 1)^2 + 1 = 19$

⑤ Write the following expressions in completed square form.

(i) $x^2 + 4x + 5$ (ii) $x^2 - 6x + 3$

(iii) $x^2 + 2x - 5$ (iv) $x^2 - 8x - 4$

TECHNOLOGY

You could use graphing software or a graphical calculator to check your work.

⑥ For each of the following equations:

(a) write it in completed square form

(b) hence write down the equation of the line of symmetry and the coordinates of the vertex

(c) sketch the curve.

(i) $y = x^2 + 4x + 9$

(ii) $y = x^2 - 4x + 9$

(iii) $y = x^2 + 4x + 3$

(iv) $y = x^2 - 4x + 3$

(v) $y = x^2 + 6x - 1$

(vi) $y = x^2 - 6x - 1$

TECHNOLOGY

You can also find maximum and minimum points using a graphical calculator which might help you to check your answers.

⑦ For each of the following:

(a) write the expression in completed square form

(b) identify the vertex and the line of symmetry and sketch the graph.

(i) $y = 2x^2 + 4x + 6$

(ii) $y = 2x^2 - 4x + 6$

(iii) $y = 3x^2 - 18x + 30$

(iv) $y = 3x^2 + 18x + 30$

⑧ For each of the following:

(a) write the expression in completed square form

(b) identify the vertex and the line of symmetry and sketch the graph.

(i) $y = -x^2 - 2x + 5$

(ii) $y = -x^2 + 2x + 5$

(iii) $y = -3x^2 - 12x - 15$

(iv) $y = -3x^2 + 12x - 15$

⑨ For each of the following:

(a) write the expression in completed square form

(b) identify the vertex and the line of symmetry and sketch the graph.

(i) $y = x^2 - 4x + 7$

(ii) $y = 2x^2 - 8x + 5$

(iii) $y = -x^2 - 8x - 7$

(iv) $y = -3x^2 + 4x - 5$

PS ⑩ The curves below all have equations of the form $y = x^2 + bx + c$. In each case find the values of b and c.

(i)

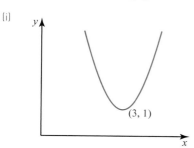

(3, 1)

(ii)

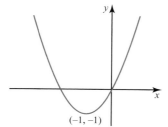

(−1, −1)

(iii)

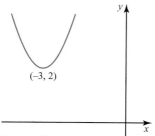

(4, 0)

(iv)

(−3, 2)

Figure 3.14

⑪ (i) Write $x^2 − 9x + 8$ in the form $(x − a)^2 + b$.

(ii) State the coordinates of the vertex of the curve $y = x^2 − 9x + 8$.

(iii) Find the points where the graph of $y = x^2 − 9x + 8$ crosses the axes.

(iv) Sketch the graph of $y = x^2 − 9x + 8$ and label the points found in (ii) and (iii).

PS ⑫ (i) Write $x^2 − 4x + 6$ in completed square form and hence state the coordinates of the vertex of the graph of $y = x^2 − 4x + 6$.

(ii) Without any further working, for what values of c is the curve $y = x^2 − 4x + c$ completely above the x axis?

⑬ (i) (a) Write $x^2 + 22x + 85$ in completed square form.

(b) Hence solve the equation $x^2 + 22x + 85 = 0$.

(ii) (a) Write $x^2 − 24x + 63$ in completed square form.

(b) Hence solve the equation $x^2 − 24x + 63 = 0$.

⑭ (i) Write $y = 3x^2 − 6x + 14$ in the form $y = a(x − p)^2 + q$.

(ii) Write down the equation of the line of symmetry and the coordinates of the vertex and hence sketch the curve, showing as much information as possible in your diagram.

PS ⑮ (i) Find two possible values of a such that $y = x^2 + ax + 9$ touches the x axis.

(ii) On the same axes, sketch the curves that you obtain using each of these values and say what you notice about them.

3 The quadratic formula

Completing the square is a powerful method because it can be used on any quadratic equation. However, it is seldom used to solve an equation because it can be generalised to give a formula which can be used instead to solve any quadratic equation.

To solve a general quadratic equation $ax^2 + bx + c = 0$ by completing the square, modify the steps in the last example.

First divide both sides by a:

$$\Rightarrow \quad x^2 + \frac{bx}{a} + \frac{c}{a} = 0$$

Subtract the constant term from both sides of the equation:

$$\Rightarrow \quad x^2 + \frac{bx}{a} = -\frac{c}{a}$$

Take the coefficient of x: $+\dfrac{b}{a}$

Halve it: $\qquad\qquad +\dfrac{b}{2a}$

Square the answer: $\qquad +\dfrac{b^2}{4a^2}$

Add it to both sides of the equation:

$$\Rightarrow \quad x^2 + \frac{bx}{a} + \frac{b^2}{4a^2} = \frac{b^2}{4a^2} - \frac{c}{a}$$

Factorise the left-hand side and tidy up the right-hand side:

$$\Rightarrow \quad \left(x + \frac{b}{2a}\right)^2 = \frac{b^2 - 4ac}{4a^2}$$

Take the square root of both sides:

$$\Rightarrow \quad x + \frac{b}{2a} = \pm \frac{\sqrt{b^2 - 4ac}}{2a}$$

$$\Rightarrow \quad x = \frac{-b \pm \sqrt{b^2 - 4ac}}{2a}$$

This important result, known as the **quadratic formula**, has significance beyond the solution of awkward quadratic equations, as you will see later. The next two examples, however, demonstrate its use as a tool for solving equations.

Example 3.13

Note

Although this quadratic formula is generally used to give an approximate solution, manipulating the surd expression gives an exact solution:

$$x = \frac{6 \pm \sqrt{36 - 24}}{6}$$

$$= 1 \pm \frac{\sqrt{3}}{3}$$

Use the quadratic formula to solve $3x^2 - 6x + 2 = 0$.

Solution

Comparing it to the form $ax^2 + bx + c = 0$ gives $a = 3$, $b = -6$, $c = 2$.

Substituting these values into the formula

$$x = \frac{-b \pm \sqrt{b^2 - 4ac}}{2a}$$

gives $x = \dfrac{6 \pm \sqrt{36 - 24}}{6}$

$$= 0.423 \text{ or } 1.577 \text{ (to 3 d.p.)}$$

Example 3.14

(i) Show that $x^2 - 2x + 2$ cannot be factorised.

(ii) What happens when you try to use the quadratic formula to solve $x^2 - 2x + 2 = 0$?

(iii) Draw the graph of $y = x^2 - 2x + 2$.

(iv) How does your graph explain your answers to (ii)?

Solution

(i) The only two whole numbers which multiply to give 2 are 2 and 1 (or −2 and −1) and they cannot be added to get −2.

(ii) Comparing $x^2 - 2x + 2$ to the form $ax^2 + bx + c = 0$ gives $a = 1$, $b = -2$ and $c = 2$.

Substituting these values in $x = \dfrac{-b \pm \sqrt{b^2 - 4ac}}{2a}$

gives $x = \dfrac{2 \pm \sqrt{4 - 8}}{2}$

$= \dfrac{2 \pm \sqrt{-4}}{2}$

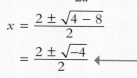

There is no real number equal to $\sqrt{-4}$ so there is no solution.

(iii)

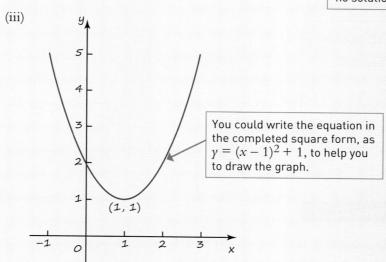

You could write the equation in the completed square form, as $y = (x - 1)^2 + 1$, to help you to draw the graph.

Figure 3.15

(iv) Sketching the graph shows that it does not cut the x axis and so there is indeed no real solution to the equation.

The discriminant

The part of the quadratic formula which determines whether or not there are real roots is the part under the square root sign. This is called the **discriminant**.

$$x = \frac{-b \pm \sqrt{b^2 - 4ac}}{2a}$$

the discriminant, $b^2 - 4ac$

The quadratic formula

Note

It is not quite true to say that a negative number has no square root. Certainly it has none among the real numbers but mathematicians have invented an imaginary number, which is denoted by i (or sometimes j) with the property $i^2 = -1$. Numbers such as $1 + i$ and $-1 - i$ (which are the solutions to the equation in Example 3.14 on the previous page) are called **complex numbers**. Complex numbers are extremely useful in both pure and applied mathematics; they are covered in Further Mathematics.

If $b^2 - 4ac > 0$, the equation has two real roots (see Figure 3.16). (These roots will be rational if $b^2 - 4ac$ is a square number.)

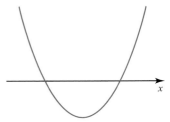

Figure 3.16

If $b^2 - 4ac < 0$, the equation has no real roots (see Figure 3.17).

Figure 3.17

If $b^2 - 4ac = 0$, the equation has one repeated root (see Figure 3.18).

Discussion points

→ Use the quadratic formula to solve the equation
$x^2 + 6x + 91 = 0$.

→ What do you notice?

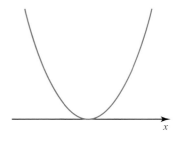

Figure 3.18

Exercise 3.3

 ① The equation $3x^2 - 5x + 1 = 0$ can be solved using the quadratic formula.

Which of the following substitutions will give the correct roots?

Explain the error in each of the incorrect ones.

(i) $\dfrac{-3 \pm \sqrt{(-5)^2 - 4 \times 1 \times 5}}{2 \times 3}$

(ii) $\dfrac{-5 + \sqrt{-5^2 - 4 \times 1 \times 3}}{2 \times 3}$

(iii) $\dfrac{5 \pm \sqrt{5^2 - 4 \times 1 \times 3}}{2}$

(iv) $\dfrac{5 \pm \sqrt{5^2 - 4 \times 1 \times 3}}{2 \times 3}$

(v) $\dfrac{5 \pm \sqrt{(-5)^2 - 4 \times 1 \times 3}}{2 \times 3}$

(vi) $5 \pm \dfrac{\sqrt{(-5)^2 - 4 \times 1 \times 3}}{2 \times 3}$

📰 TECHNOLOGY

Check your answers using equation mode if your calculator has it. This will give you the roots of the quadratic. If you are using the solve function, you will need to choose different starting values to make sure you get both roots.

② Use the quadratic formula to solve the following equations, giving your answers to 3 s.f.

(i) $x^2 + 8x + 5 = 0$

(ii) $x^2 + 2x - 4 = 0$

(iii) $x^2 - 5x - 19 = 0$

(iv) $x^2 + 3x - 1 = 0$

(v) $2x^2 - 2x - 1 = 0$

(vi) $3x^2 + x - 9 = 0$

③ Solve the following equations using the quadratic formula, giving your answers to 2 d.p.

(i) $3x^2 + 5x + 1 = 0$

(ii) $3x^2 - 5x + 1 = 0$

(iii) $2x^2 + 11x - 4 = 0$

(iv) $2x^2 + 11x + 4 = 0$

(v) $5x^2 - 10x + 1 = 0$

(vi) $5x^2 - 10x - 1 = 0$

④ For each of the following equations:

(a) find the discriminant

(b) state the number of real roots

(c) say whether or not those roots will be rational numbers.

(i) $x^2 + 6x + 4 = 0$

(ii) $x^2 - 28x + 196 = 0$

(iii) $x^2 + 6x + 12 = 0$

(iv) $2p^2 - 5p - 3$

(v) $10c^2 - 20c + 10$

(vi) $5r^2 - 15r + 25 = 0$

Remember, rational numbers are ones that can be written as a fraction.

⑤ Solve the following equations using the quadratic formula, leaving your answers in surd form.

(i) $x^2 + 1 = 4x$

(ii) $3x^2 = 1 - x$

(iii) $1 + x - 3x^2 = 0$

(iv) $1 = 5x - x^2$

(v) $8x - x^2 = 1$

(vi) $3x = 5 - 4x^2$

PS ⑥ For each of these equations find

(a) the discriminant and hence

(b) a value of c so that the equation has equal roots

(c) a range of values of c so that the equation has real roots.

(i) $x^2 - 4x + c = 0$

(ii) $2x^2 + 6x + c = 0$

(iii) $3x^2 + 4x + c = 0$

(iv) $c + 2x - 5x^2 = 0$

⑦ In each case write one of the symbols $\Rightarrow, \Leftarrow$ or $\Leftrightarrow$ between the two statements A and B.

(i) A: $x = 2$ B: $x^2 - 4 = 0$

(ii) A: $ax^2 + bx + c$ B: $b^2 - 4ac = 0$ is a perfect square

(iii) A: $b^2 - 4ac < 0$ B: The graph of $y = ax^2 + bx + c$ doesn't cross or touch the x axis

(iv) A: The equation   B: The graph of $ax^2 + bx + c = 0$ $y = ax^2 + bx + c$ has no real roots   lies completely above the x axis

(v) A: The equation   B: The expression $ax^2 + bx + c = 0$ $ax^2 + bx + c$ has no real roots   cannot be factorised

⑧ The following are all true statements. For each one,

(a) state the converse and

(b) say whether or not the converse is true.

(i) The graph of $y = ax^2 + bx + c$ touches the x axis $\Rightarrow b^2 - 4ac = 0$

(ii) $b^2 - 4ac < 0 \Rightarrow ax^2 + bx + c$ cannot be factorised.

(iii) $x = 3 \Rightarrow$ $x^2 - 9 = 0$

⑨ The height h metres of a ball at time t seconds after it is thrown up in the air is given by the expression
$$h = 1 + 15t - 5t^2.$$

(i) Find the times at which the height is 11 m.

(ii) Find the time at which the ball hits the ground.

(iii) What is the greatest height the ball reached?

⑩ The quadratic equation
$$x^2 - 4x - 1 = 2k(x - 5)$$ has two

equal roots. Calculate the possible values of k.

⑪ The height of a ball, h m, at time t seconds is given by
$$h = 6 + 20t - 4.9t^2.$$

Prove that the ball does not reach a height of 32 m.

⑫ (i) For what set of values of k does the equation
$$x^2 + (3k - 1)x + 10 + 2k = 0$$ have real roots?

(ii) Find the values of k for which it has equal roots and find these equal roots for each value of k.

LEARNING OUTCOMES

When you have completed this chapter you should be able to:

➤ work with quadratic functions and their graphs:
  ○ the discriminant of a quadratic function
  ○ the conditions for real and repeated roots
  ○ completing the square
  ○ solve quadratic equations in a function of the unknown.

KEY POINTS

1 Some quadratic equations can be solved by factorising.

2 You can sketch a quadratic graph by finding the points where the curve crosses the coordinate axes.

3 The vertex and the line of symmetry of a quadratic graph can be found by completing the square.

> You need to know the quadratic formula.

4 The quadratic formula $x = \dfrac{-b \pm \sqrt{b^2 - 4ac}}{2a}$ can be used to solve the quadratic equation $ax^2 + bx + c = 0$.

5 For the quadratic equation $ax^2 + bx + c = 0$, the discriminant is given by $b^2 - 4ac$.

  (i) If the discriminant is positive, the equation has two real roots. If the discriminant is a perfect square, these roots are rational.

  (ii) If the discriminant is zero, the equation has a repeated real root.

  (iii) If the discriminant is negative, the equation has no real roots.

FUTURE USES

- Solving simultaneous equations in which one equation is linear and one quadratic in Chapter 4.
- Finding points where a line intersects a circle in Chapter 5.
- Solving cubic equations and other polynomial equations in Chapter 7.
- You will use quadratic equations in many other areas of mathematics.

4

Equations and inequalities

Saira and her friends are going to the cinema and want to buy some packets of nuts and packets of crisps. A packet of nuts costs 40 pence more than a packet of crisps. Two packets of nuts cost the same as three packets of crisps. What is the cost of each item?

→   This is the type of question that you may find in a puzzle book. How would you set about tackling it?

→   You may think that the following question appears very similar. What happens when you try to find the answer?

A packet of nuts cost 40 pence more than a packet of crisps. Four packets of nuts and three packets of crisps cost 40p more than three packets of nuts and four packets of crisps. What is the cost of each item?

1 Simultaneous equations

Prior knowledge

You will have met this at GCSE.

Simultaneous equations – both equations linear

There are many situations which can only be described mathematically in terms of more than one variable and to find the value of each variable you need to solve two or more equations simultaneously (i.e. at the same time). Such equations are called **simultaneous equations**.

> **Discussion points**
>
> ➔ Why is one equation such as $2x + 3y = 5$ not enough to find the values of x and y? How many equations do you need?
>
> ➔ How many equations would you need to find the values of three variables x, y and z?

Example 4.1

Solution by substitution

Solve the simultaneous equations
$$x + y = 4$$
$$y = 2x + 1$$

Solution

$$x + y = 4$$

> Take the expression for y from the second equation and substitute it into the first.

$$\Rightarrow x + (2x + 1) = 4$$
$$\Rightarrow \qquad 3x = 3$$
$$\Rightarrow \qquad x = 1$$

Note

This is a useful method when y is already the subject of one of the equations.

Substituting $x = 1$ into $y = 2x + 1$ gives $y = 3$ so the solution is $x = 1, y = 3$.

Example 4.2

Solution by elimination

Solve the simultaneous equations
$$3x + 5y = 12$$
$$2x + 3y = 7$$

Solution

Multiplying the first equation by 2 and the second by 3 will give two equations each containing the term $6x$:

Multiplying $3x + 5y = 12$ by 2 $\Rightarrow 6x + 10y = 24$

Multiplying $2x + 3y = 7$ by 3 $\Rightarrow 6x + 9y = 21$

> This gives two equations containing the term $6x$.

Subtracting $\qquad\qquad\qquad\qquad y = 3$

Substituting this into either of the original equations gives $x = -1$.

The solution is $x = -1, y = 3$.

Sometimes you may need to add the two equations rather than subtracting them.

Example 4.3

Note

To find the value of the second variable you can substitute into either of the original equations – just choose the one where the numbers look 'easier' to you.

Note

It would have been equally correct to start by eliminating x by multiplying the first equation by 5 and the second by 9 and then subtracting one of the resulting equations from the other, but this would lead to slightly more complicated arithmetic.

TECHNOLOGY

You could use graphing software or a graphical calculator to do this.

Solution by elimination

Solve the simultaneous equations
$$9x - 2y = 5$$
$$5x + 3y = 11$$

Solution

Multiplying $9x - 2y = 5$ by 3 $\Rightarrow 27x - 6y = 15$

Multiplying $5x + 3y = 11$ by 2 $\Rightarrow 10x + 6y = 22$

This gives two equations each containing $6y$.

Adding $\qquad\qquad 37x \qquad = 37$

Hence $\qquad\qquad\quad x \qquad = 1$

Substituting this into $5x + 3y = 11$ gives $3y = 6$ and hence $y = 2$.

The solution is $x = 1$, $y = 2$.

A useful rule

When you have decided which variable you want to eliminate, look at its sign in each of the two equations. If in each case it has the **S**ame **S**ign, then **S**ubtract one equation from the other to eliminate it. If the signs are **D**ifferent, then a**DD** them – the **D** rule!

ACTIVITY 4.1

On a single set of axes draw the graphs of $x + y = 4$ and $y = 2x + 1$.

From your diagram, find the coordinates of the point where the two lines cross.

The coordinates of every point on the line $x + y = 4$ satisfy the equation $x + y = 4$. Similarly the coordinates of every point on the line $y = 2x + 1$ satisfy the equation $y = 2x + 1$. The point where the lines cross lies on both lines, so its coordinates satisfy both equations and therefore give the solution to the simultaneous equations.

Repeat this activity for the graphs of $2x + 3y = 7$ and $5x - 4y = 4$. How accurate is your answer?

Example 4.4

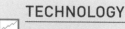

TECHNOLOGY

You may have equation mode on your calculator. You can use this to check answers. The equations must be set out like this before you use it.

Solve the simultaneous equations $2x + 3y = 7$ and $5x - 4y = 4$ using the elimination method.

Solution

Multiplying $2x + 3y = 7$ by 5 $\quad\Rightarrow 10x + 15y = 35$

Multiplying $5x - 4y = 4$ by 2 $\quad\Rightarrow 10x - 8y = 8$

Subtracting $\qquad\qquad\qquad\qquad \Rightarrow \qquad\quad 23y = 27$

Hence $y = \frac{27}{23} = 1\frac{4}{23}$ ($= 1.17$ to 2 d.p.).

Substituting gives $x = \frac{40}{23} = 1\frac{17}{23}$ ($= 1.74$ to 2 d.p.).

Note

In Activity 4.1, you probably found that you could only find an approximate value for the coordinates of the intersection point from your graphs. In Example 4.4 you can see how the same pair of simultaneous equations are solved algebraically. This gives exact fractional values for x and y.

 Note

Many calculators are able to solve simultaneous equations and you can use this facility to check your answer, but you do need to know how to solve them algebraically. You also need to be able to draw their graphs and use them to estimate the solution.

Prior knowledge

You need to be able to solve quadratic equations. This is covered in Chapter 3.

Simultaneous equations – one linear and one quadratic

Either the substitution method or the graphical method can be used to solve a pair of simultaneous equations, one of which is linear and the other quadratic. As before, the graphical method may not be completely accurate.

When using the substitution method **always** start by making one of the variables the subject of the **linear** equation. (It can be either variable – just choose the 'easiest'.)

Example 4.5

Solve the simultaneous equations $\begin{array}{l} y = x^2 + x \\ 2x + y = 4 \end{array}$ using the substitution method.

Solution

From the linear equation, $y = 4 - 2x$.

Substituting this into the quadratic equation:
$$4 - 2x = x^2 + x$$
$$\Rightarrow \quad 0 = x^2 + 3x - 4$$
$$\Rightarrow \quad (x + 4)(x - 1) = 0$$
$$\Rightarrow \quad x = -4 \text{ or } x = 1$$

Collecting all terms on the right-hand side (since that is where the x^2 term is positive) makes factorising the quadratic easier.

Substituting into $2x + y = 4$

$$x = -4 \quad \Rightarrow -8 + y = 4 \quad \Rightarrow y = 12$$
$$x = 1 \quad \Rightarrow 2 + y = 4 \quad \Rightarrow y = 2$$

The solution is $x = -4$, $y = 12$ and $x = 1$, $y = 2$.

 Note

You should always substitute the linear equation into the quadratic one, as in some situations substituting the quadratic equation into the linear one will give you incorrect extra answers. In any case, it is easier to substitute the linear one into the quadratic!

You could also solve the pair of equations from Example 4.5 by drawing their graphs, as shown in Figure 4.1.

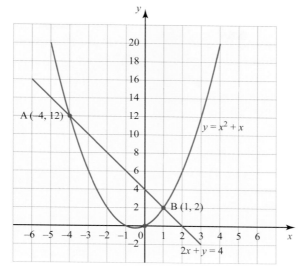

Discussion points

➜ Are there always two points where a straight line meets a quadratic graph?

➜ If not, what would happen if you try to solve the simultaneous equations graphically?

Figure 4.1

You can see that the points A and B represent the solution of the equations. At point A, $x = -4$ and $y = 12$, and at point B, $x = 1$ and $y = 2$.

Exercise 4.1

TECHNOLOGY

Answer these questions without using the equation solver on a calculator. Using graphing software or a graphical calculator is an important skill in mathematics; treat checking your answers to these questions as an opportunity to develop and practise it.

① If $x = -2$ and $y = 3$, which of these equations are true?

$y = -2x - 1$ $3x - 2y = 0$
$4x + 3y = 1$ $2x + 3y = 6$

② Which answer gives the solution to the following simultaneous equations?

$4x - y = -7$ and $3y + 2x = 7$

A $x = 1, y = 11$ B $x = -1, y = 3$
C $x = 0.5, y = 2$ D $x = -2, y = 1$

③ Solve the following pairs of simultaneous equations

     (a) by drawing accurate graphs
     (b) by an algebraic method.

(i) $x = 7$
       $2x + 3y = 20$

(ii) $x + y = 10$
       $2x - 3y = 10$

(iii) $2a + b = 8$
       $a + 2b = 1$

④ Solve the following pairs of simultaneous equations.

(i) $y = 3x - 2$ (ii) $y = 2x + 5$
    $y = x + 4$ $x + y = 8$

(iii) $x + y = 7$ (iv) $a - 3b = 7$
    $x - y = 4$ $5a + b = 3$

⑤ Solve the following pairs of simultaneous equations.

(i) $4x + 3y = 5$ (ii) $5x + 4y = 11$
   $2x - 6y = -5$ $2x - 3y = 9$

(iii) $4l - 3m = 2$ (iv) $3r - 2s = 17$
    $5l - 7m = 9$ $5r - 3s = 28$

⑥ A student wishes to spend exactly £20 at a second-hand bookshop. All the paperbacks are one price, all the hardbacks another. She can buy five paperbacks and eight hardbacks. Alternatively she can buy ten paperbacks and six hardbacks.

(i) Write this information as a pair of simultaneous equations.

(ii) Solve your equations to find the cost of each type of book.

PS ⑦ Two adults and one child pay £75 to go to the theatre. The cost for one adult and three children is also £75.

(i) Find the cost of each adult and each child.

(ii) How much would it cost for two adults and five children to go to the theatre?

M ⑧ A car journey of 380 km lasts 4 hours. Part of the journey is on a motorway at an average speed of 110 km h^{-1}, the rest is on country roads at an average speed of 70 km h^{-1}.

(i) Write this information as a pair of simultaneous equations.

(ii) Solve your equations to find how many kilometres of the journey is spent on each type of road.

PS ⑨ A taxi firm charges a fixed amount plus an additional cost per mile. A journey of five miles costs £10 and a journey of seven miles costs £13.20. How much does a journey of three miles cost?

⑩ Solve the following pairs of simultaneous equations

(a) by drawing accurate graphs

(b) by an algebraic method.

(i) $y = x^2$ (ii) $y = 3x + 1$

$\quad y = 5x - 4$ $2x^2 + y = 3$

⑪ Solve the following pairs of simultaneous equations.

(i) $y = 4 - x$

$\quad x^2 + y^2 = 10$

(ii) $x = 2y$

$\quad x^2 - y^2 + xy = 20$

(iii) $k^2 + km = 8$

$\quad m = k - 6$

(iv) $x + 2y = -3$

$\quad x^2 - 2x + 3y^2 = 11$

⑫ (i) On the same axes draw the line $y = x + 4$ and the curve $y = 4 - x^2$.

(ii) Use an algebraic method to find the coordinates of the points of intersection of the line and the curve.

⑬ (i) On the same axes draw the line $y = x + 1$ and the curve $y = (x - 2)(7 - x)$.

(ii) Find the coordinates of the points of intersection of the line and the curve.

M ⑭ A large window consists of six square panes of glass as shown in Figure 4.2. Each pane is x m by x m, and all the dividing wood is y m wide.

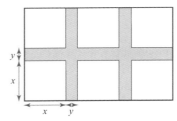

Figure 4.2

(i) Write down the total area of the whole window in terms of x and y.

(ii) Show that the total area of the dividing wood is $7xy + 2y^2$.

(iii) The total area of glass is 1.5 m^2 and the total area of dividing wood is 1 m^2. Find x, and hence find an equation for y and solve it.

⑮ The diagram shows the net of a cylindrical container of radius r cm and height h cm. The full width of the metal sheet from which the container is made is 1 m, and the shaded area is waste. The surface area of the container is 1400π cm^2.

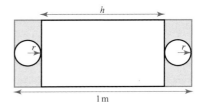

Figure 4.3

(i) Write down a pair of simultaneous equations for r and h.

(ii) Find the volume of the container, giving your answers in terms of π. (There are two possible answers.)

⑯ (i) Solve the pair of simultaneous equations $\quad y = 4 - x^2$

$\qquad\qquad\qquad\qquad y = 5 - 2x$

(ii) Sketch the line and curve given by these equations and use your sketch to explain why there is only one solution.

PS (17) (i) Try to solve the pair of simultaneous equations

$$y = 2x - 15$$
$$y = x^2 - 3x - 4$$

(ii) What happens?

(iii) Explain this result by sketching the graphs of the line and curve on the same axes.

ACTIVITY 4.2

Rachael is shopping for some cakes for a party and has decided that she will buy some muffins at 60p each and some doughnuts at 40p each. She has up to £25 to spend and needs at least 50 cakes. Letting m be the number of muffins bought and d the number of doughnuts bought, write down two inequalities which need to be satisfied by m and d. Find some possible solutions to the problem.

---

2 Inequalities
Linear inequalities

Although Mathematics is generally a precise subject, there are times when the most precise statement that you can make is to say that an answer will lie within certain limits.

The methods for linear inequalities are much the same as those for equations but you must be careful when multiplying or dividing through an inequality by a negative number.

For example: $5 > 3$ is true

Multiply both sides by -1: $-5 > -3$ is false

Multiplying or dividing by a negative number reverses an inequality, but you may prefer to avoid this as shown in the example below.

Example 4.6

Solve $2x - 15 \leqslant 5x - 3$.

Discussion point

➜ What happens if you decide to add 15 and subtract $5x$ from both sides in the first step?

Solution

Add 3 to, and subtract $2x$ from both sides $\Rightarrow -15 + 3 \leqslant 5x - 2x$

Tidy up $\Rightarrow -12 \leqslant 3x$

Divide both sides by 3 $\Rightarrow -4 \leqslant x$

Note that this implies that x is the larger of the two terms.

Make x the subject $\Rightarrow x \geqslant -4$

The solution to an inequality can also be represented as an interval on a number line:

- A solid circle at the end of a line segment means that that value is included in the solution.
- An open circle at the end of a line segment means that the value is *not* included in the solution.

Hence, $x \geqslant -4$ may be illustrated as

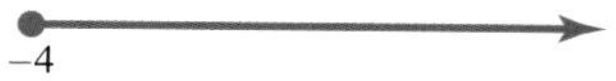

−4

Figure 4.4

Example 4.7

Solve the inequality $8 < 3x - 1 \leqslant 14$ and illustrate the solution on a number line.

Solution

$$8 < 3x - 1 \leqslant 14$$

Add 1 throughout $\quad\quad\quad \Rightarrow 9 < 3x \leqslant 15$

Divide by 3 $\quad\quad\quad\quad\quad \Rightarrow 3 < x \leqslant 5$

This is illustrated as

Figure 4.5

Notice how the different circles show that 5 is included in the solution but 3 is not.

It is useful to emphasise the line segment that you want as shown in Figure 4.5.

Quadratic inequalities

When dealing with a quadratic inequality the easiest method is to sketch the associated graph. You must be able to do this manually, but you may wish to check your graphs using a graphical calculator or graphing software.

Example 4.8

Solve (i) $\quad x^2 - 4x + 3 > 0$ $\quad\quad$ (ii) $\quad x^2 - 4x + 3 \leqslant 0$.

Solution

$$x^2 - 4x + 3 = (x - 1)(x - 3)$$

So the graph of $y = x^2 - 4x + 3$ is a $\cup$-shaped quadratic (shown by the positive coefficient of x^2) which crosses the x axis at $x = 1$ and $x = 3$.

> **Note**
>
> The graph for (i) shows that there are two distinct parts to the solution set; those values of x less than 1 together with those values greater than 5. This requires two statements, joined by 'or' or '$\cup$' to express the solution. You cannot write this solution as a single inequality.
>
> In (ii) there is only one interval that describes the solution, requiring the use of 'and' or '$\cap$'.
>
> You could also write this as $x \leqslant 3$ and $x \geqslant 1$ but the single inequality $1 \leqslant x \leqslant 3$ is more helpful as it makes it clear that x lies between two values.

(i) $x^2 - 4x + 3 > 0$

Values of x correspond to where the curve is above the axis (i.e. $y > 0$).

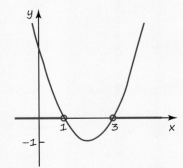

Figure 4.6

(ii) $x^2 - 4x + 3 \leqslant 0$

Values of x correspond to where the curve is on or below the axis (i.e. $y \leqslant 0$).

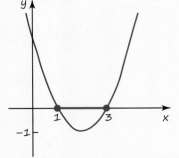

Figure 4.7

There are various ways of expressing these results:

(i) $x < 1$ or $x > 3$ $\quad\quad\quad\quad \{x : x < 1\} \cup \{x : x > 3\}$

(ii) $1 \leqslant x \leqslant 3$ $\quad\quad\quad\quad\quad\quad \{x : x \leqslant 3\} \cap \{x : x \geqslant 1\}$

If the quadratic expression can be factorised, you can solve the quadratic inequality algebraically by thinking about the sign of each factor.

Example 4.9

Solve (i) $x^2 - 4x - 5 < 0$ (ii) $x^2 - 4x - 5 \geqslant 0$

Solution

Start by factorising the quadratic expression and finding the values of x which make the expression zero.

$$x^2 - 4x - 5 = (x + 1)(x - 5)$$

$$(x + 1)(x - 5) = 0 \implies x = -1 \text{ or } x = 5$$

Now think about the sign of each factor at these values and in the three intervals into which they divide the number line:

Table 4.1

	$x < -1$	$x = -1$	$-1 < x < 5$	$x = 5$	$x > 5$
Sign of $(x + 1)$	$-$	0	$+$	$+$	$+$
Sign of $(x - 5)$	$-$	$-$	$-$	0	$+$
Sign of $(x + 1)(x - 5)$	$(-) \times (-)$ $= +$	$(0) \times (-)$ $= 0$	$(+) \times (-)$ $= -$	$(+) \times (0)$ $= 0$	$(+) \times (+)$ $= +$

(i) From the table the solution to $(x + 1)(x - 5) < 0$ is $-1 < x < 5$ which can also be written as $\{x : x > -1\} \cap \{x : x < 5\}$.

(ii) The solution to $(x + 1)(x - 5) \geqslant 0$ is $x \leqslant -1$ or $x \geqslant 5$ which can also be written as $(x : x \leqslant -1\} \cup (x : x \geqslant -5\}$.

REPRESENTING INEQUALITIES GRAPHICALLY

An inequality such as $y > 2x + 1$ describes a set of points for which the x and y coordinates satisfy the inequality. An inequality like this can be represented graphically.

Example 4.10

Show the inequalities (i) $y \leqslant x - 1$ (ii) $y > x^2 + 1$ graphically.

Solution

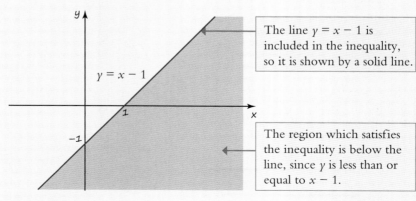

The line $y = x - 1$ is included in the inequality, so it is shown by a solid line.

The region which satisfies the inequality is below the line, since y is less than or equal to $x - 1$.

Figure 4.8

(ii)

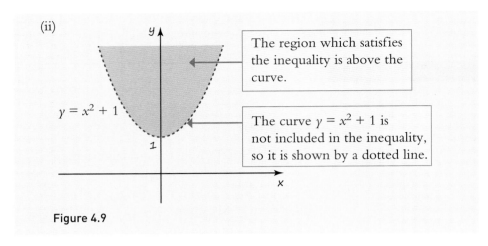

The region which satisfies the inequality is above the curve.

$y = x^2 + 1$

The curve $y = x^2 + 1$ is not included in the inequality, so it is shown by a dotted line.

Figure 4.9

Exercise 4.2

💻 TECHNOLOGY

When working through this chapter, you may wish to use a graphical calculator or graphing software to check your answers where appropriate

① Match each number line with the correct inequality.

$x > -2$ $x > -1$ and $x \leqslant -2$

$x \leqslant 2$ $-1 \leqslant x < 2$

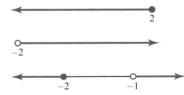

Figure 4.10

② Which inequality is represented by this number line?

Figure 4.11

A $-3 \leqslant x < -1$ B $x \leqslant -3, x > -1$

C $-3 \geqslant x > -1$ D $x < -3, x \geqslant -1$

③ The following number lines each represent an inequality. Express each inequality algebraically.

(i)

(ii)

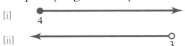

(iii)

(iv)

(v)

(vi)

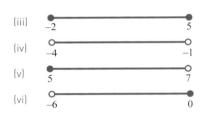

Figure 4.12

④ Illustrate the following inequalities on number lines.

(i) $x > 3$ (ii) $x \leqslant -2$

(iii) $-1 < x < 5$ (iv) $2 \leqslant x \leqslant 6$

(v) $0 < x \leqslant 7$ (vi) $-4 < x \leqslant 1$

⑤ Find the solution sets identified in each of the following graphs:

(i)

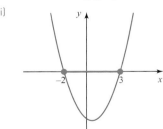

(ii)

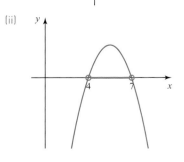

(iii)

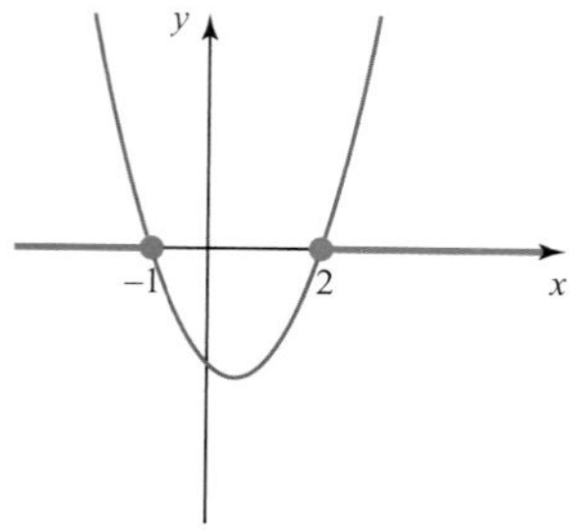

Figure 4.13

⑥ Solve the following inequalities.

(i) $2x - 5 < 9$

(ii) $4 + 3x > 13$

(iii) $2x + 5 \leqslant x + 9$

(iv) $5x - 2 \geqslant 3x - 6$

(v) $\dfrac{p + 3}{2} < 5$

(vi) $\dfrac{2 + 3s}{4} \geqslant 5$

⑦ Solve the following inequalities and represent their solutions on a number line.

(i) $2 \leqslant 5x - 8 \leqslant 12$

(ii) $7 < 9 + 2x < 15$

(iii) $4 \geqslant 3 + 2x \geqslant 0$

(iv) $5 > 2x + 3 > 1$

(v) $11 \leqslant 3x + 5 \leqslant 20$

(vi) $5 > 7 + 4x > 1$

⑧ Solve the following inequalities.

(i) $4(c - 1) > 3(c - 2)$

(ii) $d - 3(d + 2) \geqslant 2(1 + 2d)$

(iii) $\frac{1}{2}e + 3\frac{1}{2} < e$

(iv) $-f - 2f - 3 < 4(1 + f)$

(v) $5(2 - 3g) + g \geqslant 8(2g - 4)$

(vi) $3(h + 2) - 2(h - 4) > 7(h + 2)$

⑨ Solve the following inequalities by sketching by hand the curves of the functions involved and indicating the parts of the x axis which correspond to the solution.

(i) $p^2 - 5p + 4 < 0$

(ii) $x^2 + 3x + 2 \leqslant 0$

(iii) $x^2 + 3x > -2$

(iv) $3x^2 + 5x - 2 < 0$

(v) $2x^2 - 11x - 6 \geqslant 0$

(vi) $10x^2 > x + 3$

⑩ Solve the following inequalities and express your solutions in set notation using the symbols $\cup$ or $\cap$.

(i) $y^2 - 2y - 3 > 0$

(ii) $z(z - 1) \leqslant 20$

(iii) $x^2 \geqslant 4x - 3$

(iv) $(a + 2)(a - 1) > 4$

(v) $8 - 2a \geqslant a^2$

(vi) $3s^2 + 2s < 1$

⑪ Show graphically the regions represented by the following inequalities.

(i) $y > x + 2$

(ii) $y \leqslant 2x - 1$

(iii) $y \geqslant x^2 - 4$

(iv) $y < x^2 - x - 2$

PS ⑫ The equation $x^2 + kx + k = 0$ has no real roots. Find the set of possible values for k.

PS ⑬ What happens when you try to solve the following inequalities?

(i) $a^2 - 4a + 4 \leqslant 0$

(ii) $b^2 + 5 > 0$

(iii) $c^2 - 3c - 3 > 0$

PS ⑭ Jasmine is 13 years old. Her mother tells her: 'Two years ago, you were less than $\frac{1}{3}$ of my age, and in two years' time, you will be more than $\frac{1}{4}$ of my age.'

What are the possible ages for Jasmine's mother now?

LEARNING OUTCOMES

When you have completed this chapter you should be able to:

➤ solve linear simultaneous equations in two variables by:
- ○ elimination
- ○ substitution
➤ solve simultaneous equations in two variables that include one linear and one quadratic equation by substitution
➤ solve linear and quadratic inequalities in a single variable including inequalities with brackets and fractions
➤ interpret linear and quadratic inequalities in a single variable graphically, including inequalities with brackets and fractions
➤ express solutions through correct use of 'and' and 'or', or through set notation
➤ represent linear and quadratic inequalities such as $y > x + 1$ and $y > ax^2 + bx + c$ graphically.

KEY POINTS

1. Simultaneous equations may be solved by
   (i) substitution
   (ii) elimination
   (iii) drawing graphs.
2. Linear inequalities are dealt with like equations *but* if you multiply or divide by a negative number, you must reverse the inequality sign: $<$ reverses to $>$ and $\leqslant$ to $\geqslant$.
3. When solving a quadratic inequality it is helpful to start by sketching a graph.

FUTURE USES

■   The relationship between finding the intersection points of graphs and solving simultaneous equations will be developed further in Chapters 5 and 7.

Coordinate geometry

The diagram shows some scaffolding in which all of the horizontal pieces are 1 m. All the vertical pieces are 2 m.

→ How many 1 m pieces are there?

→ How many 2 m pieces are there?

→ An ant crawls along the scaffolding from point P to point Q, travelling either horizontally or vertically. How far does the ant crawl?

→ A mouse also goes from point P to point Q, travelling either horizontally or along one of the sloping pieces. How far does the mouse travel?

→ A bee flies directly from point P to point Q. How far does the bee fly?

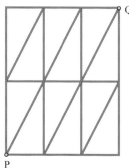

Figure 5.1

1 Working with coordinates

Coordinates are a means of describing a position relative to a fixed point, or origin.

In two dimensions you need two pieces of information; in three dimensions you need three pieces of information.

In the Cartesian system (named after René Descartes), position is given in perpendicular directions: x, y in two dimensions; x, y, z in three dimensions.

This chapter concentrates exclusively on two dimensions.

Prior knowledge

You will have met this at GCSE.

The midpoint and length of a line segment

When you know the coordinates of two points you can work out the midpoint and length of the line segment which connects them.

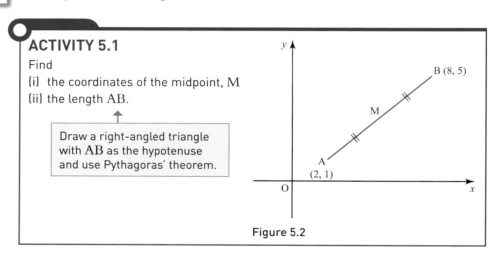

ACTIVITY 5.1

Find
(i) the coordinates of the midpoint, M
(ii) the length AB.

> Draw a right-angled triangle with AB as the hypotenuse and use Pythagoras' theorem.

Figure 5.2

You can generalise these methods to find the midpoint and length of any line segment AB.

Let A be the point (x_1, y_1) and B the point (x_2, y_2).

(i) Find the midpoint of AB.

The midpoint of two values is the mean of those values.

The mean of the x coordinates is $\dfrac{x_1 + x_2}{2}$.

The mean of the y coordinates is $\dfrac{y_1 + y_2}{2}$.

So the coordinates of the midpoint are $\left(\dfrac{x_1 + x_2}{2}, \dfrac{y_1 + y_2}{2} \right)$.

> C has the same x coordinate as B...
>
> ...and the same y coordinate as A.

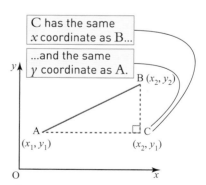

Figure 5.3

(ii) Find the length of AB.

First find the lengths of AC and AB: $AC = x_2 - x_1$

$$BC = y_2 - y_1$$

By Pythagoras' theorem: $AB^2 = AC^2 + BC^2$

$$= (x_2 - x_1)^2 + (y_2 - y_1)^2$$

So the length AB is $\sqrt{(x_2 - x_1)^2 + (y_2 - y_1)^2}$

The gradient of a line

When you know the coordinates of any two points on a straight line, then you can draw that line. The slope of a line is given by its **gradient**. The gradient is often denoted by the letter m.

Prior knowledge

You will have met this at GCSE.

Discussion point

→ Does it matter which point you call (x_1, y_1) and which (x_2, y_2)?

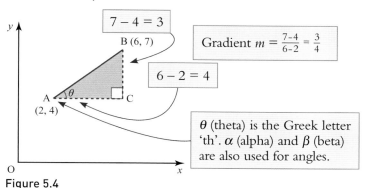

$7 - 4 = 3$

B (6, 7)

Gradient $m = \frac{7-4}{6-2} = \frac{3}{4}$

$6 - 2 = 4$

A (2, 4)

θ

C

θ (theta) is the Greek letter 'th'. α (alpha) and β (beta) are also used for angles.

Figure 5.4

In Figure 5.4, A and B are two points on the line. The gradient of the line AB is given by the increase in the y coordinate from A to B divided by the increase in the x coordinate from A to B.

In general, when A is the point (x_1, y_1) and B is the point (x_2, y_2), the gradient is

$$m = \frac{y_2 - y_1}{x_2 - x_1}$$

Gradient = $\frac{change\ in\ y}{change\ in\ x}$

When the same scale is used on both axes, $m = \tan\theta$ (see Figure 5.4). The gradient equals the tangent of the angle the line makes with the x axis.

Parallel and perpendicular lines

Prior knowledge

You will have met this at GCSE.

TECHNOLOGY

Use graphing software to generate more examples.

ACTIVITY 5.2

It is best to use squared paper for this activity.

Draw the line L_1 joining $(0, 2)$ to $(4, 4)$.

Draw another line L_2 perpendicular to L_1 from $(4, 4)$ to $(6, 0)$.
Find the gradients m_1 and m_2 of these two lines.
What is the relationship between the gradients?
Is this true for other pairs of perpendicular lines?

When you know the gradients m_1 and m_2, of two lines, you can tell at once if they are either parallel or perpendicular – see Figure 5.5.

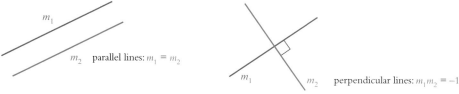

m_1

m_2 parallel lines: $m_1 = m_2$

m_1 m_2 perpendicular lines: $m_1 m_2 = -1$

Figure 5.5

⚠ Lines for which $m_1 m_2 = -1$ will only look perpendicular if the same scale has been used for both axes.

So for perpendicular lines:

$$m_1 = -\frac{1}{m_2} \text{ and likewise, } m_2 = -\frac{1}{m_1}.$$

So m_1 and m_2 are each the negative reciprocal of each other.

Example 5.1

A and B are the points (2, 5) and (6, 3) respectively (see Figure 5.6).

Find:

(i) the gradient of AB

(ii) the length of AB

(iii) the midpoint of AB

(iv) the gradient of the line perpendicular to AB.

Solution

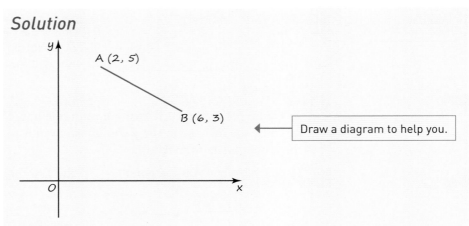

Draw a diagram to help you.

Figure 5.6

(i) Gradient $m_1 = \dfrac{y_A - y_B}{x_A - x_B}$

Gradient is difference in y coordinates divided by difference in x coordinates. It doesn't matter which point you use first, as long as you are consistent!

$= \dfrac{5 - 3}{2 - 6}$

$= -\dfrac{1}{2}$

(ii) Length $AB = \sqrt{(x_B - x_A)^2 + (y_B - y_A)^2}$

$= \sqrt{(6 - 2)^2 + (3 - 5)^2}$

$= \sqrt{16 + 4}$

$= \sqrt{20}$

(iii) Midpoint $= \left(\dfrac{x_A + x_B}{2}, \dfrac{y_A + y_B}{2} \right)$

$= \left(\dfrac{2 + 6}{2}, \dfrac{5 + 3}{2} \right)$

$= (4, 4)$

(iv) Gradient of AB: $m_{AB} = -\dfrac{1}{2}$

So gradient of perpendicular to AB is 2.

The gradient of the line perpendicular to AB is the negative reciprocal of m_{AB}.

Check: $-\dfrac{1}{2} \times 2 = -1$ ✓

Example 5.2
The points P(2, 7), Q(3, 2) and R(0, 5) form a triangle.

(i)   Use gradients to show that RP and RQ are perpendicular.

(ii) Use Pythagoras' theorem to show that PQR is right-angled.

Solution

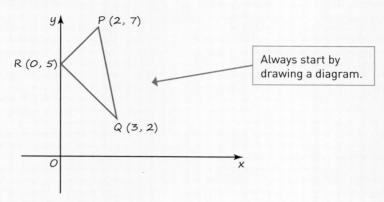

Always start by drawing a diagram.

Figure 5.7

(i)   Show that the gradients satisfy $m_1 m_2 = -1$

Gradient of RP $\dfrac{7 - 5}{2 - 0}$

Gradient of RQ $= \dfrac{2 - 5}{3 - 0} = -1$

$\Rightarrow$ product of gradients $= 1 \times (-1) = -1$

$\Rightarrow$ sides RP and RQ are at right angles.

(ii) Pythagoras' theorem states that for a right-angled triangle with hypotenuse of length a and other sides of lengths b and c, $a^2 = b^2 + c^2$.

Conversely, when $a^2 = b^2 + c^2$ for a triangle with sides of lengths a, b and c, then the triangle is right-angled and the side of length a is the hypotenuse.

$$\text{length}^2 = (x_2 - x_1)^2 + (y_2 - y_1)^2$$

$$PQ^2 = (3 - 2)^2 + (2 - 7)^2 = 1 + 25 = 26$$

$$RP^2 = (2 - 0)^2 + (7 - 5)^2 = 4 + 4 = 8$$

$$RQ^2 = (3 - 0)^2 + (2 - 5)^2 = 9 + 9 = 18$$

Since $26 = 8 + 18$, $PQ^2 = RP^2 + RQ^2$

PQ is the hypotenuse since RP and RQ are perpendicular.

$\Rightarrow$ sides RP and RQ are at right angles.

Exercise 5.1

 ① Draw axes and plot the points P(1, −2) and Q(4, 4).

Find the midpoint of PQ and plot it on the diagram.

Draw an appropriate right-angled triangle to show the length of PQ and the gradient of PQ. Use this triangle to calculate the length of PQ and its gradient.

 ② What is the gradient of the line joining A(−1, 2) and B(3, −1)?

A $\frac{3}{4}$ B $-\frac{3}{4}$ C $-\frac{4}{3}$ D 5

Justify your choice.

③ For the following pairs of points A and B, calculate:

(a) the midpoint of the line joining A to B

(b) the distance AB

(c) the gradient of the line AB

(d) the gradient of the line perpendicular to AB.

(i) A(2, 5) and B(6, 8)

(ii) A(−2, −5) and B(−6, −8)

(iii) A(−2, −5) and B(6, 8)

(iv) A(−2, 5) and B(6, −8)

④ The gradient of the line joining the point P(3, −4) to Q(q, 0) is 2. Find the value of q.

⑤ The three points X(2, −1), Y(8, y) and Z(11, 2) are collinear. Find the value of y.

> **They lie on the same straight line.**

⑥ For the points P(x, y), and Q(3x, 5y), find in terms of x and y:

(i) the gradient of the line PQ

(ii) the midpoint of the line PQ

(iii) the length of the line PQ.

⑦ The points A, B, C and D have coordinates (1, 2), (7, 5), (9, 8) and (3, 5) respectively.

(i) Find the gradients of the lines AB, BC, CD and DA.

(ii) What do these gradients tell you about the quadrilateral ABCD?

(iii) Draw an accurate diagram to check your answer to part (ii).

PS ⑧ The points A, B, and C have coordinates (−4, 2), (7, 4) and (−3, −1).

(i) Draw the triangle ABC.

(ii) Show by calculation that the triangle ABC is isosceles and name the two equal sides.

(iii) Find the midpoint of the third side.

(iv) Work out the area of the triangle ABC.

⑨ The points A, B and C have coordinates (2, 1), (b, 3) and (5, 5), where b > 3, and ∠ABC = 90°.

Find:

(i) the value of b

(ii) the lengths of AB and BC

(iii) the area of triangle ABC.

⑩ Three points A, B and C have coordinates (1, 3), (3, 5) and (−1, y). Find the value of y in each of the following cases:

(i) AB = AC

(ii) BC = AC

(iii) AB is perpendicular to BC

(iv) A, B and C are collinear.

⑪ The triangle PQR has vertices P(8, 6), Q(0, 2) and R(2, r).

Find the values of r when the triangle PQR:

(i) has a right angle at P

(ii) has a right angle at Q

(iii) has a right angle at R

(iv) is isosceles with RQ = RP.

PS ⑫ A quadrilateral has vertices A(0, 0), B(0, 3), C(6, 6) and D(12, 6).

(i) Draw the quadrilateral.

(ii) Show by calculation that it is a trapezium.

(iii) EBCD is a parallelogram. Find the coordinates of E.

PS ⑬ Show that the points with coordinates (1, 2), (8, −2), (7, 6) and (0, 10) are the vertices of a rhombus, and find its area.

PS ⑭ The lines AB and BC in Figure 5.8 are equal in length and perpendicular.

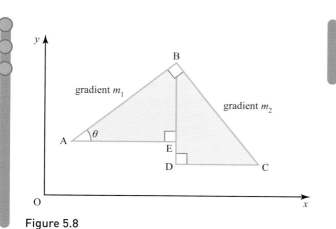

Figure 5.8

(i) Show that triangles ABE and BCD are congruent.

(ii) Hence prove that the gradients m_1 and m_2 satisfy $m_1 m_2 = -1$.

Prior knowledge
You will have met some of these ideas at GCSE.

2 The equation of a straight line

Drawing a line, given its equation

There are several standard forms for the equation of a straight line, as shown in Figure 5.9.

(a) Equations of the form $x = a$

Each point on the line has an x coordinate of 3.

All such lines are parallel to the y axis.

(b) Equations of the form $y = b$

Each point on the line has a y coordinate of 2.

All such lines are parallel to the x axis.

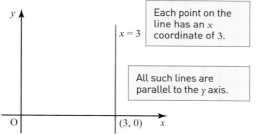

(c) Equations of the form $y = mx$

These are lines through the origin, with gradient m.

(d) Equations of the form $y = mx + c$

These lines have gradient m and cross the y axis at point $(0, c)$.

(e) Equations of the form $px + qy + r = 0$

This is often a tidier way of writing the equation.

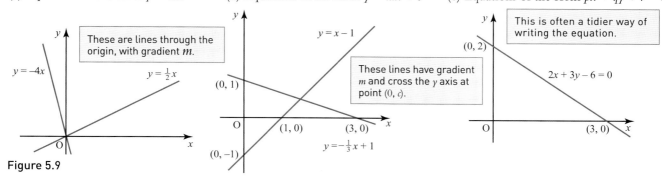

Figure 5.9

The two forms of the equation of a straight line where y is the subject, $y = mx$ and $y = mx + c$, are particularly useful as you can write down the gradient and y-intercept without further working. Use graphing software with a constant controller to explore the effect of changing the values of m and c.

Example 5.3

(i) Draw the lines (a) $y = x - 1$ and (b) $3x + 4y = 24$ on the same axes.

(ii) Are these lines perpendicular?

Solution

(i) To draw a line you need to find the coordinates of two points on it.

> Usually it is easiest to find where the line cuts the x and y axes.

(a) The line $y = x - 1$ passes through the point $(0, -1)$.

> The line is already in the form $y = mx + c$.

Substituting $y = 0$ gives $x = 1$, so the line also passes through $(1, 0)$.

(b) Find two points on the line $3x + 4y = 24$.

> Set $x = 0$ and find y to give the y-intercept.
> Then set $y = 0$ and find x to give the x-intercept.

Substituting $x = 0$ gives $4y = 24 \Rightarrow y = 6$

substituting $y = 0$ gives $3x = 24 \Rightarrow x = 8$.

So the line passes through $(0, 6)$ and $(8, 0)$.

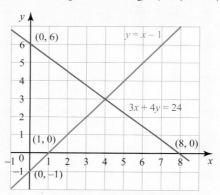

Figure 5.10

(ii) The lines look as if they are perpendicular but you need to use the gradient of each line to check.

$$\text{Gradient of } y = x - 1 \text{ is } 1.$$

$$\text{Gradient of } 3x + 4y = 24 \text{ is } -\frac{3}{4}.$$

> Rearrange the equation to make y the subject so you can find the gradient.
> $4y = -3x + 24$
> $y = -\frac{3}{4}x + 6$

Therefore the lines are not perpendicular as $1 \times \left(-\frac{3}{4}\right) \neq -1$.

Finding the equation of a line

To find the equation of a line, you need to think about what information you are given.

(i) **Given the gradient, m, and the coordinates (x_1, y_1) of one point on the line**

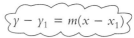

$$y - y_1 = m(x - x_1)$$

Take a general point (x, y) on the line, as shown in Figure 5.11.

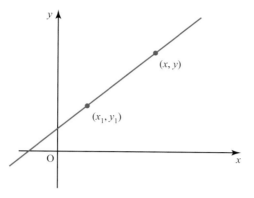

Figure 5.11

The gradient, m, of the line joining (x_1, y_1) to (x, y) is given by

$$m = \frac{y - y_1}{x - x_1}$$

$$\Rightarrow y - y_1 = m(x - x_1)$$

This is a very useful form of the equation of a straight line.

For example, the equation of the line with gradient 2 that passes through the point $(3, -1)$ can be written as $y - (-1) = 2(x - 3)$

which can be simplified to $y = 2x - 7$.

(ii) **Given the gradient, m, and the y-intercept $(0, c)$**

$$y = mx + c$$

A special case of $y - y_1 = m(x - x_1)$ is when (x_1, y_1) is the y-intercept $(0, c)$.

The equation then becomes

$$y = mx + c$$

Substituting $x_1 = 0$ and $y_1 = c$ into the equation

as shown in Figure 5.12.

When the line passes through the origin, the equation is

$$y = mx$$

The y-intercept is $(0, 0)$, so $c = 0$

as shown in Figure 5.13.

ACTIVITY 5.3

A Show algebraically that an equivalent form of

$$\frac{y - y_1}{y_2 - y_1} = \frac{x - x_1}{x_2 - x_1}$$

is

$$\frac{y - y_1}{x - x_1} = \frac{y_2 - y_1}{x_2 - x_1}.$$

B Use both forms to find the equation of the line joining $(2, 4)$ to $(5, 3)$ and show they give the same equation.

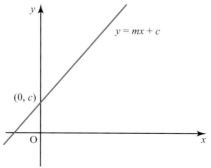

Figure 5.12

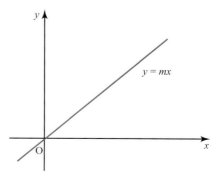

Figure 5.13

(iii) **Given two points, (x_1, y_1) and (x_2, y_2)**

The two points are used to find the gradient:

$$m = \frac{y_2 - y_1}{x_2 - x_1}$$

$$\frac{y - y_1}{y_2 - y_1} = \frac{x - x_1}{x_2 - x_1}$$

This value of m is then substituted in the equation

$$y - y_1 = m(x - x_1)$$

This gives

$$y - y_1 = \frac{y_2 - y_1}{x_2 - x_1}(x - x_1)$$

or $\dfrac{y - y_1}{y_2 - y_1} = \dfrac{x - x_1}{x_2 - x_1}$

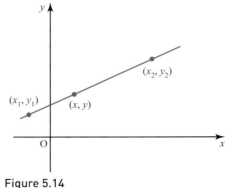

Figure 5.14

Discussion points

→ How else can you write the equation of a line?

→ Which form do you think is best?

Example 5.4

Find the equation of the line perpendicular to $4y + x = 12$ which passes through the point P(2, −5).

Solution

First rearrange $4y + x = 12$ into the form $y = mx + c$ to find the gradient.

$$4y = -x + 12$$

$$y = -\frac{1}{4}x + 3$$

> For perpendicular gradients $m_1 m_2 = -1$
> So $m_2 = -\dfrac{1}{m_1}$

So the gradient is $-\frac{1}{4}$

The negative reciprocal of $-\frac{1}{4}$ is 4.

> Check: $-\frac{1}{4} \times 4 = -1$ ✓

So the gradient of a line perpendicular to $y = -\frac{1}{4}x + 3$ is 4.

Using $y - y_1 = m(x - x_1)$ when $m = 4$ and (x_1, y_1) is $(2, -5)$

$$\Rightarrow y - (-5) = 4(x - 2)$$

$$\Rightarrow y + 5 = 4x - 8$$

$$\Rightarrow y = 4x - 13$$

> Linear means a straight line.

Straight lines can be used to model real–life situations. Often simplifying assumptions need to be made so that a linear model is appropriate.

Example 5.5

Ⓜ

The diameter of a snooker cue varies uniformly from 9 mm to 23 mm over its length of 140 cm.

> Varying uniformly means that the graph of diameter against distance from the tip is a straight line.

(i) Sketch the graph of diameter (y mm) against distance (x cm) from the tip.

(ii) Find the equation of the line.

(iii) Use the equation to find the distance from the tip at which the diameter is 15 mm.

Solution

(i) The graph passes through the points (0, 9) and (140, 23).

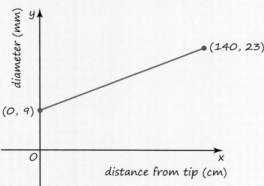

Figure 5.15

(ii) Gradient $= \dfrac{y_2 - y_1}{x_2 - x_1}$

$$= \frac{23 - 9}{140 - 0} = 0.1$$

Using the form $y = mx + c$, the equation of the line is $y = 0.1x + 9$.

(iii) Substituting $y = 15$ into the equation gives

$$15 = 0.1x + 9$$

$$0.1x = 6$$

$$x = 60$$

$\Rightarrow$ The diameter is 15 mm at a point 60 cm from the tip.

Discussion points

→ Which of these situations in Figure 5.16 could be modelled by a straight line?

→ For each straight line model, what information does the gradient of the line give you?

→ What assumptions do you need to make so that a linear model is appropriate?

→ How reasonable are your assumptions?

Interest earned on savings in a bank account against time	Height of ball dropped from a cliff against time	Profit of ice cream seller against number of sales
Tax paid against earnings	Cost of apples against mass of apples	Value of car against age of car
Mass of candle versus length of time it is burning	Distance travelled by a car against time	Mass of gold bars against volume of gold bars
Population of birds on an island against time	Mobile phone bill against number of texts sent	Length of spring against mass of weights attached

Figure 5.16

Exercise 5.2

① Here are the equations of some lines:

$y = 2x - 3$

$y + 3x = 1$

$y + 2x - 1 = 0$

$y = x + 6$

Match them to the lines on the graph.

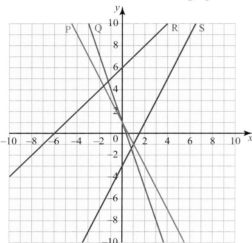

Figure 5.17

② Which of these lines has a gradient of $-\frac{1}{2}$ and a y-intercept of 3?

A $x + 3y = 2$

B $y - 2x = 3$

C $2y + x = 3$

D $2y + x - 6 = 0$

③ Sketch the following lines:

(i) $y = -2$ (ii) $x = 2$

(iii) $y = -2x$ (iv) $y = x + 2$

(v) $y = 2x + 5$ (vi) $y = 5 - 2x$

(vii) $2x - y = 5$ (viii) $y + 2x + 5 = 0$

④ Find the equations of the lines (i)–(v) in Figure 5.18.

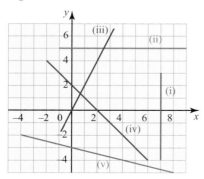

Figure 5.18

⑤ Find the equations of the lines

(i)   parallel to $y = 3x - 2$ and passing through $(0, 0)$

(ii)   parallel to $y = 3x$ and passing through $(2, 5)$

(iii)   parallel to $2x + y - 3 = 0$ and passing through $(-2, 5)$

(iv)   parallel to $3x - y - 2 = 0$ and passing through $(5, -2)$

(v)   parallel to $x + 2y = 3$ and passing through $(-2, -5)$.

⑥ Find the equations of the lines

(i)   perpendicular to $y = 3x$ and passing through $(0, 0)$

(ii)   perpendicular to $y = 2x + 3$ and passing through $(4, 3)$

(iii)   perpendicular to $2x + y = 4$ and passing through $(4, -3)$

(iv)   perpendicular to $2y = x + 5$ and passing through $(-4, 3)$

(v)   perpendicular to $2x + 3y = 4$ and passing through $(-4, -3)$.

⑦ Find the equations of the line AB in each of the following cases.

(i)   A$(3, 1)$, B$(5, 7)$

(ii)   A$(-3, -1)$, B$(-5, -7)$

(iii)   A$(-3, 1)$, B$(-5, 7)$

(iv)   A$(3, -1)$, B$(5, -7)$

(v)   A$(1, 3)$, B$(7, 5)$

PS ⑧ Show that the region enclosed by the lines

$$y = \frac{2}{3}x + 1, \quad y = 1 - \frac{3x}{2},$$

$$3y - 2x + 1 = 0 \text{ and } 2y + 3x + 5 = 0$$

forms a rectangle.

> **The perpendicular bisector is the line at right angles to AB (perpendicular) that passes though the midpoint of AB (bisects).**

⑨ Find the equation of the perpendicular bisector of each of the following pairs of points.

(i)   A$(2, 4)$ and B$(3, 5)$

(ii)   (A$(4, 2)$ and B $(5, 3)$

(iii)   A$(-2, -4)$ and B$(-3, -5)$

(iv) A(−2, 4) and B(−3, 5)

(v) A(2, −4) and B(3, −5)

⑩ A median of a triangle is a line joining one of the vertices to the midpoint of the opposite side.

In a triangle OAB, O is at the origin, A is the point $(0, 6)$ and B is the point $(6, 0)$.

(i) Sketch the triangle.

(ii) Find the equations of the three medians of the triangle.

(iii) Show that the point $(2, 2)$ lies on all three medians. (This shows that the medians of this triangle are concurrent.)

PS ⑪ A quadrilateral ABCD has its vertices at the points $(0, 0)$, $(12, 5)$, $(0, 10)$ and $(−6, 8)$ respectively.

(i) Sketch the quadrilateral.

(ii) Find the gradient of each side.

(iii) Find the length of each side.

(iv) Find the equation of each side.

(v) Find the area of the quadrilateral.

PS **M** ⑫ A firm manufacturing jackets finds that it is capable of producing 100 jackets per day, but it can only sell all of these if the charge to the wholesalers is no more than £20 per jacket. On the other hand, at the current price of £25 per jacket, only 50 can be sold per day. Assuming that the graph of price P against number sold per day N is a straight line:

(i) sketch the graph, putting the number sold per day on the horizontal axis (as is normal practice for economists)

(ii) find its equation.

Use the equation to find:

(iii) the price at which 88 jackets per day could be sold

(iv) the number of jackets that should be manufactured if they were to be sold at £23.70 each.

PS **M** ⑬ To clean the upstairs window on the side of a house, it is necessary to position the ladder so that it just touches the edge of the lean-to shed as shown in Figure 5.19 with the the top of the ladder touching the wall. The coordinates represent distances from O in metres, in the x and y directions shown.

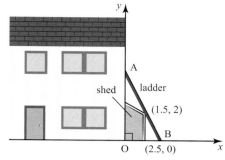

Figure 5.19

(i) Find the equation of the line of the ladder.

(ii) Find the height of the point A reached by the top of the ladder.

(iii) Find the length of the ladder to the nearest centimetre.

PS **M** ⑭ A spring has an unstretched length of 10 cm. When it is hung with a load of 80 g attached, the stretched length is 28 cm. Assuming that the extension of the spring is proportional to the load:

(i) draw a graph of extension E against load L and find its equation

> Hint: E is in place of the y axis and L is in place of the x axis.

(ii) find the extension caused by a load of 48 g

(iii) find the load required to extend the spring to a length of 20 cm.

This particular spring passes its elastic limit when it is stretched to four times its original length. (This means that if it is stretched more than that it will not return to its original length.)

(iv) Find the load which would cause this to happen.

PS ⑮

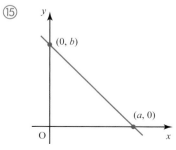

Figure 5.20

Show that the equation of the line in Figure 5.20 can be written

$$\frac{x}{a} + \frac{y}{b} = 1.$$

Prior knowledge

You will have met this at GCSE.

3 The intersection of two lines

The intersection of any two curves (or lines) can be found by solving their equations simultaneously. In the case of two distinct lines, there are two possibilities:

(i) they are parallel, or **(ii)** they intersect at a single point.

You often need to find where a pair of lines intersect in order to solve problems.

Example 5.6

The lines $y = 5x - 13$ and $2y + 3x = 0$ intersect at the point P.

Find the coordinates of P.

Solution

You need to solve the equations

$$y = 5x - 13 \qquad ①$$

and $\qquad 2y + 3x = 0 \qquad ②$

simultaneously.

Substitute equation ① into ② : $2(5x - 13) + 3x = 0$

$10x - 26 + 3x = 0$ ← Multiply out the brackets.

$13x - 26 = 0$ ← Simplify

$13x = 26$

$x = 2$

Substitute $x = 2$ into equation ① to find y. ← Don't forget to find the y coordinate.

$y = 5 \times 2 - 13$

$y = -3$

So the coordinates of P are $(2, -3)$.

Discussion point

→ The line l has equation $2x - y = 4$ and the line m has equation $y = 2x - 3$. What can you say about the intersection of these two lines?

Exercise 5.3

 ① These three lines intersect to form a triangle:

$y = x + 4$, $y = -\frac{1}{2}x - \frac{1}{2}$ and $y = -3x + 12$

The vertices of the triangle are $(5, -3)$, $(2, 6)$ and $(-3, 1)$. Work out which point is at the intersection of each pair of lines.

 ② Choose the coordinates of the point where the lines $3x - y = 9$ and $2x - 3y = 13$ intersect.

A $(2, -3)$ B $(2, 3)$

C $(3\frac{7}{11}, 1\frac{10}{11})$ D $(3\frac{7}{11}, -1\frac{10}{11})$

 ③ Find the coordinates of the point of intersection of the following pairs of lines.

(i) $y = 2x + 3$ and $y = 6x + 1$

(ii) $y = 2 - 3x$ and $2y + x = 14$

(iii) $3x + 2y = 4$ and $5x - 4y = 3$

④ (i) Find the coordinates of the points where the following pairs of lines intersect.

(a) $y = 2x - 4$ and $2y = 7 - x$

(b) $y = 2x + 1$ and $2y = 7 - x$

The lines form three sides of a square.

(ii) Find the equation of the fourth side of the square.

(iii) Find the area of the square.

PS ⑤ (i) Find the vertices of the triangle ABC whose sides are given by the lines

AB: $x - 2y = -1$

BC: $7x + 6y = 53$

AC: $9x + 2y = 11$.

(ii) Show that the triangle is isosceles.

⑥ A(0, 1), B(1, 4), C(4, 3) and D(3, 0) are the vertices of a quadrilateral ABCD.

   (i) Find the equations of the diagonals AC and BD.

   (ii) Show that the diagonals AC and BD bisect each other at right angles.

   (iii) Find the lengths of AC and BD.

   (iv) What type of quadrilateral is ABCD?

⑦ The line $y = 5x - 2$ crosses the x axis at A.

The line $y = 2x + 4$ crosses the x axis at B.

The two lines intersect at P.

PS

   (i) Find the coordinates of A and B.

   (ii) Find coordinates of the point of intersection, P.

   (iii) Find the exact area of the triangle ABP.

PS ⑧ Triangle ABC has an angle of 90° at B. Point A is on the y axis, AB is part of the line $x - 2y + 8 = 0$, and C is the point (6, 2).

   (i) Sketch the triangle.

   (ii) Find the equations of the lines AC and BC.

   (iii) Find the lengths of AB and BC and hence find the area of the triangle.

   (iv) Using your answer to (iii), find the length of the perpendicular from B to AC.

PS ⑨ Two rival taxi firms have the following fare structures:

   Firm A: fixed charge of £1 plus 40p per kilometre;

   Firm B: 60p per kilometre, no fixed charge.

   (i) Sketch the graph of price (vertical axis) against distance travelled (horizontal axis) for each firm (on the same axes).

   (ii) Find the equation of each line.

   (iii) Find the distance for which both firms charge the same amount.

   (iv) Which firm would you use for a distance of 6 km?

PS ⑩ Two sides of a parallelogram are formed by parts of the lines

$$2x - y = -9$$
$$\text{and } x - 2y = -9.$$

   (i) Show these two lines on a graph.

   (ii) Find the coordinates of the vertex where they intersect.

Another vertex of the parallelogram is the point (2, 1).

   (iii) Find the equations of the other two sides of the parallelogram.

   (iv) Find the coordinates of the other two vertices.

PS ⑪ The line with equation $5x + y = 20$ meets the x axis at A and the line with equation $x + 2y = 22$ meets the y axis at B. The two lines intersect at a point C.

   (i) Sketch the two lines on the same diagram.

   (ii) Calculate the coordinates of A, B and C.

   (iii) Calculate the area of triangle OBC where O is the origin.

   (iv) Find the coordinates of the point E such that ABEC is a parallelogram.

⑫ Figure 5.21 shows the supply and demand of labour for a particular industry in relation to the wage paid per hour.

Supply is the number of people willing to work for a particular wage, and this increases as the wage paid increases. Demand is the number of workers that employers are prepared to employ at a particular wage: this is greatest for low wages.

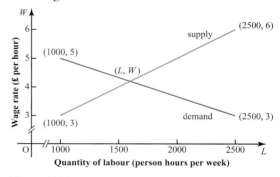

Figure 5.21

   (i) Find the equation of each of the lines.

   (ii) Find the values of L and W at which the market 'clears', i.e. at which supply equals demand.

(iii) Although economists draw the graph this way round, mathematicians would plot wage rate on the horizontal axis. Why?

 ⑬ When the market price £p of an article sold in a free market varies, so does the number demanded, D, and the number supplied, S.

In one case $D = 20 + 0.2p$ and $S = -12 + p$.

(i) Sketch both of these lines on the same graph. (Put p on the horizontal axis.)

The market reaches a state of equilibrium when the number demanded equals the number supplied.

(ii) Find the equilibrium price and the number bought and sold in equilibrium.

 ⑭ A median of a triangle is a line joining a vertex to the midpoint of the opposite side. In any triangle, the three medians meet at a point called the centroid of the triangle.

Find the coordinates of the centroid for each triangle shown in Figure 5.22.

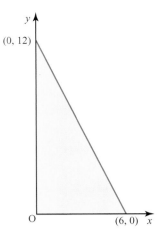

(i)

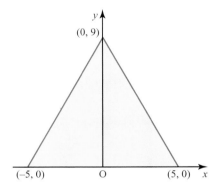

Figure 5.22

 ⑮ Find the exact area of the triangle whose sides have the equations $x + y = 4$, $y = 2x - 8$ and $x + 2y = -1$.

4 The circle

You are, of course, familiar with the circle, and have done calculations involving its area and circumference. In this section you are introduced to the equation of a circle.

The circle is defined as the **locus** of all the points in a plane which are at a fixed distance (the radius) from a given point (the centre).

> **Locus** means possible positions subject to given conditions. In two dimensions the locus can be a path or a region.

This definition allows you to find the equation of a circle.

 TECHNOLOGY

Graphing software needs to be set to equal aspect to get these graphs looking correct.

Remember, the length of a line joining (x_1, y_1) to (x_2, y_2) is given by

$$\text{length} = \sqrt{(x_2 - x_1)^2 + (y_2 - y_1)^2}$$

> This is just Pythagoras' theorem.

For a circle of radius 3, with its centre at the origin, any point (x, y) on the circumference is distance 3 from the origin.

So the distance of (x, y) from $(0, 0)$ is given by

$$\sqrt{(x - 0)^2 + (y - 0)^2} = 3$$
$$\Rightarrow \qquad x^2 + y^2 = 3^2$$
$$\Rightarrow \qquad x^2 + y^2 = 9$$

Squaring both sides.

This is the equation of the circle in Figure 5.23.

The circle in Figure 5.24 has a centre $(9, 5)$ and radius 4, so the distance between any point on the circumference and the centre $(9, 5)$ is 4.

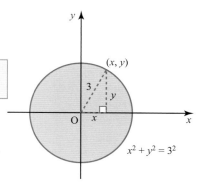

Figure 5.23

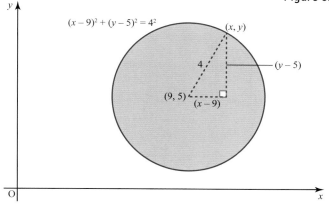

Figure 5.24

The equation of this circle in Figure 5.24 is:

$$\sqrt{(x - 9)^2 + (y - 5)^2} = 4$$
$$\Rightarrow (x - 9)^2 + (y - 5)^2 = 16.$$

ACTIVITY 5.4

Sophie tries to draw the circle $x^2 + y^2 = 9$ on her graphical calculator.
Explain what has gone wrong for each of these outputs.

(i)

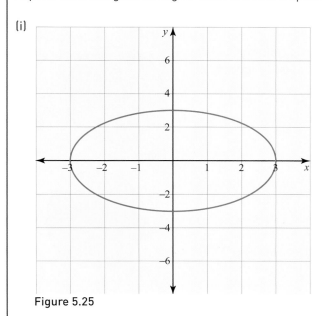

Figure 5.25

(ii)

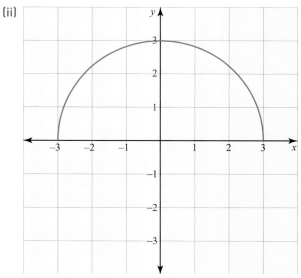

Figure 5.26

Note

In the form shown in the activity, the equation highlights some of the important characteristics of the equation of a circle. In particular:

(i) the coefficients of x^2 and y^2 are equal

(ii) there is no xy term.

> **ACTIVITY 5.5**
>
> Show that you can rearrange $(x - a)^2 + (y - b)^2 = r^2$ to give
> $x^2 + y^2 - 2ax - 2by + (a^2 + b^2 - r^2) = 0$

These results can be generalised to give the equation of a circle as follows:

centre $(0, 0)$, radius r: $x^2 + y^2 = r^2$

centre (a, b), radius r: $(x - a)^2 + (y - b)^2 = r^2$.

Example 5.7

Find the centre and radius of the circle $x^2 + y^2 - 6x + 10y - 15 = 0$.

Solution

You need to rewrite the equation so it is in the form $(x - a)^2 + (y - b)^2 = r^2$.

$$x^2 - 6x \quad + \quad y^2 + 10y \quad - 15 = 0$$

Complete the square on the terms involving x …

$$(x - 3)^2 - 9 + (y + 5)^2 - 25 - 15 = 0$$

… then complete the square on the terms involving y.

$$(x - 3)^2 \quad + \quad (y + 5)^2 \quad = 49$$

So the centre is $(3, -5)$ and the radius is 7. $\leftarrow$ $7^2 = 49$

Prior knowledge

You will have met these at GCSE.

Circle geometry

There are some properties of a circle that are useful when solving coordinate geometry problems.

1   The angle in a semicircle is a right angle (see Figure 5.27).

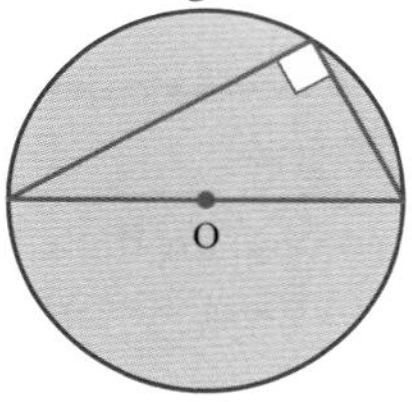

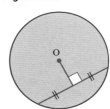

Figure 5.27

2   The perpendicular from the centre of a circle to a chord bisects the chord (see Figure 5.28).

Figure 5.28

3   The tangent to a circle at a point is perpendicular to the radius through that point (see Figure 5.29).

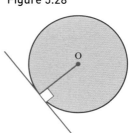

Figure 5.29

Discussion points

→ How can you prove these results?

→ State the converse of each of these results. ◄

> The converse of 'p implies q' is 'q implies p'.

The converse of each of the three circle properties above is also true.

The next three examples use these results in coordinate geometry.

Example 5.8

A circle has a radius of 5 units, and passes through the points $(0, 0)$ and $(0, 8)$.

Sketch the two possible positions of the circle and find their equations.

Solution

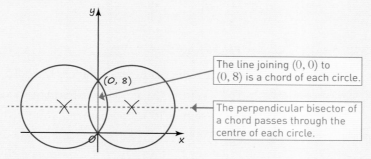

> The line joining $(0, 0)$ to $(0, 8)$ is a chord of each circle.

> The perpendicular bisector of a chord passes through the centre of each circle.

Figure 5.30

The midpoint of the chord is $(0, 4)$.

The equation of the bisector is $y = 4$. ◄

So the centre of the circle lies on the line $y = 4$.

> The chord is along the y axis, so the perpendicular bisector passes through $(0, 4)$ and is parallel to the x axis.

Let the centre be the point $(a, 4)$. ◄

> The radius of the circle is 5 and the circle passes through the origin ...

Using Pythagoras' theorem $a^2 + 16 = 25$ ◄

> ... so the distance between the centre $(a, 4)$ and the origin is 5.

$\Rightarrow \qquad\qquad\qquad\qquad\qquad a^2 = 9$

$\Rightarrow \qquad\qquad\qquad\qquad a = 3 \text{ or } a = -3.$

The two possible equations are therefore

$(x - 3)^2 + (y - 4)^2 = 25$ and
$(x + 3)^2 + (y - 4)^2 = 25.$ ◄

> $(x - (-3))^2 + (y - 4)^2 = 25$

Example 5.9

(i) Show that OB is a diameter of the circle which passes through the points O(0, 0), A(2, 6) and B(8, 4).

(ii) Find the equation of the circle.

Solution

(i)

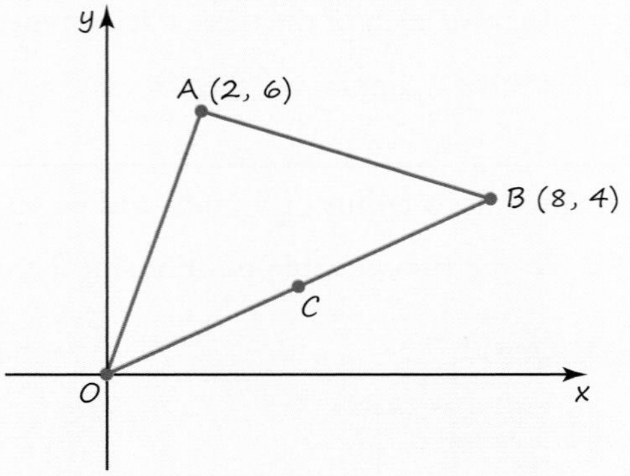

A (2, 6)

B (8, 4)

Always draw a sketch.

C

O

Figure 5.31

If OB is the diameter of the circle, and A lies on the circle then $\angle OAB$ is 90°.

The angle in a semicircle is a right angle.

So to show OB is the diameter you need to show that OA and AB are perpendicular.

$$\text{Gradient of OA} = \frac{6}{2} = 3$$

$$\text{Gradient of AB} = \frac{6-4}{2-8} = \frac{2}{-6} = -\frac{1}{3}$$

$$\text{Product of gradients} = 3 \times -\frac{1}{3} = -1$$

by the converse of the theorem that the angle in a semicircle is a right angle.

$\Rightarrow$ Lines OA and AB are perpendicular so angle OAB = 90°.

$\Rightarrow$ OAB is the angle in a semicircle where OB is the diameter, as required.

(ii) The centre C of the circle is the midpoint of OB.

$$C = \left(\frac{0+8}{2}, \frac{0+4}{2}\right) = (4, 2)$$

To find the equation of a circle you need the centre and radius.

The radius of the circle, CO $= \sqrt{4^2 + 2^2} = \sqrt{20}$.

So the radius, $r = \sqrt{20} \Rightarrow r^2 = 20$

Hence the equation of the circle is $(x - 4)^2 + (y - 2)^2 = 20$.

Example 5.10

Figure 5.32 shows the circle $x^2 + y^2 = 25$.

The point P(4, 3) lies on the circle and the tangent to the circle at P cuts the coordinate axes at the points Q and R.

Find:

(i) the equation of the tangent to the circle at P

(ii) the exact area of triangle OQR.

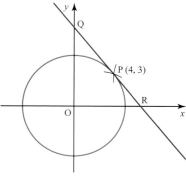

Figure 5.32

Solution

(i) The gradient of OP is $\frac{3}{4}$.

> To find the equation of the tangent you need the gradient.

> OP is the radius of the circle ...

So the gradient of the tangent is $-\frac{4}{3}$.

> ... so the gradient of the tangent is the negative reciprocal of the gradient of the radius.

> ... the tangent and radius meet at right angles ...

The equation of the tangent at P(4, 3) is

$$y - 3 = -\frac{4}{3}(x - 4)$$

$$\Rightarrow \qquad 3y - 9 = 16 - 4x$$

$$\Rightarrow 4x + 3y - 25 = 0$$

> Area is $\frac{1}{2}$ × base × height.

(ii) OQR forms a right-angled triangle.

Find Q:

> The base is the x coordinate of R and the height is the y coordinate of Q.

$$3y - 25 = 0$$

$$\Rightarrow \qquad y = \frac{25}{3}$$

> Substitute $x = 0$ into the tangent equation to find Q.

Find R:

$$4x - 25 = 0$$

> Substitute $y = 0$ into the tangent equation to find R.

$$\Rightarrow \qquad x = \frac{25}{4}$$

Area of triangle OQR is

$$\frac{1}{2} \times \frac{25}{4} \times \frac{25}{3} = \frac{625}{24} \text{ square units.}$$

> Exact means leave your answer as a fraction (or a surd).

Exercise 5.4

① Find the equations of the following circles.

(i) centre (2, 3), radius 1

(ii) centre (2, −3), radius 2

(iii) centre (−2, 3), radius 3

(iv) centre (−2, −3), radius 4

② For each of the following circles state

(a) the coordinates of the centre

(b) the radius.

(i) $x^2 + y^2 = 1$

(ii) $x^2 + (y - 2)^2 = 2$

(iii) $(x - 2)^2 + y^2 = 3$

(iv) $(x + 2)^2 + (y + 2)^2 = 4$

(v) $(x - 2)^2 + (y + 2)^2 = 5$

③ The equation of a circle is $(x - 3)^2 + (y + 2)^2 = 26$.

Complete the table to show whether each point lies inside the circle, outside the circle or on the circle.

Point	Inside	Outside	On
(3, −2)	✓		
(−2, −5)			
(6, −6)			
(4, 3)			
(0, 2)			
(−2, −3)			

④ Draw the circles $(x - 4)^2 + (y - 5)^2 = 16$ and $(x - 3)^2 + (y - 3)^2 = 4$.

In how many points do they intersect?

PS ⑤ Sketch the circle $(x + 2)^2 + (y - 3)^2 = 16$, and find the equations of the four tangents to the circle which are parallel to the coordinate axes.

⑥ Find the coordinates of the points where each of these circles crosses the axes.

(i) $x^2 + y^2 = 25$

(ii) $(x - 4)^2 + (y + 5)^2 = 25$

(iii) $(x + 6)^2 + (y - 8)^2 = 100$

PS ⑦ Find the equation of the circle with centre $(1, 7)$ passing through the point $(-4, -5)$.

⑧ Show that the equation $x^2 + y^2 + 2x - 4y + 1 = 0$ can be written in the form $(x + 1)^2 + (y - 2)^2 = r^2$, where the value of r is to be found.

Hence give the coordinates of the centre of the circle, and its radius.

PS ⑨ Draw the circle of radius 4 units which touches the positive x and y axes, and find its equation.

⑩ $A(3, 5)$ and $B(9, -3)$ lie on a circle. Show that the centre of the circle lies on the line with equation $4y - 3x + 14 = 0$.

⑪ For each of the following circles find

(a)   the coordinates of the centre

(b)   the radius.

(i) $x^2 + y^2 - 6x - 2y - 6 = 0$

(ii) $x^2 + y^2 + 2x + 6y - 6 = 0$

(iii) $x^2 + y^2 - 2x + 8y + 8 = 0$

PS ⑫ A circle passes through the points $A(3, 2)$, $B(5, 6)$ and $C(11, 3)$.

(i)   Calculate the lengths of the sides of the triangle ABC.

(ii)   Hence show that AC is a diameter of this circle. State which theorems you

have used, and in each case whether you have used the theorem or its converse.

(iii)   Calculate the area of triangle ABC.

⑬ (i)   Find the midpoint, C, of AB where A and B are $(1, 8)$ and $(3, 14)$ respectively. Find also the distance AC.

(ii)   Hence find the equation of the circle which has AB as a diameter.

PS ⑭ $A(1, -2)$ is a point on the circle $(x - 3)^2 + (y + 1)^2 = 5$.

(i)   State the coordinates of the centre of the circle and hence find the coordinates of the point B, where AB is a diameter of the circle.

(ii) $C(2, 1)$ also lies on the circle. Use coordinate geometry to verify that angle ACB = 90°.

⑮ The tangent to the circle $x^2 + (y + 4)^2 = 25$ at the point $(-4, -1)$ intersects the x axis at A and the y axis at B. Find the exact area of the triangle AOB.

PS ⑯ A circle passes through the points $(2, 0)$ and $(8, 0)$ and has the y axis as a tangent. Find the two possible equations for the circle.

PS ⑰ $A(6, 3)$ and $B(10,1)$ are two points on a circle with centre $(11, 8)$.

(i)   Calculate the distance of the chord AB from the centre of the circle.

(ii)   Find the equation of the circle.

PS ⑱ $A(6, 6)$, $B(6, -2)$ and $C(-1, -1)$ are three points on a circle.

Find the equation of the circle.

Prior knowledge

You need to be able to

- solve a quadratic equation
- use the discriminant to determine the number of roots of a quadratic equation

These are covered in Chapter 3.

5 The intersection of a line and a curve

When a line and a curve are in the same plane, the coordinates of the point(s) of intersection can be found by solving the two equations simultaneously.

There are three possible situations.

(i) All points of intersection are distinct (see Figure 5.33).

> Distinct means they are separate points.

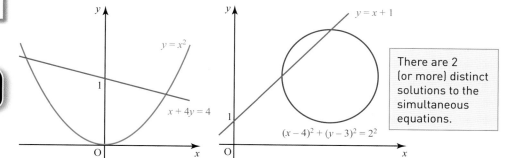

> There are 2 (or more) distinct solutions to the simultaneous equations.

Figure 5.33

> Coincident means they are in the same position.

(ii) The line is a tangent to the curve at one (or more) point(s) (see Figure 5.34). In this case, each point of contact corresponds to two (or more) coincident points of intersection. It is possible that the tangent will also intersect the curve somewhere else (as in Figure 5.34b).

(a)

(b)

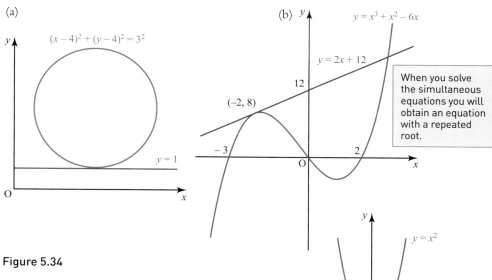

> When you solve the simultaneous equations you will obtain an equation with a repeated root.

Figure 5.34

(iii) The line and the curve do not meet (see Figure 5.35).

> When you try to solve the simultaneous equations you will obtain an equation with no real roots. So there is no point of intersection.

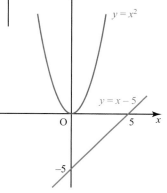

Figure 5.35

Example 5.11

A circle has equation $x^2 + y^2 = 8$.

For each of the following lines, find the coordinates of any points where the line intersects the circle.

(i) $y = x$ (ii) $y = x + 4$ (iii) $y = x + 6$

Solution

(i) Substituting $y = x$ into $x^2 + y^2 = 8$ gives ⟵ | Simplify. |

$$x^2 + x^2 = 8$$
$$2x^2 = 8$$
$$x^2 = 4$$
$$x = \pm 2 \quad ⟵ \boxed{\text{Don't forget the negative square root!}}$$

Since $y = x$ then the coordinates are $(-2, -2)$ and $(2, 2)$. ⟵ | The line intersects the circle twice. |

(ii) Substituting $y = x + 4$ into $x^2 + y^2 = 8$ gives

$$x^2 + (x + 4)^2 = 8 \quad ⟵ \boxed{\text{Multiply out the brackets.}}$$
$$\Rightarrow x^2 + x^2 + 8x + 16 = 8$$
$$\Rightarrow \quad 2x^2 + 8x + 8 = 0$$
$$\Rightarrow \quad \quad x^2 + 4x + 4 = 0 \quad ⟵ \boxed{\text{Divide by 2.}}$$
$$\Rightarrow \quad \quad \quad (x + 2)^2 = 0$$
$$x = -2 \quad ⟵ \boxed{\text{This is a repeated root, so } y = x + 4 \text{ is a tangent to the circle.}}$$

When $x = -2$ then $y = -2 + 4 = 2$

So the coordinates are $(-2, 2)$.

(iii) Substituting $y = x + 6$ into $x^2 + y^2 = 8$ gives

$$x^2 + (x + 6)^2 = 8$$
$$\Rightarrow x^2 + x^2 + 12x + 36 = 8$$
$$\Rightarrow \quad 2x^2 + 12x + 28 = 0$$

Check the discriminant: $b^2 - 4ac$

$$12^2 - 4 \times 2 \times 28 = -80$$

Since the discriminant is less than 0, the equation has no real roots.

So the line $y = x + 6$ does not meet the circle.

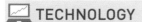 **TECHNOLOGY**

Use a graphing package or calculator to check the solutions.

Note

This example shows you an important result. When you are finding the intersection points of a line and a quadratic curve, or two quadratic curves, you obtain a quadratic equation.

If the discriminant is:

- positive – there are **two** points of intersection
- zero – there is **one repeated** point of intersection
- negative – there are **no** points of intersection.

The intersection of two curves

The same principles apply to finding the intersection of two curves, but it is only in simple cases that it is possible to solve the equations simultaneously using algebra (rather than a numerical or graphical method).

Example 5.12

Sketch the circle $x^2 + y^2 = 16$ and the curve $y = x^2 - 4$ on the same axes. Find the coordinates of any points of intersection.

Discussion points

➜ How else could you solve the simultaneous equations in the example?

➜ Which method is more efficient in this case?

Solution

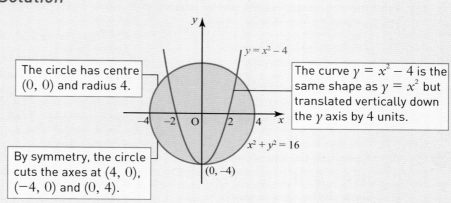

The circle has centre $(0, 0)$ and radius 4.

The curve $y = x^2 - 4$ is the same shape as $y = x^2$ but translated vertically down the y axis by 4 units.

By symmetry, the circle cuts the axes at $(4, 0)$, $(-4, 0)$ and $(0, 4)$.

Figure 5.36

Substituting $y = x^2 - 4$ into $x^2 + y^2 = 16$ gives

$$x^2 + (x^2 - 4)^2 = 16$$

$$\Rightarrow x^2 + x^4 - 8x^2 + 16 = 16 \quad \leftarrow \boxed{\text{Simplify.}}$$

$$\Rightarrow \qquad\qquad x^4 - 7x^2 = 0$$

$$\Rightarrow \qquad\qquad x^2(x^2 - 7) = 0 \quad \leftarrow \boxed{\text{Factorise.}}$$

$$\Rightarrow \qquad\qquad x^2 = 0 \Rightarrow x = 0 \text{ (twice)}.$$

$$\text{or} \Rightarrow \qquad x^2 = 7 \Rightarrow x = \pm\sqrt{7} \quad \leftarrow \boxed{\begin{array}{l}\text{Don't forget the negative} \\ \text{square root.}\end{array}}$$

Substitute back into $y = x^2 - 4$ to find the y coordinates.

$$x = 0 \qquad \Rightarrow y = -4$$

$$x = \pm\sqrt{7} \Rightarrow y = 7 - 4 = 3$$

So the points of intersection are $(-\sqrt{7}, 3)$, $(\sqrt{7}, 3)$ and $(0, -4)$ (twice).

Exercise 5.5

① Which of these points are on both of $x^2 + y^2 = 13$ and $x + 2y = 4$?

A $(2, 1)$ B $(2, 3)$
C $(3.6, 0.2)$ D $(-0.4, 2.2)$

② Show that the line $y = 3x + 1$ crosses the curve $y = x^2 + 3$ at $(1, 4)$ and find the coordinates of the other point of intersection.

③ Find the coordinates of the points where the line $y = 2x - 1$ cuts the circle $(x - 2)^2 + (y + 1)^2 = 5$.

④ Find the coordinates of the points of intersection of the line $2y = x - 5$ and the circle $(x + 1)^2 + (y - 2)^2 = 20$.

What can you say about this line and the circle?

PS ⑤ (i) Show that the line $x + y = 6$ is a tangent to the circle $x^2 + y^2 = 18$.

(ii) Show the line and circle on a diagram.

(iii) Find the point of contact of the tangent parallel to the line $x + y = 6$, and the equation of the tangent.

⑥ (i) Find the coordinates of the points of intersection of the line $y = 2x$ and the curve $y = x^2 + 6x - 5$.

(ii) Show also that the line $y = 2x$ does not cross the curve $y = x^2 + 6x + 5$.

⑦ Find the coordinates of the points A and B where the line $x - 3y + 15 = 0$ cuts the circle $x^2 + y^2 + 2x - 6y + 5 = 0$.

Also find the coordinates of the points where the line $y = x + 1$ meets the curve $y = x^3 - 3x^2 + 3x + 1$.

M ⑧ Figure 5.37 shows the cross-section of a goldfish bowl. The bowl can be thought of as a sphere with its top removed and its base flattened.

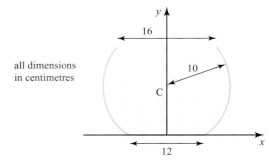

all dimensions in centimetres

Figure 5.37

Assume the base is on the x axis and the y axis is a line of symmetry.

(i) Find the height of the bowl.

(ii) Find the equation of the circular part of the cross-section.

(iii) The bowl is filled with water to a depth of 12 cm.

Find the area of the surface of the water.

PS ⑨ The line $y = 1 - x$ intersects the circle $x^2 + y^2 = 25$ at two points A and B.

(i) Find the coordinates of the points and the distance AB.

(ii) Is AB a diameter of the circle?

Give a reason for your answer.

PS ⑩ (i) Find the value of k for which the line $2y = x + k$ forms a tangent to the curve $y^2 = 2x$.

(ii) Hence find the coordinates of the point where the line $2y = x + k$ meets the curve for the value of k found in part (i).

PS ⑪ The equation of a circle is $(x + 2)^2 + y^2 = 8$ and the equation of a line is $x + y = k$, where k is a constant.

Find the values of k for which the line forms a tangent to the curve.

⑫ The equations of two circles are given below.

$$(x + 1)^2 + (y - 2)^2 = 10$$

and $(x - 1)^2 + (y - 3)^2 = 5$

The two circles intersect at the points A and B.

Find the area of the triangle AOB where O is the origin.

PS

Integer point circles

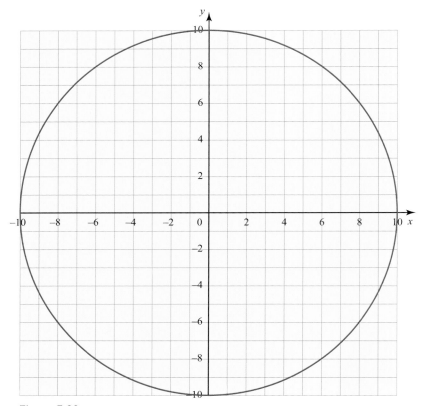

Figure 5.38

Look at the circle in Figure 5.38. Its equation is $x^2 + y^2 = 100$.

It goes through the point $(6, 8)$. Since both 6 and 8 are integers, this is referred to as an **integer point** in this question. This is not the only integer point this circle goes through; another is $(-10, 0)$ and there are others as well.

(i)   How many integer points are there **inside** the circle?

(ii) How many circles are there with equations of the form $x^2 + y^2 = N$, where $0 < N < 100$, that pass through at least one integer point?

      How many of these circles pass through at least 12 integer points?

(iii) Devise and explain at least one method to find the equation of a circle with radius greater than 10 units that passes through at least 12 integer points.

1   **Problem specification and analysis**

Parts (i) and (ii) of the problem are well defined and so deal with them first. Start by thinking about possible strategies. There are several quite different approaches, based on geometry or algebra. You may decide to try more than one and see how you get on.

Part (iii) is more open ended. You have to devise and explain at least one method. Leave this until you get to the last stage of the problem solving cycle. By then your earlier work may well have given you some insight into how to go about it.

2   **Information collection**

In this problem there will probably be a large amount of trial and error in your data collection. As well as collecting information, you will be trying out different possible approaches.

There are a number of cases that you could try out and so you need to be on the lookout for patterns that will cut down on your work. You have to think carefully about how you are going to record your findings systematically.

3   **Processing and representation**

The work you need to do at this stage will depend on what you have already done at the Information collection stage.

You may have already collected all the information you need to answer parts (i) and (ii) by just counting up the numbers. Alternatively, however, you may have found some patterns that will help you to work out the answers.

You then need to find a good way to present your answers. Think of someone who is unfamiliar with the problem. How are you going to show such a person what you have found in a convincing way?

4   **Interpretation**

So far you have been looking at parts (i) and (ii) of the problem. They are well defined and all the answers are numbers.

In part (iii), you are now expected to interpret what you have been doing by finding not just numbers but also a method, so that you can continue the work with larger circles.

To give a good answer you will almost certainly need to use algebra but you will also need to explain what you are doing in words.

The wording of the questions suggests there is more than one method and that is indeed the case. So a really good answer will explore the different possibilities.

LEARNING OUTCOMES

When you have completed this chapter you should be able to:

➤ understand and use the equation of a straight line including the forms:
  ○ $y - y_1 = m(x - x_1)$
  ○ $ax + by + c = 0$
➤ understand and use conditions on the gradient for two lines to be:
  ○ parallel
  ○ perpendicular
➤ use straight line models in a variety of contexts
➤ understand and use the coordinate geometry of the circle:
  ○ the equation of a circle in the form $(x - a)^2 + (y - b)^2 = r^2$
  ○ completing the square to find the centre and radius of a circle
➤ use the properties that:
  ○ the angle in a semicircle is a right angle
  ○ the perpendicular from the centre to a chord bisects the chord
  ○ the radius of a circle at a given point on its circumference is perpendicular to the tangent to the circle at that point.

KEY POINTS

1   For a line segment $A(x_1, y_1)$ and $B(x_2, y_2)$ (Figure 5.39) then:

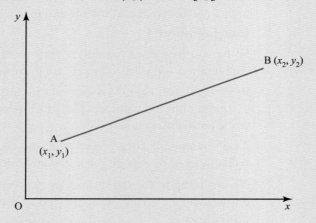

Figure 5.39

■ the gradient of AB is $\dfrac{y_2 - y_1}{x_2 - x_1}$

These are a formule you need to know.

■ the midpoint is $\left(\dfrac{x_1 + x_2}{2}, \dfrac{y_1 + y_2}{2} \right)$

■ the distance AB is $\sqrt{(x_2 - x_1)^2 + (y_2 - y_1)^2}$.

Using Pythagoras' theorem.

2 Two lines are parallel $\Leftrightarrow$ their gradients are equal.
3 Two lines are perpendicular $\Leftrightarrow$ the product of their gradients is -1.
4 The equation of a straight line may take any of the following forms:
   - line parallel to the y axis: $\qquad\qquad\quad x = a$
   - line parallel to the x axis: $\qquad\qquad\quad y = b$
   - line through the origin with gradient m: $\; y = mx$
   - line through $(0, c)$ with gradient m: $\qquad y = mx + c$
   - line through (x_1, y_1) with gradient m: $y - y_1 = m(x - x_1)$.
5 The equation of a circle is
   - centre $(0, 0)$, radius r: $x^2 + y^2 = r^2$
   - centre (a, b), radius r: $(x - a)^2 + (y - b)^2 = r^2$.

> These are a formule you need to know.

6 The angle in a semicircle is a right angle (Figure 5.40).
7 The perpendicular from the centre of a circle to a chord bisects the chord (Figure 5.40).
8 The tangent to a circle at a point is perpendicular to the radius through that point (Figure 5.41).

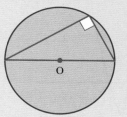

Figure 5.40

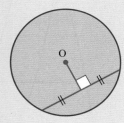

Figure 5.41

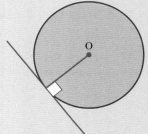

Figure 5.42

9 To find the points of intersection of two curves, you solve their equations simultaneously.

FUTURE USES

- The process of finding the equations of tangents and normal to curves will be used in Chapter 10 on Differentiation.
- The use of a straight line graph to calculate gradient and y-intercept will be developed in Chapter 13 with exponentials and logarithms.

PRACTICE QUESTIONS: PURE MATHEMATICS 1

MP ① (i) Prove that $\sqrt{2\frac{2}{3}} = 2\sqrt{\frac{2}{3}}$. [2 marks]

(ii) Show that $\dfrac{\sqrt{3}+1}{\sqrt{3}-1} = \sqrt{3}+2$. [2 marks]

PS ② (i) Solve the equation $2^{3x} = 4^{x+4}$. [3 marks]

(ii) Which of the following is a counter-example to $10^x > 2^x$?
$x = 10, x = 1, x = 0.1, x = 0$ [1 mark]

MP ③ Do not use a calculator in this question.

(i) Figure 1 shows the curves $y = x^2 - 4x + 1$ and $y = 7 - x^2$.

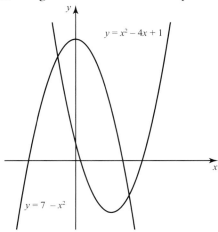

Figure 1

Find the coordinates of their points of intersection. [5 marks]

(ii) Prove that $y = -2x$ is a tangent to $y = x^2 - 4x + 1$ and
state the coordinates of the point of contact. [4 marks]

MP ④ Do not use a calculator in this question.

(i) Write $x^2 + 6x + 7$ in the form $(x + a)^2 + b$. [3 marks]

(ii) State the coordinates of the turning point of
$y = x^2 + 6x + 7$ and whether it is a minimum or maximum. [3 marks]

(iii) Sketch the curve $y = x^2 + 6x + 7$ and solve the
inequality $x^2 + 6x + 7 > 0$. [4 marks]

MP ⑤ Figure 2 shows a circle with centre C which passes through
the points A(2, 4) and B(−1, 1).

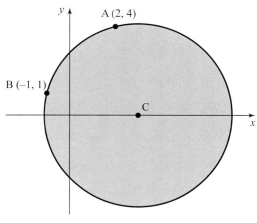

Figure 2

(i) AB is a chord of the circle. Show that the centre of the circle must lie on the line $x + y = 3$, Explain your reasoning at each step. [7 marks]

(ii) The centre of the circle also lies on the x axis. Find the equation of the circle. [5 marks]

PS ⑥ Figure 3 shows an equilateral triangle, ABC with A and B on the x axis and C on the y axis.

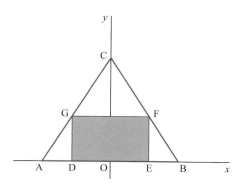

Figure 3

Each side of triangle ABC measures 4 units.

(i) Find the coordinates of points A, B and C in exact form. [4 marks]

(ii) Show that the equation of line BC can be written as $y = \sqrt{3}(2 - x)$. [2 marks]

A rectangle DEFG is drawn inside the triangle, as also shown.

D and E lie on the x axis, G on AC and F on BC.

(iii) Find the greatest possible area of rectangle DEFG. [7 marks]

M ⑦ Table 1 shows a spreadsheet with the information about typical stopping distances for cars from the Highway Code. Figure 4 has been drawn using the spreadsheet.

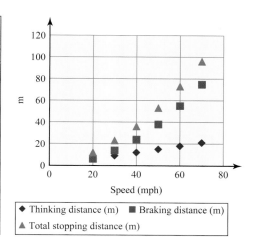

	A	B	C	D
1	Speed (mph)	Thinking distance (m)	Braking distance (m)	Total stopping distance (m)
2	20	6	6	12
3	30	9	14	23
4	40	12	24	36
5	50	15	38	53
6	60	18	55	73
7	70	21	75	96

Table 1

Figure 4

(i) (a) What feature of the scatter diagram suggests that the thinking distance is directly proportional to speed? [1 mark]

    (b)   What does this tell you about the thinking time for different speeds? Comment, with a brief explanation, on whether this is a reasonable modelling assumption.   **[2 marks]**

    (c)   Write down a formula connecting the speed, x mph and the thinking distance d m.   **[1 mark]**

(ii) The spreadsheet gives the following linear best fit model for the total stopping distance, y m, in terms of the speed, x mph.

$$y = 1.6771x - 26.38$$

    (a)   Use the model to find the total stopping distance for a speed of 10 mph.   **[1 mark]**

    (b)   Explain why this is not a suitable model for total stopping distance.   **[1 mark]**

(iii) The spreadsheet gives the following quadratic best fit model for the total stopping distance.

$$y = 0.0157x^2 + 0.2629x + 0.6$$

The values for total stopping distance using this model are shown in Table 2.

Speed (mph)	20	30	40	50	60	70
Quadratic model (m)		22.617	36.236	52.995	72.894	95.933

Table 2

    (a)   Calculate the missing value for 20 mph.   **[1 mark]**

    (b)   Give one possible reason why the model does not give exactly the same total stopping distances as those listed in the Highway Code.   **[1 mark]**

6 Trigonometry

I must go down to the seas again, to the lonely sea and the sky,

And all I ask is a tall ship and a star to steer her by,

John Masefield
(1878–1967)

There are many situations in which you can measure *some* sides or angles of a triangle, but it is either impossible or impractical to measure them all.

What information would you need to answer each of the questions below, and how would you obtain it?

→   How could someone on the bullet train in the picture above estimate how far they are away from Mount Fuji?

→   How can the sailors on the boat be sure that they are a safe distance from the lighthouse?

1 Trigonometric functions

Prior knowledge

You will have met this at GCSE.

You need to be able to use surds such as $\sqrt{2}$ and $\sqrt{3}$ – see Chapter 2.

The simplest definitions of the trigonometric functions are given in terms of the ratios of the sides of a right-angled triangle, for values of the angle θ between 0° and 90°.

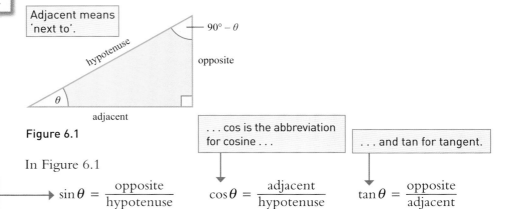

Adjacent means 'next to'.

Figure 6.1

... cos is the abbreviation for cosine ...

... and tan for tangent.

In Figure 6.1

sin θ is the abbreviation for sine θ...

$$\sin\theta = \frac{\text{opposite}}{\text{hypotenuse}} \qquad \cos\theta = \frac{\text{adjacent}}{\text{hypotenuse}} \qquad \tan\theta = \frac{\text{opposite}}{\text{adjacent}}$$

You will see from the triangle in Figure 6.1 that

$$\sin\theta = \cos(90° - \theta) \quad \text{and} \quad \cos\theta = \sin(90° - \theta).$$

Special cases

Certain angles occur frequently in mathematics and you will find it helpful to learn the values of their trigonometric functions.

(i)   The angles 30° and 60°

In Figure 6.2, triangle ABC is an equilateral triangle with side length 2 units, and AD is a line of symmetry.

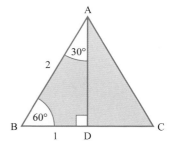

Figure 6.2

Using Pythagoras' theorem

$$AD^2 + 1^2 = 2^2 \Rightarrow AD = \sqrt{3}.$$

From triangle ABD,

$$\sin 60° = \frac{\sqrt{3}}{2} \qquad \cos 60° = \frac{1}{2} \qquad \tan 60° = \sqrt{3};$$

For the angle 60° at B,
opposite = AD = $\sqrt{3}$
adjacent = BD = 1
hypotenuse = AB = 2.

$$\sin 30° = \frac{1}{2} \qquad \cos 30° = \frac{\sqrt{3}}{2} \qquad \tan 30° = \frac{1}{\sqrt{3}}$$

For the angle 30° at A,
opposite = BD = 1
adjacent = AD = $\sqrt{3}$
hypotenuse = AB = 1.

Example 6.1

Without using a calculator, find the value of $\cos 60° \sin 30° + \cos^2 30°$.

> Note that $\cos^2 30°$ means $(\cos 30°)^2$.

Solution

$\cos 60° = \frac{1}{2}$, $\sin 30° = \frac{1}{2}$ and $\cos 30° = \frac{\sqrt{3}}{2}$

So $\cos 60° \sin 30° + \cos^2 30° = \frac{1}{2} \times \frac{1}{2} + \left(\frac{\sqrt{3}}{2}\right)^2$

$= \frac{1}{4} + \frac{3}{4}$

$= 1$

(ii)   The angle 45°

In Figure 6.3, triangle PQR is a right-angled isosceles triangle, in which sides PR and QR both have a length of 1 unit.

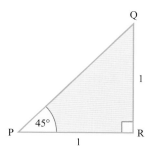

> Opposite = QR = 1
> Adjacent = PR = 1
> Hypotenuse = PQ = $\sqrt{2}$

Figure 6.3

Using Pythagoras' theorem

$$PQ^2 = 1^2 + 1^2 = 2 \Rightarrow PQ = \sqrt{2}.$$

This gives

$$\sin 45° = \frac{1}{\sqrt{2}} \qquad \cos 45° = \frac{1}{\sqrt{2}} \qquad \tan 45° = 1$$

(iii)   The angles 0° and 90°

Although you cannot have an angle of 0° in a triangle (because one side would be lying on top of another), you can still imagine what it might look like. In Figure 6.4, the hypotenuse has length 1 unit and the angle at X is very small.

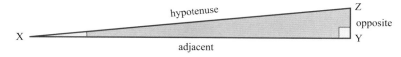

Figure 6.4

Imagine the angle at X becoming smaller and smaller until it is zero; you can deduce that

$$\sin 0° = \frac{0}{1} = 0 \qquad \cos 0° = \frac{1}{1} = 1 \qquad \tan 0° = \frac{0}{1} = 0$$

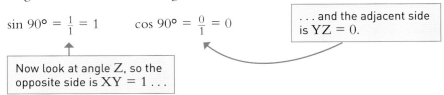

If the angle at X is 0°, then the angle at Z is 90°, and so

$$\sin 90° = \frac{1}{1} = 1 \qquad \cos 90° = \frac{0}{1} = 0$$

... and the adjacent side is YZ = 0.

Now look at angle Z, so the opposite side is XY = 1 ...

However, when you come to find tan 90°, there is a problem. The triangle suggests this has value $\frac{1}{0}$, but you cannot divide by zero.

If you look at the triangle XYZ, you will see that what we actually did was to draw it with angle X not zero but just very small, and to argue:

'We can see from this what will happen if the angle becomes smaller and smaller so that it is effectively zero.'

In this case we are looking at the limits of the values of sin **θ**, cos **θ** and tan **θ** as the angle **θ** approaches zero. The same approach can be used to look again at the problem of tan 90°.

If the angle X is not quite zero, then the side ZY is also not quite zero, and tan Z is 1 (XY is almost 1) divided by a very small number and so tan Z is large. The smaller the angle X, the smaller the side ZY and so the larger the value of tan Z. We conclude that in the limit when angle X becomes zero and angle Z becomes 90°, tan Z is infinitely large, and so we say

as Z → 90°, tan Z → ∞ (infinity)

Read these arrows as 'tends to'.

You can see this happening in the table of values given here.

Table 6.1

Z	tan Z
80°	5.67
89°	57.29
89.9°	527.96
89.99°	5279.6
89.999°	52796

Use your calculator to check these values.

Note

You will learn more about limits in Chapters 10 and 11 on differentiation and integration.

When Z actually equals 90°, you say that tan Z is **undefined**.

Example 6.2

In the diagram, angles ADB and CBD are right angles, AB = 2a and BC = 3a. Find the angle θ.

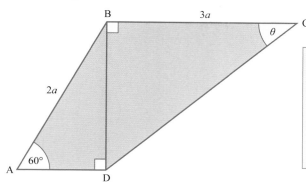

Figure 6.5

> There is not enough information in triangle CBD to work out θ. We can use triangle ADB to find BD and, because it is common to both triangles, we can then work out θ.

Solution

First, find an expression for BD.

In triangle ABD, $\dfrac{\text{BD}}{\text{AB}} = \sin 60°$ $\quad\boxed{\text{AB} = 2a}$

$$\Rightarrow \text{BD} = 2a \sin 60°$$
$$= 2a \times \frac{\sqrt{3}}{2}$$
$$= \sqrt{3}a$$

In triangle BCD, $\tan\theta = \dfrac{\text{BD}}{\text{BC}}$

$$= \frac{\sqrt{3}a}{3a}$$
$$= \frac{1}{\sqrt{3}}$$
$$\Rightarrow \boldsymbol{\theta} = 30°$$

Exercise 6.1

① Mel writes $\sin\theta \geq 1$.

Is Mel's statement always true, sometimes true or never true? Justify your answer.

②

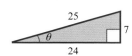

Figure 6.6

What is the value of $\cos\theta$?

A $\dfrac{7}{25}$ B $\dfrac{7}{24}$ C $\dfrac{24}{25}$ D $\dfrac{25}{24}$

③ In the triangle PQR, PQ = 17 cm, QR = 15 cm and PR = 8 cm.

(i) Show that the triangle is right-angled.

(ii) Write down the values of $\sin Q$, $\cos Q$ and $\tan Q$, leaving your answers as fractions.

(iii) Use your answers to part (ii) to show that

(a) $\sin^2 Q + \cos^2 Q = 1$

(b) $\tan Q = \dfrac{\sin Q}{\cos Q}$.

> **Exact** means leave your answer in surd form (e.g. $\sqrt{2}$) or as a fraction.

④ Without using a calculator, find the **exact** values of:

(i) $\tan 45° + \cos 60°$

(ii) $\tan^2 30°$

(iii) $\sin 90° - \cos 60°$

(iv) $\dfrac{\sin 60°}{\cos 60°}$

(v) $\tan^2 60°$

(vi) $\cos 30°(\tan^2 60° + \sin 0°)$

⑤ Without using a calculator, show that:

(i) $\sin 60° \cos 30° + \cos 60° \sin 30° = 1$

(ii) $\sin^2 30° + \sin^2 45° = \sin^2 60°$

(iii) $3\sin^2 30° = \cos^2 30°$

(iv) $\sin^2 45° + \cos^2 45° = 1$

PS ⑥ In Figure 6.7, $AB = 10\,cm$, $\angle BAC = 30°$, $\angle BCD = 45°$ and $\angle BDC = 90°$.

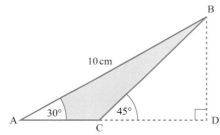

Figure 6.7

(i) Find the length of BD.

(ii) Show that $AC = 5(\sqrt{3} - 1)$ cm.

⑦ In Figure 6.8, $OA = 1\,cm$, $\angle AOB = \angle BOC = \angle COD = 30°$ and $\angle OAB = \angle OBC = \angle OCD = 90°$.

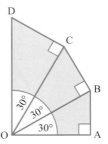

Figure 6.8

(i) Find the length of OD giving your answer in the form $a\sqrt{3}$.

(ii) Show that the perimeter of OABCD is $\dfrac{5}{3}(1 + \sqrt{3})$ cm.

<u>Prior knowledge</u>

You will have met this at GCSE.

2 Trigonometric functions for angles of any size

Unless given in the form of bearings, angles are measured from the x axis (see Figure 6.9). Anticlockwise is taken to be positive and clockwise to be negative.

Figure 6.9

Is it possible to extend the use of the trigonometric functions to angles greater than 90°, like $\sin 120°$, $\cos 275°$ or $\tan 692°$? The answer is yes – provided you change the definition of sine, cosine and tangent to one that does not require

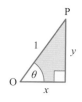

Figure 6.10

Unit length means the length is 1.

ACTIVITY 6.1

Draw x and y axes. For each of the four quadrants formed, work out the sign of $\sin\theta$, $\cos\theta$ and $\tan\theta$, from the definitions above.

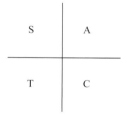

Figure 6.12

the angle to be in a right-angled triangle. It is not difficult to extend the definitions, as follows.

First look at the right-angled triangle in Figure 6.10 which has hypotenuse of unit length.

This gives rise to the definitions:

$$\sin\theta = \frac{y}{1} = y \qquad \cos\theta = \frac{x}{1} = x \qquad \tan\theta = \frac{y}{x}$$

Now think of the angle θ being situated at the origin, as in Figure 6.11, and allow θ to take any value. The vertex marked P has coordinates (x, y) and can now be anywhere on the unit circle.

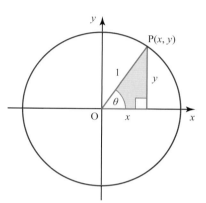

You can now see that the definitions above can be applied to *any* angle θ, whether it is positive or negative, and whether it is less than or greater than 90°:

$$\sin\theta = y \qquad \cos\theta = x \qquad \tan\theta = \frac{y}{x}.$$

For some angles, x or y (or both) will take a negative value, so the sign of $\sin\theta$, $\cos\theta$ and $\tan\theta$ will vary accordingly.

Figure 6.11

> **Note**
>
> Some people use this diagram (called a **CAST** diagram; Figure 6.12) to help them remember when sin, cos and tan are positive, and when they are negative. **A** means all positive in this quadrant, **S** means only sin positive, so cos and tan negative, etc.

Identities involving $\sin\theta$, $\cos\theta$ and $\tan\theta$

Since $\tan\theta = \frac{y}{x}$ and $y = \sin\theta$ and $x = \cos\theta$ it follows that

$$\tan\theta = \frac{\sin\theta}{\cos\theta}.$$

Prior knowledge

You need to be familiar with the idea of proof from Chapter 1.

It would be more accurate here to use the identity sign, $\equiv$, since the relationship is true for all values of θ except when $\cos\theta = 0$.

Remember you can't divide by 0.

$$\tan\theta \equiv \frac{\sin\theta}{\cos\theta}, \cos\theta \neq 0$$

So $\theta \neq 90°, 270°$ and so on.

An **identity** is different from an equation since an equation is only true for certain values of the variable, called the **solution** of the equation. For example, $\tan\theta = 1$ is an equation: it is true when $\theta = 45°$ or $225°$, but not when it takes any other value in the range $0° \leqslant \theta \leqslant 360°$.

By contrast, an identity is true for all values of the variable, for example

$$\tan 30° \equiv \frac{\sin 30°}{\cos 30°} \qquad \tan 72° \equiv \frac{\sin 72°}{\cos 72°} \qquad \tan(-339°) \equiv \frac{\sin(-339°)}{\cos(-339°)}$$

and so on for all values of the angle θ except when $\cos\theta = 0$.

Check this for yourself on your calculator.

In this book, as in mathematics generally, we often use an equals sign where it would be more correct to use an identity sign. The identity sign is kept for situations where we really want to emphasise that the relationship is an identity and not an equation.

The unit circle has radius 1.

Another useful identity can be found by applying Pythagoras' theorem to any point P(x, y) on the unit circle See Figure 6.11.

$$y^2 + x^2 \equiv OP^2$$

$$(\sin\theta)^2 + (\cos\theta)^2 \equiv 1$$

This is written as

$$\sin^2\theta + \cos^2\theta \equiv 1$$

You can use the identities $\tan\theta \equiv \dfrac{\sin\theta}{\cos\theta}$ and $\sin^2\theta + \cos^2\theta \equiv 1$ to prove other identities are true.

Example 6.3

Prove the identity $\cos^2\theta - \sin^2\theta \equiv 2\cos^2\theta - 1$.

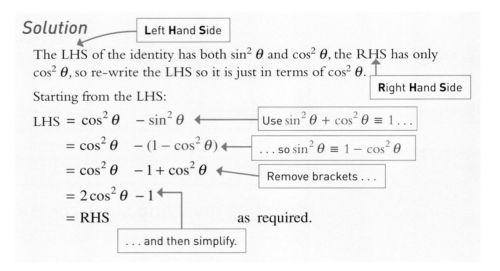

Solution   — Left Hand Side

The LHS of the identity has both $\sin^2\theta$ and $\cos^2\theta$, the RHS has only $\cos^2\theta$, so re-write the LHS so it is just in terms of $\cos^2\theta$.

Right Hand Side

Starting from the LHS:

$$\begin{aligned}
\text{LHS} &= \cos^2\theta \quad - \sin^2\theta \quad\quad\quad \text{Use } \sin^2\theta + \cos^2\theta \equiv 1\ldots\\
&= \cos^2\theta \quad - (1 - \cos^2\theta) \quad\quad \ldots\text{so } \sin^2\theta \equiv 1 - \cos^2\theta\\
&= \cos^2\theta \quad - 1 + \cos^2\theta \quad\quad \text{Remove brackets}\ldots\\
&= 2\cos^2\theta \quad - 1\\
&= \text{RHS} \quad\quad\quad\quad\quad\quad\quad \text{as required.}
\end{aligned}$$

$\ldots$ and then simplify.

When one side of the identity looks more complicated than the other side it can help to work with this side until you end up with the same as the simpler side.

Example 6.4

Prove the identity $\dfrac{\cos\theta}{1-\sin\theta} - \dfrac{1}{\cos\theta} \equiv \tan\theta$.

Solution

The LHS of this identity is more complicated, so manipulate the LHS until you end up with $\tan\theta$.

Write the LHS as a single fraction: $\dfrac{\cos\theta}{1-\sin\theta} - \dfrac{1}{\cos\theta}$

| Since $\sin^2\theta + \cos^2\theta \equiv 1$ $\Rightarrow \cos^2\theta \equiv 1 - \sin^2\theta\dots$ |

$\equiv \dfrac{\cos^2\theta - (1-\sin\theta)}{\cos\theta(1-\sin\theta)}$

$\equiv \dfrac{\cos^2\theta + \sin\theta - 1}{\cos\theta(1-\sin\theta)}$

| $\dots$ so replace $\cos^2\theta$ with $1 - \sin^2\theta$. |

$\equiv \dfrac{1 - \sin^2\theta + \sin\theta - 1}{\cos\theta(1-\sin\theta)}$

$\equiv \dfrac{\sin\theta - \sin^2\theta}{\cos\theta(1-\sin\theta)}$ ← Simplify

$\equiv \dfrac{\sin\theta(1-\sin\theta)}{\cos\theta(1-\sin\theta)}$ ← Factorise

$\equiv \dfrac{\sin\theta}{\cos\theta}$ ← Cancel

$\equiv \tan\theta$ as required

TECHNOLOGY

When using graphing software to draw these graphs, check the software is set to degrees.

The sine and cosine graphs

In Figure 6.13, angles have been drawn at intervals of 30° in the unit circle, and the resulting y coordinates plotted relative to the axes on the right. They have been joined with a continuous curve to give the graph of $\sin\theta$ for $0° \leqslant \theta \leqslant 360°$.

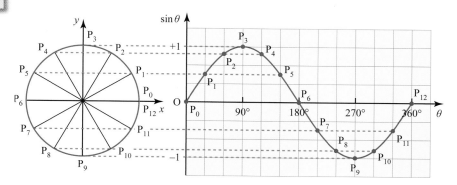

Figure 6.13

The angle 390° gives the same point P_1 on the circle as the angle 30°, the angle 420° gives point P_2 and so on. You can see that for angles from 360° to 720° the sine wave will simply repeat itself, as shown in Figure 6.14. This is true also for angles from 720° to 1080° and so on.

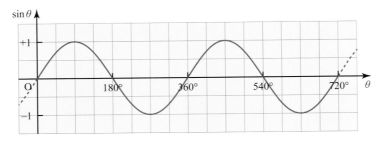

Figure 6.14

Since the curve repeats itself every 360° the sine function is described as **periodic**, with **period** 360°.

In a similar way you can transfer the x coordinates on to a set of axes to obtain the graph of $\cos\theta$. This is most easily illustrated if you first rotate the circle through 90° anticlockwise.

Figure 6.15 shows the circle in this new orientation, together with the resulting graph.

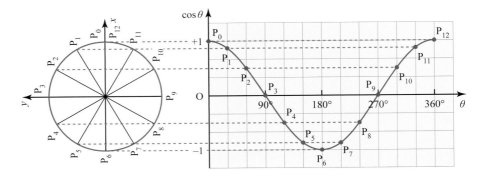

Figure 6.15

For angles in the interval $360° < \theta < 720°$, the cosine curve will repeat itself. You can see that the cosine function is also periodic with a period of 360°.

Notice that the graphs of $\sin\theta$ and $\cos\theta$ have exactly the same shape. The cosine graph can be obtained by translating the sine graph 90° to the left, as shown in Figure 6.16.

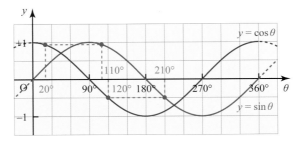

Discussion point

What do the graphs of $\sin\theta$ and $\cos\theta$ look like for negative angles?

Figure 6.16

From the graphs it can be seen that, for example

$$\cos 20° = \sin 110°, \qquad \cos 90° = \sin 180°, \qquad \cos 120° = \sin 210°, \text{ etc.}$$

In general

$$\cos\theta \equiv \sin(\theta + 90°).$$

The tangent graph

You can use the fact that $\tan\theta = \frac{y}{x}$ and, since $\sin\theta° = y$ and $\cos\theta° = x$ then $\tan\theta° = \frac{\sin\theta}{\cos\theta}$,

Look at the unit circle in Figure 6.11 on page 105

to work out the value of $\tan\theta$ for any angle.

You have already seen that $\tan\theta$ is undefined for $\theta = 90°$. This is also the case for all other values of θ for which $\cos\theta = 0$, namely $270°, 450°, \ldots$, and $-90°, -270°, \ldots$

The graph of $\tan\theta$ is shown in Figure 6.17. The dotted lines at $\theta = \pm 90°$ and at $\theta = 270°$ are **asymptotes**. They are not actually part of the curve. Its branches get closer and closer to them without ever quite reaching them.

Note

The graph of $\tan\theta$ is periodic, like those for $\sin\theta$ and $\cos\theta$, but in this case the period is $180°$.

These are asymptotes.

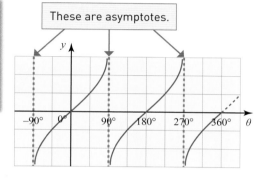

Figure 6.17

Discussion point

Describe the symmetry of each of your graphs.

ACTIVITY 6.2

Draw the graphs of $y = \sin\theta$, $y = \cos\theta$ and $y = \tan\theta$ for values of θ between $-90°$ and $450°$.

These graphs are very important. Keep them in your mind, because you will often use them, for example, when solving trigonometric equations.

Example 6.5

Given that $\sin\theta = \frac{4}{7}$ and θ is **obtuse** find the exact value of

(i) $\cos\theta$

(ii) $\tan\theta$

An **obtuse** angle is between $90°$ and $180°$.

Solution

(i)   Use the identity $\sin^2\theta + \cos^2\theta \equiv 1$

$$\Rightarrow \cos^2\theta \equiv 1 - \sin^2\theta$$

$$\Rightarrow \cos^2\theta = 1 - \left(\frac{4}{7}\right)^2$$

$$\text{So } \cos^2\theta = 1 - \left(\frac{16}{49}\right) = \left(\frac{33}{49}\right)$$

So $\cos\theta = \pm\sqrt{\frac{33}{49}} = \pm\frac{\sqrt{33}}{7}$

Now you need to decide on the correct sign for $\cos\theta$.

Look at the graph of $y = \cos\theta$ in Figure 6.18: it shows that $\cos\theta$ is negative for any angle between $90°$ and $180°$.

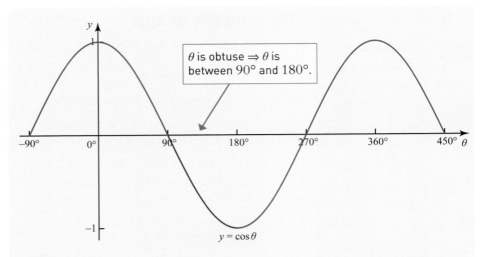

Figure 6.18

Or look at the CAST diagram in Figure 6.19: if θ *is* obtuse then it lies in the second quadrant and only sin is positive there.

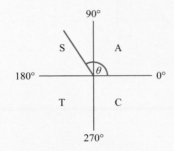

Figure 6.19

Hence $\cos\theta = -\dfrac{\sqrt{33}}{7}$

If the question had said that θ is acute (i.e. θ is between 0° and 90°) then the answer would be $\cos\theta = +\dfrac{\sqrt{33}}{7}$.

(ii) Use the identity $\tan\theta \equiv \dfrac{\sin\theta}{\cos\theta}$

$$\Rightarrow \tan\theta = \frac{\frac{4}{7}}{-\frac{\sqrt{33}}{7}} = -\frac{4}{\sqrt{33}}$$

Example 6.9

Solve the equation $\cos\theta + 2 = 2\sin^2\theta$ for $0° \leqslant \theta \leqslant 360°$.

Solution

You can't solve the equation as it stands, as it involves both sin and cos. You need to rewrite it so it is in terms of just one trigonometric function.

You need to use the identity $\sin^2\theta + \cos^2\theta \equiv 1 \Rightarrow \sin^2\theta \equiv 1 - \cos^2\theta$

$$\cos\theta + 2 = 2\sin^2\theta$$
$$\Rightarrow \quad \cos\theta + 2 = 2(1 - \cos^2\theta)$$
$$\Rightarrow \quad \cos\theta + 2 = 2 - 2\cos^2\theta$$

This is a quadratic in $\cos\theta$.

$$\Rightarrow \quad 2\cos^2\theta + \cos\theta = 0$$
$$\Rightarrow \quad \cos\theta\,(2\cos\theta + 1) = 0$$

Always factorise, never cancel . . .

$$\Rightarrow \cos\theta = 0 \text{ or } 2\cos\theta + 1 = 0$$

. . . because if you cancel you will lose the roots for when $\cos\theta = 0$.

$$\cos\theta = 0 \Rightarrow \theta = 90° \text{ or } 270°$$
$$\cos\theta = -\tfrac{1}{2} \Rightarrow \theta = 120°$$
$$\text{or} \qquad\qquad\qquad \theta = -120° + 360° = 240°$$

Find the first value from your calculator . . .

. . . then use the graph in Figure 6.24 to find the other possible values of θ.

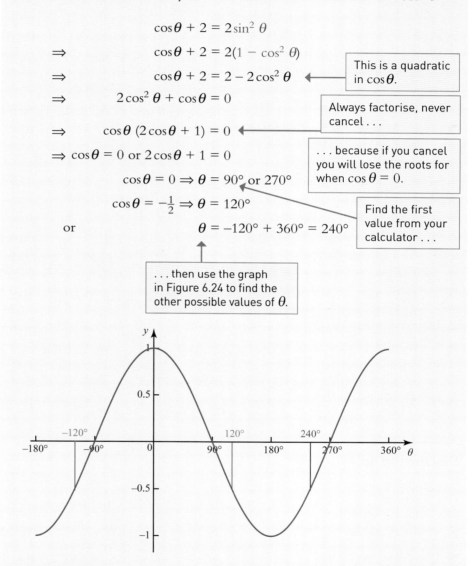

Figure 6.24

The values of θ are 120° or 240°.

Example 6.10

Solve (i) $\cos(\theta + 70°) = 0.5$ for $0° \leqslant \theta \leqslant 360°$

       (ii) $\sin 2\theta = \sqrt{3}\cos 2\theta$ for $0° \leqslant \theta \leqslant 360°$

Solution

(i) Let $\theta + 70° = x$

Add 70° to each part of the inequality.

Change the interval: $\quad 70° \leqslant \theta + 70° \leqslant 430° \Rightarrow 70° \leqslant x \leqslant 430°$

$\cos x = 0.5 \Rightarrow x = 60°$ This is out of range.

or $\qquad\qquad x = 360° - 60° = 300°$

or $\qquad\qquad x = 360° + 60° = 420°$

Once you have found all the values for x, you can use them to find the values of θ.

As $\theta + 70° = x$ then $\theta + 70° = 300° \Rightarrow \theta = 230°$

or $\qquad\qquad\qquad\qquad \theta + 70° = 420° \Rightarrow \theta = 350°$

so $\qquad\qquad\qquad\qquad\qquad \theta = 230°$ or $\theta = 350°$

(ii) $\sin 2\theta = \sqrt{3}\cos 2\theta \Rightarrow \dfrac{\sin 2\theta}{\cos 2\theta} = \sqrt{3}$

Use the identity $\tan x \equiv \dfrac{\sin x}{\cos x}$ to re-write the equation.

$\Rightarrow \tan 2\theta = \sqrt{3}$

θ lies between 0° and 360°, so 2θ lies between 0° and 720°.

Let $2\theta = x$

Change the interval: $\quad 0° \leqslant 2\theta \leqslant 720° \Rightarrow 0° \leqslant x \leqslant 720°$

$\tan x = \sqrt{3} \Rightarrow x = 60°$

or $\qquad\qquad x = 60° + 180° = 240°$

or $\qquad\qquad x = 240° + 180° = 420°$

Keep adding 180° until you have all the roots for x in the interval $0° \leqslant x \leqslant 720°$.

or $\qquad\qquad x = 420° + 180° = 600°$

Find **all** the values for x, **before** you work out the values for θ.

As $2\theta = x$ then $2\theta = 60° \Rightarrow \theta = 30°$

or $\qquad\qquad\qquad 2\theta = 240° \Rightarrow \theta = 120°$

or $\qquad\qquad\qquad 2\theta = 420° \Rightarrow \theta = 210°$

or $\qquad\qquad\qquad 2\theta = 600° \Rightarrow \theta = 300°$

so $\qquad\qquad\qquad \theta = 30°, 120°, 210°$ or $300°$

Discussion point

Use graphing software or a graphical calculator to sketch the graph of $y = \tan 2x$.

➜ How does this graph show that there are four roots to the equation $\tan 2\theta = \sqrt{3}$ in the range $0° \leqslant \theta \leqslant 360°$?

You will find out more about functions such as $y = \tan 2x$ and $y = \cos(\theta + 70°)$ in Chapter 8.

Exercise 6.3

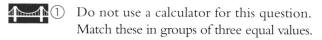 ① Do not use a calculator for this question. Match these in groups of three equal values.

$\sin 60°$	$\tan 60°$	$\cos 120°$
$-\tan 120°$	$-\sin 150°$	$\tan 150°$
$-\cos 210°$	$-\tan 210°$	$\tan 240°$
$-\sin 300°$	$-\cos 300°$	$\tan 330°$

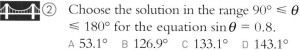 ② Choose the solution in the range $90° \leqslant \theta \leqslant 180°$ for the equation $\sin\theta = 0.8$.

   A 53.1°   B 126.9°   C 133.1°   D 143.1°

③ Solve the following equations for $0° \leqslant x \leqslant 360°$.

  (i) $\tan x = -\sqrt{3}$ (ii) $\tan x = \dfrac{1}{\sqrt{3}}$

  (iii) $\cos x = \dfrac{1}{2}$ (iv) $\cos x = -\dfrac{1}{2}$

  (v) $\sin x = \dfrac{\sqrt{2}}{2}$ (vi) $\sin x = -\dfrac{1}{\sqrt{2}}$

④ Solve the following equations for $0° \leqslant x \leqslant 360°$. Give your answers correct to 1 decimal place.

  (i) $\sin x = 0.6$ (ii) $\sin x = -0.6$

  (iii) $\cos x = 0.6$ (iv) $\cos x = -0.6$

  (v) $\tan x = 0.6$ (vi) $\tan x = -0.6$

⑤ In this question all the angles are in the interval $-180°$ to $180°$. Give your answers correct to 1 d.p.

  (i)   Given that $\sin\alpha < 0$ and $\cos\alpha = 0.7$, find α.

  (ii)   Given that $\tan\beta = 0.4$ and $\cos\beta < 0$, find β.

  (iii)   Given that $\sin\gamma = 0.8$ and $\tan\gamma > 0$, find γ.

⑥ Solve the following equations for $0° \leqslant x \leqslant 360°$.

  (i) $2\cos x - 1 = 0$ (ii) $3\sin x + 2 = 1$

  (iii) $\tan^2 x = 3$ (iv) $\cos^2 x = 1$

  (v) $3\tan^2 x - 10\tan x + 3 = 0$

  (vi) $\cos^2 x - x - 2 = 0$

⑦ Solve the following equations for $0° \leqslant x \leqslant 360°$.

  (i) $2\sin^2\theta = \cos\theta + 1$

  (ii) $\sin^2 x = 1 - \cos x$

  (iii) $1 - \cos^2 x = 2\sin x$

  (iv) $\sin^2 x = 2\cos^2 x$

  (v) $2\sin^2 x = 3\cos x$

  (vi) $4\cos\theta + 5 = 4\sin^2\theta$

⑧ Solve the following equations for $0° \leqslant \theta \leqslant 360°$.

  (i) $\cos(\theta - 10°) = \dfrac{1}{2}$

  (ii) $\tan(\theta + 20°) = \dfrac{\sqrt{3}}{3}$

  (iii) $\sin(\theta + 15°) = \dfrac{\sqrt{2}}{2}$

  (iv) $\tan 3\theta = \dfrac{\sqrt{3}}{3}$ (v) $\sin\frac{1}{2}\theta = \dfrac{\sqrt{3}}{2}$

  (vi) $\cos 2\theta = -\dfrac{1}{2}$ (vii) $\cos 3\theta = \dfrac{1}{2}$

  (viii) $\sin 2\theta = -\dfrac{\sqrt{3}}{2}$ (ix) $\tan 3\theta = 0$

⑨ Solve $\tan 5\theta = 1$ for $0° \leqslant \theta \leqslant 90°$.

PS ⑩ Figure 6.25 shows part of the curves $y = \cos x$ and $y = \tan x$ which intersect at the points A and B. Find the coordinates of A and B.

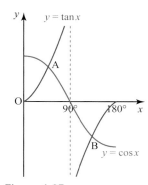

Figure 6.25

⑪ Solve the following equations for $0° \leqslant x \leqslant 360°$.

  (i) $\cos x = 2\sin x$

  (ii) $\tan x + \dfrac{1}{\tan x} = 2$

  (iii) $\dfrac{\tan x}{2} = \dfrac{1}{\tan x}$

⑫ Solve the following equations for $0° \leqslant \theta \leqslant 360°$.

  (i) $\sin(2\theta - 30°) = \dfrac{\sqrt{3}}{2}$

  (ii) $1 - 2\tan 2\theta = 0$

(iii) $\cos 3\theta - \sin 3\theta = 0$

(iv) $\sqrt{3} \sin 2\theta + \cos 2\theta = 0$

(v) $4\sin^2 \frac{1}{2}\theta - 4\sin \frac{1}{2}\theta + 1 = 0$

(vi) $\cos^2 2\theta - \cos 2\theta - 2 = 0$

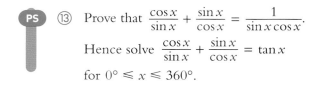

⑬ Prove that $\dfrac{\cos x}{\sin x} + \dfrac{\sin x}{\cos x} = \dfrac{1}{\sin x \cos x}$.

Hence solve $\dfrac{\cos x}{\sin x} + \dfrac{\sin x}{\cos x} = \tan x$

for $0° \leqslant x \leqslant 360°$.

Prior knowledge

You will have met this idea at GCSE.

4 Triangles without right angles

Now that you have moved away from the simple trigonometric definitions in terms of right-angled triangles, it is time to look at other triangles, not just those with an angle of 90°.

There is a standard notation used for triangles. The vertices are labelled with capital letters, and the side opposite a vertex is labelled with the corresponding lower case letter, as in Figure 6.26.

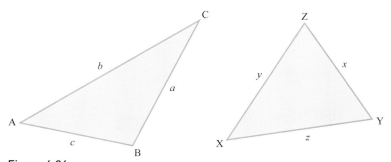

Figure 6.26

There are two rules which connect the values of the sides and angles of any triangle: the sine rule and the cosine rule.

The sine rule

For any triangle ABC

$$\frac{a}{\sin A} = \frac{b}{\sin B} = \frac{c}{\sin C}$$

Proof

For the triangle ABC, CD is the perpendicular from C to AB (extended if necessary). There are two cases to consider, as shown in Figure 6.27.

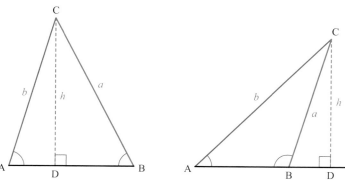

Figure 6.27

Case (i)	Case (ii)
In $\triangle ACD$: $\quad \sin A = \dfrac{h}{b}$	$\sin A = \dfrac{h}{b}$
$\Rightarrow h = b \sin A \quad$ ①	$\Rightarrow h = b \sin A \quad$ ①
In $\triangle BCD$: $\quad \sin B = \dfrac{h}{a}$	$\sin(180° - B) = \dfrac{h}{a}$
$\Rightarrow h = a \sin B \quad$ ②	But $\sin(180° - B) = \sin B$
	$\Rightarrow h = a \sin B \quad$ ②

Combining ① and ② gives

$$a \sin B = b \sin A \qquad \text{or} \qquad \frac{a}{\sin A} = \frac{b}{\sin B}$$

In the same way, starting with a perpendicular from A would give $\dfrac{b}{\sin B} = \dfrac{c}{\sin C}$.

Example 6.11

Find the side AB in the triangle shown in Figure 6.28.

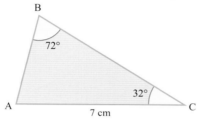

Figure 6.28

Solution

Using the sine rule:

$$\frac{b}{\sin B} = \frac{c}{\sin C}$$

$$\Rightarrow \frac{7}{\sin 72°} = \frac{c}{\sin 32°}$$

$$\Rightarrow \frac{7 \sin 32°}{\sin 72°} = c$$

$$c = 3.900...$$

> Do the calculation entirely on your calculator, and only round the final answer.

Side AB = 3.9 cm (to 1 d.p.)

Using the sine rule to find an angle

The sine rule may also be written as

$$\frac{\sin A}{a} = \frac{\sin B}{b} = \frac{\sin C}{c}$$

and this form is easier to use when you need to find an angle.

Be careful because sometimes there are two possible answers, as the next example demonstrates.

Example 6.12

Find the angle Z in the triangle XYZ, given that $Y = 27°$, $y = 4\,cm$ and $z = 5\,cm$.

Solution

The sine rule in $\triangle XYZ$ is $\dfrac{\sin X}{x} = \dfrac{\sin Y}{y} = \dfrac{\sin Z}{z}$

$$\Rightarrow \frac{\sin Z}{5} = \frac{\sin 27°}{4}$$

$$\sin Z = \frac{5 \sin 27°}{4} = 0.567488\ldots$$

$Z = 34.6°$ (to 1 d.p.) or $Z = 180° - 34.6° = 145.4°$ (to 1 d.p.)

Both solutions are possible, as shown in Figure 6.29.

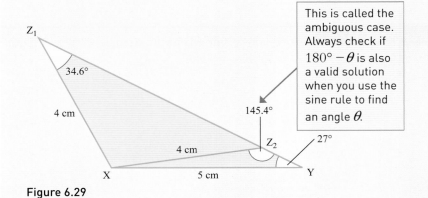

This is called the ambiguous case. Always check if $180° - \theta$ is also a valid solution when you use the sine rule to find an angle θ.

> **Note**
>
> Sometimes only one solution is possible since the second would give a triangle with angle sum greater than 180°.

Figure 6.29

The cosine rule

Sometimes it is not possible to use the sine rule with the information you have about a triangle, for example when you know all three sides but none of the angles.

An alternative approach is to use the cosine rule, which can be written in two forms:

$$a^2 = b^2 + c^2 - 2bc \cos A \qquad \text{or} \qquad \cos A = \frac{b^2 + c^2 - a^2}{2bc}$$

Like the sine rule, the cosine rule can be applied to any triangle.

Proof

For the triangle ABC, CD is the perpendicular from C to AB (extended if necessary). As with the sine rule, there are two cases to consider. Both are shown in Figure 6.30.

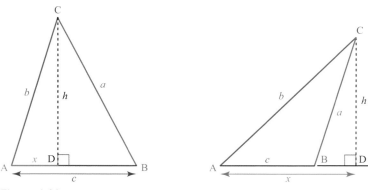

Figure 6.30

In $\triangle ACD$ $b^2 = x^2 + h^2$ ①

$$\cos A = \frac{x}{b} \Rightarrow x = b \cos A \quad ②$$

In $\triangle BCD$, for case (i) $a^2 = (c - x)^2 + h^2$ `Using Pythagoras' theorem`

for case (ii) $a^2 = (x - c)^2 + h^2$

In both cases this expands to give

$$a^2 = c^2 - 2cx + x^2 + h^2$$
$$= c^2 - 2cx + x^2 + b^2 \quad \text{Using } b^2 = x^2 + h^2 \quad ①$$
$$= c^2 - 2cb \cos A + b^2 \quad \text{Using } x = b \cos A \quad ②$$

So $a^2 = b^2 + c^2 - 2bc \cos A$, as required.

Discussion point

→ Show that you can rearrange the formula to give

$$\cos A = \frac{b^2 + c^2 - a^2}{2bc}$$

Use this form of the cosine rule when you need to find a missing angle.

Often you will need to choose between the sine and cosine rules, or possibly use both, as in Example 6.13 overleaf.

You can use Table 6.2 to help you decide which rule to use.

Table 6.2

Triangle	You know	You want	Use
	3 sides	Any angle $a^2 = b^2 + c^2 - 2bc \cos A$	Cosine rule
	2 sides + **included** angle	3rd side $\cos A = \dfrac{b^2 + c^2 - a^2}{2bc}$	
	2 angles + 1 side	Any side $\dfrac{a}{\sin A} = \dfrac{b}{\sin B}$	Sine rule
	2 sides + 1 angle	Any angle $\dfrac{\sin A}{a} = \dfrac{\sin B}{b}$	

Don't forget that once you know 2 angles in a triangle, you can use 'angles sum to 180°' to find the third angle!

Example 6.13

A ship S is 5 km from a lighthouse H on a bearing of 152°, and a tanker T is 7 km from the lighthouse on a bearing of 069°.

Find the distance and bearing of the ship from the tanker.

Solution

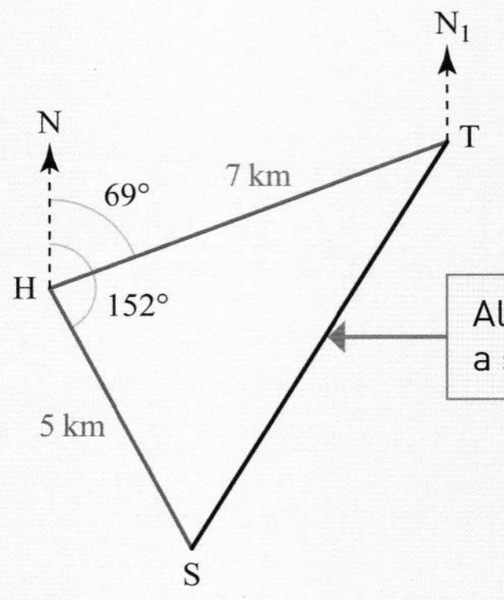

Figure 6.31

The bearing (or compass bearing) is the direction measured as an angle clockwise from North.

Always start with a sketch.

$\angle THS = 152° - 69° = 83°$

You know 2 sides and the angle between them, so use the cosine rule to find the 3rd side.

Using the cosine rule:

$$h^2 = s^2 + t^2 - 2st\cos H$$
$$= 7^2 + 5^2 - 2 \times 7 \times 5 \times \cos 83°$$
$$= 49 + 25 - 8.530\ldots$$
$$= 65.469\ldots$$
$$h = 8.091\ldots$$

Store the exact value on your calculator; never round until the end!

Therefore the distance of the ship from the tanker is 8.09 km (to 2 d.p.).

Using the sine rule

$$\frac{\sin T}{t} = \frac{\sin H}{h}$$
$$\frac{\sin T}{5} = \frac{\sin 83°}{8.0\ldots}$$
$$\sin T = 0.6\ldots$$
$$T = 37.8\ldots°$$

From the diagram: $\angle N_1TH = 180° - 69° = 111°$

It follows that the bearing of the ship from the tanker (the reflex angle N_1TS) is

$$360° - 111° - 37.8\ldots° = 211° \text{ (to the nearest degree).}$$

Angles of elevation and depression

Some problems may refer to an **angle of elevation** or an **angle of depression**.

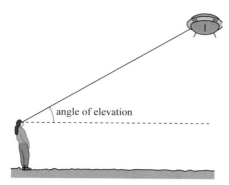

Figure 6.32

The angle of elevation is the angle between the horizontal and a direction **above** the horizontal.

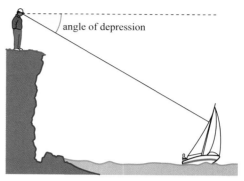

Figure 6.33

The angle of depression is the angle between the horizontal and a direction **below** the horizontal.

Exercise 6.4

 ①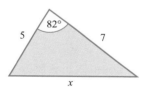

Figure 6.34

Someone has used the cosine rule to work out x and the result was 0.746.

How do you know it must be wrong?

What error did they make in their calculation?

② Use the sine rule to find the length x in each of the following triangles.

(i)

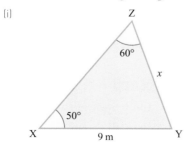

Figure 6.35

(ii)

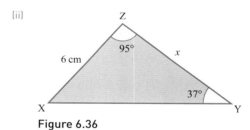

Figure 6.36

③ Use the cosine rule to find the length x in each of the following triangles.

(i)

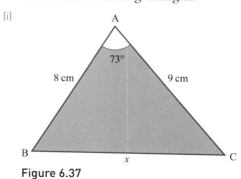

Figure 6.37

PS

(ii)

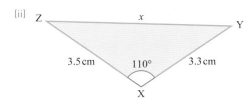

Figure 6.38

④ Use the sine rule to find the angle **θ** in each of the following triangles.

(i)

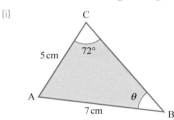

Figure 6.39

(ii)

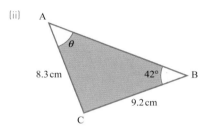

Figure 6.40

⑤ Use the cosine rule to find the angle **θ** in each of the following triangles.

(i)

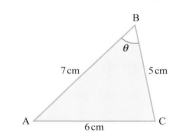

Figure 6.41

(ii)

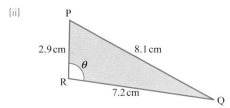

Figure 6.42

⑥ In the quadrilateral ABCD shown, find:

(i) AC

(ii) ∠ADC.

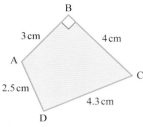

Figure 6.43

⑦ Bradley is 4.5 km due south of Applegate, and Churchgate is 6.7 km from Applegate on a bearing 079°.

PS

How far is it from Bradley to Churchgate?

PS ⑧ From a ship S, two other ships P and Q are on bearings of 315° and 075° respectively. The distance PS = 7.4 km and QS = 4.9 km.

Find the distance PQ.

PS ⑨ A man sets off from A and walks 3 km on a bearing of 045° to a point B. He then walks 1 km on a bearing of 322° to a point C.

Find the distance AC.

⑩ From two points A and B on level ground, the angles of elevation of the top, P, of a lighthouse are found to be 27.6° and 46.2°. The distance between A and B is 34 m.

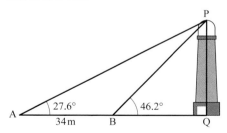

Figure 6.44

Find:

(i) ∠APB   (ii) AP

(iii) the height of the lighthouse.

⑪ The diagram shows three survey points, P, Q and R which are on a north–south line on level ground. From P, the bearing of the foot of a statue S is 104°, and from Q it is 081°. Point P is 400 m north of Q and Q is 300 m north of R.

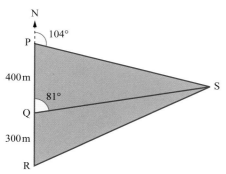

Figure 6.45

(i)   Find the distance PS.

(ii) Find the distance RS.

(iii) Find the bearing of S from R.

⑫ At midnight, a ship sailing due north passes two lightships, A and B, which are 10 km apart in a line due east from the ship. Lightship A is closer to the ship than B. At 2 a.m. the bearings of the lightships are 149° and 142°.

(i)   Draw a sketch to show A and B and the positions of the ship at midnight and 2 a.m.

(ii) Find the distance of the ship from A at 2 a.m.

(iii) Find the speed of the ship.

⑬ A ship travelling with a constant speed and direction is sighted from a lighthouse. At this time it is 2.7 km away, on a bearing of 042°. Half an hour later it is on a bearing of 115° at a distance of 7.6 km from the same lighthouse.

Find its speed in kilometres per hour.

⑭ A helicopter leaves a point P and flies on a bearing of 162° for 3 km to a point Q, and then due west to a point R which is on a bearing of 220° from P.

(i)   Find PR.

(ii) Find RQ.

The flight takes 4 minutes.

(iii) Find the speed of the helicopter in kilometres per hour.

⑮ A walker at point P can see the spires of St. Mary's (M) and St. George's (G) on bearings of 340° and 015° respectively. She then walks 500 m due north to a point Q and notes the bearings as 315° to St. Mary's and 030° to St. George's. Take the measurements as exact and assume they are all made in the same horizontal plane.

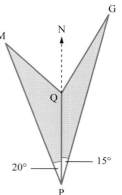

Figure 6.46

(i)   Copy the diagram (Figure 6.46) and fill in the information given.

(ii) Calculate the distance PM to the nearest metre.

(iii) Given that the distance PG is 966 m, calculate MG, the distance between the spires, to the nearest metre.

⑯ In Figure 6.47, PQRS is a square of side a and PQT is an equilateral triangle.

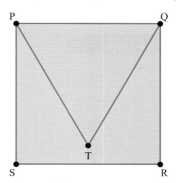

Figure 6.47

(i)   Find the lengths of the sides of the triangle QRT in terms of a.

(ii) Hence prove that

$$\sin 15° = \tfrac{1}{2}\sqrt{2 - \sqrt{3}}.$$

5 The area of a triangle

As well as finding the unknown sides and angles of a triangle, if you are given enough information you can also find its area. A particularly useful rule is that for any triangle (using the same notation as before)

$$\text{area} = \tfrac{1}{2}bc\sin A.$$

Proof

Figure 6.48 shows a triangle, ABC, the height of which is denoted by h.

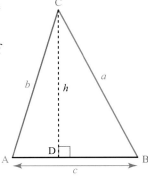

Figure 6.48

> **Note**
>
> In this case point C was taken to be the top of the triangle, AB the base. Taking the other two points, in turn, to be the top give the equivalent results:
>
> $\text{area} = \tfrac{1}{2}ca\sin B$ and
>
> $\text{area} = \tfrac{1}{2}ab\sin C.$

Using the fact that the area of a triangle is equal to half its base times its height:

$$\text{area} = \tfrac{1}{2}ch \qquad \text{①}$$

In triangle ACD

$$\sin A = \frac{h}{b}$$

$$\Rightarrow h = b\sin A$$

Substituting in ① gives

$$\text{area} = \tfrac{1}{2}\,bc\sin A.$$

> **ACTIVITY 6.3**
>
> 1 Write down an expression for the area of triangle ABC in Figure 6.48 in terms of
>   (i) $\sin A$ (ii) $\sin B$ (iii) $\sin C$.
> 2 Explain why all of your expressions are equal.
> 3 Equate your three expressions and divide through by $\tfrac{1}{2}abc$. What rule have you just proved?

Example 6.14

Find the area of triangle ABC shown in Figure 6.49.

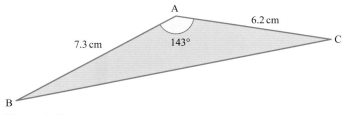

Figure 6.49

Solution

$$\begin{aligned}
\text{Area} &= \tfrac{1}{2}bc\sin A \\
&= \tfrac{1}{2} \times 7.3 \times 6.2 \times \sin 143° \\
&= 13.6 \text{ cm}^2 \text{ (to 3 s.f.)}.
\end{aligned}$$

Exercise 6.5

 ① Which expression will **not** calculate the area of a triangle? (with the usual notation)

$\frac{1}{2} pq \sin R$ $\qquad$ $\frac{1}{2} bc \sin A$

$\frac{1}{2} xy \sin Z$ $\qquad$ $\frac{1}{2} ab \sin A$

 ②

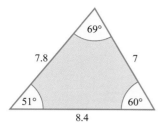

Figure 6.50

Which calculation will give the area of the triangle?

A $\frac{1}{2} \times 7 \times 8.4 \times \sin 69°$

B $\frac{1}{2} \times 7.8 \times 8.4 \times \sin 51°$

C $\frac{1}{2} \times 7 \times 7.8 \times \sin 60°$

D $7 \times 8.4 \times \sin 60°$

③ Find the area of each of the following triangles.

(i)

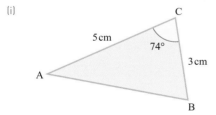

(ii)

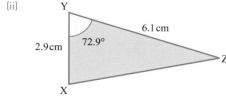

(iii)

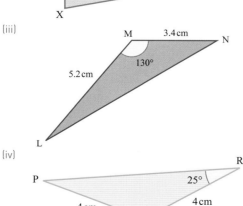

(iv)

④ A triangular flower bed has two sides of length 3 m and the angle between them is 30°. Find

(i) the area of the bed

(ii) the amount of top soil (in m³) which would be needed to cover the bed evenly to a depth of 15 cm.

⑤ In the triangle XYZ, XY = 4.6 cm and YZ = 6.9 cm.

The area of the triangle is 7.3 cm².

Find two possible values for ∠XYZ.

⑥ An icosahedron is a solid with 20 faces, each of which is an equilateral triangle.

Find the surface area of an icosahedron whose edges are all 3 cm.

⑦ Figure 6.51 shows the end wall of a bungalow. The roof has one face inclined at 30° to the horizontal and the other inclined at 58° to the horizontal. The lengths are as shown in Figure 6.51.

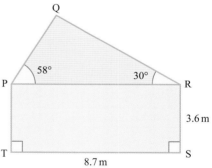

Figure 6.51

Find

(i) the length PQ

(ii) the length QR

(iii) the area of the end wall, TPQRS

(iv) the height of Q above TS.

⑧ A field is in the shape of a regular pentagon of side 80 m.

Calculate the area of the field.

⑨ The area of an irregularly shaped plot of land is often calculated by dividing it into triangles and taking appropriate measurements.

Using the information given in Figure 6.52, calculate the area of the plot ABCDE.

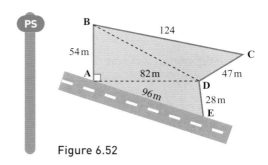

Figure 6.52

LEARNING OUTCOMES

When you have completed this chapter you should be able to:

➤ understand and use the definitions of sine, cosine and tangent for all arguments
➤ understand and use the sine rule
➤ understand and use the cosine rule
➤ understand and use the area of a triangle in the form $\frac{1}{2}\,ab\sin C$
➤ understand and use the sine, cosine and tangent functions including their:
   ○ graphs   ○ symmetries   ○ periodicity
➤ understand and use $\tan \theta = \dfrac{\sin \theta}{\cos \theta}$
➤ understand and use $\sin^2 \theta + \cos^2 \theta = 1$
➤ solve simple trigonometric equations in a given interval:
   ○ linear equations
   ○ quadratic equations in sin, cos and tan
   ○ equations involving multiples of the unknown angle.

KEY POINTS

1 The point (x, y) at angle $\boldsymbol{\theta}$ on the unit circle centre $(0, 0)$ has coordinates $(\cos \boldsymbol{\theta}, \sin \boldsymbol{\theta})$ for all $\boldsymbol{\theta}$.

2 Steps in solving a trigonometric equation

	Use your calculator to find the 1st root	Find a 2nd root for sin or cos	Find all other roots in range
sin	θ_1 Figure 6.53	$\theta_2 = 180° - \theta_1$ Figure 6.54	$\theta_1 \pm 360°$ and $\theta_2 \pm 360°$
cos	θ_1 Figure 6.55	$\theta_2 = -\theta_1$ Figure 6.56	$\theta_1 \pm 360°$ and $\theta_2 \pm 360°$
tan	θ Figure 6.57		$\theta \pm 180°$

You need to know these results.

3 The following identities are true for all values of θ.

(i) $\tan\theta \equiv \dfrac{\sin\theta}{\cos\theta}$, $\cos\theta \neq 0$

(ii) $\sin^2\theta + \cos\theta^2 \equiv 1$

4 For any triangle ABC

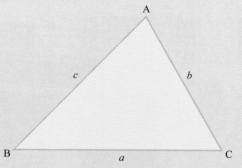

Figure 6.58

(i) $\dfrac{a}{\sin A} = \dfrac{b}{\sin B} = \dfrac{c}{\sin C}$ (sine rule)

(ii) $a^2 = b^2 + c^2 - 2bc\cos A$ (cosine rule)

(iii) area $= \dfrac{1}{2}ab\sin C$

FUTURE USES

■ You will look at transforming the graphs of trigonometric functions using translations, stretches and reflections in Chapter 8.

■ Work on trigonometry will be extended in the A-level textbook. You will learn more about solving trigonometric equations and proving identities.

■ You will also learn how to find the gradient of the curve of trigonometric functions and the area under their graphs.

7 Polynomials

A brilliant mathematician, Ramanujan was largely self-taught, being too poor to afford a university education. He left India at the age of 26 to work with G. H. Hardy in Cambridge on number theory, but fell ill in the English climate and died six years later in 1920. On one occasion when Hardy visited him in hospital, Ramanujan asked about the registration number of the taxi he came in. Hardy replied that it was 1729, an uninteresting number; Ramanujan's instant response is quoted here on the left.

→ Find the two pairs of cubes referred to in the quote.

→ How did you find them?

1 Polynomial expressions

Notation

A polynomial such as $x^3 + 3x^2 - 2x + 11$ will often be referred to as $f(x)$, meaning a function of x. The letters x and f could be replaced by any other letter, for example you could write

$g(y) = 3y^2 - 1$.

Notice that a polynomial does not contain terms involving $\sqrt{x}$, $\frac{1}{x}$, x^{-3} etc. Apart from the constant term, all the others are multiples of x raised to a **positive** integer power.

You have already met quadratic expressions, like $x^2 - 5x + 6$, and solved quadratic equations, such as $x^2 - 5x + 6 = 0$. Quadratic expressions have the form $ax^2 + bx + c$ where x is a variable and a, b and c are constants with $a \neq 0$.

An expression of the form $ax^3 + bx^2 + cx + d$, which includes a term in x^3, is called a **cubic** in x. Examples of cubic expressions are

$$x^3 + 3x^2 - 2x + 11, \qquad 3y^3 - 1 \qquad \text{and} \qquad 4z^3 - 2z$$

Similarly a **quartic** expression in x, such as $x^4 - 4x^3 + 6x^2 - 4x + 1$, contains a term in x^4; a **quintic** expression contains a term in x^5 and so on.

All these expressions are called **polynomials**. The **order** of a polynomial is the highest power of the variable it contains, so a quadratic is a polynomial of order 2, a cubic has order 3 and so on.

Operations with polynomials

Addition and subtraction

Polynomials are added by collecting like terms, for example you add the **coefficients** of x^3 together (i.e. the numbers multiplying x^3), the coefficients of x^2 together and so on as in the following example.

Example 7.1

Add $(5x^4 - 3x^3 - 2x)$ to $(7x^4 + 5x^3 + 3x^2 - 2)$.

Solution

$$(5x^4 - 3x^3 - 2x) + (7x^4 + 5x^3 + 3x^2 - 2)$$
$$= (5 + 7)x^4 + (-3 + 5)x^3 + 3x^2 - 2x - 2$$
$$= 12x^4 + 2x^3 + 3x^2 - 2x - 2$$

Alternatively this can be set out as:

$7x^4$	$+5x^3$	$+3x^2$		-2
$+ 5x^4$	$-3x^3$		$-2x$	
$12x^4$	$+2x^3$	$+3x^2$	$-2x$	-2

> Leave a space on the top row where there is no term in x.

Subtraction follows the same pattern.

Example 7.2

Subtract $(5x^4 - 3x^3 - 2x)$ from $(7x^4 + 5x^3 + 3x^2 - 2)$.

Solution

> Make sure you get these the right way round.

$(7x^4 + 5x^3 + 3x^2 - 2) - (5x^4 - 3x^3 - 2x)$

> Be careful with signs!

$= (7 - 5)x^4 + (5 - (-3))x^3 + 3x^2 + 2x - 2$

$= 2x^4 + 8x^3 + 3x^2 + 2x - 2$

Alternatively:

> ! Be careful of the signs when subtracting. You may find it easier to change the signs on the bottom line and then go on as if you were adding.

$$
\begin{array}{rrrrr}
7x^4 & +5x^3 & +3x^2 & & -2 \\
-5x^4 & -3x^3 & & -2x & \\
\hline
2x^4 & +8x^3 & +3x^2 & +2x & -2
\end{array}
$$

> Remember you are subtracting each term on the bottom line.

Multiplication

When you multiply two polynomials you multiply each term of the first by each term of the second. Then you add all the resulting terms. Remember that when you multiply powers of x, you add the indices: $x^3 \times x^2 = x^5$.

Example 7.3

Multiply $(x^3 + 3x - 2)$ by $(x^2 - 2x - 4)$.

Solution

Method 1

	x^3	$3x$	-2
x^2	x^5	$3x^3$	$-2x^2$
$-2x$	$-2x^4$	$-6x^2$	$+4x$
-4	$-4x^3$	$-12x$	$+8$

> e.g. $3x^3$ comes from multiplying x^2 by $3x$.

Adding like terms from the body of the table gives

$(x^3 + 3x - 2)(x^2 - 2x - 4)$

$\quad = x^5 - 2x^4 + (3 - 4)x^3 + (-2 - 6)x^2 + (4 - 12)x + 8$

$\quad = x^5 - 2x^4 - x^3 - 8x^2 - 8x + 8$

Method 2

This may also be set out as follows, multiplying each term in the second bracket by each term in the first one:

$(x^3 + 3x - 2) \times (x^2 - 2x - 4)$

$\quad = x^3(x^2 - 2x - 4) + 3x(x^2 - 2x - 4) - 2(x^2 - 2x - 4)$

$\quad = x^5 - 2x^4 - 4x^3 + 3x^3 - 6x^2 - 12x - 2x^2 + 4x + 8$

$\quad = x^5 - 2x^4 + x^3(-4 + 3) + x^2(-6 - 2) + x(-12 + 4) + 8$

$\quad = x^5 - 2x^4 - x^3 - 8x^2 - 8x + 8$

Discussion points

→ If you multiply a linear expression by a quadratic expression, what is the order of the resulting polynomial?

→ What is the order of the polynomial obtained by multiplying two polynomials of degree m and n respectively?

→ A third method of multiplying polynomials uses long multiplication, like this.

			$+x^3$		$+3x$	-2	
				x^2	$-2x$	-4	
x^5		$+3x^3$	$-2x^2$				A
	$-2x^4$		$-6x^2$	$+4x$			B
	$-4x^3$			$-12x$	$+8$		C
$-x^5$	$-2x^4$	$-x^3$	$-8x^2$	$-8x$	$+8$		D

Figure 7.1

Explain what has been done at each of the lines A, B, C and D.

→ How has the multiplication been laid out?

You may need to multiply out an expression involving three brackets, such as $(x + 1)(x - 2)(2x + 3)$. In this case start by multiplying two of the brackets and then multiply by the third, as in the following example.

Example 7.4

Write $(x + 1)(x - 2)(2x + 3)$ as a cubic polynomial.

Solution

$$(x + 1)(x - 2)(2x + 3) = (x + 1)(2x^2 + 3x - 4x - 6)$$
$$= (x + 1)(2x^2 - x - 6)$$
$$= x(2x^2 - x - 6) + 1(2x^2 - x - 6)$$
$$= 2x^3 - x^2 - 6x + 2x^2 - x - 6$$
$$= 2x^3 + x^2 - 7x - 6$$

> It doesn't matter which two brackets you multiply out first.

Polynomial curves

It is useful to be able to look at a polynomial and to visualise immediately the general shape of its curve. The most important clues to the shape of the curve are the order of the polynomial and the sign of the highest power of the variable.

Turning points

A turning point is a place where a curve changes from increasing (curve going up) to decreasing (curve going down), or vice versa. A **turning point** may be described as a **maximum** (change from increasing to decreasing) or a **minimum** (change from decreasing to increasing). In general, the curve of a polynomial of order n has $n - 1$ turning points as shown in Figure 7.2.

There are some polynomials for which not all the turning points materialise,

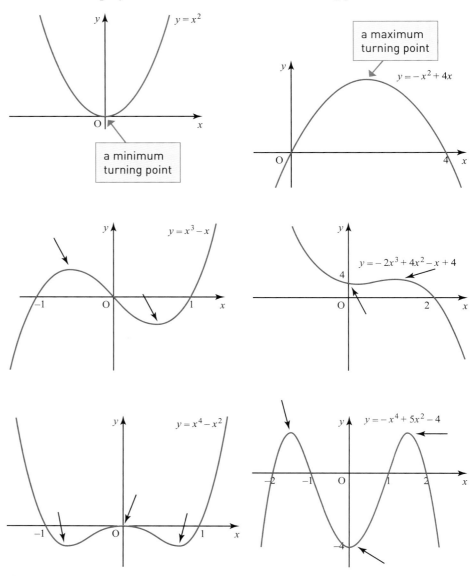

Figure 7.2

as in the case of $y = x^4 - 4x^3 + 5x^2$ (whose curve is shown in Figure 7.3). To be accurate, you say that the curve of a polynomial of order n has *at most* $n - 1$ turning points.

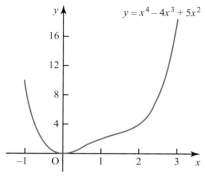

Figure 7.3

Behaviour for large x (positive and negative)

What can you say about the value of a polynomial for large positive values and large negative values of x? As an example, look at

$$f(x) = x^3 + 2x^2 + 3x + 9$$

and take 1000 as a large number.

$$\begin{aligned} f(1000) &= 1\,000\,000\,000 + 2\,000\,000 + 3000 + 9 \\ &= 1\,002\,003\,009 \end{aligned}$$

So as $x \to \infty$, $f(x) \to \infty$.

Similarly,

$$\begin{aligned} f(-1000) &= -1\,000\,000\,000 + 2\,000\,000 - 3000 + 9 \\ &= -998\,002\,991 \end{aligned}$$

So as $x \to -\infty$, $f(x) \to -\infty$.

> ### Note
>
> 1 The term x^3 makes by far the largest contribution to the answers. It is the **dominant** term.
>   For a polynomial of order n, the term in x^n is dominant as $x \to \pm\infty$.
>
> 2 In both cases the answers are extremely large numbers. You will probably have noticed already that away from their turning points, polynomial curves quickly disappear off the top or bottom of the page.
>   For all polynomials as $x \to \pm\infty$, either $f(x) \to +\infty$ or $f(x) \to -\infty$.

When investigating the behaviour of a polynomial of order n as $x \to \pm\infty$, you need to look at the term in x^n and ask two questions.

    (i)   Is n even or odd?

    (ii) Is the coefficient of x^n positive or negative?

According to the answers, the curve will have one of the four types of shape illustrated in Figure 7.4.

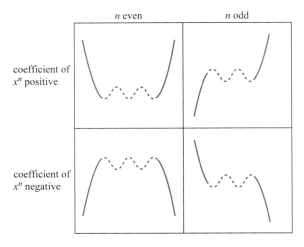

n even   *n* odd

coefficient of x^n positive

coefficient of x^n negative

Figure 7.4

ACTIVITY 7.1

Find an equation to fit each of the four cases shown in Figure 7.4, and check the shape of its curve using a graphical calculator or graphing software.

Discussion point

→ Explain why this is the case.

Intersections with the x axis and the y axis

The constant term in the polynomial gives the value of y where the curve intersects the y axis. So $y = x^8 + 5x^6 + 17x^3 + 23$ crosses the y axis at the point $(0, 23)$. Similarly, $y = x^3 + x$ crosses the y axis at $(0, 0)$, the origin, since the constant term is zero.

When the polynomial is given, or known, in factorised form you can see at once where it crosses the x axis. The curve $y = (x - 2)(x - 8)(x - 9)$, for example, crosses the x axis at $x = 2$, $x = 8$ and $x = 9$. Each of these values makes one of the brackets equal to zero, and so $y = 0$.

Example 7.5

Sketch the curve $y = x^3 - 3x^2 - x + 3 = (x + 1)(x - 1)(x - 3)$.

Solution

Since the polynomial is of order 3, the curve has up to two turning points. The term in x^3 has a positive coefficient $(+1)$ and 3 is an odd number, so the general shape is as shown in Figure 7.5.
The actual equation
$$y = x^3 - 3x^2 - x + 3 = (x + 1)(x - 1)(x - 3)$$
tells you that the curve:

- crosses the y axis at $(0, 3)$
- crosses the x axis at $(-1, 0)$, $(1, 0)$ and $(3, 0)$.

This is enough information to sketch the curve.

Figure 7.5

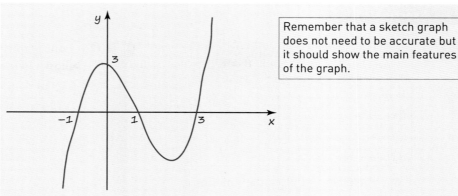

Remember that a sketch graph does not need to be accurate but it should show the main features of the graph.

Figure 7.6

In this example the polynomial $x^3 - 3x^2 - x + 3$ has three factors, $(x + 1)$, $(x - 1)$ and $(x - 3)$. Each of these corresponds to an intersection with the x axis, and to a root of the equation $x^3 - 3x^2 - x + 3 = 0$. A cubic polynomial cannot have more than three factors of this type, since the highest power of x is 3, so it cannot cross the x axis more than three times. A cubic polynomial may, however, cross the x axis fewer than three times, as in the case of $f(x) = x^3 - x^2 - 4x + 6$ (see Figure 7.7).

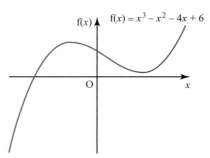

Figure 7.7

> ### Note
>
> This illustrates an important result. If $f(x)$ is a polynomial of degree n, the curve with equation $y = f(x)$ crosses the x axis at most n times, and the equation $f(x) = 0$ has at most n roots.

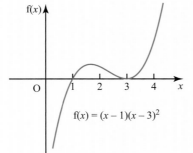
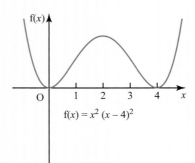

Figure 7.8

Discussion point

➜ What happens to the curve of a polynomial if it has a factor of the form $(x - a)^3$? Or $(x - a)^4$?

An important case occurs when the polynomial function has one or more repeated factors, as in Figure 7.8. In such cases the curves touch the x axis at points corresponding to the repeated roots.

Polynomial expressions

Exercise 7.1

 ①

	Order		
	3	4	5
$x \to \infty$ $f(x) \to \infty$			
$x \to \infty$ $f(x) \to -\infty$			

Copy the table and place the functions in the correct cell.

P: $f(x) = x - x^5 + 1$

Q: $f(x) = 3x^4 - x^2 + 2$

R: $f(x) = 4x^3 - 2x + 3$

S: $f(x) = 1 - x^2 - x^3$

T: $f(x) = x^4$

U: $f(x) = 11 - 2x - x^4$

 ② What is the constant term, when $x^2 - 3x + 2$ and $2x^2 + 4x - 1$ are multiplied together?

A 2   B −2   C 1   D −1

 ③ State the orders of the following polynomials:

(i) $x^3 + 3x^2 - 4x$

(ii) x^4

(iii) $3 + 2x - x^2$.

 ④ In this question $f(x) = x^3 + x^2 + 3x + 2$ and $g(x) = x^3 - x^2 - 3x - 2$.
Find

(i) $f(x) + g(x)$

(ii) $f(x) - g(x)$

(iii) $g(x) - f(x)$.

⑤ In this question $f(x) = x^3 - x^2$, $g(x) = 3x^2 + 2x + 1$ and $h(x) = x^3 + 5x^2 + 7x + 9$.
Find

(i) $f(x) + g(x) + h(x)$

(ii) $g(x) + h(x) - f(x)$

(iii) $2h(x) - g(x)$

(iv) $h(x) - f(x)$.

⑥ (i) Multiply $(x^3 + 3x^2 + 3x + 1)$ by $(x + 1)$.

(ii) Multiply $(x^2 + 2x - 3)$ by $(x - 2)$.

PS ⑦ Four graphs A, B, C and D are shown below.

A
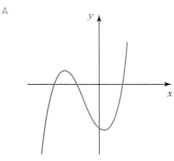

B
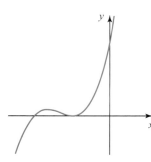

C
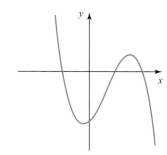

D
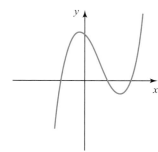

Figure 7.9

Match the graphs to the equations below.

(i) $y = (x - 1)(x + 1)(x - 2)$

(ii) $y = (x - 1)(x + 1)(x + 2)$

(iii) $y = (x - 1)(x + 1)(2 - x)$

(iv) $y = (x - 1)^2(x + 2)$

138

⑧ Sketch the following curves:

    (i) $y = x - 1$

    (ii) $y = (x - 1)(x - 2)$

    (iii) $y = (x - 1)(x - 2)(x - 3)$

    (iv) $y = x(x - 1)(x - 2)(x - 3)$

⑨ (i) Simplify
$$(x^2 + 1)(x - 1) - (x^2 - 1)(x + 1).$$
    (ii) Simplify
$$(x^2 + 1)(x^2 + 4) - (x^2 - 1)(x^2 - 4).$$

> **Hint: Expand two brackets first.**

⑩ (i) Expand $(x + 2)(x - 3)(x + 1)$.

    (ii) Expand $(2x - 1)(3 - x)(x + 1)$.

⑪ Sketch the following curves:

    (i) $y = x^2(x - 3)$

    (ii) $y = (x + 1)^2(2 - x)$

    (iii) $y = (x + 2)^2(x - 4)^2$

PS ⑫ Anna plots her results from a physics experiment and joins the points with a smooth curve. Her graph is shown in Figure 7.10.

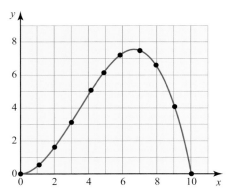

Figure 7.10

Anna wants to find an equation for her graph. She thinks a polynomial equation will fit the data.

(i) First, Anna writes down the equation $y = x(10 - x)$. Explain why this equation does not fit the data.

(ii) Anna decides to try an equation of the form $y = kx(10 - x)$. Find an approximate value for k.

(iii) Anna's friend Fatima says that an equation of the form $y = px^2(10 - x)$ would be a better fit for the data. Explain why Fatima's equation is sensible.

(iv) Find an approximate value for p in Fatima's equation.

⑬ A curve has equation
$$f(x) = (x - 2)(2x + 1)(x - 3).$$

(i) Sketch the curve.

(ii) Write the equation in the form $f(x) = ax^3 + bx^2 + cx + d$ where a, b, c, d are constants to be determined.

⑭ Suggest equations for the following curves.

(i)

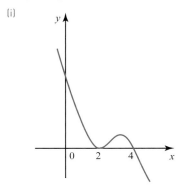

Figure 7.11

(ii)

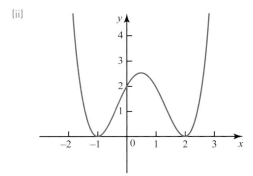

Figure 7.12

Are these suggestions the only possibilities? Try to find some alternatives.

2 Dividing polynomials

You can divide one polynomial by another, to give a lower order polynomial.

> ### Discussion points
> → If you divide a polynomial of degree 4 by a linear factor, what is the degree of the resulting polynomial?
> → What is the degree of the polynomial obtained by dividing a polynomial of degree m by a polynomial of degree n?

Two methods for dividing a polynomial by a linear expression are shown in the next example.

Example 7.6

Divide $2x^3 - 3x^2 + x - 6$ by $x - 2$.

Solution

(i) Long division

One method is to set out the polynomial division rather like arithmetic long division.

> found by dividing $2x^3$ (the first term in $2x^3 - 3x^2 + x - 6$) by x (the first term in $x - 2$)

$$
\begin{array}{r}
2x^2 \\
x - 2 \overline{)2x^3 - 3x^2 + x - 6} \\
2x^3 - 4x^2
\end{array}
$$

> $2x^3 \times (x - 2)$

Now subtract $2x^3 - 4x^2$ from $2x^3 - 3x^2$, bring down the next term (i.e. x) and repeat the method above:

> $x^2 \div x$

$$
\begin{array}{r}
2x^2 + x \\
x - 2 \overline{)2x^3 - 3x^2 + x - 6} \\
\underline{2x^3 - 4x^2} \\
x^2 + x \\
x^2 - 2x
\end{array}
$$

> $x \times (x - 2)$

Continuing gives:

> This is the answer.

$$
\begin{array}{r}
2x^2 + x + 3 \\
x - 2 \overline{)2x^3 - 3x^2 + x - 6} \\
\underline{2x^3 - 4x^2} \\
x^2 + x \\
\underline{x^2 - 2x} \\
3x - 6 \\
\underline{3x - 6} \\
0.
\end{array}
$$

> The final remainder of zero means that $x - 2$ divides exactly into $2x^3 - 3x^2 + x - 6$.

Thus $(2x^3 - 3x^2 + x - 6) \div (x - 2) = 2x^2 + x + 3$.

> Notice that the result of dividing a cubic expression by a linear expression is a quadratic expression.

(ii) By inspection

Alternatively, if you know that there is no remainder, you can use 'inspection'.

You can write $2x^3 - 3x^2 + x - 6 = (x - 2) \times$ quadratic expression.

By thinking about multiplying out, you can find the quadratic expression 'by inspection'.

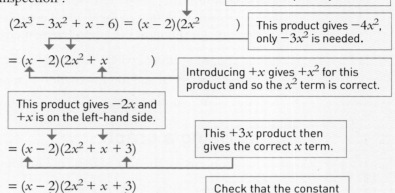

Needed to give the $2x^3$ term when multiplied by the x.

$$(2x^3 - 3x^2 + x - 6) = (x - 2)(2x^2 \qquad)$$

This product gives $-4x^2$, only $-3x^2$ is needed.

$$= (x - 2)(2x^2 + x \qquad)$$

Introducing $+x$ gives $+x^2$ for this product and so the x^2 term is correct.

This product gives $-2x$ and $+x$ is on the left-hand side.

$$= (x - 2)(2x^2 + x + 3)$$

This $+3x$ product then gives the correct x term.

$$= (x - 2)(2x^2 + x + 3)$$

Check that the constant term (-6) is correct.

Exercise 7.2

Note: In each of the following questions the divisor will be a factor of the given expression – there will be no remainder.

① Copy and complete:

(i) $x^2 - 2x - 3 = (x - 3)(x + \square)$

(ii) $x^3 - 3x^2 - 10x + 24 = (x^2 - x - 12)$
$(\square x - \square)$

(iii) $2x^3 + x^2 - 7x - 6 = (x + 1)$
$(2x^2 - \square x - \square)$

② (i) Divide $(x^2 + 2x - 3)$ by $(x - 1)$.

(ii) Divide $(x^2 + 2x - 3)$ by $(x + 3)$.

(iii) Divide $(x^3 + 2x^2 - 3x)$ by $(x - 1)$.

③ (i) Divide $(3x^2 - 2x - 1)$ by $(x - 1)$.

(ii) Divide $(5x^2 + x - 4)$ by $(x + 1)$.

(iii) Divide $(2x^2 - 3x - 2)$ by $(x - 2)$.

④ (i) Divide $(6x^2 + x - 2)$ by $(3x + 2)$.

(ii) Divide $(12x^2 - 25x + 12)$ by $(3x - 4)$.

(iii) Divide $(8x^2 - 10x - 3)$ by $(4x + 1)$.

PS ⑤ (i) Divide $(x^3 + 4x^2 + 5x + 6)$ by $(x + 3)$.

(ii) Explain how you can use your answer to (i) to divide 1456 by 13.

(iii) Explain how you can use your answer to (i) to divide 256 by 8.

⑥ (i) Divide $(x^3 + x^2 - 4x - 4)$ by $(x - 2)$.

(ii) Divide $(2x^3 - x^2 - 5x + 10)$ by $(x + 2)$.

⑦ (i) Divide $(x^3 + x - 2)$ by $(x - 1)$

(ii) Divide $(2x^3 - 10x^2 + 3x - 15)$ by $(x - 5)$.

⑧ (i) Given that $(x - 2)$ is a factor of $(x^3 + x^2 - 6x)$, divide $(x^3 + x^2 - 6x)$ by $(x - 2)$.

(ii) Given that $(x - 1)$ is a factor of $(x^4 + x^2 - 2)$, divide $(x^4 + x^2 - 2)$ by $(x - 1)$.

(iii) Given that $(x - 5)$ is a factor of $(2x^3 - 10x^2 + 3x - 15)$, divide $(2x^3 - 10x^2 + 3x - 15)$ by $(x - 5)$.

⑨ (i) Divide $(x^4 + x^2 - 2)$ by $(x + 1)$.

(ii) Divide $(3x^4 - 4x^3 - 2x^2 - 8)$ by $(x - 2)$.

3 Polynomial equations

Prior knowledge

You need to be able to solve quadratic equations, which are covered in Chapter 3.

You have already met the formula

$$x = \frac{-b \pm \sqrt{b^2 - 4ac}}{2a}$$

for the solution of the quadratic equation $ax^2 + bx + c = 0$.

Unfortunately there is no such simple formula for the solution of a cubic equation, or indeed for any higher power polynomial equation. So you use one (or more) of three possible methods:

- finding where the graph of the expression cuts the x axis
- trial and improvement (often called a numerical method)
- using algebra.

Using a graph

You can see the first of these methods by taking the equation

$$x^3 - 5x^2 + 5x + 1 = 0$$

as an example.

Start by making a table of values.

Table 7.1

x	-1	0	1	2	3	4
x^3	-1	0	$+1$	$+8$	$+27$	$+64$
$-5x^2$	-5	0	-5	-20	-45	-80
$+5x$	-5	0	$+5$	$+10$	$+15$	$+20$
$+1$	$+1$	$+1$	$+1$	$+1$	$+1$	$+1$
y	-10	$+1$	$+2$	-1	-2	$+5$

Then plot the graph.

You can see that the roots are roughly -0.2, 1.7 and 3.4.

You can be more accurate by taking values of x that are closer together in your table of values, say every 0.5 or every 0.1 instead of the present 1, but the roots will still only be approximate.

Using a graph is an easy way to solve a polynomial equation if you don't need to know the roots very accurately.

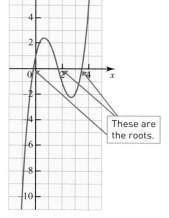

These are the roots.

Figure 7.13

Trial and improvement

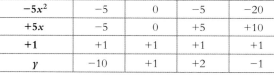

ACTIVITY 7.2

In Figure 7.13, you saw that one of the points where the curve $y = x^3 - 5x^2 + 5x + 1$ crossed the x axis, i.e. there is a root of the equation, was in the interval $1 < x < 2$. Using trial and improvement (the table function on a graphical calculator or a spreadsheet will help with this) find this root to 3 decimal places. Remember that you need to go as far as the fourth decimal place to make sure that your final approximation is correct to three decimal places.

Using algebra

Sometimes you will be able to spot one (or more) of the roots.

So, for example, if you want to solve the equation $x^3 - 5x^2 + 7x - 2 = 0$, and you make a table of values, you get

Table 7.2

x	-1	0	1	2	3	4
y	-15	-2	1	0	-3	6

You can see at once that one of the roots is $x = 2$ because substituting 2 in the polynomial gives 0.

That tells you that $(x - 2)$ is a factor of $x^3 - 5x^2 + 7x - 2$. This is an example of the **factor theorem**.

You can state the factor theorem more formally as follows.

> If $x - a$ is a factor of f(x), then f$(a) = 0$ and a is a root of the equation f$(x) = 0$. And conversely
>
> If f$(a) = 0$, then $(x - a)$ is a factor of f(x).

Example 7.7

 **TECHNOLOGY**

Table mode on a calculator may be useful here. Remember to include negative and positive values of x.

Discussion point

➔ Why was it not necessary to calculate f(4) in this instance?

Given that f$(x) = x^3 - 6x^2 + 11x - 6$:

(i)   find f(0), f(1), f(2), f(3) and f(4)

(ii) factorise $x^3 - 6x^2 + 11x - 6$

(iii) solve the equation $x^3 - 6x^2 + 11x - 6 = 0$

(iv) sketch the curve whose equation is f$(x) = x^3 - 6x^2 + 11x - 6$.

Solution

(i)   f$(0) = 0^3 - 6 \times 0^2 + 11 \times 0 - 6 = -6$

      f$(1) = 1^3 - 6 \times 1^2 + 11 \times 1 - 6 = 0$

      f$(2) = 2^3 - 6 \times 2^2 + 11 \times 2 - 6 = 0$

      f$(3) = 3^3 - 6 \times 3^2 + 11 \times 3 - 6 = 0$

      f$(4) = 4^3 - 6 \times 4^2 + 11 \times 4 - 6 = 6$

(ii) Since f(1), f(2) and f(3) all equal 0, it follows that $(x - 1)$, $(x - 2)$ and $(x - 3)$ are all factors. This tells you that

$$x^3 - 6x^2 + 11x - 6 = (x - 1)(x - 2)(x - 3) \times \text{constant}$$

By checking the coefficient of the term in x^3, you can see that the constant must be 1, and so

$$x^3 - 6x^2 + 11x - 6 = (x - 1)(x - 2)(x - 3)$$

(iii) $x = 1, 2$ or 3.

(iv)

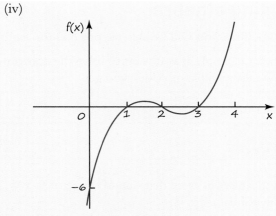

Figure 7.14

Discussion point

➔ How does the constant term help you to decide which values of x to try when you are looking for a root?

Factorising a polynomial

When a polynomial factorises it is possible to find at least one factor by first spotting a root, as in the previous example.

To check whether an integer root exists for any polynomial equation, look at the constant term. Decide which whole numbers divide into it and check them – you do not have to check every integer.

Example 7.8

Given that $f(x) = x^3 - x^2 - 3x + 2$
(i) find a linear factor of $f(x)$
(ii) solve the equation $f(x) = 0$.

Solution

(i) Start by looking systematically for a value of x where $f(x) = 0$.

$$f(1) = 1^3 - 1^2 - 3(1) + 2 = -1$$
$$f(-1) = (-1)^3 - (-1)^2 - 3(-1) + 2 = 3$$
$$f(2) = 2^3 - 2^2 - 3(2) + 2 = 0$$

So $(x - 2)$ is a factor.

(ii) Using this information, and using long division:

$$
\begin{array}{r}
x^2 + x - 1 \\
x - 2 \overline{\smash{)}x^3 - x^2 - 3x + 2} \\
\underline{x^3 - 2x^2} \\
x^2 - 3x \\
\underline{x^2 - 2x} \\
-x + 2 \\
\underline{-x + 2} \\
0
\end{array}
$$

$f(x) = 0$ therefore becomes $(x - 2)(x^2 + x - 1) = 0$
$\Rightarrow$ either $x - 2 = 0$ or $x^2 + x - 1 = 0$

Using the quadratic formula on $x^2 + x - 1 = 0$ gives

$$x = \frac{-1 \pm \sqrt{1 - 4 \times 1 \times (-1)}}{2}$$

$$= \frac{-1 \pm \sqrt{5}}{2}$$

$$= -1.618 \text{ or } 0.618 \text{ (to 3 d.p.)}$$

The complete solution is $x = -1.618, 0.618$ or 2.

Example 7.9

(i)   Show that $(x - 1)$ is a linear factor of $2x^3 - 5x^2 - 6x + 9$.

(ii) Factorise $2x^3 - 5x^2 - 6x + 9 = 0$.

(iii) Hence sketch the curve $f(x) = 2x^3 - 5x^2 - 6x + 9$.

Solution

(i) $f(1) = 2(1)^3 - 5(1)^2 - 6(1) + 9$

$\qquad = 0$

so $(x - 1)$ is a factor.

(ii) $2x^3 - 5x^2 - 6x + 9 = (x - 1)(2x^2 - 3x - 9)$

$\qquad\qquad\qquad\qquad\quad = (x - 1)(2x + 3)(x - 3)$

> Using inspection – you could use long division if you prefer.

(iii) The curve is a cubic with a positive x^3 term which crosses the x axis when $x = 1$, 3 and -1.5.

When $x = 0$, $y = 9$

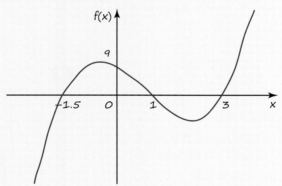

Figure 7.15

Example 7.10

Find the value of a given that $x - 3$ is a factor of the expression $x^3 - 2x^2 + ax + 6$.

Solution

$(x - 3)$ is a factor of the expression means that $x = 3$ is a root of the equation $x^3 - 2x^2 + ax + 6 = 0$.

Substituting $x = 3$ into the equation gives

$\qquad 27 - 18 + 3a + 6 = 0$

$\Rightarrow \qquad\qquad 15 + 3a = 0$

$\Rightarrow \qquad\qquad\qquad a = -5$

Exercise 7.3

① (i) Sort the following into two sets of equivalent statements.

$f(1) = 0$ $\qquad$ $f(-1) = 0$
$x + 1$ is a factor of $f(x)$
$x - 1$ is a factor of $f(x)$
when $x = 1, f(x) = 0$
when $x = -1, f(x) = 0$

(ii) Now complete a similar set for '$x + 2$ is a factor of $f(x)$'

② You know that $f(1) = 3, f(2) = 1$, $f(-1) = 0$ and $f(-2) = -2$.

Which of the following is a factor of $f(x)$?

A $x + 1$ $\quad$ B $x - 1$ $\quad$ C $x + 2$ $\quad$ D $x - 2$

③ Show that $x - 1$ is a factor of $3x^3 + 2x^2 - 5$.

④ Show that $x + 2$ is not a factor of $x^3 + x^2 + x - 3$.

⑤ Show that $x = 2$ is a root of the equation $x^4 - 5x^2 + 2x = 0$ and write down another integer root.

⑥ The polynomial $p(x) = x^3 - 6x^2 + 9x + k$ has a factor $x - 4$.

Find the value of k.

⑦ Show that neither $x = 1$ nor $x = -1$ is a root of $x^4 - 2x^3 + 3x^2 - 8 = 0$.

⑧ You are given that $f(x) = x^3 - 19x + 30$.

(i) Calculate $f(0)$ and $f(3)$. Hence write down a factor of $f(x)$.

(ii) Find p and q such that $f(x) \equiv (x - 2)(x^2 + px + q)$.

(iii) Solve the equation $x^3 - 19x + 30 = 0$.

(iv) Without further calculation draw a sketch of $y = f(x)$.

⑨ Find the possible values of a, if $(x - 2)$ is a factor of $x^3 + x^2 - 5ax + 2a^2$.

⑩ The polynomial $f(x) = x^3 + ax^2 + bx + 6$ has factors $(x - 1)$ and $(x - 2)$. Find the values of a and b and the other linear factor.

⑪ Show that $x = 3$ is a root of the equation $2x^3 - 7x^2 + 2x + 3 = 0$ and hence solve the equation.

⑫ The expression $x^4 + px^3 + qx + 36$ is exactly divisible by $(x - 2)$ and $(x + 3)$. Find and simplify two simultaneous equations for p and q and hence find p and q.

⑬ Factorise $x^3 + x^2 - 14x - 24$.

⑭ Figure 7.16 shows sketches of the graphs of $y = f(x)$ and $y = g(x)$ where $f(x) = x^3 + 2x^2 - 7x + 4$ and $g(x) = \dfrac{(x + 4)^2}{4}$.

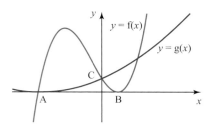

Figure 7.16

(i) Factorise $f(x)$.

(ii) Find the coordinates of the points A, B and C.

(iii) Find the coordinates of the other point of intersection of the two curves.

⑮ Figure 7.17 shows an open rectangular tank with a square base of side x metres and a volume of $18\,\text{m}^3$.

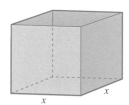

Figure 7.17

(i) Write down an expression in terms of x for the height of the tank.

(ii) Show that the surface area of the tank is $\left(x^2 + \dfrac{72}{x}\right)\text{m}^2$.

(iii) Given that the surface area is $33\,\text{m}^2$, show that $x^3 - 33x + 72 = 0$.

(iv) Solve $x^3 - 33x + 72 = 0$ and hence find possible dimensions for the tank.

LEARNING OUTCOMES

When you have completed this chapter you should be able to:

➤ Manipulate polynomials algebraically:
  ○ expanding brackets
  ○ collecting like terms
  ○ using factorisation
  ○ using simple algebraic division
➤ use the factor theorem
➤ sketch curves of polynomials
➤ interpret the algebraic solution of a polynomial graphically
➤ Use intersection points of graphs to solve equations.

KEY POINTS

1 A polynomial in x has terms in positive integer powers of x and may also have a constant term.
2 The order of a polynomial in x is the highest power of x which appears in the polynomial.
3 The factor theorem states that if $(x - a)$ is a factor of a polynomial $f(x)$ then $f(a) = 0$ and a is a root of the equation $f(x) = 0$.
4 The curve of a polynomial function of order n, has up to $(n - 1)$ turning points.
5 The behaviour of the curve of a polynomial of order n, for large positive values of n, depends on whether n is even or odd, and whether the coefficient of the term in x^n is positive or negative.

Graphs and transformations

Mathematics is the art of giving the same name to different things.

Henri Poincaré (1854–1912)

The photograph shows the Tamar Railway bridge (the Royal Albert bridge). You can model the spans of this bridge by two curves – one dashed and one solid, as in Figure 8.1.

Discussion point

➜ How would you set about trying to fit equations to these two curves?

Assume the horizontal ⟨...⟩ vertical scales are the s⟨...⟩ for the bridge.

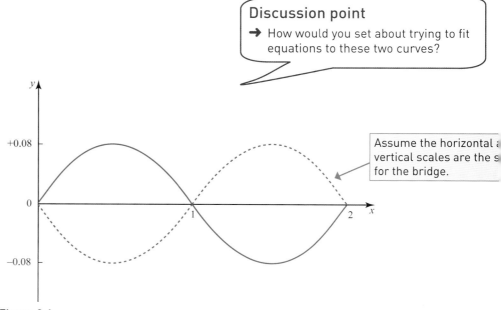

Figure 8.1

1 The shapes of curves

You can always plot a curve, point by point, if you know its equation. Often, however, all you need is a general idea of its shape and a sketch will do.

When you sketch a curve the most important thing is to get its overall **shape** right. You have already met some important curves: quadratics, polynomials and those of trigonometric functions, and know their shapes. In this chapter you will meet the curves of some further functions. Exercise 8.1 covers all the types of curves and so some of the questions revisit earlier work.

In this chapter you will learn to sketch the graphs of some other functions.

Function notation

There are several different but equivalent ways of writing a function. For example, the function which maps x on to x^2 can be written in any of the following ways.

$$y = x^2 \qquad\qquad f(x) = x^2 \qquad\qquad f : x \rightarrow x^2$$

It is often helpful to represent a function graphically.

Example 8.1

Two functions are given as $f(x) = x^2 + 4$ and $g(x) = 3x^2$.

(i) Find (a) $f(-3), f(0)$ and $f(3)$

      (b) $g(-3), g(0)$ and $g(3)$.

(ii) Solve $g(x) = 12$.

(iii) Sketch the functions $y = f(x)$ and $y = g(x)$ on the same axes.

(iv) (a) Solve the equation $f(x) = g(x)$.

     (b) Hence find the coordinates of the points where the two curves intersect.

Solution

(i) (a) $f(-3) = (-3)^2 + 4 = 13$ ← Substitute $x = -3$ into $x^2 + 4$

       $f(0) = 0^2 + 4 = 4$

       $f(3) = 3^2 + 4 = 13$

   (b) $g(-3) = 3 \times (-3)^2 = 27$ ← Substitute $x = -3$ into $3x^2$

       $g(0) = 3 \times 0^2 = 0$

       $g(3) = 3 \times (3)^2 = 27$

(ii) $g(x) = 12 \Rightarrow 3x^2 = 12$

             Substitute $g(x) = 3x^2$

        $\Rightarrow x^2 = 4$

  So $x = \pm 2$

(iii)

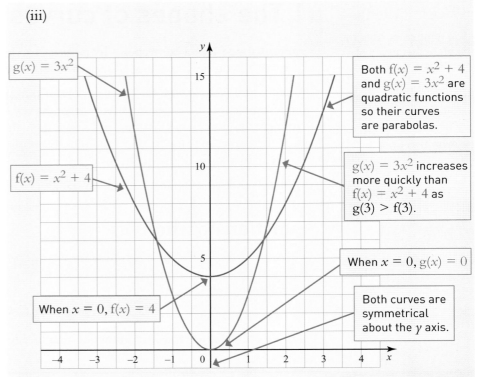

Figure 8.2

(iv) (a) $f(x) = g(x)$

> Rearrange the equation to make x^2 the subject.

$$\Rightarrow x^2 + 4 = 3x^2$$

$$\Rightarrow 2x^2 = 4$$

$$\Rightarrow x^2 = 2$$

$$\Rightarrow x = \pm\sqrt{2}$$

(b) The curves intersect at $x = \sqrt{2}$ and $-\sqrt{2}$

When $x = \sqrt{2}$ then $y = g(x) = 3 \times (\sqrt{2})^2 = 3 \times 2 = 6$

When $x = -\sqrt{2}$ then $y = g(x) = 3 \times (-\sqrt{2})^2 = 3 \times 2 = 6$

So the curves intersect at $(\sqrt{2}, 6)$ and $(-\sqrt{2}, 6)$.

> Check
> $f(\sqrt{2}) = (\sqrt{2})^2 + 4 = 6 ✓$

ACTIVITY 8.1

(i) The graphs below in Figure 8.3 show the curves $y = x^2$, $y = x^3$, $y = x^4$ and $y = x^5$, but not in that order.

Match the graphs to the equations.

A
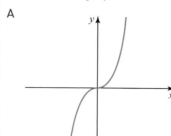

B
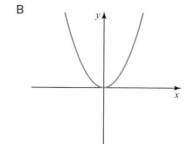

C
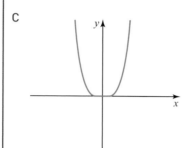

D
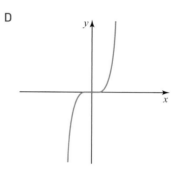

Figure 8.3

(ii) What do the graphs of $y = x^n$ look like when n is (a) even and (b) odd?

(iii) What do the curves of $y = -x^n$ look like?

Prior knowledge

In Chapter 7 you met other polynomial curves, and you saw how you can sketch the graphs of polynomials in factorised form.

Curves of the form $y = \dfrac{1}{x^n}$ (for $x \neq 0$)

Not all curves are polynomials. Figure 8.4 shows two reciprocal functions.

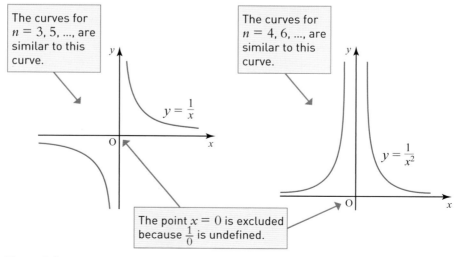

The curves for $n = 3, 5, \ldots,$ are similar to this curve.

$y = \dfrac{1}{x}$

The curves for $n = 4, 6, \ldots,$ are similar to this curve.

$y = \dfrac{1}{x^2}$

The point $x = 0$ is excluded because $\dfrac{1}{0}$ is undefined.

Figure 8.4

An important feature of the curves of reciprocal functions is that they approach both the x and the y axes ever more closely but never actually reach them. These lines are described as **asymptotes** to the curves. Asymptotes may be vertical (e.g. the y axis), horizontal, or lie at an angle, when they are called oblique.

Asymptotes are usually marked on graphs as dotted lines but in Figure 8.4 the lines are already there, being coordinate axes. The curves have different branches which never meet. A curve with different branches is said to be **discontinuous**, whereas one with no breaks, like $y = x^2$, is **continuous**.

> You can draw a continuous curve without lifting your pencil. When you draw a discontinuous curve you need to lift your pencil as you cross the asymptote.

Graphs showing proportion

Graphs can show proportion in the following ways:

Direct proportion: When two quantities y and x are in **direct proportion** then as one increases the other increases at the same rate. You say 'y is directly proportional to x' and write this as $y \propto x$. On a graph this is shown with a straight line through the origin: $y = kx$.

Inverse proportion: When two quantities are in **inverse proportion** then as one quantity increases, the other quantity decreases at the same rate. You say 'y is inversely proportional to x' and write this as $y \propto \dfrac{1}{x}$. On a graph this is shown as the curve $y = \dfrac{k}{x}$.

> Note: the product of y and x is a constant, k.

Other proportion: y can be directly or inversely proportional to a power or root of x. For example, if y is directly proportional to the square of x then you would write $y \propto x^2$ and so $y = kx^2$.

> In this case, the graph of y against x^2 is a straight line through the origin.

Example 8.2

The quantity y is inversely proportional to x^4.

When $x = \frac{1}{2}$, $y = 8$.

(i) Find the equation connecting y and x.

(ii) Sketch the curve of y against x.

Label clearly any asymptotes with their equations.

(iii) Sketch the curve of y against x^4.

Label clearly any asymptotes with their equations.

Solution

(i) $y \propto \dfrac{1}{x^4}$

$\Rightarrow y = \dfrac{k}{x^4}$

When $x = \frac{1}{2}$, $y = 8$

$\Rightarrow 8 = \dfrac{k}{\left(\frac{1}{2}\right)^4}$

> Substitute in $x = \frac{1}{2}$ and $y = 8$.

$\Rightarrow k = 8 \times \dfrac{1}{16} = \dfrac{1}{2}$

So $y = \dfrac{1}{2x^4}$

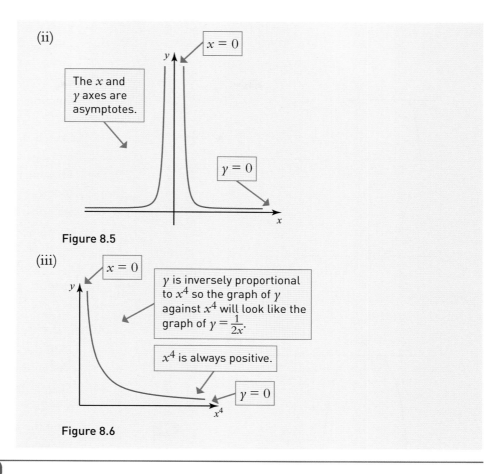

Figure 8.5

Figure 8.6

Exercise 8.1

 ① Sort the functions into the cells of the table. k is a positive integer.

	No negative values	Some negative values
Continuous		
Discontinuous		

$y = kx$ $y = \dfrac{k}{x}$ $y = \dfrac{k}{x^2}$

$y = kx^2$ $y = kx^3$ $y = \dfrac{k}{x^3}$

$y = k\sqrt{x}$

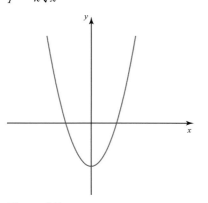 ②

Figure 8.7

Which function is shown by the curve in Figure 8.7?

A $y = -x^2$ B $y = x^2$

C $y = \dfrac{1}{x^2}$ D $y = x^2 - 3$

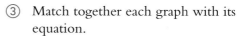 ③ Match together each graph with its equation.

A $y = -x^4$

B $y = 3 + x^2$

C $y = -x^3$

D $y = x^5$

E $y = \sqrt{x}$

(i)

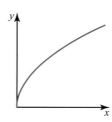

(ii)

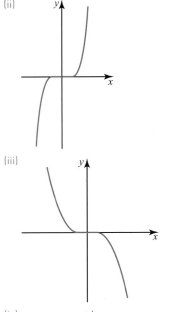

(iii)

(iv)

(v)

Figure 8.8

Sketch the curves in Questions 4–6 by hand, marking clearly the values of x and y where they cross the coordinate axes and label the equations of any asymptotes.

TECHNOLOGY

You can use a graphical calculator or graphing software to check your results.

④ (i) $y = \dfrac{1}{x}$ (ii) $y = \dfrac{1}{x^2}$

  (iii) $y = \dfrac{1}{x^3}$ (iv) $y = \dfrac{1}{x^4}$

⑤ (i) $y = (x - 1)(x + 3)$
  (ii) $y = (x + 1)(x + 2)(x + 3)$
  (iii) $y = (2x + 1)(2x - 3)$
  (iv) $y = x(2x + 1)(2x - 3)$

⑥ (i) $y = (x + 1)^2(3 - x)$
  (ii) $y = (3x - 4)(4x - 3)^2$
  (iii) $y = (x + 2)^2(x - 4)^2$
  (iv) $y = (x - 3)^2(4 + x)^2$

⑦ Two functions are given as $f(x) = 2x^2 - x$ and $g(x) = x^3$.
  (i) Sketch the functions $y = f(x)$ and $y = g(x)$ on the same axes.
  (ii) How many points of intersection are there?
  (iii) (a) Solve the equation $f(x) = g(x)$.
    (b) Hence find the coordinates of the points where the two curves intersect.

⑧ The quantity y is directly proportional to $\sqrt{x}$ for $x > 0$.

  When $x = \frac{1}{2}$, $y = 1$.
  (i) Find the equation connecting y and x.
  (ii) Sketch the curve of y against x.

⑨ (i) Sketch the graph of $y = \dfrac{1}{x}$ and $y = 2x + 1$ on the same axes.
  (ii) Find the coordinates of their points of intersection.
  (iii) Isobel says that both graphs show a proportional relationship. Is Isobel correct?

    Give a reason for your answer.

⑩ Look at this curve.

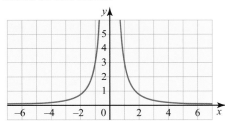

Figure 8.9

  y is inversely proportional to the square of x.
  (i) Find the equation of the curve.
  (ii) Find the value of y when $x = 10$.
  (iii) Find the values of x when $y = \frac{3}{4}$.

⑪ The quantity y is inversely proportional to a power of x for $x > 0$.
When $x = 1$, $y = \sqrt{2}$ and when $x = 8$, $y = \frac{1}{2}$.

(i) Find the equation connecting y and x.

(ii) Sketch the curve of y against x.

PS ⑫ How many times will the graphs of $y = \dfrac{2}{x}$ and $y = \dfrac{1}{x^3}$ intersect?
Find the exact coordinates of any points of intersection.

Prior knowledge

You will have met some of this at GCSE.

2 Using transformations to sketch curves

Another approach to sketching curves is to start from a known curve and apply a transformation to it.

Here are some words associated with this work.

- The original curve or shape is called the **object**.
- The new curve or shape is called its **image**.
- Going from an object to its image is described as **mapping**.
- A **transformation** describes the type of mapping you are using.

Some of this work will already be familiar to you. Now it is being described in more mathematical language.

This section looks at the following types of transformation:

- translations
- reflections in the x or y axis
- stretches.

Translations

 ACTIVITY 8.2

A Using a graphical calculator or graphing software, draw the graphs of
  $$y = x^2, \qquad y = x^2 + 3 \qquad \text{and} \qquad y = x^2 - 2$$
  on the same axes.
  What do you notice?
  How can the second and third graphs be obtained from the first one?
  How could your findings be generalised?

B Repeat part A using the graphs of $y = x^2$, $y = (x - 2)^2$ and $y = (x + 3)^2$.

Figure 8.10 shows the graphs of $y = x^2$ and $y = x^2 + 3$. For any given value of x, the y coordinate for the second curve is 3 units more than the y coordinate for the first curve.

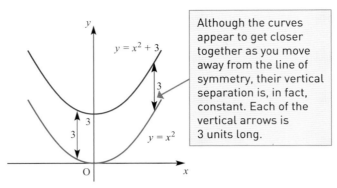

Figure 8.10

You can see that the graphs of $y = x^2 + 3$ and $y = x^2$ are exactly the same shape, but $y = x^2$ has been translated by 3 units in the positive y direction to obtain $y = x^2 + 3$.

Similarly, $y = x^2 - 2$ is a translation of $y = x^2$ by 2 units in the negative y direction. ◄———— So –2 units in the positive y direction.

In general, for any function f(x), the curve $y = $ f(x) $+ s$ is a translation of $y = $ f(x) by s units in the positive y direction.

What about the relationship between the graphs of $y = x^2$ and $y = (x - 2)^2$?

Figure 8.11 shows the graphs of these two functions.

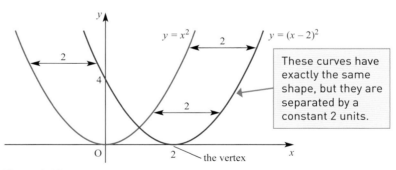

Figure 8.11

Note

The last result may be written as:

the graph of $y = $ f(x) when translated by $\binom{t}{s}$ results in the graph $y - s = $ f($x - t$).

Writing the result in this form shows the symmetry of the results.

Translating $y = $ f(x) by t in the positive x direction is equivalent to replacing x with $x - t$.

Translating $y = $ f(x) by s in the positive y direction is equivalent to replacing y with $y - s$.

You may find it surprising that $y = x^2$ moves in the positive x direction when 2 is subtracted from x. It happens because x must be 2 units larger if $(x - 2)$ is to give the same output that x did in the first mapping. ———► For $y = (x - 2)^2$ when $x = 2$, $y = 0$

Notice that the axis of symmetry of the curve $y = (x - 2)^2$ is the line $x = 2$.

In general, the curve with equation $y = $ f(x) maps onto $y = $ f($x - t$) by a translation of t units in the positive x direction.

Combining these results, $y = $ f(x) maps onto $y = $ f($x - t$) $+ s$ by a translation of s units in the positive y direction and t units in the positive x direction.

This translation is represented by the vector $\binom{t}{s}$.

Example 8.3

Figure 8.12 shows the graph of $y = f(x)$.

Sketch the graph of each of these functions.

(i) $y = f(x) + 2$

(ii) $y = f(x - 1)$

(iii) $y = f(x + 1) - 2$

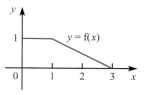

Figure 8.12

Solution

(i)

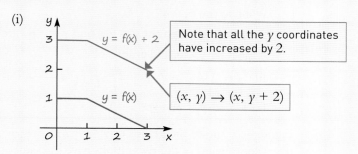

Note that all the y coordinates have increased by 2.

$(x, y) \rightarrow (x, y + 2)$

Figure 8.13

(ii)

Note that all the x coordinates have increased by 1.

So $(x, y) \rightarrow (x + 1, y)$

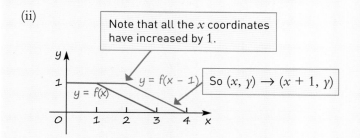

Figure 8.14

(iii)

Note that all the x coordinates have decreased by 1 and all the y coordinates have decreased by 2.

So $(x, y) \rightarrow (x - 1, y - 2)$

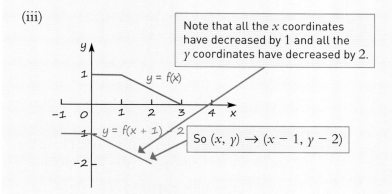

Figure 8.15

Reflections

T | **ACTIVITY 8.3**

A Using a graphical calculator or graphing software, draw the graphs of

(i) $y = x^2$ and $y = -x^2$

(ii) $y = \dfrac{1}{x^2}$ and $y = -\dfrac{1}{x^2}$

on the same axes.

What do you notice?

How are the graphs of $y = f(x)$ and $y = -f(x)$ related?

B Repeat part A for the graphs of

(i) $y = 2 + \dfrac{1}{x}$ and $y = 2 + \dfrac{1}{(-x)}$ ← | This is the same as $y = 2 - \dfrac{1}{x}$

(ii) $y = x^3 + 1$ and $y = (-x)^3 + 1$

Figure 8.16 shows the graphs of $y = x^2$ and $y = -x^2$. For any particular value of x, the y coordinates of the two graphs have the same magnitude but opposite signs.

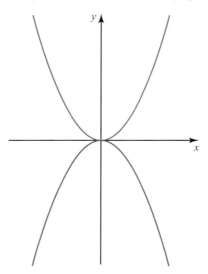

Figure 8.16

The graphs are reflections of each other in the x axis.

In general, starting with the graph of $y = f(x)$ and replacing $f(x)$ by $-f(x)$ gives a reflection in the x axis. This is the equivalent of replacing y by $-y$ in the equation.

Figure 8.17 shows the graph of $y = 2x + 1$, a straight line with gradient 2 passing through $(0, 1)$. The graph of $y = 2(-x) + 1$ (which can be written as $y = -2x + 1$) is a straight line with gradient -2, and as you can see it is a reflection of the line $y = 2x + 1$ in the y axis.

In general, starting with the graph of $y = f(x)$ and replacing x by $(-x)$ gives a reflection in the y axis.

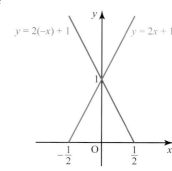

Figure 8.17

(c)

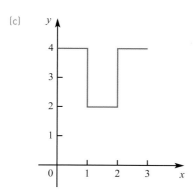

Figure 8.32

(d)

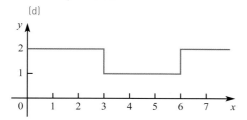

Figure 8.33

(ii) Sketch the graphs of:
- (a) $y = g(x + 2) + 3$
- (b) $y = g(2x)$
- (c) $y = -2g(x)$

⑨ Figure 8.34 shows the function f(x).

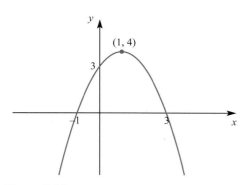

Figure 8.34

(i) Write down the coordinates of the vertex and of the points of intersection with axes for the following curves.
- (a) $f(-x)$
- (b) $-f(x)$
- (c) $f(3x)$
- (d) $-\frac{1}{2}f(x)$

(ii) The maximum point of the curve $f(x + a)$ lies on the y axis.

Find the value of a.

⑩ In each of the parts of Figure 8.35, the curve drawn with a dashed line is obtained as a mapping of the curve $y = f(x)$ using a single transformation.

In each case, write down the equation of the image (dashed) in terms of f(x).

(i)

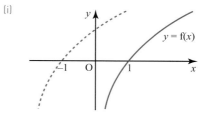

(ii)

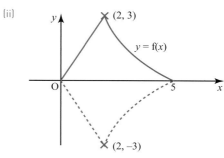

(iii)

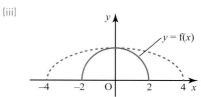

(iv)

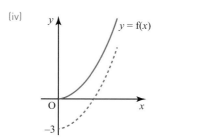

(v)

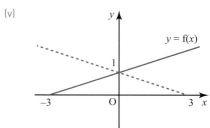

(vi)

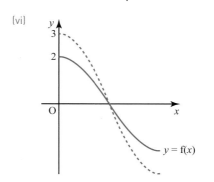

Figure 8.35

Prior knowledge

You need to be able to complete the square. This is covered in Chapter 3.

3 Using transformations

You now know how to sketch the curves of many functions. Frequently, curve sketching can be made easier by relating the equation of the function to that of a standard function of the same form. This allows you to map the points on the standard curve to equivalent points on the curve you need to draw.

Example 8.6

(i) Write the equation $y = x^2 - 4x - 1$ in the form $y = (x + p)^2 + q$.

(ii) Show how the graph of $y = x^2 - 4x - 1$ can be obtained from the graph of $y = x^2$.

(iii) Sketch the graph.

Solution

(i) Complete the square:

$$y = x^2 - 4x - 1$$

> Take the coefficient of x: -4; halve it: -2 then square it: $+4$. Add it and subtract it.

$$= \underline{x^2 - 4x + 4} \quad - 1 - 4$$
$$y = \quad (x - 2)^2 \quad - 1 - 4$$

So the equation is $y = (x - 2)^2 - 5$.

> It is always a good idea to check your results using a graphical calculator whenever possible.

(ii) The curve $y = x^2$ maps to the curve

$y = (x - 2)^2 - 5$ by a translation of $\begin{pmatrix} 2 \\ -5 \end{pmatrix}$.

(iii)

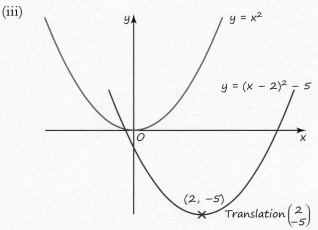

Figure 8.36

Discussion points

You have now met two ways of looking at the graph of a quadratic function: completing the square and using translations.

→ What are the advantages and disadvantages of each way?

→ Would you be justified in finding the values of s and t by substituting, for example, $x = 0$ and $x = 1$ in the identity $x^2 - 2x + 5 \equiv (x - t)^2 + s$?

Using the technique of completing the square, any quadratic expression of the form $y = x^2 + bx + c$ can be written as $y = (x - t)^2 + s$, so its graph can be sketched by relating it to the graph of $y = x^2$.

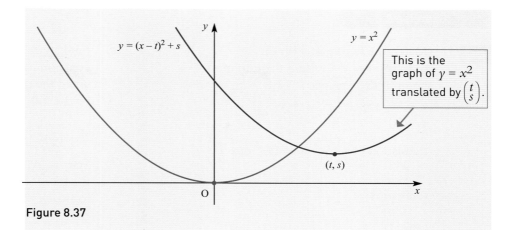

Figure 8.37

Example 8.7

Figure 8.38 shows the graph of $y = x^3$ translated by $\begin{pmatrix} -1 \\ 4 \end{pmatrix}$. Show that the resulting graph has the equation $y = x^3 + 3x^2 + 3x + 5$.

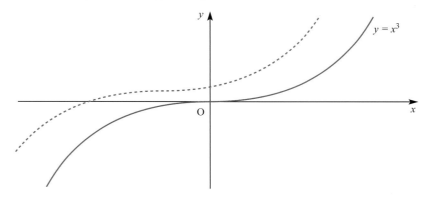

Figure 8.38

Solution

After a translation of $\begin{pmatrix} -1 \\ 4 \end{pmatrix}$, the resulting graph has the equation

$$= (x - (-1))^3 + 4$$
$$\Rightarrow y = (x + 1)^3 + 4$$
$$\Rightarrow y = (x + 1)(x^2 + 2x + 1) + 4$$
$$\Rightarrow y = (x^3 + 2x^2 + x + x^2 + 2x + 1) + 4$$
$$\Rightarrow y = (x^3 + 3x^2 + 3x + 1) + 4$$
$$\Rightarrow y = x^3 + 3x^2 + 3x + 5, \text{ as required.}$$

Using transformations

Exercise 8.3

① Match the functions with the translation of $y = x^2$ and the minimum point of their graph.

$y = (x + 5)^2 - 1$ $y = (x - 1)^2 - 5$
$y = (x + 1)^2 + 5$

translation of $\begin{pmatrix} -5 \\ -1 \end{pmatrix}$ translation of $\begin{pmatrix} -1 \\ 5 \end{pmatrix}$

translation of $\begin{pmatrix} 1 \\ -5 \end{pmatrix}$

$(-5, -1)$ $(-1, 5)$ $(1, -5)$

② Which point is the minimum point of the curve of $y = (x + 2)^2 - 3$?

A $(2, 3)$ B $(2, -3)$
C $(-2, 3)$ D $(-2, -3)$

③ (i) Starting with the graph of $y = x^3$, find the equation of the graph resulting from the following translations.

(a) $\begin{pmatrix} 0 \\ 4 \end{pmatrix}$ (b) $\begin{pmatrix} -3 \\ 0 \end{pmatrix}$ (c) $\begin{pmatrix} 3 \\ -4 \end{pmatrix}$

> You do not need to multiply out any brackets.

(ii) Starting with the graph of $y = x^3$, find the equation of the graph resulting from the following one-way stretches.

(a) Scale factor 2 parallel to the y axis

(b) Scale factor 2 parallel to the x axis

(c) Scale factor 3 parallel to the y axis

(d) Scale factor $\frac{1}{2}$ parallel to the x axis.

④ Draw the line $y = 2x + 3$.
Find the equation obtained when the line is translated through $\begin{pmatrix} 4 \\ 1 \end{pmatrix}$:

(i) by a graphical method

(ii) by transforming the equation.

⑤ (i) On one diagram, sketch both lines $y = 3x$ and $y = 3x - 3$.

(ii) $y = 3x$ can be transformed to $y = 3x - 3$ by a translation of $\begin{pmatrix} 0 \\ a \end{pmatrix}$. What is the value of a?

(iii) $y = 3x$ can be transformed to $y = 3x - 3$ by a translation of $\begin{pmatrix} b \\ 0 \end{pmatrix}$. What is the value of b?

⑥ Find, in the form $y = mx + c$, the equation of the resulting straight line when the graph of $y = 2x$ is translated by the following vectors.

(i) $\begin{pmatrix} 0 \\ -10 \end{pmatrix}$ (ii) $\begin{pmatrix} -5 \\ 0 \end{pmatrix}$ (iii) $\begin{pmatrix} -5 \\ -10 \end{pmatrix}$.

⑦ Starting with the graph of $y = x^2$, state the transformations which can be used to sketch the following curves.

(i) $y = -2x^2$ (ii) $y = x^2 - 2$
(iii) $y = (x - 2)^2$ (iv) $y = (x - 2)^2 - 2$

⑧ Starting with the graph of $y = x^2$, state the transformations which can be used to sketch each of the following curves.

In each case there are two possible answers. Give both.

(i) $y = 4x^2$ (ii) $3y = x^2$

⑨ Starting with the graph of $y = x^2$, give the vectors for the translations which can be used to sketch the following curves.

State also the equation of the line of symmetry of each of these curves.

(i) $y = x^2 + 4$

(ii) $y = (x + 4)^2$

(iii) $y = x^2 - 3$

(iv) $y = (x - 3)^2$

(v) $y = (x - 4)^2 + 3$

(vi) $y = (x + 3)^2 + 4$

(vii) $y = x^2 - 4x$

(viii) $y = x^2 - 4x + 3$

⑩ (i) Show that $x^2 + 6x + 5$ can be written in the form $(x + a)^2 + b$, where a and b are constants to be determined.

(ii) Sketch the graph of $y = x^2 + 6x + 5$, giving the equation of the axis of symmetry and the coordinates of the vertex.

⑪ A cubic graph has the equation $y = 3(x - 1)(x + 2)(x + 5)$.
Write in similar form the equation of the graph after a translation of $\begin{pmatrix} 3 \\ 0 \end{pmatrix}$.

⑫ A quantity y is inversely proportional to the square of x.

When $x = 2$, $y = 2$.

(i) Find the equation connecting x and y.

(ii) Sketch the curve of $f(x) = y$. Clearly label any asymptotes.

(iii) Sketch the curves of

(a) $-f(x)$ (b) $f(x) + 2$

(c) $f(x - 3) + 2$

Label each curve with its equation.

Clearly label any asymptotes and the coordinates of any points of intersection with the coordinate axes.

(iv) Another function is given as $g(x) = x - 1$.

(a) Solve $f(x - 3) + 2 = g(x)$.

(b) Hence find the coordinates of the points of intersection of $y = x - 1$ and

$$y = \frac{8}{(x - 3)^2} + 2$$

⑬ The graph of $y = x^2 + 8x + 3$ is translated by $\begin{pmatrix} -6 \\ 4 \end{pmatrix}$.

Find, in the form $y = ax^2 + bx + c$, the equation of the resulting graph.

⑭ (i) Given that $f(x) = x^2 - 6x + 11$, find values of p and q such that $f(x) = (x - p)^2 + q$.

(ii) On the same set of axes, sketch and label the following curves.

(a) $y = f(x)$

(b) $y = f(x + 4)$

(c) $-f(x)$

Find the equation of each of these curves.

⑮ Sketch the following curves, and state the equations of the asymptotes in each case.

(i) $y = \dfrac{1}{x - 1} + 2$

(ii) $y = \dfrac{1}{(x - 1)^2} + 2$

Prior knowledge

- Graphs of trigonometric functions
- Solving trigonometric equations

These are covered in Chapter 6.

4 Transformations and graphs of trigonometric functions

You can use transformations to help you sketch graphs of trigonometric functions. The transformations give you important information about

- where the curve cuts the axes
- the period
- the maximum and minimum values of the function.

Example 8.8

Starting with the curve $y = \cos x$, show how transformations can be used to sketch these curves.

(i) $y = \cos 3x$ (ii) $y = 3 + \cos x$

(iii) $y = \cos (x - 60°)$ (iv) $y = 2 \cos x$

Solution

(i)

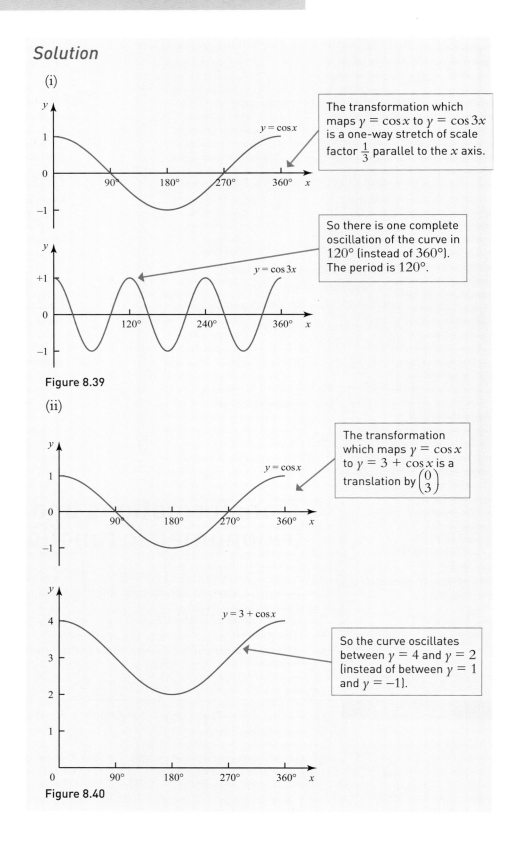

The transformation which maps $y = \cos x$ to $y = \cos 3x$ is a one-way stretch of scale factor $\frac{1}{3}$ parallel to the x axis.

So there is one complete oscillation of the curve in $120°$ (instead of $360°$). The period is $120°$.

Figure 8.39

(ii)

The transformation which maps $y = \cos x$ to $y = 3 + \cos x$ is a translation by $\begin{pmatrix} 0 \\ 3 \end{pmatrix}$

So the curve oscillates between $y = 4$ and $y = 2$ (instead of between $y = 1$ and $y = -1$).

Figure 8.40

(iii)

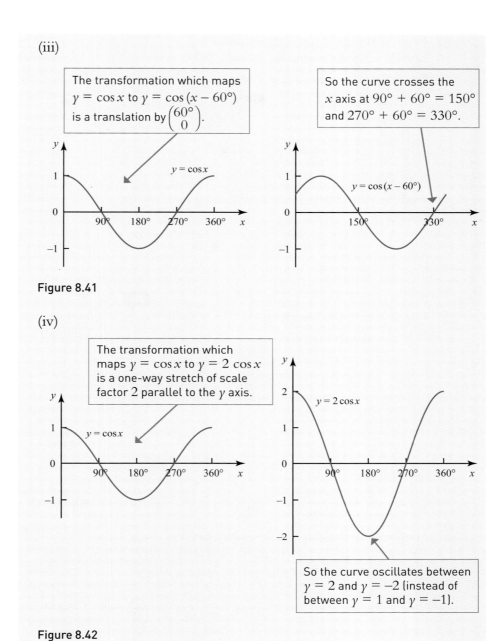

The transformation which maps $y = \cos x$ to $y = \cos(x - 60°)$ is a translation by $\begin{pmatrix} 60° \\ 0 \end{pmatrix}$.

So the curve crosses the x axis at $90° + 60° = 150°$ and $270° + 60° = 330°$.

Figure 8.41

(iv)

The transformation which maps $y = \cos x$ to $y = 2\cos x$ is a one-way stretch of scale factor 2 parallel to the y axis.

So the curve oscillates between $y = 2$ and $y = -2$ (instead of between $y = 1$ and $y = -1$).

Figure 8.42

Exercise 8.4

 ① Sort the functions into the correct cell in the table.

	Maximum value of the function		
	1	2	3
Maximum at $x = 90°$			
Maximum not at $x = 90°$			

P: $y = \sin x$

Q: $y = \sin x + 2$

R: $y = \cos(x - 90°)$

S: $y = 3\sin 2x$

T: $y = 2\sin(x - 180°)$

U: $y = \sin(x + 90°)$

 ② Which function does **not** oscillate between $y = 3$ and $y = -3$?

A $y = 3\sin(x + 90°)$ B $y = 3\sin 3x$

C $y = \sin x + 3$ D $y = 3\sin \dfrac{x}{3}$

③ Find the maximum and minimum values of each of the following functions.

Express your answers in the form

$a \leqslant f(x) \leqslant b$.

(i) $f(x) = \sin 3x$

(ii) $f(x) = 3 \cos x$

(iii) $f(x) = \cos x - 1$

(iv) $f(x) = 3 + \sin x$

④ Find the period of each of the following functions.

(i) $f(x) = \cos (x - 30°)$

(ii) $f(x) = 2 \tan x$

(iii) $f(x) = \sin \frac{1}{2}x$

(iv) $f(x) = \sin 3x$

⑤ For each of the following curves

(a) sketch the curve

(b) identify the curve as being the same as one of the following:

$y = \pm\sin x, \quad y = \pm\cos x,$
or $y = \pm\tan x.$

(i) $y = \sin (x + 360°)$

(ii) $y = \sin (x + 90°)$

(iii) $y = \tan (x - 180°)$

(iv) $y = \cos (x - 90°)$

(v) $y = \cos (x + 180°)$

⑥ Starting with the graph of $y = \sin x$, state the transformations which can be used to sketch each of the following curves.

(i) $y = \sin (x - 90°)$

(ii) $y = \sin 2x$

(iii) $2y = \sin x$

(iv) $y = \sin \frac{x}{2}$

(v) $y = 2 + \sin x$

(vi) $y = -\sin x$

Sketch each curve for $0° \leqslant x \leqslant 360°$ and label any intersections with the axes.

⑦ Starting with the graph of $y = \tan x$

(a) find the equation of the graph

(b) sketch the graph

after the following transformations.

(i) Translation of $\begin{pmatrix} 0 \\ 4 \end{pmatrix}$

(ii) Translation of $\begin{pmatrix} -30° \\ 0 \end{pmatrix}$

(iii) One-way stretch with scale factor 2 parallel to the x axis.

For all, clearly label any asymptotes and any intersections with the axes.

⑧ For each of these curves find the coordinates of the point where the curve:

(a) first intersects the x axis for $x > 0$

(b) intersects the y axis.

(i) $y = \sin (x + 60°)$

(ii) $y = \tan (x - 60°)$

(iii) $y = \frac{1}{2}\cos x$

(iv) $y = \sin \frac{1}{2}x$

(v) $y = 1 + \cos x$

PS ⑨ The curve below intersects the y axis at P and has a maximum at Q. It has a minimum at $(225°, -1)$.

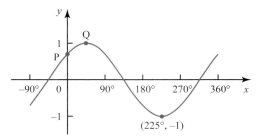

Figure 8.43

(i) Find two possible equations for this curve.

(ii) Hence find the coordinates of the points marked P and Q.

⑩ The graph of $y = \sin x$ is stretched with scale factor 4 parallel to the y axis.

(i) State the equation of the new graph.

(ii) Find the exact value of y on the new graph when $x = 240°$.

PS ⑪ Here are 5 statements.

State whether each statement is TRUE for all values of x in degrees, or FALSE.

Draw suitable graphs to explain your answers.

(i) $\sin 2x = 2 \sin x$

(ii) $\cos (x - 90°) = \sin x$

(iii) $\tan (x + 180°) = \tan x$

(iv) $\cos x = -\sin x$

(v) $\sin (x + 360°) = \sin (x - 360°)$

LEARNING OUTCOMES

When you have completed this chapter you should be able to:

➤ understand and use graphs of functions
➤ sketch curves defined by simple equations including $y = \frac{a}{x}$ and $y = \frac{a}{x^2}$
➤ include vertical and horizontal asymptotes
➤ interpret algebraic solution of equations graphically
➤ use intersection points of graphs to solve equations
➤ understand and use proportional relationships and their graphs
➤ understand the effect of simple transformations on the graph of $y = f(x)$
➤ sketch the graphs of:
  ○ $y = af(x)$
  ○ $y = f(x) + a$
  ○ $y = f(x + a)$
  ○ $y = f(ax)$

KEY POINTS

1 The curves $y = \frac{a}{x}$ and $y = \frac{a}{x^2}$ for $x \neq 0$ show proportional relationships.

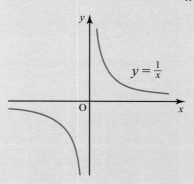 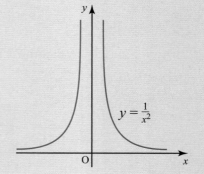

Figure 8.44

2 Transformations of the graphs of the function $y = f(x)$:

Function	Transformation
$f(x - t) + s$	Translation $\begin{pmatrix} t \\ s \end{pmatrix}$
$-f(x)$	Reflection in x axis
$f(-x)$	Reflection in y axis
$af(x)$	One-way stretch, parallel to y axis, scale factor a
$f(ax)$	One-way stretch, parallel to x axis, scale factor $\frac{1}{a}$

9 The binomial expansion

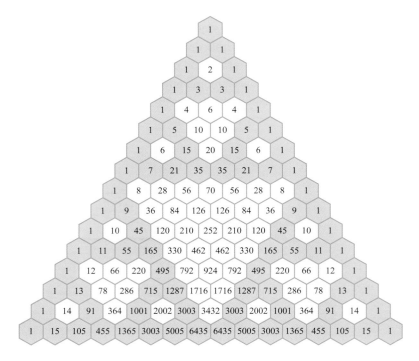

Figure 9.1 below shows part of a town in which the streets form a grid.

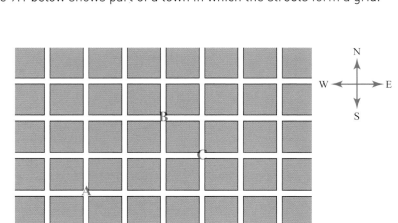

Figure 9.1

→ How many different routes are there from A to B, travelling along the streets and always going East or North?

→ How can you describe each route?

→ How many different routes are there from A to C?

→ Try to find a quick way to work out the number of routes from any one point to another.

1 Binomial expansions

A special type of polynomial is produced when a binomial (i.e. two–part) expression such as $(x + 1)$ is raised to a power. The resulting polynomial is often called a **binomial expansion**.

The simplest binomial expansion is $(x + 1)$ itself. This and other powers of $(x + 1)$ are given below.

$$
\begin{array}{llllllllllll}
(x + 1)^1 = & & & & & & 1x & + & 1 \\
(x + 1)^2 = & & & & 1x^2 & + & 2x & + & 1 \\
(x + 1)^3 = & & & 1x^3 & + & 3x^2 & + & 3x & + & 1 \\
(x + 1)^4 = & & 1x^4 & + & 4x^3 & + & 6x^2 & + & 4x & + & 1 \\
(x + 1)^5 = & 1x^5 & + & 5x^4 & + & 10x^3 & + & 10x^2 & + & 5x & + & 1
\end{array}
$$

> It will soon become obvious why these 1's have been included.

Figure 9.2

If you look at the coefficients you will see that they form a pattern.

> The 1 at the vertex completes the pattern.

> These numbers are called **binomial coefficients**.

$$
\begin{array}{ccccccccccc}
 & & & & & (1) & & & & & \\
 & & & & 1 & & 1 & & & & \\
 & & & 1 & & 2 & & 1 & & & \\
 & & 1 & & 3 & & 3 & & 1 & & \\
 & 1 & & 4 & & 6 & & 4 & & 1 & \\
1 & & 5 & & 10 & & 10 & & 5 & & 1
\end{array}
$$

Figure 9.3

This is called **Pascal's triangle** or **the Chinese triangle**. Each number is obtained by adding the two above it, for example

$$
\begin{array}{ccc}
4 & + & 6 \\
\text{gives} & 10 &
\end{array}
$$

Figure 9.4

The pattern of coefficients is very useful. It enables you to write down the expansions of other **binomial expressions**. For example,

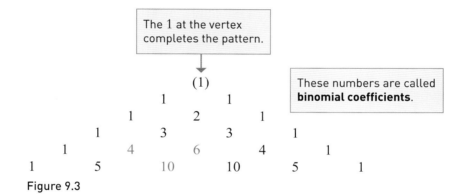

> Notice how in each term the sum of the powers of x and y is the same as the power of $(x + y)$.

> This is a binomial expression.

> These numbers are called binomial coefficients.

Figure 9.5

Example 9.1

Write out the binomial expansion of $(x + 2)^4$.

Solution

The binomial coefficients for power 4 are   1   4   6   4   1.

In each term, the sum of the powers of x and 2 must equal 4.

So the expansion is

$$1 \times x^4 \;+\; 4 \times x^3 \times 2^1 \;+\; 6 \times x^2 \times 2^2 \;+\; 4 \times x^1 \times 2^3 + 1 \times 2^4$$

i.e. x^4 $+ \; 8x^3$ $+ \; 24x^2$ $+ \; 32x$ $+ \; 16$

Example 9.2

Write out the binomial expansion of $(2a - 3b)^5$.

The expression $(2a - 3b)$ is treated as $(2a + (-3b))$.

So the expansion is

$$1 \times (2a)^5 + 5 \times (2a)^4 \times (-3b)^1 + 10 \times (2a)^3 \times (-3b)^2 + 10 \times (2a)^2 \times (-3b)^3$$
$$+ \, 5 \times (2a)^1 \times (-3b)^4 + 1 \times (-3b)^5$$

i.e. $32a^5 - 240a^4b + 720a^3b - 1080a^2b^3 + 810ab^4 - 243b^5$

Historical note

Blaise Pascal has been described as the greatest might-have-been in the history of mathematics. Born in France in 1623, he was making discoveries in geometry by the age of 16 and had developed the first computing machine before he was 20.

Pascal's triangle (and the binomial theorem) had actually been discovered by Chinese mathematicians several centuries earlier, and can be found in the works of Yang Hui (around 1270 A.D.) and Chu Shi-kie (in 1303 A.D.). Pascal is remembered for his application of the triangle to elementary probability, and for his study of the relationship of binomial coefficients.

Pascal was also interested in Physics and spent some time exploring the subject of pressure. He postulated the idea of a vacuum and sought to confirm it through his own experiments. The unit for pressure is named after him, a pascal is a pressure of one newton per square metre.

Pascal died at the early age of 39.

Finding binomial coefficients

The formula for a binomial coefficient

What happens if you need to find the coefficient of x^{17} in the expansion of $(x + 2)^{25}$? To write out 25 rows of Pascal's triangle would take a long while! Clearly you need a formula that gives binomial coefficients.

The first thing you need is a notation for identifying binomial coefficients. It is usual to denote the power of the binomial expansion by n, and the position in the row of binomial coefficients by r, where r can take any value from 0 to n. So for row 5 of Pascal's triangle

$n = 5$:   1   5   10   10   5   1

            $r = 0$ $r = 1$ $r = 2$ $r = 3$ $r = 4$ $r = 5$

Figure 9.6

The general binomial coefficient corresponding to values of n and r is written as $_nC_r$ and said as 'n C r'. You may also see it written as ^nC_r.

Our preferred notation is $\binom{n}{r}$. Thus $_5C_3 = \binom{5}{3} = 10$.

Before you can find a formula for the general binomial coefficient $\binom{n}{r}$, you must be familiar with the term **factorial**.

The quantity '8 factorial', written 8!, is

$$8! = 8 \times 7 \times 6 \times 5 \times 4 \times 3 \times 2 \times 1 = 40\,320.$$

Similarly, $12! = 12 \times 11 \times 10 \times 9 \times 8 \times 7 \times 6 \times 5 \times 4 \times 3 \times 2 \times 1$
$$= 479\,001\,600,$$

and $n! = n \times (n - 1) \times (n - 2) \times \cdots \times 1$, where n is a positive integer.

> ! Note that 0! is defined to be 1. You will see the need for this when you use the formula for $\binom{n}{r}$.

ACTIVITY 9.1

The table shows an alternative way of laying out Pascal's triangle.

		Column (r)							
		0	**1**	**2**	**3**	**4**	**5**	**6**	**...**	**r**
Row	**1**	1	1							
(n)	**2**	1	2	1						
	3	1	3	3	1					
	4	1	4	6	4	1				
	5	1	5	10	10	5	1			
	6	1	6	15	20	15	6	1		
	...	...	...	...	...	...	...	...	...	
	...	...	...	...	...	...	...	...	...	...
	n	1	n	?	?	?	?	?	?	?

Show that $\begin{pmatrix} n \\ r \end{pmatrix} = \dfrac{n!}{r!(n-r)!}$ following the procedure below.

The numbers in column 0 are all 1.

(i) Multiply each number in column 0 by the row number, n. Which column does this give you?

(ii) Now multiply each number in column 1 by $\dfrac{n-1}{2}$. What column does this give you? Repeat the process to find the relationship between each number in column 2 and the corresponding one in column 3.

(iii) What about column 4? And column 5?

(iv) Use, and extend, the rules you have found to complete rows 7 and 8.

(v) Show that repeating the process leads to
$$\begin{pmatrix} n \\ r \end{pmatrix} = \frac{n(n-1)(n-2)\cdots(n-r+1)}{1 \times 2 \times 3 \times \cdots \times r} \text{ for } r \geqslant 1$$

(vi) Show that this can also be written as
$$\begin{pmatrix} n \\ r \end{pmatrix} = \frac{n!}{r!(n-r)!}$$

(vii) You know from Pascal's triangle that $\begin{pmatrix} 5 \\ 2 \end{pmatrix}$ is 15. Check that the formula gives the same answer.

(viii) Use this formula to find $\begin{pmatrix} 5 \\ 5 \end{pmatrix}$. What does this tell you about 0!?

Example 9.3

Use the formula $\begin{pmatrix} n \\ r \end{pmatrix} = \dfrac{n!}{r!(n-r)!}$ to calculate:

(i) $\begin{pmatrix} 5 \\ 0 \end{pmatrix}$ (ii) $\begin{pmatrix} 5 \\ 1 \end{pmatrix}$ (iii) $\begin{pmatrix} 5 \\ 2 \end{pmatrix}$ (iv) $\begin{pmatrix} 5 \\ 3 \end{pmatrix}$

(v) $\begin{pmatrix} 5 \\ 4 \end{pmatrix}$ (vi) $\begin{pmatrix} 5 \\ 5 \end{pmatrix}$

Solution

(i) $\begin{pmatrix} 5 \\ 0 \end{pmatrix} = \dfrac{5!}{0!(5-0)!} = \dfrac{120}{1 \times 120} = 1$

(ii) $\begin{pmatrix} 5 \\ 1 \end{pmatrix} = \dfrac{5!}{1!4!} = \dfrac{120}{1 \times 24} = 5$

(iii) $\begin{pmatrix} 5 \\ 2 \end{pmatrix} = \dfrac{5!}{2!3!} = \dfrac{120}{2 \times 6} = 10$

(iv) $\begin{pmatrix} 5 \\ 3 \end{pmatrix} = \dfrac{5!}{3!2!} = \dfrac{120}{6 \times 2} = 10$

> **Note**
>
> You can see that these numbers, 1, 5, 10, 10, 5, 1, are row 5 of Pascal's triangle.

(v) $\begin{pmatrix} 5 \\ 4 \end{pmatrix} = \dfrac{5!}{4!1!} = \dfrac{120}{24 \times 1} = 5$

(vi) $\begin{pmatrix} 5 \\ 5 \end{pmatrix} = \dfrac{5!}{5!0!} = \dfrac{120}{120 \times 1} = 1$

Most scientific calculators have factorial buttons, e.g. $\boxed{x!}$. Many also have $\boxed{{}_nC_r}$ buttons. Find out how best to use your calculator to find binomial coefficients, as well as practising non–calculator methods.

Example 9.4

Find the coefficient of x^{17} in the expansion of $(x + 2)^{25}$.

Solution

$$(x + 2)^{25} = \binom{25}{0}x^{25} + \binom{25}{1}x^{24}2^1 + \binom{25}{2}x^{23}2^2 + \cdots$$

$$+ \binom{25}{8}x^{17}2^8 + \cdots + \binom{25}{25}2^{25}$$

So the required term is $\binom{25}{8}x^{17}2^8$

$$\binom{25}{8} = \frac{25!}{8!17!} = \frac{25 \times 24 \times 23 \times 22 \times 21 \times 20 \times 19 \times 18 \times 17!}{8! \times 17!}$$

$$= \frac{25 \times 24 \times 23 \times 22 \times 21 \times 20 \times 19 \times 18}{8!}$$

$$= 1\,081\,575$$

So the coefficient of x^{17} is $1\,081\,575 \times 2^8 = 276\,883\,200$.

> **Note**
>
> Notice how 17! was cancelled in working out $\binom{25}{8}$.
>
> Factorials become large numbers very quickly and you should keep a lookout for such opportunities to simplify calculations.

The expansion of $(1 + x)^n$

When deriving the result for $\binom{n}{r}$, you found the binomial coefficients in the form

$$1 \qquad n \qquad \frac{n(n - 1)}{2!} \qquad \frac{n(n - 1)(n - 2)}{3!} \qquad \frac{n(n - 1)(n - 2)(n - 3)}{4!} \qquad \cdots$$

This form is commonly used in the expansion of expressions of the type $(1 + x)^n$.

$$(1 + x)^n = 1 + nx + \frac{n(n - 1)x^2}{1 \times 2} + \frac{n(n - 1)(n - 2)x^3}{1 \times 2 \times 3} + \frac{n(n - 1)(n - 2)(n - 3)x^4}{1 \times 2 \times 3 \times 4}$$

$$+ \frac{n(n - 1)x^{n-2}}{1 \times 2} + nx^{n-1} + 1x^n$$

Example 9.5

Use the binomial expansion to write down the first four terms of $(1 + x)^9$.

Solution

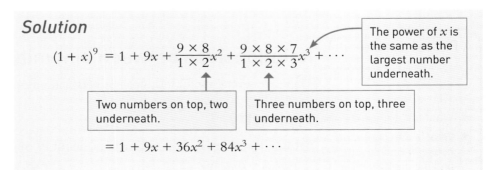

$$(1 + x)^9 = 1 + 9x + \frac{9 \times 8}{1 \times 2}x^2 + \frac{9 \times 8 \times 7}{1 \times 2 \times 3}x^3 + \cdots$$

Two numbers on top, two underneath.

Three numbers on top, three underneath.

The power of x is the same as the largest number underneath.

$$= 1 + 9x + 36x^2 + 84x^3 + \cdots$$

Example 9.6

Use the binomial expansion to write down the first four terms of $(1 - 3x)^7$.

Simplify the terms.

Solution

Think of $(1 - 3x)^7$ as $(1 + (-3x))^7$. Keep the brackets while you write out the terms.

$$(1 + (-3x))^7 = 1 + 7(-3x) + \frac{7 \times 6}{1 \times 2}(-3x)^2 + \frac{7 \times 6 \times 5}{1 \times 2 \times 3}(-3x)^3$$

$$= 1 - 21x + 189x^2 - 945x^3 + \cdots$$ ← Note how the signs alternate.

Example 9.7

(i) Write down the binomial expansion of $(1 - x)^4$.

(ii) Using $x = 0.03$ and the first three terms of the expansion find an approximate value for $(0.97)^4$.

(iii) Use your calculator to find the percentage error in your answer.

 **TECHNOLOGY**

You can use graphing software to gain understanding of the approximation. Type $y = (1 - x)^4$ and $y = 1 - 4x$, $y = 1 - 4x + 6x^2$ etc. and compare the graphs.

Solution

(i) $(1 - x)^4 = (1 + (-x))^4$

$$= 1 + 4(-x) + 6(-x)^2 + 4(-x)^3 + (-x)^4$$

$$= 1 - 4x + 6x^2 - 4x^3 + x^4$$

(ii) $(0.97)^4 \approx 1 - 4(0.03) + 6(0.03)^2$ ← $0.97 = 1 - 0.03$

$$= 0.8854$$

(iii) $(0.97)^4 = 0.885\,292\,81$ ← the result using a calculator

$$\text{Error} = 0.8854 - 0.885\,292\,81$$

$$= 0.000\,107\,19$$

$$\text{Percentage error} = \frac{\text{error}}{\text{true value}} \times 100$$

$$= \frac{0.000\,107\,19}{0.885\,292\,81} \times 100$$

$$= 0.0121\%$$

Discussion points

→ In this question, is 0.885 282 81 an exact value?

→ What about the final answer of 0.0121%?

A Pascal puzzle

$1.1^2 = 1.21$
$1.1^3 = 1.331$
$1.1^4 = 1.4641$

→ What is 1.1^5?

→ What is the connection between your results and the coefficients in Pascal's triangle?

Relationships between binomial coefficients

There are several useful relationships between binomial coefficients.

Symmetry

Because Pascal's triangle is symmetrical about its middle, it follows that

$$\binom{n}{r} = \binom{n}{n - r}.$$

Adding terms

You have seen that each term in Pascal's triangle is formed by adding the two above it.

This is written formally as

$$\binom{n}{r} + \binom{n}{r+1} = \binom{n+1}{r+1}.$$

Sum of terms

You have seen that

$$(x + y)^n = \binom{n}{0}x^n + \binom{n}{1}x^{n-1}y + \binom{n}{2}x^{n-2}y^2 + \cdots + \binom{n}{n}y^n.$$

Substituting $x = y = 1$ gives

$$2^n = \binom{n}{0} + \binom{n}{1} + \binom{n}{2} + \cdots + \binom{n}{n}.$$

Thus the sum of the binomial coefficients for power n is 2^n.

The binomial theorem and its applications

The binomial expansions covered in the last few pages can be stated formally as the binomial theorem for positive integer powers:

$$(a + b)^n = \sum_{r=0}^{n} \binom{n}{r} a^{n-r} b^r \text{ for } n \in \mathbb{Z}^+, \text{ where } \binom{n}{r} = \frac{n!}{r!(n-r)!} \text{ and } 0! = 1.$$

> ### Note
>
> Notice the use of the summation symbol, $\sum$.
>
> $\displaystyle\sum_{r=0}^{n} \binom{n}{r} a^{n-r} b^r$ reads 'the sum of $\binom{n}{r} a^{n-r} b^r$ for values of r from
>
> 0 to n'. So it means
>
> $$\underset{r=0}{\binom{n}{0}a^n} + \underset{r=1}{\binom{n}{1}a^{n-1}b} + \underset{r=2}{\binom{n}{2}a^{n-2}b^2} + \cdots + \underset{r=k}{\binom{n}{k}a^{n-k}b^k} + \cdots + \underset{r=n}{\binom{n}{n}b^n},$$

Exercise 9.1

 ① Jon wrote down the third term in the expansion of $(1 - 2x)^5$ in ascending powers of x.

He wrote $-10 \times 1 \times 2x^2$. Explain his error.

 ② Which answer gives the term in x^3 in the expansion of $(1 - 2x)^5$?

A $10 \times 1^2 \times (-2x)^3$ B $10 \times 1^2 \times -2x^3$

C $1^2 \times (-2x)^3$ D $10 \times 1^2 \times (2x)^3$

 ③ Write out the following binomial expansions.

(i) $(1 + x)^3$ (ii) $(x + 1)^5$

(iii) $(x + 2)^4$

④ Write out the following binomial expansions.

(i) $(3 + x)^3$ (ii) $(2x - 3)^4$

(iii) $(x + 2y)^5$

⑤ Without using a calculator or tables, calculate the following binomial coefficients. Check your answers using your calculator.

(i) $\binom{6}{2}$ (ii) $\binom{5}{0}$ (iii) $\binom{12}{9}$

(iv) $\binom{7}{7}$ (v) $\binom{9}{2}$

⑥ Find

(i) the coefficient of x^4 in the expansion of $(1 + x)^7$

(ii) the coefficient of x^3 in the expansion of $(1 - x)^{10}$

(iii) the coefficient of x^8 in the expansion of $(2 - x)^{10}$

(iv) the coefficient of x^4 in the expansion of $(x + 3)^6$

(v) the value of the term in the expansion of $\left(x - \dfrac{1}{x}\right)^8$ which is independent of x.

⑦ Simplify $(1 + 2x)^4 - (1 - x)^4$.

⑧ Expand $(1 - 3x)^3$ and use your result to expand $(2 + x)(1 - 3x)^3$.

PS ⑨ (i) Expand $(x - 2)^3$.

(ii) Find the values of x for which $(x - 2)^3 = x^3 - 8$.

⑩ (i) Expand $(1 - 2x)^3$.

(ii) Hence expand $(1 + 3x)(1 - 2x)^3$.

⑪ (i) Expand $(1 + x)^3$.

(ii) Hence write down and simplify the expansion of $(1 + y - y^2)^3$ by replacing x by $(y - y^2)$.

⑫ (i) Write down the first six rows of a triangle of numbers built up by the same method as Pascal's, but starting with the row

$$1 \qquad 2$$

instead of

$$1 \qquad 1$$

for $n = 1$.

(Each row starts with 1 and ends with 2.)

(ii) Investigate the row sums of this triangle.

(iii) Repeat this starting with 1   3.

(iv) Generalise your result.

2 Selections

An important application of binomial coefficients occurs when you want to find the number of possible ways of selecting objects, as in the two examples that follow.

Example 9.8

Child Prodigy

Little Gary Forest looks like any other toddler but all the evidence points to him being a budding genius.

Recently Gary's mother gave him five bricks to play with, each showing one of the numbers 1, 2, 3, 4 and 5. Without hesitation Gary sat down and placed them in the correct order.

Figure 9.7

What is the probability that Gary chose the bricks at random and just happened by chance to get them in the right order?

Solution

One way to answer this question is to work out the number of possible ways of placing the five bricks in line.

To start with there are five bricks to choose from, so there are five ways of choosing brick 1. Then there are four bricks left and so there are four ways of choosing brick 2, and so on.

The total number of ways is

$$5 \times 4 \times 3 \times 2 \times 1 = 5! = 120$$

Brick 1 Brick 2 Brick 3 Brick 4 Brick 5

Only one of these is the order 1, 2, 3, 4, 5, so the probability of Gary selecting it at random is $\frac{1}{120}$.

> Number of possible outcomes.

In the previous example Gary selected all the possible number bricks and put them in order. The next example is about him choosing just A, B and C out of 26 alphabet blocks and placing them in order.

Example 9.9

In order to provide Gary with a more challenging test, he is given a set of 26 alphabet bricks to play with and asked to find the bricks displaying the first three letters of the alphabet and put them in order.

If Gary selects the blocks at random, what is the probability that his selection is correct?

Solution

The first block can be chosen in 26 ways.
The second block can be chosen in 25 ways.
The third block can be chosen in 24 ways.

Thus the total number of ways of placing three blocks A, B, C, in the first three positions is $26 \times 25 \times 24 = 15\,600$. So if Gary selects the blocks at random, the probability that his selection is correct is $\frac{1}{15\,600}$.

In this example attention is given to the order in which the blocks are put down. The solution required a **permutation** of three objects from twenty-six.

In general the number of permutations, $_n\mathrm{P}_r$, of r objects from n is given by

$$_n\mathrm{P}_r = n \times (n-1) \times (n-2) \times \cdots \times (n-r+1).$$

This can be written more compactly as

$$_n\mathrm{P}_r = \frac{n!}{(n-r)!}$$

Combinations

It is often the case that you are not concerned with the order in which items are chosen, only with which ones are picked.

Discussion point

→ Show that the number of combinations of 6 objects from 59 can be written as $\dfrac{59!}{6!\,53!}$ and find its value.

Discussion point

→ How can you prove this general result?

To take part in the National Lottery you fill in a ticket by selecting 6 numbers out of a possible 59 (numbers 1, 2, . . . , 59). When the draw is made a machine selects six numbers at random. If they are the same as the six on your ticket, you win the jackpot.

The probability of a single ticket winning the jackpot is often said to be 1 in 45 million. How can you work out this figure?

The key question is, 'How many ways are there of choosing 6 numbers out of 59?' If the order mattered, the answer would be $_{59}P_6$, or $59 \times 58 \times 57 \times 56 \times 55 \times 54$.

However, the order does not matter. 1, 3, 15, 19, 31 and 48 is the same as 15, 48, 31, 1, 19, 3 and 3, 19, 48, 1, 15, 31 and lots more. For each set of six numbers there are 6! arrangements that all count as being the same.

So, the number of ways of selecting six balls, given that the order does not matter, is

$$\frac{59 \times 58 \times 57 \times 56 \times 55 \times 54}{6!}.$$

This is called the number of combinations of 6 objects from 59.

Returning to the National Lottery, it follows that the probability of your one ticket winning the jackpot is $\dfrac{1}{45\,057\,474}$ or about 1 in 45 million.

This example illustrates an important general result.

The number of ways of selecting r objects from n, when the order does not matter, is given by the binomial coefficients

$$\binom{n}{r} = \frac{n!}{r!(n-r)!}.$$

Example 9.10

A School Governors' committee of five people is to be chosen from eight applicants. How many different selections are possible?

Solution

Number of selections $= \dbinom{8}{5} = \dfrac{8!}{5! \times 3!} = \dfrac{8 \times 7 \times 6}{3 \times 2 \times 1} = 56.$

Example 9.11

In how many ways can a committee of four people be selected from four applicants?

Solution

Common sense says that there is only one way to make the committee which is by appointing all applicants. However, if we work from the formula

$\dbinom{4}{4} = \dfrac{4!}{4! \times 0!}$ and this must $= 1$.

To achieve the answer 1 requires the convention that 0! is taken to be 1.

Example 9.12

A cricket team consisting of 6 batters, 4 bowlers and 1 wicket-keeper is to be selected from a group of 18 cricketers comprising 9 batters, 7 bowlers and 2 wicket-keepers. How many different teams can be selected?

Solution

The batters can be selected in $\binom{9}{6}$ ways.

The bowlers can be selected in $\binom{7}{4}$ ways.

The wicket-keeper can be selected in $\binom{2}{1}$ ways.

Therefore total number of possible teams

$$= \binom{9}{6} \times \binom{7}{4} \times \binom{2}{1}$$

$$= \frac{9!}{3!6!} \times \frac{7!}{3!4!} \times \frac{2!}{1!1!}$$

$$= \frac{9 \times 8 \times 7}{3 \times 2 \times 1} \times \frac{7 \times 6 \times 5}{3 \times 2 \times 1} \times 2$$

$$= 5880 \text{ ways.}$$

Exercise 9.2

 ① For each of these events, decide whether the order matters or not.

P: Choosing 10 books to take to read on holiday.

Q: Choosing a basketball team from a squad of 20.

R: Choosing 4 digits for a PIN.

S: Arranging 10 books on a shelf.

T: Choosing 6 different numbers to make a target number

 ② How many ways can you choose 3 people from a group of 8 people?

A 24　　B 56　　C 336　　D 512

③ How many different four letter words can be formed from the letters A, B, C, D if letters cannot be repeated? (The words do not need to mean anything.)

④ To win the jackpot in a lottery a contestant must correctly select six numbers from the numbers 1 to 30 inclusive. What is the probability that a contestant wins the jackpot with one selection of six numbers?

⑤ How many different ways can eight books be arranged in a row on a shelf?

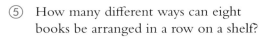 **PS** ⑥ A group of 5 computer programmers is to be chosen to form the night shift from a set of 14 programmers. The 5 must include the shift leader. In how many ways can the programmers be chosen?

 ⑦ In a 60-metre hurdles race there are five runners, one from each of the nations Austria, Belgium, Canada, Denmark and Estonia.

(i) How many different finishing orders are there?

(ii) What is the probability of predicting the finishing order by choosing first, second, third, fourth and fifth at random?

 PS ⑧ My brother Matt decides to put together a rock band from amongst his year at school. He wants a lead singer, a guitarist, a keyboard player and a drummer. He invites applications and gets 7 singers, 5 guitarists, 4 keyboard players and 2 drummers. In how many ways can Matt put the group together?

9. Jay has a CD player which can play CDs in 'shuffle' mode. If a CD is played in 'shuffle' mode the tracks are selected in a random order with a different track selected each time until all the tracks have been played. Jay plays a 14-track CD in 'shuffle' mode.

   (i) In how many different orders could the tracks be played?

   (ii) What is the probability that 'shuffle' mode will play the tracks in the normal set order listed on the CD?

10. A factory advertises for four employees. Eight men and five women apply.

    (i) How many different selections of employees are possible from these applicants?

    (ii) How many of these selections include no men?

    (iii) What is the probability that only women applicants are successful?

    (iv) What assumption did you make in part (iii)?

11. A committee of four is to be selected from five boys and four girls. The members are selected at random.

    (i) How many different selections are possible?

    (ii) What is the probability that the committee will be made up of

       (a) all girls?

       (b) more boys than girls?

12. Baby Imran has a set of alphabet blocks. His mother often uses the blocks I, M, R, A and N to spell Imran's name.

    (i) One day she leaves him playing with these five blocks. When she comes back into the room Imran has placed them in the correct order to spell his name. What is the probability

of Imran placing the blocks in this order? (He is only 18 months old so he certainly cannot spell!)

    (ii) A couple of days later she leaves Imran playing with all 26 of the alphabet blocks. When she comes back into the room she again sees that he has placed the five blocks I, M, R, A and N in the correct order to spell his name. What is the probability of him choosing the five correct blocks and placing them in this order?

13. In a 'Goal of the season' competition, participants pay an entry fee of ten pence. They are then asked to rank ten goals in order of quality. The organisers select their 'correct' order at random. They offer £100 000 to anybody who matches their order. There are no other prizes.

    (i) What is the probability of a participant's order being the same as that of the organisers?

    PS (ii) How much does a participant expect to win or lose with each entry?

    (iii) Five million people enter the competition. How much profit do the organisers expect to make?

PS 14. At a small branch of the MidWest bank the manager has a staff of 12, consisting of 5 men and 7 women including a Mr Brown and a Mrs Green. The manager receives a letter from head office saying that 4 of his staff are to be made redundant. In the interests of fairness, the manager selects the 4 staff at random.

    (i) How many different selections are possible?

    (ii) How many of these selections include both Mr Brown and Mrs Green?

    (iii) Write down the probability that both Mr Brown and Mrs Green will be made redundant.

LEARNING OUTCOMES

When you have completed this chapter you should be able to:

➤ understand and use the binomial expansion of $(a + bx)^n$ for positive integer n

➤ understand and use the notation $n!$ and $_nC_r$.

KEY POINTS

1 $n! = n \times (n-1) \times (n-2) \times \cdots \times 3 \times 2 \times 1;\ 0! = 1$

2 Binomial coefficients, denoted by $_nC_r$ or $\binom{n}{r}$, can be found
   - using Pascal's triangle
   - using your calculator (or tables)
   - using the formula $\binom{n}{r} = \dfrac{n!}{r!(n-r)!}$

3 The binomial expansion of $(1 + x)^n$ may also be written

$$(1 + x)^n = 1 + nx + \frac{n(n-1)}{2!}x^2 + \frac{n(n-1)(n-2)}{3!}x^3 + \cdots + nx^{n-1} + x^n$$

4 This can be generalised as

$$(a + b)^n = a^n + \binom{n}{1}a^{n-1}b + \binom{n}{2}a^{n-2}b^2 + \cdots + \binom{n}{r}a^{n-r}b^r + \cdots + b^n,\ n \in \mathbb{Z}^+$$

   where $\binom{n}{r} = {}^nC_r = \dfrac{n!}{r!(n-r)!}$

5 The number of ways of arranging n unlike objects in a line is $n!$.

6 The number of permutations of r objects from n is

$$_nP_r = \frac{n!}{(n-r)!}$$

7 The number of combinations of r objects from n is

$$\binom{n}{r} = \frac{n!}{r!(n-r)!}$$

8 For permutations the order matters. For combinations it does not.

FUTURE USES

- The binomial expansion forms the basis of the probability distribution called the binomial distribution. This is covered in Chapter 17.

PS ① Figure 1 shows the graph of a polynomial function.

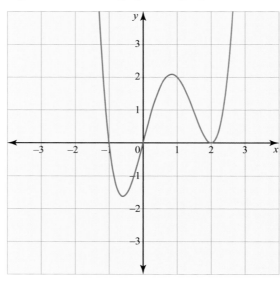

Figure 1

(i)   State, with a reason, whether the order of the polynomial is odd or even. [1 mark]

(ii) Give a possible equation for the graph. [4 marks]

T MP ② Holly is using a graph-drawing program to investigate graphs of the form $y = x^3 - ax^2 + ax - 1$ for different values of a. Figure 2 is a print-out from her program showing six members of the family.

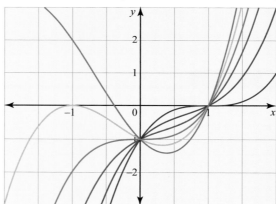

Figure 2

Holly makes the following three conjectures from her investigation.

A: All the members of this family of curves meet at the same two points.

B: When $a = 3$ the graph is a translation to the right of $y = x^3$.

C: When $a = -1$ the graph has repeated zeros.

(i)   Prove that conjecture A is true. [2 marks]

(ii)   Prove that conjecture B is true. [2 marks]

(iii) Prove that conjecture C is true. [3 marks]

M ③ A mission to Mars is being planned. The distance to be travelled from Earth to Mars is calculated as 1.65×10^{11} metres. This figure is used in an initial model to estimate how long it will take to make the journey to Mars, using the following relationship.

> journey time (in seconds) × maximum speed of spacecraft (in m s^{-1})
> = 1.65×10^{11}

(i) State one assumption used in constructing this model. [1 mark]

(ii) Sketch a graph of the relationship between journey time and maximum speed of the spacecraft. Put maximum speed on the horizontal axis. You need not show any scales on the axes. [1 mark]

A spacecraft with a maximum speed of $11\,000$ m s^{-1} is loaded with enough supplies for 175 days.

(iii) Does the initial model suggest that the supplies are sufficient for the journey to Mars? [3 marks]

(iv) State, with reasons, whether you would trust this model if you were an astronaut going on this mission. [1 mark]

④ Figure 3 shows the graphs of $y = 2\sin^2 x$ and $y = 1 - \cos x$.

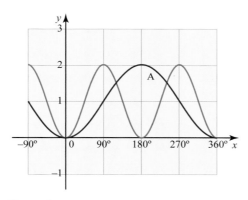

Figure 3

(i) Which curve is labelled A? Explain how you know. [1 mark]

(ii) Use a non-calculator method to solve the equation $2\sin^2 x = 1 - \cos x$ in the interval $0° \leqslant x \leqslant 360°$. [5 marks]

⑤ Figure 4 shows a triangle ABC with some sides and angles labelled.

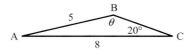

Figure 4

(i) Find the two possible values of θ. [3 marks]

(ii) Sketch the two triangles which correspond to the two answers to part (i) and find the area of the larger triangle. [5 marks]

⑥ **Do not use a calculator in this question.**

Each part-question below has a diagram in which four curves are drawn on the same axes, and labelled A, B, C, D. The equations of the curves are given next to each diagram. Match the correct label to the correct equation.

(i)

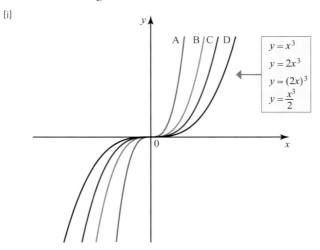

$y = x^3$

$y = 2x^3$

$y = (2x)^3$

$y = \dfrac{x^3}{2}$

Figure 5 [3 marks]

(ii)

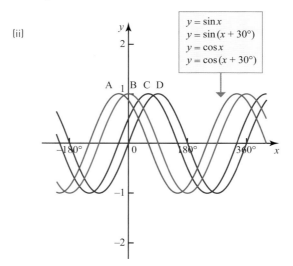

$y = \sin x$

$y = \sin(x + 30°)$

$y = \cos x$

$y = \cos(x + 30°)$

Figure 6 [3 marks]

(iii)

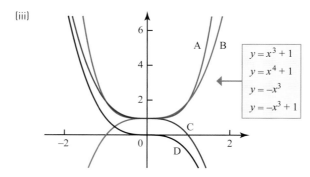

$y = x^3 + 1$

$y = x^4 + 1$

$y = -x^3$

$y = -x^3 + 1$

Figure 7 [3 marks]

⑦ Figure 8 shows a triangle ABC with some sides and angles marked. R is the area of the square on the side of the triangle with length x.

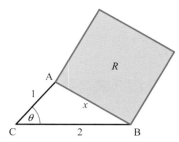

Figure 8

(i)   Find x when $\theta = 20°$. [2 marks]

(ii) Show that, for $0°< \theta <180°$, $R = 5 - 4\cos\theta$ [2 marks]

(iii) Obtain the values of θ for which $R < 4$. [3 marks]

PS ⑧ Figure 9 shows the graphs of $y = (2 + x)^3$ and $y = (1 - x)^3$.

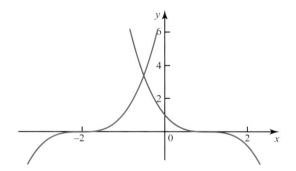

Figure 9

(i)   Obtain the binomial expansions of $(2 + x)^3$ and $(1 - x)^3$. [4 marks]

(ii) Show that

  (a) $y = (2 + x)^3 + (1 - x)^3$ is a quadratic function and find the equation of its line of symmetry [3 marks]

  (b) $y = (2 + x)^3 + (1 - x)^3$ is a cubic function. Show that it has a root at $x = -\frac{1}{2}$ and no other roots. [5 marks]

10 Differentiation

The photo shows a ride at an amusement park. To ensure that the ride is absolutely safe, designers of the ride need to know the gradient of the curve at every point along it. What is meant by the gradient of a curve?

To understand what this means, think about the log on a log-flume, shown in Figure 10.1. If you draw the straight line passing along the bottom of the log, then this line is a tangent to the curve at the point of contact.

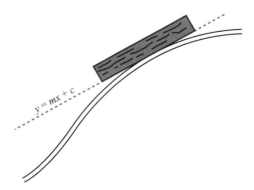

$$y = mx + c$$

Figure 10.1 A log flume

The gradient of a curve at any point on the curve is defined as the gradient of the tangent to the curve at the point and is the rate of change of the function, at that point.

ACTIVITY 10.1

Sketch the curve $y = x^2$. Then discuss the following questions.

(i)   For what values of x is the gradient:

   (a)   positive?

   (b)   negative?

   (c)   zero?

(ii)   For positive values of x, what happens to the gradient as x gets bigger?

(iii)   For negative values of x, what happens to the gradient as x gets more negative?

(iv)   What can you say about the gradient of the curve at the points where $x = 2$ and $x = -2$?

   Now generalise this result for the points $x = a$ and $x = -a$, where a is any constant.

(v)   Repeat this for the curve $y = x^3$.

1 The gradient of the tangent as a limit

Suppose you want to find the gradient of the curve $y = x^2$ at the point where $x = 1$. One approach is to draw the graph and the tangent by hand. You can then find the gradient of your tangent. This gives you an approximate value for the gradient of the curve.

> It is an approximate value not an exact value because you drew the tangent by hand.

Another approach is to find the gradients of chords.

> Remember, a chord is a straight line joining two points on a curve.

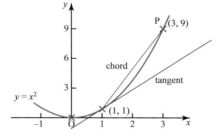

Figure 10.2

The gradient of the chord that joins the points $(1,1)$ and $(3,9)$ is $\dfrac{9 - 1}{3 - 1} = 4$.

Although you can see from Figure 10.2 that this is not the same as the gradient of the tangent, you can also see that as you move point P closer to $(1,1)$, the gradient of the chord becomes closer to the gradient of the tangent (Figure 10.3).

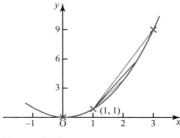

Figure 10.3

TECHNOLOGY

You can set up a spreadsheet to do the arithmetic here. Copy and paste can be used effectively to find the answers at other points on the graph.

ACTIVITY 10.2

Complete the table to find the values of y_2 and the gradients of the chords of the curve $y = x^2$ as you move point P closer to $(1,1)$, and discuss the following questions:

1 Do your gradients seem to be getting closer and closer to a particular number? (This is called **converging** or **tending to a limit**.)

2 What do you think the gradient of $y = x^2$ is at the point $(1,1)$?

(x_1, y_1)	(x_2, y_2)	Gradient $= \dfrac{y_2 - y_1}{x_2 - x_1}$
$(1,1)$	$(3,9)$	4
$(1,1)$	$(2, \quad)$	
$(1,1)$	$(1.5, \quad)$	
$(1,1)$	$(1.25, \quad)$	
$(1,1)$	$(1.1, \quad)$	
$(1,1)$	$(1.01, \quad)$	
$(1,1)$	$(1.001, \quad)$	

You can repeat Activity 10.2 at different points on the graph of $y = x^2$, to find the gradients of a series of chords that converge to a tangent at that point. You could use a spreadsheet to do the calculations, or you could use graphing software. You will find that there is a pattern.

Example 10.1

The gradient of the curve $y = x^2$ at different points is given below.

Point	Gradient of curve
$(-3,9)$	-6
$(-2,-4)$	-4
$(-1,1)$	2
$(0,0)$	0
$(1,1)$	2
$(2,4)$	4
$(3,9)$	6

This is the limit of the gradients of the chords.

The formula for the gradient is called the gradient function. For the curve $y = f(x)$, the gradient function can be written as $f'(x)$.

(i) Describe the pattern, and suggest a formula for the gradient.

(ii) Use your answer to part (i) to

(a) write down the gradient of $y = x^2$ at the point $(10, 100)$

(b) find the point on $y = x^2$ which has gradient 5.

Solution

(i) The gradient appears to be twice the x-value, so the gradient is $2x$. ◄

> You can use function notation to write this as; for $f(x) = x^2$, $f'(x) = 2x$.

(ii)

    (a) $x = 10 \Rightarrow$ gradient $= 2 \times 10 = 20$ ◄

> You can write this as $f'(10) = 20$.

    (b) $2x = 5$

         $\Rightarrow x = \dfrac{5}{2}$

When $x = \dfrac{5}{2}, y = \left(\dfrac{5}{2}\right)^2 = \dfrac{25}{4}$

The point is $\left(\dfrac{5}{2}, \dfrac{25}{4}\right)$.

Exercise 10.1

① Some of these calculations work out the gradient of a chord of $y = x^2$. Which ones?

$$\dfrac{4.2 - 4}{2.1 - 2} \qquad \dfrac{9 - 8.41}{3 - 2.9} \qquad \dfrac{16 - 4}{-4 - -2}$$

$$\dfrac{3.4225 - 2.89}{1.85 - 1.7} \qquad \dfrac{-9 - 9.61}{-3 - -3.1}$$

② (i) Copy and complete the table below for the curve $y = x^3$. You may use either a spreadsheet to look at the limit of the gradient of chords, or graphing software to look at the gradient of the tangent to the curve at the given point.

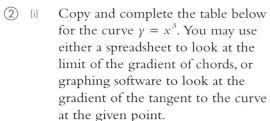

(x_1, y_1)	Gradient of $y = x^3$ at (x_1, y_1)
$(-3, \quad)$	
$(-2, \quad)$	
$(-1, \quad)$	
$(0, \quad)$	
$(1, \quad)$	
$(2, \quad)$	
$(3, \quad)$	

(ii) Plot a graph to show the gradient of $y = x^3$, with your values for x_1 on the horizontal axis, and your values for the gradient on the vertical axis.

(iii) For $f(x) = x^3$, suggest a formula for $f'(x)$.

③ (i) Using either a spreadsheet or graphing software, copy and complete the table below for the curve $y = x^4$.

(x_1, y_1)	Gradient of $y = x^4$ at x_1
$(-3, \quad)$	
$(-2, \quad)$	
$(-1, \quad)$	
$(0, \quad)$	
$(1, \quad)$	
$(2, \quad)$	
$(3, \quad)$	

(ii) Plot a graph to show the gradient function of $y = x^4$, with your values for x_1 on the horizontal axis, and your values for the gradient on the vertical axis.

(iii) Use your graph and your knowledge of the shape of graphs and graph transformations, to suggest a formula for $f'(x)$, where $f(x) = x^4$.

④ (i) From your answers to questions 2(iii) and 3(iii), predict the formula $f'(x)$ for where $f(x) = x^5$.

(ii) Using either a spreadsheet or graphing software, find the gradient of the curve $y = x^5$ at $x = 2$.

(iii) Does this result confirm your prediction?

⑤ (i) Make predictions for $f'(x)$ in the cases

(a) $f(x) = x^6$

(b) $f(x) = x^7$

(ii) Suggest an expression for $f'(x)$ in the case $f(x) = x^n$ where n is a positive integer.

2 Differentiation using standard results

In the work above, you investigated the gradient of a curve $y = f(x)$ by looking at the gradient of a chord joining two points that are close together. If one point has coordinates $(x, f(x))$ and the second point has coordinates $(x + h, f(x + h))$, then the gradient of the chord is given by

$$\frac{f(x + h) - f(x)}{h}.$$

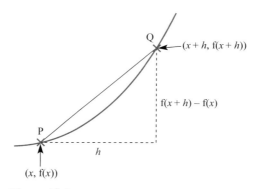

Figure 10.4

The gradient of the tangent is the limit of the gradient of the chord, as h tends to zero. This is written as

$$\lim_{h \to 0}\left(\frac{f(x + h) - f(x)}{h}\right).$$

You will look at formalising this idea in the final section of this chapter.

An alternative notation can be used as follows. The notation δx is used to represent a small change in x, and the notation δy to represent a small change in y. Figure 10.5 shows a chord PQ of a curve, in which the points P and Q are close together.

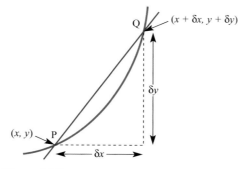

Figure 10.5

In the diagram, the gradient of the chord is $\dfrac{\delta y}{\delta x}$. In the limit as the points come together, the gradient is written $\dfrac{\mathrm{d}y}{\mathrm{d}x}$. This will be explored further in the final section of this chapter, from first principles 'Finding the gradient'. The gradient function $\dfrac{\mathrm{d}y}{\mathrm{d}x}$ or $\text{f}'(x)$ is sometimes called the **derivative** of y with respect to x, and when you find it you have **differentiated** y with respect to x.

In the previous exercise, you saw that, when n is a positive integer,

$$\text{f}(x) = x^n \Rightarrow \text{f}'(x) = nx^{n-1}.$$

The same result can be written as $y = x^n \Rightarrow \dfrac{\mathrm{d}y}{\mathrm{d}x} = nx^{n-1}$.

This can be extended to functions of the type $y = kx^n$ for any constant k, to give

$$y = kx^n \Rightarrow \dfrac{\mathrm{d}y}{\mathrm{d}x} = knx^{n-1}.$$

When $n = 1$, this result becomes $y = kx \Rightarrow \dfrac{\mathrm{d}y}{\mathrm{d}x} = k$.

Another important result is that $y = c \Rightarrow \dfrac{\mathrm{d}y}{\mathrm{d}x} = 0$.

Example 10.2

Differentiate $y = 3x^4$, and hence find the gradient of the curve at the point where $x = -2$.

Solution

$$y = 3x^4 \Rightarrow \dfrac{\mathrm{d}y}{\mathrm{d}x} = 3 \times 4x^{4-1} = 12x^3$$

At $x = -2$, the gradient is $12(-2)^3 = -96$.

Sums and differences of functions

Many of the functions you will meet are sums or differences of simpler ones. For example, the function $3x^2 + 4x^3$ is the sum of the functions $3x^2$ and $4x^3$.

To differentiate a function such as this you differentiate each part separately and then add the results together.

Example 10.3

Differentiate $y = 3x^2 + 4x^3$.

Solution

$$\frac{dy}{dx} = 6x + 12x^2$$

Note

This illustrates a general result:

$$y = f(x) + g(x) \Rightarrow$$

$$\frac{dy}{dx} = f'(x) + g'(x)$$

Example 10.4

Find the coordinates of the points on the curve with equation $y = x^3 - 6x^2 + 5$ where the value of the gradient is -9.

Solution

The gradient at any point on the curve is given by $\frac{dy}{dx} = 3x^2 - 12x$.

You need to find the points at which $\frac{dy}{dx} = -9$.

$$3x^2 - 12x = -9$$

$$3x^2 - 12x + 9 = 0 \quad \longleftarrow \boxed{\text{Rearrange equation.}}$$

$$x^2 - 4x + 3 = 0 \quad \longleftarrow \boxed{\text{Divide by 3.}}$$

$$(x - 1)(x - 3) = 0 \quad \longleftarrow \boxed{\text{Factorise.}}$$

$$\Rightarrow x = 1, x = 3$$

When $x = 1$, $y = 1^3 - 6(1)^2 + 5 = 0$

When $x = 3$, $y = 3^3 - 6(3)^2 + 5 = -22$

Therefore the gradient is -9 at the points $(1, 0)$ and $(3, -22)$.

Example 10.5

Differentiate $y = (2x + 1)(3x - 4)$.

Solution

$$y = (2x + 1)(3x - 4) \quad \longleftarrow \boxed{\begin{array}{l}\text{You must multiply out the} \\ \text{brackets before differentiating.}\end{array}}$$

$$= 6x^2 - 5x - 4$$

$$\frac{dy}{dx} = 12x - 5$$

Differentiation with other variables

Although a curve often has x on the horizontal axis and y on the vertical axis, there is nothing special about these letters. If, for example, your curve represented time t on the horizontal axis and velocity v on the vertical axis, then the relationship may be referred to as $v = f(t)$, i.e. v is a function of t, and the gradient function is given by $\frac{dv}{dt} = f'(t)$. Just as $\frac{dy}{dx}$ is the rate of change of y with respect to x, $\frac{dv}{dt}$ is the rate of change of v with respect to t, and the process is called differentiating v with respect to t.

Example 10.6

Differentiate the following functions:

(i) $\quad x = t^2 - 3t + 2$ $\qquad$ (ii) $\quad z = 4p^8 + 2p^6 - p^3 + 5$

Solution

(i) $\quad \dfrac{dx}{dt} = 2t - 3$ $\qquad$ (ii) $\quad \dfrac{dz}{dp} = 32p^7 + 12p^5 - 3p^2$

Example 10.7

Differentiate the function $g(r) = \dfrac{4\pi r^3}{3}$ with respect to r.

Solution

$g'(r) = \dfrac{4\pi}{3} \times 3r^2$

$= 4\pi r^2$

> Because the question has been given in function notation, use the function notation $g'(r)$ for the derivative.

> Remember that π is just another number.

Exercise 10.2

① These expressions are either $f(x)$ or $f'(x)$. Match pairs of functions and their derivatives.

$12 \quad 6x \quad 8x \quad 12x \quad 3x^2$
$4x^2 \quad 6x^2 \quad 12x^2 \quad 2x^3 \quad 4x^3$

② What is the derivative of $y = x^3 - 3x^2$?
A 0 $\quad$ B $6x^2$ $\quad$ C $3x^2 - 6x$ $\quad$ D $3x^2 - 3x$

③ Differentiate the following functions:

(i) $\quad y = x^7$

(ii) $\quad y = x^{11}$

(iii) $\quad y = 2x^7 - 3x^{11}$

④ Differentiate the following functions:

(i) $\quad V = x^3$

(ii) $\quad x = 2t^2 - 5t + 6$

(iii) $\quad z = 3l^5 - l^2 + 5l - 36$

⑤ Find the gradient of the following curves at the point where $x = 1$.

(i) $\quad y = x^2 - 3x$

(ii) $\quad y = x^3 + 6x + 10$

(iii) $\quad y = x^4 - 7x^2 + 10x$

⑥ Find the gradient of the following curves at the point where $x = -2$.

(i) $\quad y = 3x^2 + 4x - 7$

(ii) $\quad y = \frac{1}{2}x^3 - 2x + 9$

(iii) $\quad y = \frac{1}{3}x^3 - \frac{2}{3}x^2 + 5x$

⑦ Find the coordinates of the points on the curve $y = 2x^3 - 9x^2 - 12x + 7$ where the gradient is 12.

PS

⑧ (i) Multiply out $\left(x^2 + 3\right)(5x - 1)$.

(ii) Use your answer to (i) to differentiate $y = \left(x^2 + 3\right)(5x - 1)$ with respect to x.

(iii) Markus has written

$y = \left(x^2 + 3\right)(5x - 1) \Rightarrow$

$\dfrac{dy}{dx} = 2x \times 5 = 10x.$

What mistake has he made?

⑨ (i) Simplify $\dfrac{2x^2 + 3x}{x}$.

(ii) Use your answer to (i) to differentiate $y = \dfrac{2x^2 + 3x}{x}$.

⑩ Differentiate the following with respect to x:

(i) $\quad y = (x + 2)\left(x^3 - 4\right)$

(ii) $\quad y = \dfrac{6x^3 - 3x^2 + 4x}{2x}$.

⑪ (i) Sketch the curve $y = x^2 - 6x$.

(ii) Differentiate $y = x^2 - 6x$.

(iii) Show that the point $(3, -9)$ lies on the curve $y = x^2 - 6x$ and find the gradient of the curve at this point.

(iv) What does your answer to (iii) tell you about the curve at the point $(3, -9)$?

⑫ (i) Sketch the curve $y = x^2 - 4$.

(ii) Write down the coordinates of the points, P and Q, where the curve crosses the x axis.

(iii) Differentiate $y = x^2 - 4$.

(iv) Find the gradient of the curve at P and at Q.

(v) On your sketch, draw the tangents to P and Q, and write down the coordinates of the point where they meet.

⑬ (i) Factorise $x^3 - 6x^2 + 11x - 6$, and hence write down the coordinates of the points where the curve $y = x^3 - 6x^2 + 11x - 6$ cuts the x axis.

(ii) Sketch the curve $y = x^3 - 6x^2 + 11x - 6$.

(iii) Differentiate $y = x^3 - 6x^2 + 11x - 6$.

(iv) Show that the tangents to the curve at two of the points at which it cuts the x axis are parallel.

⑭ (i) Sketch, on the same axes, the line with equation $y = 2x + 5$ and the curve with equation $y = 4 - x^2$ for $-3 \leqslant x \leqslant 3$.

(ii) Show that the point $(-1, 3)$ lies on both curves.

(iii) Differentiate $y = 4 - x^2$ and hence find its gradient at $(-1, 3)$.

(iv) Do you have sufficient evidence to decide whether the line $y = 2x + 5$ is a tangent to the curve $y = 4 - x^2$?

(v) Is the line joining $\left(2\frac{1}{2}, 0\right)$ to $(0, 5)$ a tangent to the curve $y = 4 - x^2$?

PS ⑮ (i) Sketch the curve $y = x^2 + 3x - 1$.

(ii) Differentiate $y = x^2 + 3x - 1$.

(iii) Find the coordinates of the point on the curve $y = x^2 + 3x - 1$ at which it is parallel to the line $y = 5x - 1$.

(iv) Is the line $y = 5x - 1$ a tangent to the curve $y = x^2 + 3x - 1$? Give reasons for your answer.

⑯ The equation of a curve is $y = ax^3 + bx + 4$, where a and b are constants. The point $(2, 14)$ lies on the **PS** curve and its gradient at this point is 21. Find the values of a and b.

PS ⑰ (i) Sketch, on the same axes, the curves with equations $y = x^2 - 9$ and $y = 9 - x^2$ for $-4 \leqslant x \leqslant 4$.

(ii) The tangents to $y = x^2 - 9$ at $(2, -5)$ and $(-2, -5)$, and the tangents to $y = 9 - x^2$ at $(2, 5)$ and $(-2, 5)$ are drawn to form a quadrilateral. Describe this quadrilateral, giving reasons for your answer.

Prior knowledge

You need to know how to find the equation of a straight line, given its gradient and a point on the line (covered in Chapter 5). You also need to know the relationship between the gradients of two perpendicular lines (covered in Chapter 5).

3 Tangents and normals

You know that the gradient of a curve at any point is the same as the gradient of the tangent to the curve at that point. Now that you know how to find the gradient of a curve at any point by differentiation, you can use this to find the equation of the tangent at any point on the curve.

Example 10.8

Find the equation of the tangent to the curve $y = x^2 - 2x + 3$ at the point $(2, 3)$.

Solution

$$\frac{dy}{dx} = 2x - 2$$

> First work out $\frac{dy}{dx}$.

When $x = 2$, $\frac{dy}{dx} = 2 \times 2 - 2 = 2$.

Therefore $m = 2$.

> Now substitute your value of x into the expression $\frac{dy}{dx}$ to find the gradient m of the tangent at that point.

The equation of the tangent is given by $y - y_1 = m(x - x_1)$

In this case $x_1 = 2, y_1 = 3, m = 2$ so

$$y - 3 = 2(x - 2)$$
$$\Rightarrow y = 2x - 1$$

This is the equation of the tangent.

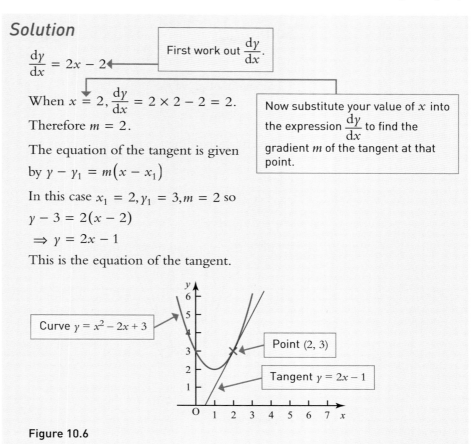

Curve $y = x^2 - 2x + 3$

Point $(2, 3)$

Tangent $y = 2x - 1$

Figure 10.6

The **normal** to a curve at a particular point is the straight line which is at right angles to the tangent at that point.

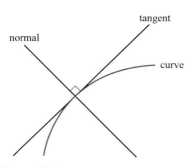

Figure 10.7

Remember that for perpendicular lines, $m_1 m_2 = -1$. So if the gradient of the tangent is m_1, the gradient, m_2, of the normal is given by

$$m_2 = -\frac{1}{m_1}.$$

This enables you to find the equation of the normal at any point on a curve.

Example 10.9

Find the equation of the normal to the curve $y = x^2 - 2x + 3$ at the point $(2, 3)$.

Solution

See Example 10.8.

The gradient m_1 of the tangent at the point $(2, 3)$ is $m_1 = 2$.

The gradient m_2 of the normal to the curve at this point is given by

$$m_2 = -\frac{1}{m_1} = -\frac{1}{2}$$

The equation of the normal is given by $y - y_1 = m(x - x_1)$ and in this case

$x_1 = 2, y_1 = 3, m = -\frac{1}{2}$ so

$$y - 3 = -\frac{1}{2}(x - 2)$$
$$x + 2y - 8 = 0$$

Tidying this up by multiplying both sides by 2 and rearranging.

This is the equation of the normal.

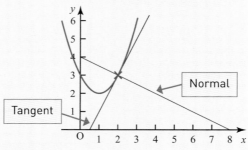

Normal

Tangent

Figure 10.8

Exercise 10.3

 ① Copy and complete the following table.

$f(x)$	$f'(x)$	Gradient of tangent when $x = 2$	Gradient of normal when $x = 2$
$f(x) = x^2 - 3x$		1	
$f(x) = 2x^3$			$-\frac{1}{24}$
$f(x) = x^4 - 3x^2$			

② (i) Draw the curve $y = x^2 + 2$, using the same scale for both axes.

(ii) Find $\frac{dy}{dx}$.

(iii) Find the gradient of the tangent to the curve $y = x^2 + 2$ at the point $(1, 3)$.

(iv) Find the gradient of the normal to the curve $y = x^2 + 2$ at the point $(1, 3)$.

(v) Show the tangent and normal to the curve $y = x^2 + 2$ at the point $(1, 3)$ on your graph.

③ (i) Draw the curve $y = 6x - x^2$, using the same scale for both axes.

(ii) Find $\frac{dy}{dx}$.

(iii) Find the gradient of the tangent to the curve $y = 6x - x^2$ at the point $(2, 8)$, and show this tangent on your graph.

(iv) Find the gradient of the normal to the curve $y = 6x - x^2$ at the point $(5, 5)$, and show this normal on your graph.

④ (i) Draw the curve $y = 1 - x^2$, using the same scale on both axes.

(ii) Find $\frac{dy}{dx}$.

(iii) Find the equation of the tangent to the curve $y = 1 - x^2$ at the point $(2, -3)$, and show this tangent on your graph.

(iv) Find the equation of the normal to the curve $y = 1 - x^2$ at the point $(1, 0)$, and show this normal on your graph.

(v) Find the coordinates of the point of intersection of the tangent and the normal found in (iii) and (iv).

PS ⑤ (i) Draw the curve $y = x^2 + x - 6$, using the same scale on both axes.

(ii) Find $\frac{dy}{dx}$.

(iii) Find the equation of the tangent to the curve $y = x^2 + x - 6$ at the point where $x = 2$, and show this tangent on your graph.

(iv) Find the equation of the normal to the curve $y = x^2 + x - 6$ at the point where $x = 1$, and show this normal on your graph.

(v) Find the area of the triangle formed by the tangent, the normal and the x axis.

⑥ (i) Sketch the curve $y = 6 - x^2$.

(ii) Find the gradient of the curve at the points $(-1, 5)$ and $(1, 5)$.

(iii) Find the equations of the tangents to the curve at these points, and show these tangents on your sketch.

(iv) Find the coordinates of the point of intersection of these two tangents.

⑦ (i) Factorise $x^3 - 4x^2$ and hence sketch the curve $y = x^3 - 4x^2$.

(ii) Differentiate $y = x^3 - 4x^2$.

(iii) Find the gradient of $y = x^3 - 4x^2$ at the point $(2, -8)$.

(iv) Find the equation of the tangent to the curve $y = x^3 - 4x^2$ at the point $(2, -8)$, and show this tangent on your sketch.

(v) Find the coordinates of the other point at which this tangent meets the curve.

PS ⑧ (i) Find the equation of the tangent to the curve $y = 2x^3 - 15x^2 + 42x$ at $P(2, 40)$.

(ii) Using your expression for $\frac{dy}{dx}$, find the coordinates of the point Q on the curve at which the tangent is parallel to the one at P.

(iii) Find the equation of the normal at Q.

⑨ The curve $y = x^4 - 4x^2$ meets the x axis at three points. The triangle T has vertices that are the points of intersection of the tangents to the curve at these three points. Draw a diagram to show the triangle T, and find the area of T.

PS

4 Increasing and decreasing functions, and turning points

Prior knowledge

You need to be able to solve inequalities. This is covered in Chapter 4.

If the gradient of a function is positive between the points $x = a$ and $x = b$, the function is described as an **increasing function** in the interval $a < x < b$. Similarly, if the gradient is negative in a particular interval, it is a **decreasing function** in that interval.

ACTIVITY 10.3

(i) Sketch the following functions.

(a) $y = \dfrac{1}{x}$

(b) $y = x$

(c) $y = x^3$

(d) $y = 2^x$

(ii) State whether the function is increasing for all values of x, increasing for some values of x or increasing for no values of x.

(iii) For each of the categories above, find another example of a suitable function.

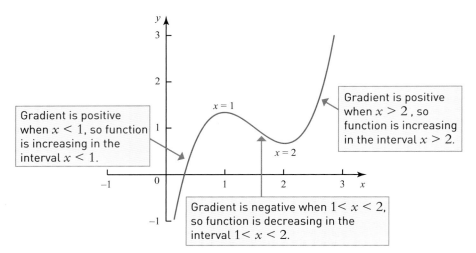

Gradient is positive when $x < 1$, so function is increasing in the interval $x < 1$.

Gradient is positive when $x > 2$, so function is increasing in the interval $x > 2$.

$x = 1$

$x = 2$

Gradient is negative when $1 < x < 2$, so function is decreasing in the interval $1 < x < 2$.

Figure 10.9

If a function is described as an increasing function without specifying an interval for x, then this means that the gradient of the function is positive for all values of x.

Example 10.10

Find the range of values of x for which the function $y = x^2 - 6x$ is a decreasing function. Confirm your answer by sketching the graph of $y = x^2 - 6x$, and write down the range of values of x for which the function $y = x^2 - 6x$ is an increasing function.

Solution

$$y = x^2 - 6x \Rightarrow \frac{dy}{dx} = 2x - 6 \qquad \boxed{\text{Differentiate}}$$

$$y \text{ decreasing} \Rightarrow \frac{dy}{dx} < 0$$

$$\Rightarrow 2x - 6 < 0 \qquad \longleftarrow \boxed{\text{Set up your inequality.}}$$

$$\Rightarrow 2x < 6$$

$$\Rightarrow x < 3 \qquad \longleftarrow \boxed{\text{Solve your inequality.}}$$

The graph of $y = x^2 - 6x$ is shown below.

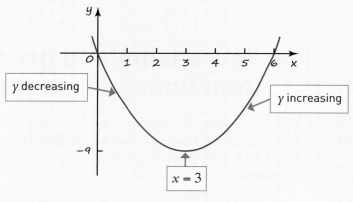

y decreasing

y increasing

$x = 3$

Figure 10.10

y is increasing for $x > 3$.

Example 10.11

Show that $y = x^3 + x$ is an increasing function for all values of x.

Solution

$$y = x^3 + x \Rightarrow \frac{dy}{dx} = 3x^2 + 1.$$

Since $x^2 \geqslant 0$ for all real values of x, $\frac{dy}{dx} \geqslant 1$

$\Rightarrow y = x^3 + x$ is an increasing function.

ACTIVITY 10.4

Figure 10.11 shows the graph of $y = x^3 - 3x$.

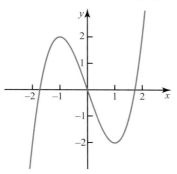

Figure 10.11

It has two turning points, one maximum and one minimum. Discuss the shape of the curve, and copy and complete the table below.

	What are the coordinates of the turning point?	Is the gradient just to the left of the turning point positive or negative?	What is the gradient at the turning point?	Is the gradient just to the right of the turning point positive or negative?
Maximum				
Minimum				

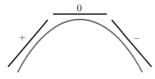

Figure 10.12

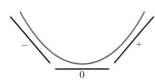

Figure 10.13

Note that although there is a maximum point at $(-1, 2)$, 2 is not the highest point on the curve. Such a point is therefore sometimes referred to as a **local maximum**. Likewise, there is a minimum point at $(1, -2)$ but $y = -2$ is not the lowest point of the graph, so this point is a **local minimum**.

At any maximum point:

- the gradient $\frac{dy}{dx}$ is zero at the turning point
- the gradient is positive just to the left of the maximum and negative just to the right of it.

At any minimum point:

- the gradient $\frac{dy}{dx}$ is zero at the turning point
- the gradient is negative just to the left of the maximum and positive just to the right of it.

Example 10.12

Find the turning points on the curve $y = -x^3 + 12x + 4$ and determine whether they are maximum points or minimum points. Hence sketch the curve.

Solution

The gradient of this curve is given by $\dfrac{dy}{dx} = -3x^2 + 12$.

When $\dfrac{dy}{dx} = 0, -3x^2 + 12 = 0.$ ← | The turning points are the points where the gradient is zero.

$$-3(x^2 - 4) = 0$$
$$-3(x + 2)(x - 2) = 0$$
$$\Rightarrow x = -2 \text{ or } x = 2$$

| Substitute the x values into the original equation to find the y values.

When $x = -2, y = -(-2)^3 + 12(-2) + 4 = -12$.

When $x = 2, y = -2^3 + 12(2) + 4 = 20.$

Therefore there are turning points at $(-2, -12)$ and $(2, 20)$.

Now look at the gradients near these points.

For $x = -2$: $x = -2.1 \Rightarrow \dfrac{dy}{dx} = -3(-2.1)^2 + 12 = -1.23$

$\qquad\qquad x = -1.9 \Rightarrow \dfrac{dy}{dx} = -3(-1.9)^2 + 12 = 1.17$

| Substitute a value of x just to the left and a value of x just to the right of your turning point into your expression for $\dfrac{dy}{dx}$.

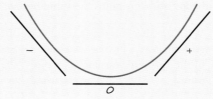

| Notice that the actual values of the gradients at nearby points are not important in this process. It is the signs (+ or –) of the gradients that matter.

Figure 10.14

So there is a minimum point at $(-2, -12)$.

For $x = 2$: $x = 1.9 \Rightarrow \dfrac{dy}{dx} = -3(1.9)^2 + 12 = 1.17$

$\qquad\qquad x = 2.1 \Rightarrow \dfrac{dy}{dx} = -3(2.1)^2 + 12 = -1.23$

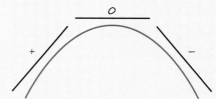

Figure 10.15

Therefore there is a maximum point at $(2, 20)$.

The sketch can now be drawn.

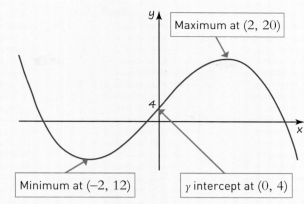

Figure 10.16

Exercise 10.4

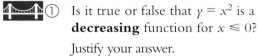 ① Is it true or false that $y = x^2$ is a **decreasing** function for $x \leqslant 0$?

Justify your answer.

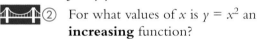 ② For what values of x is $y = x^2$ an **increasing** function?

A all x B $x > 0$

C $x < 0$ D all x except $x = 0$

③ The curve shown below has turning points at $x = -1$ and at $x = 3$. Write down the values of x for which it shows:

(i) an increasing function

(ii) a decreasing function.

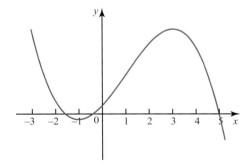

Figure 10.17

④ Sketch the graphs of the following functions, and hence write down the values of x for which the function is increasing in each case.

(i) $y = x^2$

(ii) $y = (x - 3)^2$

(iii) $y = (x + 5)^2$

⑤ (i) Use differentiation to find the range of values of x for which $y = x^2 + 2x - 3$ is an increasing function.

(ii) Solve the quadratic equation $x^2 + 2x - 3 = 0$.

(iii) Sketch the graph of $y = x^2 + 2x - 3$ marking on the coordinates of the points where the curve crosses the axes.

(iv) Explain how your sketch confirms your answer to part (i).

⑥ Use differentiation to find the coordinates of any turning points on the curve $y = x^3 - 12x + 2$ and determine their nature.

⑦ (i) Use differentiation to find the range of values of x for which the function $y = x^2 - 4x - 7$ is a decreasing function.

(ii) By writing the function $y = x^2 - 4x - 7$ in the form $y = (x + a)^2 + b$, find the coordinates of the turning point of the curve $y = x^2 - 4x - 7$.

(iii) Sketch the curve $y = x^2 - 4x - 7$ marking on the coordinates of the turning point.

(iv) Explain how your sketch confirms your answer to part (i).

⑧ (i) Given that $y = x^3 + 3x^2 - 24x - 7$, find $\dfrac{\mathrm{d}y}{\mathrm{d}x}$.

(ii) By solving the appropriate quadratic inequality, find the range of values of x for which $y = x^3 + 3x^2 - 24x - 7$ is a decreasing function.

⑨ Given that $y = x^2(x - 2)^2$

   (i) Multiply out the right-hand side, then find $\frac{dy}{dx}$.

   (ii) Find the coordinates and nature of any turning points.

   (iii) Sketch the curve. Mark the coordinates of the turning point and the points of intersection with the axes.

PS ⑩ Find the range of values of x for which $y = -x^3 + 15x^2 - 63x + 47$ is a decreasing function.

PS ⑪ Show that $f(x) = -3x^3 - x + 40$ is a decreasing function for all values of x.

⑫ (i) Given that $y = x^3 - 6x^2 + 15x + 3$, find $\frac{dy}{dx}$.

   (ii) By writing your expression for $\frac{dy}{dx}$ in the form $a(x + b)^2 + c$, show that $y = x^3 - 6x^2 + 15x + 3$ is an increasing function for all values of x.

⑬ Show that the function $f(x) = x^5 - 10x^3 + 50x - 21$ is an increasing function for all values of x.

PS

5 Sketching the graphs of gradient functions

To understand how the gradient of a curve behaves, it can be helpful to plot or sketch the gradient on a graph of its own. You still have the x values on the horizontal axis, but on the vertical axis you put the values of the gradient instead of the y values. The resulting curve is called the graph of the gradient function.

To sketch the graph of the gradient function of a curve

- look for the **stationary points** on the curve, where the gradient is zero
- then look for the regions in which the function is *increasing,* where the gradient is positive
- and the regions in which the function is *decreasing,* where the gradient is negative.

Example 10.13

Sketch the graph of the gradient function for the function shown in Figure 10.18.

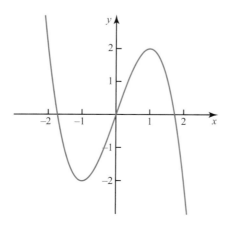

Figure 10.18

Solution

First, look for the stationary points on the graph, where the gradient is zero. These are the point where $x = -1$ and the point where $x = 1$. Therefore the gradient curve crosses the x axis at two points $(-1, 0)$ and $(1, 0)$.

> You are not interested in the y values, just in the x values and the gradient at these x values.

Now, use these two points to split up the graph into the three regions $x < -1, -1 < x < 1$ and $x > 1$, see Figure 10.19.

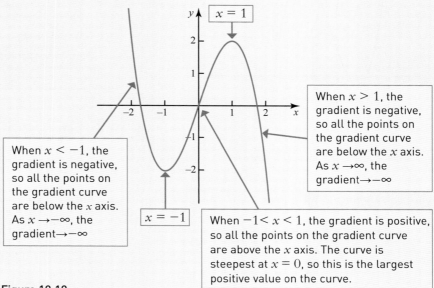

$x = 1$

When $x < -1$, the gradient is negative, so all the points on the gradient curve are below the x axis. As $x \rightarrow -\infty$, the gradient$\rightarrow -\infty$

When $x > 1$, the gradient is negative, so all the points on the gradient curve are below the x axis. As $x \rightarrow \infty$, the gradient$\rightarrow -\infty$

$x = -1$

When $-1 < x < 1$, the gradient is positive, so all the points on the gradient curve are above the x axis. The curve is steepest at $x = 0$, so this is the largest positive value on the curve.

Figure 10.19

You can also see that the gradient of the curve is the same at $x = a$ and $x = -a$ for all values of a, therefore the gradient curve is symmetrical about the vertical axis.

Putting this information together, you can sketch the graph of the gradient function, as in Figure 10.20.

Discussion points

➜ What is the gradient of the straight line $2y - 5x + 4 = 0$? What would the gradient graph of this straight line look like?
➜ Generalise this to describe the gradient graph of the straight line $y = mx + c$.
➜ What would the graph of the gradient function of the straight line $y = 3$ look like? What about the straight line $x = 3$?

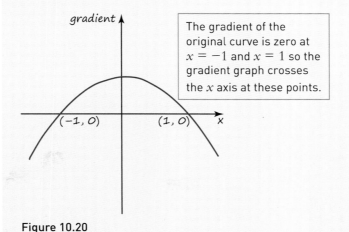

The gradient of the original curve is zero at $x = -1$ and $x = 1$ so the gradient graph crosses the x axis at these points.

Figure 10.20

Exercise 10.5

 ① Explain why the curve in Figure 10.21 could be the gradient function of the curve in Figure 10.22.

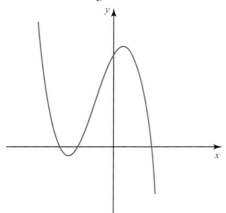

Figure 10.21

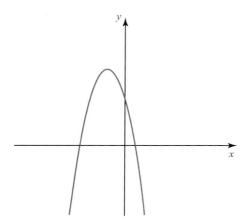

Figure 10.22

②

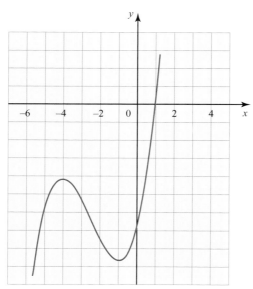

Figure 10.23

Which of the diagrams A, B, C or D shows the gradient function for the curve in Figure 10.23?

A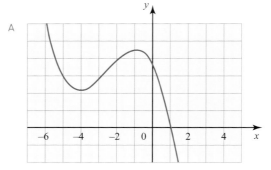

B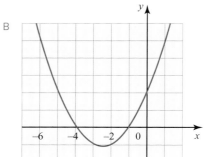

C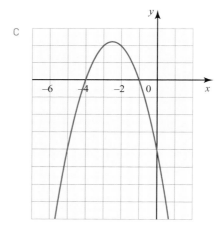

D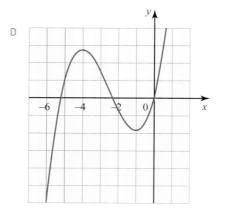

Figure 10.24

③ On the same set of axes, sketch the graphs of the gradient functions of the following straight lines.

(i) $y = 3x + 1$

(ii) $y = -5x - 1$

(iii) $2y - x + 3 = 0$

(iv) $4y + 3x + 7 = 0$

④ Sketch the curve $y = x^3$, and decide which of the following statements is true.

(i)   The gradient is always positive.

(ii) The gradient is always negative.

(iii) The gradient is never positive.

(iv) The gradient is never negative.

Explain your answer.

⑤

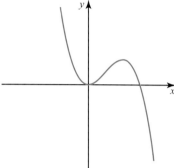

Figure 10.25

Which of the diagrams A, B, C or D shows the gradient function for the curve in Figure 10.25? Explain your answer.

A

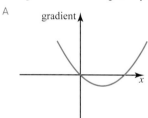

B

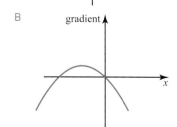

C

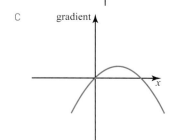

D

Figure 10.26

⑥

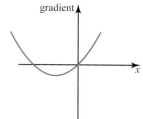

Figure 10.27

(i)   For the function in Figure 10.27, write down the coordinates of the stationary points.

(ii) For what values of x is this an increasing function?

(iii) For what values of x is this a decreasing function?

(iv) Sketch the graph of the gradient function.

⑦ (i)   Decide which, if any, of the three symbols $\Rightarrow$, $\Leftarrow$ or $\Leftrightarrow$ makes each of the statements below true for values of x at points on the curve shown in Figure 10.28.

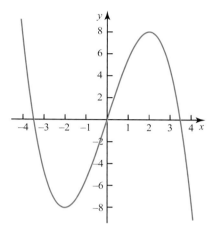

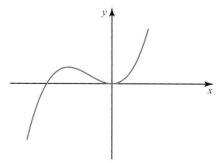

Figure 10.28

(a) $x = 0$ ☐ the gradient is zero

(b) $x < 0$ ☐ the gradient is negative

(c) $x > 0$ ☐ the gradient is positive

(d) $x < 0$ ☐ the gradient is positive

(ii) Sketch the graph of the gradient function.

⑧ For both of the curves below, sketch the graph of the gradient function.

(i)

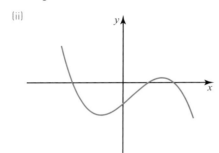

Figure 10.29

(ii)

Figure 10.30

⑨ (i) Sketch the gradient function of the curve in Figure 10.31.

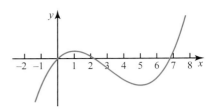

Figure 10.31

(ii) The curve given in (i) has equation $y = x^3 - 9x^2 + 15x$. By differentiating and sketching the graph of the resulting function, confirm your answer to (i).

⑩ The graph in Figure 10.32 shows the gradient function of a curve. On the same axes, sketch three different possibilities for the original curve, and explain how your three graphs are related to each other.

PS

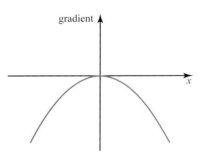

Figure 10.32

Prior knowledge

You need to know what is meant by negative and fractional indices. This is covered in Chapter 2.

6 Extending the rule

ACTIVITY 10.5

(i) You are going to investigate whether the result $y = x^n \Rightarrow \dfrac{dy}{dx} = nx^{n-1}$ is also true for the curve $y = \sqrt{x}$.

(a) Find the gradient of the curve at $x = 4$. You can do this either by using graphing software to find the gradient of the tangent to the curve, or by using a spreadsheet to find the limit of the gradients of a series of chords.

(b) By writing $y = \sqrt{x}$ in index form, calculate the value of the gradient at $x = 4$ using the result $y = x^n \Rightarrow \dfrac{dy}{dx} = nx^{n-1}$. Does it give the same answer?

(c) Repeat using a different point on the curve.

(ii) Now investigate whether the result is true for other curves where n is a negative integer or a fraction, for example $y = \dfrac{1}{x}, y = \dfrac{1}{x^2}, y = x^{\frac{3}{2}}$.

The results of the activity above suggest that the rule $y = kx^n \Rightarrow \dfrac{dy}{dx} = knx^{n-1}$ is true for all real values of n.

Example 10.14

Given that $y = x\sqrt{x}$, find $\dfrac{dy}{dx}$, and hence find the gradient of the curve $y = x\sqrt{x}$ at the point $(9, 27)$.

Solution

$y = x\sqrt{x} = x^1 \times x^{\frac{1}{2}} = x^{\frac{3}{2}}$ ← Write the function as a power of x.

$\dfrac{dy}{dx} = \dfrac{3}{2}x^{\frac{3}{2}-1} = \dfrac{3}{2}x^{\frac{1}{2}}$

When $x = 9$, $\dfrac{dy}{dx} = \dfrac{3}{2} \times 9^{\frac{1}{2}} = \dfrac{3}{2} \times 3 = \dfrac{9}{2}$

So the gradient of the curve $y = x\sqrt{x}$ at the point $(9, 27)$ is $\dfrac{9}{2}$.

Example 10.15

Differentiate $y = \left(\dfrac{1}{x} + x\right)\left(\dfrac{1}{x^2} + x\right)$.

Solution

$y = \left(\dfrac{1}{x} + x\right)\left(\dfrac{1}{x^2} + x\right)$

$\quad = \dfrac{1}{x} \times \dfrac{1}{x^2} + \dfrac{1}{x} \times x + x \times \dfrac{1}{x^2} + x \times x$ ← Start by expanding the brackets.

$\quad = \dfrac{1}{x^3} + 1 + \dfrac{1}{x} + x^2$

$\quad = x^{-3} + 1 + x^{-1} + x^2$ ← Now write as powers of x.

$\quad = x^2 + 1 + x^{-1} + x^{-3}$

$\dfrac{dy}{dx} = 2x - x^{-2} - 3x^{-4}$

$\quad = 2x - \dfrac{1}{x^2} - \dfrac{3}{x^4}$ ← Rewrite as fractions instead of as powers of x.

Example 10.16

Differentiate $x = \dfrac{t^2 + 2\sqrt{t}}{t}$.

Solution

$x = \dfrac{t^2}{t^1} + \dfrac{2t^{\frac{1}{2}}}{t^1}$

$\quad = t + 2t^{-\frac{1}{2}}$ ← Simplify before you differentiate.

$\dfrac{dx}{dt} = 1 - t^{-\frac{3}{2}}$

$\quad = 1 - \dfrac{1}{\sqrt{t^3}}$

Extending the rule

Exercise 10.6

① Copy and complete these linear sequences:

(i) $\square$, $\square$, $-\frac{1}{2}$, $\square$, $\frac{3}{2}$, $\square$

(ii) $\square$, $\square$, $\square$, $\frac{1}{3}$, $\frac{4}{3}$, $\square$

② $y = x^{-\frac{1}{2}}$. Which is the **incorrect** version of $\frac{dy}{dx}$?

A $-\frac{1}{2}x^{-\frac{3}{2}}$ B $-\frac{1}{2\sqrt{x}}$

C $-\frac{1}{2x\sqrt{x}}$ D $-\frac{1}{2x^{\frac{3}{2}}}$

③ Differentiate the following functions.

(i) $y = 4x^{-5}$

(ii) $y = 6x^{\frac{1}{2}}$

(iii) $y = x^{-\frac{2}{3}}$

④ Differentiate the following functions.

(i) $z = 7t^{-3}$

(ii) $x = 4t^{-\frac{1}{2}}$

(iii) $p = 6r^{-2} + 3r - 7$

⑤ (i) Copy and complete the following working to show that
$$y = \sqrt{x} \Rightarrow \frac{dy}{dx} = \frac{1}{2\sqrt{x}}$$

$y = \sqrt{x} = $ ← **Write as a power of x.**

$\Rightarrow \frac{dy}{dx} = $ ← **Use $y = x^n \rightarrow \frac{dy}{dx} = nx^{n-1}$.**

$\Rightarrow \frac{dy}{dx} = \frac{1}{2} \times$ ← **Rewrite the power of x as a fraction.**

$\Rightarrow \frac{dy}{dx} = \frac{1}{2\sqrt{x}}$

(ii) Now demonstrate the following results showing every stage in your working.

(a) $y = \frac{1}{\sqrt{x}} \Rightarrow \frac{dy}{dx} = -\frac{1}{2x\sqrt{x}}$

(b) $y = \frac{1}{2x^2} \Rightarrow \frac{dy}{dx} = -\frac{1}{x^3}$

⑥ Find the gradient of the following curves at the point where $x = 4$.

(i) $y = \frac{2}{x}$ (ii) $y = x\sqrt{x} - x + 4$

⑦ Differentiate the following functions.

(i) $y = x^2\left(x\sqrt{x} - \frac{1}{x^3}\right)$

(ii) $y = \dfrac{x\sqrt{x} - \dfrac{1}{x^3}}{x^2}$

⑧ (i) Write down the coordinates of the point where the curve $y = \frac{1}{x} + 2$ crosses the x axis.

(ii) Sketch the curve.

(iii) Find $\frac{dy}{dx}$.

(iv) Find the gradient of the curve at the point where it crosses the x axis.

⑨ The graph of $y = \frac{4}{x^2} + x$ is shown in Figure 10.33.

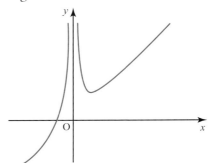

Figure 10.33

(i)   Differentiate $y = \frac{4}{x^2} + x$.

(ii) Show that the point $(-2, -1)$ lies on the curve.

(iii) Find the gradient of the curve at $(-2, -1)$.

(iv) Find the coordinates of the turning point of the curve.

⑩ (i) Sketch the curve with equations $y = \frac{1}{x^2} + 1$ for $-3 \leqslant x \leqslant 3$.

(ii) Show that the point $\left(\frac{1}{2}, 5\right)$ lies on the curve.

(iii) Differentiate $y = \frac{1}{x^2} + 1$ and find its gradient at $\left(\frac{1}{2}, 5\right)$.

(iv) Find the equation of the tangent to the curve $y = \frac{1}{x^2} + 1$ at the point $\left(\frac{1}{2}, 5\right)$, and show this tangent on your sketch.

(11) The graph of $y = \sqrt{x} - 1$ is shown in Figure 10.34.

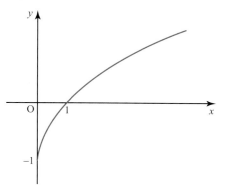

Figure 10.34

(i) Differentiate $y = \sqrt{x} - 1$.

(ii) Find the coordinates of the point on the curve $y = \sqrt{x} - 1$ at which the tangent is parallel to the line $y = 2x - 1$.

(iii) Is the line $y = 2x - 1$ a tangent to the curve $y = \sqrt{x} - 1$? Give reasons for your answer.

(12) The graph of $y = 3x - \dfrac{1}{x^2} + 2$ is shown in Figure 10.35. At the point P, $x = -1$.

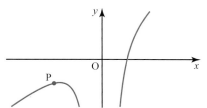

Figure 10.35

(i) Find the y coordinate at P.

(ii) Find the gradient function $\dfrac{dy}{dx}$.

(iii) Find the value of the gradient at P.

(iv) Find the equation of the tangent to the curve at P.

(13) The graph of $y = x^2 + \dfrac{1}{x}$ is shown in Figure 10.36. The point marked Q is (1, 2).

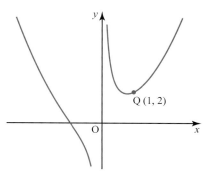

Figure 10.36

(i) Find the gradient function $\dfrac{dy}{dx}$.

(ii) Find the gradient of the tangent at Q.

(iii) Show that the equation of the normal to the curve at Q can be written as $x + y = 3$.

(iv) At what other points does the normal cut the curve?

(14) The curve $y = a\sqrt{x} + \dfrac{b}{x} + \dfrac{c}{\sqrt{x}}$ passes through the point $(1, -2)$ where it has gradient -3. It also passes through the point $(4, -1)$. Find the gradient of the curve at this point.

7 Higher order derivatives

Discussion points

→ Is $\dfrac{d^2y}{dx^2}$ the same as $\left(\dfrac{dy}{dx}\right)^2$?

→ How do you think $\dfrac{d^3y}{dx^3}$ is found? What do you think it is called, and what rate of change is it?

At a minimum point, the gradient is zero and changing from negative to positive; the rate of change of the gradient with respect to x is positive. At a maximum point, the gradient is zero and changing from positive to negative; the rate of change of the gradient is negative.

The rate of change of the gradient with respect to x is $\dfrac{d}{dx}\left(\dfrac{dy}{dx}\right)$, and is found by differentiating the function for a second time.

It is written as $\dfrac{d^2y}{dx^2}$ or $f''(x)$ and called the **second derivative**. ← $\dfrac{d^2y}{dx^2}$ is said as 'd two y by d x squared'.

Given that $y = x^5 - 2x^2 + 3$, find $\dfrac{d^2y}{dx^2}$.

Example 10.17

Solution

$$\frac{dy}{dx} = 5x^4 - 4x$$

$$\frac{d^2y}{dx^2} = 20x^3 - 4$$

Differentiate $\frac{dy}{dx}$ to find $\frac{d^2y}{dx^2}$.

You can often use the second d[...]re of a turning point.

If $\frac{d^2y}{dx^2}$ is positive at the turning point, then the rate of change of the gradient is positive, so the gradient is going from negative to positive, and therefore the turning point is a minimum.

If $\frac{d^2y}{dx^2}$ is negative then the gradient is going from positive to negative, and so it is a maximum. However, if $\frac{d^2y}{dx^2}$ is zero at a turning point, then it is not possible to use this method to identify the nature of the turning point, and you must use the method of looking at the sign of the gradient just to each side of the turning point.

Example 10.18

Given that $y = 2x^3 + 3x^2 - 12x$

(i) find $\frac{dy}{dx}$, and find the values of x for which $\frac{dy}{dx} = 0$

(ii) find the coordinates of each of the turning points

(iii) find the value of $\frac{d^2y}{dx^2}$ at each turning point and hence determine its nature

(iv) sketch the curve given by $y = 2x^3 + 3x^2 - 12x$.

Solution

(i) $\frac{dy}{dx} = 6x^2 + 6x - 12$

Set $\frac{dy}{dx}$ equal to zero to find turning points.

$\frac{dy}{dx} = 0 \Rightarrow 6x^2 + 6x - 12 = 0$

$6(x^2 + x - 2) = 0$

$6(x + 2)(x - 1) = 0$

$\Rightarrow x = -2$ or $x = 1$

(ii) At $x = -2$, $y = 2(-2)^3 + 3(-2)^2 - 12(-2) = 20$

The point is $(-2, 20)$.

Substitute x values into original equation to find the y values.

At $x = 1$, $y = 2(1)^3 + 3(1)^2 - 12(1) = -7$

The point is $(1, 7)$.

(iii) $\frac{d^2y}{dx^2} = 12x + 6$

Differentiate $\frac{dy}{dx}$ to find $\frac{d^2y}{dx^2}$.

At $x = -2$, $\frac{d^2y}{dx^2} = -18 < 0$ and so $(-2, 20)$ is a maximum.

At $x = 1$, $\frac{d^2y}{dx^2} = 18 > 0$ and so $(1, -7)$ is a minimum.

(iv)

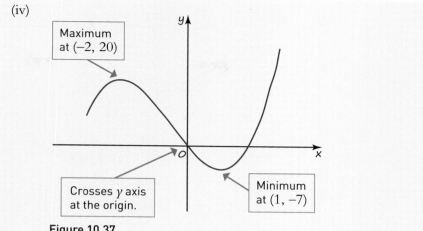

Maximum at (−2, 20)

Crosses y axis at the origin.

Minimum at (1, −7)

Figure 10.37

Example 10.19

Find the coordinates of the turning points of the curve $y = 4x^5 + 5x^4$ and identify their nature. Hence sketch the curve.

Solution

Set $\dfrac{dy}{dx}$ equal to zero to find turning points.

$$\frac{dy}{dx} = 20x^4 + 20x^3$$

$$\frac{dy}{dx} = 0 \Rightarrow 20x^4 + 20x^3 = 0$$

$$20x^3(x+1) = 0$$

Factorise and solve the equation.

$$\Rightarrow x = -1 \text{ or } x = 0$$

At $x = -1$, $y = -4 + 5 = 1$

The point is $(-1, 1)$.

At $x = 0$, $y = 0$

Substitute x values into original equation to find the y values.

The point is $(0, 0)$.

$$\frac{d^2y}{dx^2} = 80x^3 + 60x^2$$

Differentiate $\dfrac{dy}{dx}$ to find $\dfrac{d^2y}{dx^2}$.

At $x = -1$, $\dfrac{d^2y}{dx^2} = -20 < 0$, therefore $(-1, 1)$ is a maximum point.

At $x = 0$, $\dfrac{d^2y}{dx^2} = 0$, therefore you cannot use this method to determine the nature of this turning point.

Substitute a value of x just to the left and a value of x just to the right of your turning point into your expression for $\dfrac{dy}{dx}$.

At $x = -0.1$, $\dfrac{dy}{dx} = 20(-0.1)^4 + 20(-0.1)^3 = -0.018$

At $x = 0.1$, $\dfrac{dy}{dx} = 20(0.1)^4 + 20(0.1)^3 = 0.022$

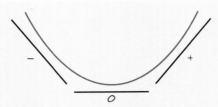

Figure 10.38

Therefore $(0, 0)$ is a minimum point.

You can now sketch the curve.

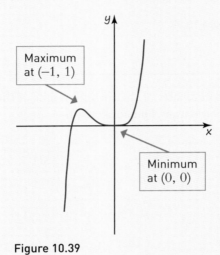

Maximum at $(-1, 1)$

Minimum at $(0, 0)$

Figure 10.39

Discussion points

The curve $y = x^3 - 3x^2 + 3x$ is shown in Figure 10.40.

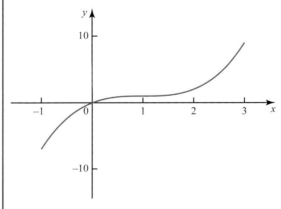

Figure 10.40

➜ What is the value of $\dfrac{\mathrm{d}y}{\mathrm{d}x}$ at $x = 1$?

➜ A stationary point that is not a turning point is called a **point of inflection**. Give examples of other curves that have points of inflection.

Note

You will learn more about points of inflection in the Year 2 textbook.

① These expressions are either $f(x), f'(x)$ or $f''(x)$.

Sort them into two groups of three functions and their first and second derivatives:

$2x^3 - 2x$ $6x^2 - 2$ $6x^2 - 3$ $12x$

$2x^3 - 3x$ $\frac{1}{2}x^4 - x^2 + 3$

 ② What is the second derivative of the function $f(x) = 4x^3 - x^2$?

A 10   B 12   C $24x - 2$ D $12x^2 - 2x$

③ Find $\frac{dy}{dx}$ and $\frac{d^2y}{dx^2}$ for the following functions.

(i) $y = x^4$

(ii) $y = 3x^5 - 2x^3 + x^2 - 7$

④ Find $\frac{dy}{dx}$ and $\frac{d^2y}{dx^2}$ for the following functions.

(i) $y = x^2 - \frac{1}{x}$

(ii) $y = 2x\sqrt{x}$

⑤ Given the curve $y = x^2 + 2x - 8$

(i)   find $\frac{dy}{dx}$ and use this to find the coordinates of the turning point of the curve

(ii)   confirm your answer to (i) by completing the square

(iii)   find $\frac{d^2y}{dx^2}$ and use this to determine whether the turning point is a maximum or minimum. Does your answer agree with what you know about the shape of quadratic graphs?

(iv)   sketch the curve, marking on your sketch the coordinates of the turning point and the points of intersection with the axes.

⑥ Given that $y = x^4 - 16$

(i)   find $\frac{dy}{dx}$

(ii)   find the coordinates of any turning points and determine their nature

(iii)   sketch the curve, marking on your sketch the coordinates of the turning point and the points of intersection with the axes.

⑦ Given that $y = (x - 1)^2(x - 3)$

(i)   multiply out the right-hand side, then find $\frac{dy}{dx}$ and $\frac{d^2y}{dx^2}$

(ii)   find the coordinates and nature of any turning points

(iii)   sketch the curve, marking on your sketch the coordinates of the turning point and the points of intersection with the axes.

⑧ For the function $y = x - 4\sqrt{x}$

(i)   find $\frac{dy}{dx}$ and $\frac{d^2y}{dx^2}$

(ii)   find the coordinates of the turning point and determine its nature.

⑨ The equation of a curve is $y = 6\sqrt{x} - x\sqrt{x}$. Find the x coordinate of the turning point and show that the turning point is a maximum.

⑩ The curve $y = px^3 + qx^2$ has a turning point at $(1, -1)$.

(i)   Find the values of p and q.

(ii)   Find the nature of the turning point.

(iii)   Find the coordinates and nature of any other turning points of the curve.

(iv)   Sketch the curve, marking on your sketch the coordinates of the turning points and the points of intersection with the axes.

8 Practical problems

There are many situations in which you need to find the maximum or minimum value of an expression.

Example 10.20

A stone is projected vertically upwards with a speed of 30 ms^{-1}. Its height, h m, above the ground after t seconds ($t < 6$) is given by $h = 30t - 5t^2$.

(i)   Find $\dfrac{dh}{dt}$ and $\dfrac{d^2h}{dt^2}$.

(ii) Find the maximum height reached.

(iii) Sketch the graph of h against t.

Solution

(i) $\dfrac{dh}{dt} = 30 - 10t$

     $\dfrac{d^2h}{dt^2} = -10$

(ii) For a turning point, $\dfrac{dh}{dt} = 0$

$$30 - 10t = 0$$
$$\Rightarrow \quad 10t = 30$$
$$\Rightarrow \quad t = 3$$

At $t = 3$, $h = 30 \times 3 - 5 \times 3^2 = 45$

Because $\dfrac{d^2h}{dt^2}$ is negative, the turning point is a maximum (as you would expect from the situation).
Maximum height is 45 m.

(iii)

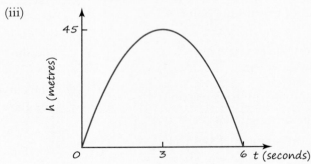

Figure 10.41

> **Note**
>
> For a position–time graph, such as this one, the gradient, $\dfrac{dh}{dt}$, is the velocity and $\dfrac{d^2h}{dt^2}$ is the acceleration.

Example 10.21

A cylindrical can with volume $16\pi \text{ cm}^3$ is to be made from a thin sheet of metal. The can has a lid. Its height is h cm and its radius r cm (Figure 10.42). Find the minimum surface area of metal needed.

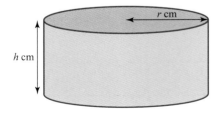

Figure 10.42

Solution

The surface area of the can is given by
$$A = 2\pi rh + 2\pi r^2.$$

> You can 'unfold' the cylinder into a rectangle of length $2\pi r$ and width h with a circle for the base and another for the lid, as shown in Figure 10.43.

Figure 10.43

Since the volume of a cylinder is given by the formula $\pi r^2 h$,
$$\pi r^2 h = 16\pi \Rightarrow h = \frac{16}{r^2}$$

> You need to write the expression for A in terms of just one variable, so that you can differentiate it.

So, $A = 2\pi rh + 2\pi r^2 = 2\pi r\left(\frac{16}{r^2}\right) + 2\pi r^2 = \frac{32\pi}{r} + 2\pi r^2$

At the turning point, $\frac{dA}{dr} = 0$
$$\frac{dA}{dr} = -\frac{32\pi}{r^2} + 4\pi r = 0.$$
$$\Rightarrow 4\pi r = \frac{32\pi}{r^2}$$
$$\Rightarrow r^3 = 8$$
$$\Rightarrow r = 2$$

Therefore the radius is $2\,cm$ at the turning point.

When $r = 2$,
$$A = \frac{32\pi}{r} + 2\pi r^2$$
$$= \frac{32\pi}{2} + 2\pi \times 2^2 = 24\pi.$$
$$\frac{d^2A}{dr^2} = \frac{64\pi}{r^3} + 4\pi$$

> To find the nature of the turning point, differentiate again.

At $r = 2, \frac{d^2A}{dr^2} = 12\pi > 0.$

Since this is positive, the turning point is a minimum.

Therefore the minimum surface area to make a closed can of height $h\,cm$ is $24\pi\,cm^2$.

Exercise 10.8

 ① Match the descriptions to the expressions.

(A) x is 10 less than y

(B) the sum of x and y is 10

(C) the mean of x and y is 10

(D) y is 10 less than x.

$x + y = 10 \qquad y - 10 = x$
$x = y + 10 \qquad x + y = 20$

 ② x is 10 times as large as y. Which of the following describes this?

A $x = 10y$ B $y = 10x$

C $y = x + 10$ D $x = y + 10$

PS

③ The sum of two numbers, x and y, is 20.

(i) Write down an expression for y in terms of x.

(ii) Write down an expression for P, the product of the two numbers, in terms of x.

(iii) Find $\frac{dP}{dx}$ and $\frac{d^2P}{dx^2}$.

(iv) Hence find the maximum value of the product, and show that it is indeed a maximum.

PS ④ The product of two numbers, x and y, is 400.

  (i) Write down an expression for y in terms of x.

  (ii) Write down an expression for S, the sum of the two numbers, in terms of x.

  (iii) Find $\dfrac{dS}{dx}$ and $\dfrac{d^2S}{dx^2}$.

  (iv) Hence find the minimum value of the sum of the two numbers. You must show that it is a minimum.

⑤ The sum of two numbers, x and y, is 8.

  (i) Write down an expression for y in terms of x.

PS (ii) Write down an expression for S, the sum of the squares of these two numbers, in terms of x.

  (iii) Find $\dfrac{dS}{dx}$ and $\dfrac{d^2S}{dx^2}$.

  (iv) Prove that the minimum value of the sum of their squares is 32.

M ⑥ A farmer wants to construct a temporary

PS rectangular enclosure of length x m and width y m for her prize bull while she works in the field. She has 120 m of fencing and wants to give the bull as much room to graze as possible.

  (i) Write down an expression for y in terms of x.

  (ii) Write down an expression for the area, A, to be enclosed in terms of x.

  (iii) Sketch the graph of A against x.

  (iv) Find $\dfrac{dA}{dx}$ and $\dfrac{d^2A}{dx^2}$.

  (v) Hence show that the maximum on your graph is indeed the greatest possible area for the bull.

PS ⑦ Charlie wants to add an extension with

M a floor area of $18\,\text{m}^2$ to the back of his house. He wants to use the minimum possible number of bricks, so he wants to know the smallest perimeter he can use. The dimensions, in metres, are x and y as shown in Figure 10.44.

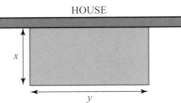

Figure 10.44

  (i) Write down an expression for the area in terms of x and y.

  (ii) Write down an expression, in terms of x and y, for the total length, T, of the outside walls.

  (iii) Show that $T = 2x + \dfrac{18}{x}$.

  (iv) Find $\dfrac{dT}{dx}$ and $\dfrac{d^2T}{dx^2}$.

  (v) Find the dimensions of the extension that give a minimum value of T, and confirm that it is a minimum.

PS ⑧ A piece of wire 16 cm long is cut into two

M pieces. One piece is $8x$ cm long and is bent to form a rectangle measuring $3x$ cm by x cm. The other piece is bent to form a square of side y cm (Figure 10.45).

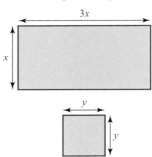

Figure 10.45

  (i) Find y in terms of x, and hence find an expression for the area of the square.

  (ii) Show that the combined area of the rectangle and the square is $A\,\text{cm}^2$ where $A = 7x^2 - 16x + 16$.

  (iii) Find the value of x for which A has its minimum value.

  (iv) Find the minimum value of A.

⑨ A rectangle has a perimeter of 32 cm. Prove that the maximum possible area of the rectangle is $64\,\text{cm}^2$.

PS

⑩ An open box is to be made from a rectangular sheet of card measuring 16 cm by 10 cm, by cutting four equal squares, each of side x cm, from the corners and folding up the flaps. Find the largest volume of the box.

⑪ Figure 10.46 shows a right-angled triangle with an area of 8 cm².

Figure 10.46

(i) Write down an expression for y in terms of x.

(ii) Write down an expression for the sum, S, of the squares of these two numbers in terms of x.

(iii) Find the least value of the sum of their squares by considering $\dfrac{dS}{dx}$ and $\dfrac{d^2 S}{dx^2}$.

(iv) Hence write down the shortest possible length for the hypotenuse.

 ⑫ A piece of wire 30 cm long is going to be made into two frames for blowing bubbles. The wire is to be

cut into two parts. One part is bent into a circle of radius r cm and the other part is bent into a square of side x cm.

(i) Write down an expression for the perimeter of the circle in terms of r, and hence write down an expression for r in terms of x.

(ii) Show that the combined area, A, of the two shapes can be written as
$$A = \frac{(4 + \pi)x^2 - 60x + 225}{\pi}.$$

(iii) Find the lengths that must be cut if the area is to be a minimum.

⑬ A cylinder with a base but no lid has a surface area of 40 cm². Find the maximum possible value for the volume of the cylinder giving your answer in terms of π. Show that this value is a maximum.

⑭ A ship is to make a voyage of 100 km at a constant speed of v km h^{-1}. The running cost of the ship is $£\left(0.8v^2 + \dfrac{2000}{v}\right)$ per hour. Find, to the nearest pound, the minimum cost of the voyage, and confirm that this is a minimum.

9 Finding the gradient from first principles

To establish the result that $y = x^n \Rightarrow \dfrac{dy}{dx} = nx^{n-1}$, you used the method of finding the limit of the gradients of a series of chords. Although this was more formal than the method of drawing a tangent and measuring its gradient, it was still somewhat experimental. The result was a sensible conclusion rather than a proven fact. This section formalises and extends this method.

Suppose you want to find the gradient at the point $P(2,4)$ on the curve $y = x^2$. Take a second point Q close to the point P on the curve. Let the x coordinate of Q be $2 + h$. Then the y coordinate will be $(2 + h)^2$.

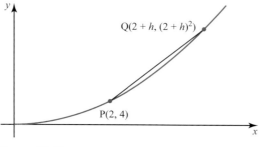

Figure 10.47

The gradient of the chord joining P and Q is

$$\frac{y_2 - y_1}{x_2 - x_1} = \frac{(2+h)^2 - 4}{2 + h - 2}$$

$$= \frac{4 + h^2 + 4h - 4}{h} = h + 4.$$

Because you want Q to move closer and closer to P, you want h to become smaller and smaller. This is described as letting h tend towards zero, and the value that you find is called the limit as h tends to zero.

$$\underset{h \to 0}{\text{Lim}}(h + 4) = 4$$

and so the gradient of $y = x^2$ at the point P(2, 4) is 4.

Example 10.22

Find from first principles the gradient of the curve $y = 2x^2 - 3x$ at the point where $x = 1$.

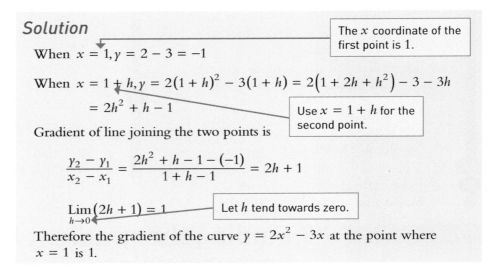

Solution

> The x coordinate of the first point is 1.

When $x = 1, y = 2 - 3 = -1$

When $x = 1 + h, y = 2(1+h)^2 - 3(1+h) = 2(1 + 2h + h^2) - 3 - 3h$

$$= 2h^2 + h - 1$$

> Use $x = 1 + h$ for the second point.

Gradient of line joining the two points is

$$\frac{y_2 - y_1}{x_2 - x_1} = \frac{2h^2 + h - 1 - (-1)}{1 + h - 1} = 2h + 1$$

$$\underset{h \to 0}{\text{Lim}}(2h + 1) = 1$$

> Let h tend towards zero.

Therefore the gradient of the curve $y = 2x^2 - 3x$ at the point where $x = 1$ is 1.

To find the gradient at any point on any curve, use the general curve $y = f(x)$ and the general point $P(x, f(x))$. The x coordinate of the second point Q is $x + h$, and so the y coordinate is $f(x + h)$ (see Figure 10.48).

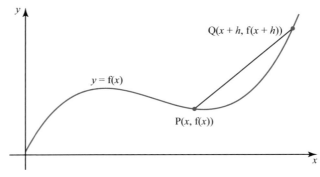

Figure 10.48

Then the gradient of the chord is $\dfrac{y_2 - y_1}{x_2 - x_1} = \dfrac{f(x + h) - f(x)}{x + h - x} = \dfrac{f(x + h) - f(x)}{h}.$

Moving Q closer and closer to P (letting h tend towards zero), the gradient function of the curve is the limit as h tends to zero of $\frac{f(x+h)-f(x)}{h}$.

So the derivative is given by the formula:

$$f'(x) = \lim_{h\to 0}\left(\frac{f(x+h)-f(x)}{h}\right)$$

Example 10.23

Show from first principles that the derivative of x^3 is $3x^2$.

Solution

$$f(x) = x^3$$

$$f'(x) = \lim_{h\to 0}\left(\frac{f(x+h)-f(x)}{h}\right) \quad \leftarrow \boxed{\text{Standard formula}}$$

$$= \lim_{h\to 0}\left(\frac{(x+h)^3 - x^3}{h}\right) \quad \leftarrow \boxed{\begin{array}{l}\text{In this case}\\ f(x)=x^3, f(x+h)=(x+h)^3\end{array}}$$

$$= \lim_{h\to 0}\left(\frac{x^3 + 3x^2h + 3xh^2 + h^3 - x^3}{h}\right) \quad \leftarrow \boxed{\text{Expand } (x+h)^3}$$

$$= \lim_{h\to 0}\left(3x^2 + 3xh + h^2\right) \quad \leftarrow \boxed{\text{Simplify}}$$

$$= 3x^2 \quad \leftarrow \boxed{\begin{array}{l}\text{Now } h \text{ is not in the denominator, you can}\\ \text{evaluate at } h = 0.\end{array}}$$

The derivative of x^3 is $3x^2$.

Example 10.24

Find, from first principles, the derivative of $x^3 - 6x + 1$.

Solution

$$f(x) = x^3 - 6x + 1$$

$$f'(x) = \lim_{h\to 0}\left(\frac{f(x+h)-f(x)}{h}\right) \quad \leftarrow \boxed{\text{Standard formula}}$$

$$= \lim_{h\to 0}\left(\frac{(x+h)^3 - 6(x+h) + 1 - \left(x^3 - 6x + 1\right)}{h}\right)$$

$$= \lim_{h\to 0}\left(\frac{x^3 + 3x^2h + 3xh^2 + h^3 - 6x - 6h + 1 - x^3 + 6x - 1}{h}\right)$$

$$= \lim_{h\to 0}\left(\frac{3x^2h + 3xh^2 + h^3 - 6h}{h}\right)$$

$$= \lim_{h\to 0}\left(3x^2 + 3xh + h^2 - 6\right)$$

$$= 3x^2 - 6$$

ACTIVITY 10.6

(i) Find, from first principles, the gradient of the line $y = 4x$. Does your result agree with what you know about the gradient of this line?

(ii) Now generalise this result to find from first principles the gradient of the general straight line $y = mx + c$. Does your result agree with what you know about the gradient of straight lines?

Exercise 10.9

① For each expression, work out the limit as $h \to 0$.

(i) $5 + h$

(ii) $\dfrac{3h + h^2}{h}$

(iii) $\dfrac{6h - h^3}{h}$

(iv) $\dfrac{2h^2 - 3h^3}{h}$

② Given that $f(x) = x^2 + 3x - 1$, find $f(2 + h)$.

A $h^2 + 3h + 1$ B $h^2 + 7h + 9$

C $h^2 + 3h + 9$ D $h^2 + 7h + 6$

③ $P(3, 9)$ and $Q\left(3 + h, (3 + h)^2\right)$ are two points on the curve $y = x^2$.

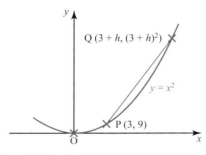

Figure 10.49

Find an expression in terms of h for the gradient of the chord joining P and Q. Simplify your answer.

④ A and B are two points on the curve $y = x^3$. At A, $x = 1$ and at B, $x = 1 + h$.

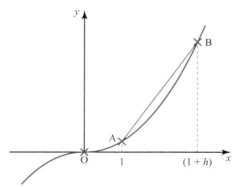

Figure 10.50

(i) What is the y coordinate of point A?

(ii) Write down an expression in terms of h for the y coordinate at point B.

(iii) Find an expression in terms of h for the gradient of the chord joining A and B, simplifying your answer.

⑤ P and Q are two points on the curve $y = 2x^2 - 4$. At P, $x = -1$ and at Q, $x = -1 + h$.

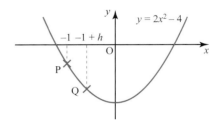

Figure 10.51

(i) What is the y coordinate of point P?

(ii) Write down an expression in terms of h for the y coordinate at point Q.

(iii) Show that the gradient of the chord PQ is $2h - 4$.

(iv) Write down the limit of $2h - 4$ as $h \to 0$.

(v) Check your answer by using differentiation to find the gradient of the curve $y = 2x^2 - 4$ at the point where $x = -1$.

⑥ P and Q are two points on the curve $y = x^2 + 3x$. At P, $x = 3$ and at Q, $x = 3 + h$.

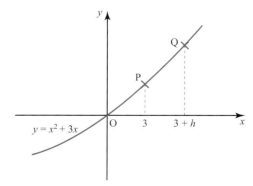

Figure 10.52

(i) What is the y coordinate of P?

(ii) Write down an expression in terms of h for the y coordinate at point Q.

(iii) Find an expression for the gradient of the chord joining P and Q, simplifying your answer.

(iv) Using your answer to (iii), and taking the limit as $h \to 0$, find the gradient of the curve $y = x^2 + 3x$ at the point where $x = 3$.

⑦ The points (x_1, y_1) and (x_2, y_2), where $x_2 = x_1 + h$, lie on the curve $y = x^3$.

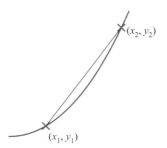

Figure 10.53

(i) Write down expressions for y_1 and y_2 in terms of x_1 and h.

(ii) Show that the gradient of the chord joining (x_1, y_1) and (x_2, y_2) is given by $3x^2 + 3xh + h^2$.

(iii) Write down the limit of the gradient of this chord as $h \to 0$.

⑧ The points (x_1, y_1) and (x_2, y_2), where $x_2 = x_1 + h$, lie on the curve $y = 2x^3 + 1$.

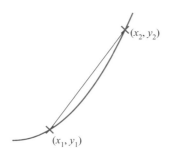

Figure 10.54

(i) Write down expressions for y_1 and y_2 in terms of x_1 and h.

(ii) Show that the gradient of the chord joining (x_1, y_1) and (x_2, y_2) is given by $6x^2 + 6xh + 2h^2$.

(iii) Write down the limit of the gradient of this chord as $h \to 0$.

⑨ The points (x_1, y_1) and (x_2, y_2), where $x_2 = x_1 + h$, lie on the curve $y = x^2 + 5x$.

(i) Write down an expression for y_1 in terms of x_1 and h.

(ii) Show that
$y_2 = x_1^2 + x_1(2h + 5) + h^2 + 5h$.

(iii) Hence find an expression for the gradient of the chord joining (x_1, y_1) and (x_2, y_2), simplifying your answer as far as possible.

(iv) Using your answer to (ii), and taking the limit as $h \to 0$, find the derivative of $x^2 + 5x$.

⑩ The points (x_1, y_1) and (x_2, y_2), where $x_2 = x_1 + h$, lie on the curve $y = x^2 - x - 6$.

(i) Write down expressions for y_1 and y_2 in terms of x_1 and h.

(ii) Hence find an expression for the gradient of the chord joining (x_1, y_1) and (x_2, y_2), simplifying your answer as far as possible.

(iii) Using your answer to (ii), and taking the limit as $h \to 0$, find the derivative of $x^2 - x - 6$.

⑪ Use differentiation from first principles to prove that the gradient function of the curve $y = x^3 + x - 3$ is $y = 3x^2 + 1$.

⑫ Use differentiation from first principles to prove that the derivative of $2x^2 - x$ is $4x - 1$.

⑬ By using differentiation from first principles, and the binomial expansion of $(x + h)^4$, show that the derivative of $f(x) = x^4$ is $f'(x) = 4x^3$.

Proofs

Tom is making an enclosure for his bull. He has a long stone wall with a right angle at one corner, and 60 m of fencing that he can place how he wants.

He considers two possible shapes for the enclosure, rectangular and triangular, as shown in Figure 10.55.

Which will be the better shape for Tom to use?

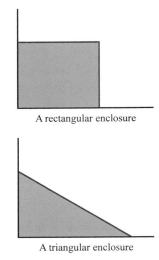

A rectangular enclosure

A triangular enclosure

Figure 10.55

The answer to this question is not obvious so you need to follow the problem solving cycle.

1 **Problem specification and analysis**

   Although the problem has been described in outline, it has not actually been fully specified. You have not been told how to decide whether one enclosure is 'better' than another.

   An obvious way is to say that the area should be as large as possible; that is, a larger area is better.

   This makes it clear that you want to find the largest area that can be enclosed for each shape, using two stone walls and 60 m of fencing.

   Before you start you will need to decide what the important dimensions are in each case and to represent them by letters.

2 **Information collection**

   For each shape (the rectangle and the triangle), try some different dimensions and work out the area. Do this systematically.

   You should start to see patterns; when that happens you will have enough information to be ready to go on to the next stage.

3 **Processing and representation**

   At this stage you may find it helpful in each case to draw a graph to show you how the area of the enclosure varies with one of the dimensions.

   Then you can find the maximum values of the areas for the two shapes, and compare them.

4 **Interpretation**

   You now have the information to advise Tom.

In this case, you should also be prepared to **prove** your results:

(A) Use calculus to prove the result you found for the rectangular field.

(B) Find a way to prove your result for the triangular field. It can be done using calculus but there are other ways you can prove it, too.

Finally, comment on which shape will suit the bull best.

LEARNING OUTCOMES

When you have completed this chapter you should be able to:

➤ understand and use the derivative of f(x) as the gradient of the tangent to the graph of $y = f(x)$ at a general point (x, y)

➤ understand and use the gradient of the tangent as a limit

➤ interpret the derivative of f(x) as a rate of change

➤ sketch the gradient function for a given curve

➤ understand and use second derivatives

➤ understand and use differentiation from first principles for small positive integer powers of x

➤ understand and use the second derivative as the rate of change of the gradient

➤ differentiate x^n, for rational values of n, and related constant multiples, sums and differences

➤ apply differentiation to find:
  ○ gradients
  ○ tangents and normals
  ○ maxima and minima and stationary points

➤ identify where functions are increasing or decreasing.

KEY POINTS

1 The rate of change of y with respect to x can be written as $\dfrac{dy}{dx}$. This is also called the derived function or derivative. In function notation, the derivative of $f(x)$ is written as $f'(x)$.

2 $y = c \Rightarrow \dfrac{dy}{dx} = 0$ $\qquad$ $y = kx \Rightarrow \dfrac{dy}{dx} = k$

$y = kx^n \Rightarrow \dfrac{dy}{dx} = nkx^{n-1}$ $\qquad$ $y = f(x) + g(x) \Rightarrow \dfrac{dy}{dx} = f'(x) + g'(x)$

3 Tangent and normal at (x_1, y_1):

gradient of tangent $m_1 = f'(x_1)$, and equation of tangent is
$y - y_1 = m_1(x - x_1)$

gradient of normal $m_2 = -\dfrac{1}{m_1}$, and equation of normal is $y - y_1 = m_2(x - x_1)$

4 $f(x)$ is increasing for $a < x < b$ if $f'(x) > 0$ for $a < x < b$

$f(x)$ is decreasing for $a < x < b$ if $f'(x) < 0$ for $a < x < b$

5 At a stationary point $\dfrac{dy}{dx} = 0$.

One method for determining the nature of a stationary point is to look at the sign of the gradient on either side of it.

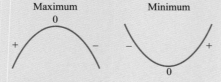

Figure 10.56

6 The second derivative is written $\dfrac{d^2y}{dx^2}$ or $f''(x)$, and is the rate of change of $\dfrac{dy}{dx}$ with respect to x.

A second method for determining the nature of the stationary point is to consider the sign of $\dfrac{d^2y}{dx^2}$.

If $\dfrac{d^2y}{dx^2} > 0$ the point is a minimum, and if $\dfrac{d^2y}{dx^2} < 0$ the point is a maximum.

If $\dfrac{d^2y}{dx^2} = 0$, it is not possible to use this method to determine the nature of the stationary point.

7 $f'(x) = \underset{h \to 0}{\text{Lim}} \left(\dfrac{f(x + h) - f(x)}{h} \right)$.

FUTURE USES

- In Chapter 21 you will see how differentiation is used in the study of motion.
- You will learn to differentiate a wider variety of functions in the A-level mathematics textbook.
- Gradient functions form the basis of differential equations, which will be covered in the A-level mathematics textbook.

11 Integration

Many small make a great.
Geoffrey Chaucer
(c.1343–1400)

This family of curves are all vertical translations of each other. Look at this family of curves and discuss the following points:

➜ How can you tell that the curves are parallel to each other?

➜ Give the equations of three other curves that are parallel to these, in this family of curves.

➜ What is the general equation of this family of curves?

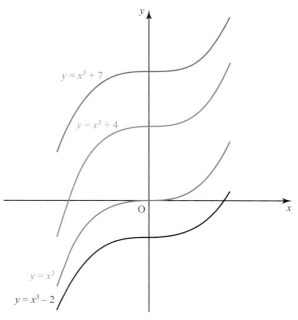

Figure 11.1

1 Integration as the reverse of differentiation

▼ **Notation**

The integral sign, $\int$

$\dfrac{dy}{dx} = 2x \Rightarrow y = x^2 + c$

can be expressed using the integral sign, as

$\int 2x\, dx = x^2 + c$

The dx denotes that x is the variable.

The solution can be read as 'the integral of $2x$ with respect to x is $x^2 + c$'.

This notation is explained more fully later in this chapter.

ACTIVITY 11.1

(i) A curve has gradient function, $\dfrac{dy}{dx} = 2x$. Give three possibilities for the equation of the curve.

(ii) Sketch your three curves, and explain why they are all parallel.

(iii) Write down the general equation of the family of parallel curves for which $\dfrac{dy}{dx} = 2x$.

(iv) Write down general equations for the family of curves for which:

(a) $\dfrac{dy}{dx} = 3x^2$ (b) $\dfrac{dy}{dx} = x^2$ (c) $\dfrac{dy}{dx} = 4x^3$ (d) $\dfrac{dy}{dx} = x^3$

(v) What about the family of curves for which $\dfrac{dy}{dx} = x^n$? Does your equation work for every value of n? Explain your answer.

The equation $\dfrac{dy}{dx} = 2x$ is an example of a **differential equation**.

Its solution is $y = x^2 + c$.

The procedure for finding the right hand side is called **integration** and c is called the **constant of integration** or the **arbitrary constant**.

Different values of c give a family of parallel curves.

Discussion point

You know that $y = x \Rightarrow \dfrac{dy}{dx} = 1$, and therefore $\int 1\, dx = x + c$.

Explain how this follows the rule for integrating x^n given above.

Rule for integrating x^n

You will have seen from Activity 11.1 that integration is the reverse of differentiation.

$$\int x^n\, dx = \frac{x^{n+1}}{n+1} + c, \quad n \neq -1$$

To integrate a power of x, add one to the power and divide by the new power.

Example 11.1

Find y for the following gradient functions.

Use differentiation to check your answers.

(i) $\dfrac{dy}{dx} = 6x^2$ (ii) $\dfrac{dy}{dx} = 4x$ (iii) $\dfrac{dy}{dx} = -3$ (iv) $\dfrac{dy}{dx} = 6x^2 + 4x - 3$

Solution

(i) $\dfrac{dy}{dx} = 6x^2$

     $\Rightarrow y = \dfrac{6x^3}{3} + c$

     $= 2x^3 + c$

   Check: $y = 2x^3 + c \Rightarrow \dfrac{dy}{dx} = 6x^2$ ← correct

Discussion point

Amelia says 'The answer to part (iv) of Example 11.1 should be $2x^3 + 2x^2 - 3x + 3c$, from adding parts (i), (ii) and (iii).' Why is Amelia wrong?

(ii) $\quad \dfrac{dy}{dx} = 4x$

$$\Rightarrow y = \frac{4x^2}{2} + c$$

$$= 2x^2 + c$$

Check: $y = 2x^2 + c \Rightarrow \dfrac{dy}{dx} = 4x$ ⟵ correct

(iii) $\quad \dfrac{dy}{dx} = -3$

$$\Rightarrow y = -3x + c$$

Check: $y = -3x + c \Rightarrow \dfrac{dy}{dx} = -3$ ⟵ correct

(iv) $\quad \dfrac{dy}{dx} = 6x^2 + 4x - 3$

$$\Rightarrow y = \frac{6x^3}{3} + \frac{4x^2}{2} - 3x + c$$

$$= 2x^3 + 2x^2 - 3x + c$$

Check: $y = 2x^3 + 2x^2 - 3x + c \Rightarrow \dfrac{dy}{dx} = 6x^2 + 4x - 3$ ⟵ correct

In this chapter, the functions you integrate are all polynomials. Sometimes, as in the next two examples, you are not given the function in this form and so need to do some preliminary work before you can do the integration.

Example 11.2

Find the following integrals.

(i) $\quad \displaystyle\int \big(x^3 - 2\big)^2 \, dx$

(ii) $\quad \displaystyle\int \dfrac{7x^7 - 4x^3 + 8x}{2x} \, dx$

Solution

(i) $\quad \displaystyle\int \big(x^3 - 2\big)^2 dx = \int \big(x^3 - 2\big)\big(x^3 - 2\big) \, dx$ ⟵ First expand the brackets.

$$= \int \big(x^6 - 4x^3 + 4\big) \, dx$$ ⟵ Now integrate.

$$= \frac{x^7}{7} - x^4 + 4x + c$$

(ii) $\quad \displaystyle\int \dfrac{7x^7 - 4x^3 + 8x}{2x} \, dx = \int \left(\frac{7x^7}{2x} - \frac{4x^3}{2x} + \frac{8x}{2x} \right) dx$ ⟵ First put the expression in polynomial form.

$$= \int \left(\frac{7}{2}x^6 - 2x^2 + 4 \right) dx$$

$$= \frac{1}{2}x^7 - \frac{2}{3}x^3 + 4x + c$$ ⟵ Now integrate.

Sometimes you are given more information about a problem and this enables you to find the numerical value of the arbitrary constant.

If you have been given the gradient function of a curve, knowing the coordinates of one point on the curve allows you to find the equation of the curve.

ACTIVITY 11.2

The gradient function of a curve is given by $\dfrac{dy}{dx} = 3x^2$.

(i) Find the equation of the family of solution curves, and sketch four possible solution curves on the same axes.

(ii) Write down the equations of the curves with the following y intercepts.

(a) $y = 3$ (b) $y = 12$ (c) $y = -7$

(iii) Work out the equations of the curves that pass through the following points:

(a) $(1, 3)$ (b) $(-1, 1)$ (c) $(2, 10)$

Example 11.3

Given that $\dfrac{dy}{dx} = 6x^2 - 2x$

(i) find the equation of the family of solution curves

(ii) find the equation of the particular curve that passes through the point $(-1, 1)$.

Solution

(i) $y = \displaystyle\int (6x^2 - 2x)\, dx$ ← First integrate.

$= \dfrac{6x^3}{3} - \dfrac{2x^2}{2} + c$

$= 2x^3 - x^2 + c$

The equation of family of solution curves is $y = 2x^3 - x^2 + c$.

(ii) When $x = -1$, $y = 1$ ← Substitute values into general equation.

$1 = 2(-1)^3 - (-1)^2 + c$

$1 = -3 + c$ ← Simplify and rearrange to find the value of c.

$c = 4$

The curve is $y = 2x^3 - x^2 + 4$.

Example 11.4

The gradient function of a curve is $\dfrac{dy}{dx} = 4x - 12$. The minimum y value is 16.

(i) Use the gradient function to find the value of x at the minimum point.

(ii) Find the equation of the curve.

Solution

(i) At the minimum, the gradient of the curve must be zero.

$\dfrac{dy}{dx} = 4x - 12 = 0$

$\Rightarrow x = 3$

(ii) $\dfrac{dy}{dx} = 4x - 12 \Rightarrow y = \displaystyle\int (4x - 12)\, dx$

$= 2x^2 - 12x + c$ ← First find the general curve.

When $x = 3$, $y = 16$ → Now work out the value of c.

$$16 = 2(3)^2 - 12(3) + c = -18 + c \Rightarrow c = 34$$

The equation of the curve is $y = 2x^2 - 12x + 34$.

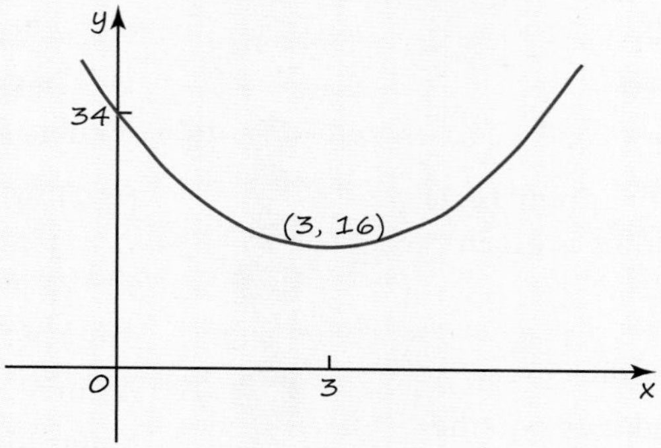

34

(3, 16)

O   3

Figure 11.2

Exercise 11.1

① A family of curves is described by $y = x^2 + 3x + c$.

What is the equation of the curve that passes through $(2, 5)$?

② Which expression is $\int x^3 \, dx$ equal to?

A $\dfrac{x^4}{4}$ B $x^4 + c$

C $\dfrac{x^4}{4} + c$ D $\dfrac{x^4}{3} + c$

③ Find the following integrals.

(i) $\displaystyle\int x^6 \, dx$ (ii) $\displaystyle\int 2x^7 \, dx$

(iii) $\displaystyle\int \left(x^6 + 2x^7\right) dx$ (iv) $\displaystyle\int 5x^6 \, dx$.

How do your answers to (iii) and (iv) relate to your answers to (i) and (ii)?

④ For both of the gradient functions below, find the equation of the family of solution curves, and in each case sketch three members of this family.

(i) $\dfrac{dy}{dx} = 8x^3$ (ii) $\dfrac{dy}{dx} = 2x - 1$

⑤ Given that $f'(x) = 3x - \frac{1}{2}x^2$, find a general expression for $f(x)$.

⑥ Given the gradient function $\dfrac{dy}{dx} = 5$.

(i) Find the equation of the family of solution curves.

(ii) Find the equation of the curve which passes through the point $(1, 8)$.

(iii) Sketch the graph of this particular curve.

⑦ The gradient function of a curve is $\dfrac{dy}{dx} = 4x$ and the curve passes through the point $(1, -5)$.

(i) Find the equation of the curve.

(ii) Find the value of y when $x = -1$.

(iii) Sketch the curve.

⑧ The curve C passes through the point $(2, 10)$ and its gradient at any point is given by $\dfrac{dy}{dx} = 6x^2$.

(i) Find the equation of the curve C.

(ii) Show that the point $(1, -4)$ lies on the curve.

(iii) Sketch the curve.

⑨ Given that $\dfrac{dy}{dx} = 12x^2 + 4x$

(i) find an expression for y in terms of x

(ii) find the equation of the curve with gradient function $\dfrac{dy}{dx}$ and which passes through $(1, 9)$

(iii) hence show that $(-1, 1)$ also lies on the curve.

⑩ Find the following integrals.

(i) $\int (x-1)(x+3)\, \mathrm{d}x$

(ii) $\int \left(\dfrac{x^2 + 7x - 2x^5}{x} \right) \mathrm{d}x$

⑪ Find the following integrals.

(i) $\int (2z-3)(3z-1)\, \mathrm{d}z$

(ii) $\int t\left(2t^3 - 4t + 1\right) \mathrm{d}t$

⑫ A curve passes through the point $(4,1)$ and its gradient at any point is given by $\dfrac{\mathrm{d}y}{\mathrm{d}x} = 2x - 6$.

(i) Find the equation of the curve.

(ii) Sketch the curve and state whether it passes under, over or through the point $(1, 4)$.

⑬ The gradient function of a curve is $3x^2 - 3$. The curve has two turning points. One is a maximum with a y value of 5 and the other is a minimum with a y value of 1.

(i) Find the value of x at each turning point.

(ii) Find the equation of the curve.

(iii) Sketch the curve.

⑭ A cubic function has a y intercept at $(0,2)$. When $x = 1$, $y = 3$, $\dfrac{\mathrm{d}y}{\mathrm{d}x} = 2$ and $\dfrac{\mathrm{d}^2 y}{\mathrm{d}x^2} = 0$. Find the equation of the original curve, and sketch the curve.

⑮ A curve $y = \mathrm{f}(x)$ passes through the points $(0, 1)$ and $(1, -8)$, and has a turning point at $x = -1$. Given that $\dfrac{\mathrm{d}^3 y}{\mathrm{d}x^3} = 24x + 24$, find the equation of the curve.

Discussion points

→ What do you get if you subtract the value of $x^2 + c$ at $x = 2$ from the value of $x^2 + c$ at $x = 5$?

→ Why is there no c in your answer?

→ Would this happen if you use any pair of values for x?

2 Finding areas

You know that $\int 2x\, \mathrm{d}x = x^2 + c$.

This is called an **indefinite integral**.

The answer

■ is a function of the variable (in this case x)

■ contains an arbitrary constant, c.

You will see later that a common application of integration is to find the area under a curve. When finding areas, and in other applications of integration, you often need to find the difference between the value of an integral at two points.

When you find the difference between the values of an integral at two points, it is called a **definite integral**. When you find a definite integral, the answer is a number and there is no arbitrary constant.

A definite integral is written like this: $\int_2^5 2x\, \mathrm{d}x$.

> The numbers 5 and 2 are called the **limits**.

When you work out a definite integral, you use square brackets to show the integrated function.

$$\int_2^5 2x\, \mathrm{d}x = \left[x^2 \right]_2^5$$

> There is no need to include the arbitrary constant, as this will cancel out.

$$= 5^2 - 2^2$$

> Substitute the upper limit $x = 5$.

$$= 25 - 4$$

> Substitute the lower limit $x = 2$.

$$= 21$$

Figure 11.3 shows the area, A, bounded by the curve $y = f(x)$, the x axis, and the lines $x = a$ and $x = b$.

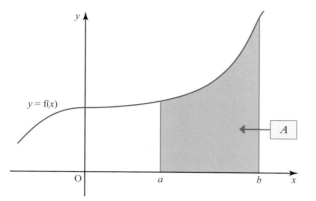

Figure 11.3

You can find an approximation for the area using rectangular strips. The narrower the rectangles are, the better this approximation will be.

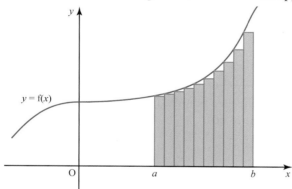

Figure 11.4

 **TECHNOLOGY**

You may be able to use graphing software to show these rectangles to estimate the area. Increase the number of rectangles for better approximations. Integration will give exact values.

Each of these rectangles has a height of y (which depends on the value of x) and a width of δx, where δx is a way of writing 'a small increase in x'. The area of a rectangle is denoted by δA, where $\delta A = y \times \delta x$, as in Figure 11.5.

The total area, A, is approximately the sum of the areas of the rectangles.

So $A \approx \sum \delta A$

Figure 11.5

To show that the rectangles cover the area from $x = a$ to $x = b$, this can be written as

$$\sum_{x=a}^{x=b} \delta A.$$

So $A = \displaystyle\sum_{x=a}^{x=b} \delta A \approx \sum_{x=a}^{x=b} y \, \delta x$

This approximation for the area becomes more accurate as the strips get thinner, i.e. as δx gets smaller. In the limiting case, as $\delta x \to 0$, the two sides of the equation are equal.

In this limiting case:

- the $\sum$ sign is written as $\displaystyle\int$

  > So the $\int$ sign is a form of the letter S.

- δx is written as $\mathrm{d}x$

  > δx is small, $\mathrm{d}x$ is infinitesimally small.

- so $A = \displaystyle\int_a^b y \, \mathrm{d}x.$

  > The limits can be written as $x = a$ and $x = b$, but it is usual to write them as just a and b.

Therefore to find the area under a curve you use definite integration.

The theorem that states this connection between areas and the reverse process of differentiation is called the **Fundamental Theorem of Calculus**.

Making the link with differentiation

$A(x)$ = area beneath the graph of f(x) between 0 and x.

Consider a rectangular strip, width h and height f(x).

We can write its area in two ways

Area $\approx h$ f(x)

Area also = $A(x + h) - A(x)$

So h f(x) $\approx A(x + h) - A(x)$

$$f(x) \approx \frac{A(x + h) - A(x)}{h}$$

The error is the area above the blue dotted line. As $h \to 0$ this tends to zero so that:

$$f(x) = \lim_{h \to 0} \frac{A(x + h) - A(x)}{h}$$

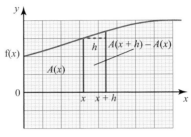

Figure 11.6

You will recognise this from Chapter 10, differentiating from first principles.

So $A'(x) = $ f(x)

And so $A(x) = \int$ f(x) dx

Example 11.5

Find the area bounded by the curve $y = x^2 + 2$, the lines $x = -2$ and $x = 1$, and the x axis.

Solution

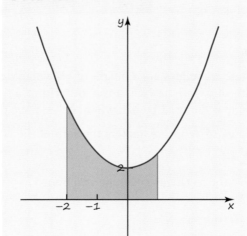

Figure 11.7

Area $= \int_{-2}^{1} \left(x^2 + 2\right) dx$ ⟵ Integrate y using the two x values as your limits.

$$= \left[\frac{x^3}{3} + 2x\right]_{-2}^{1}$$

$$= \left(\frac{1}{3} + 2\right) - \left(\frac{-8}{3} - 4\right)$$

$$= 9$$

So the area is 9 square units.

Exercise 11.2

 ① $\left[\dfrac{4x^3}{3} - 3x \right]_2^4$ has been worked out

incorrectly. Explain what was done wrong.

$$= \left(\frac{4 \times 4^3}{3} - 3 \times 4 \right) - \left(\frac{4 \times 2^3}{3} - 3 \times 2 \right)$$

$$= \frac{256}{3} - 12 - \frac{32}{3} - 6$$

$$= \frac{224}{3} - 18$$

$$= 56\frac{2}{3}$$

② Evaluate the following definite integrals.

(i) $\displaystyle\int_0^2 4x^3 \,dx$

(ii) $\displaystyle\int_1^3 6x^2 \,dx$

(iii) $\displaystyle\int_3^4 2x \,dx$

③ Evaluate the following definite integrals.

(i) $\displaystyle\int_1^2 x^2 \,dx$

(ii) $\displaystyle\int_2^3 x^2 \,dx$

(iii) $\displaystyle\int_1^3 x^2 \,dx$

Explain how your answer to (iii) relates to your answers to (i) and (ii).

④ Evaluate the following definite integrals.

(i) $\displaystyle\int_0^2 x^4 \,dx$

(ii) $\displaystyle\int_0^2 5x^4 \,dx$

Explain how your answer to (i) relates to your answer to (ii).

⑤ (i) Evaluate

(a) $\displaystyle\int_{-3}^{-2} x^2 \,dx$

(b) $\displaystyle\int_2^3 x^2 \,dx$

(ii) Sketch the curve $y = x^2$ and use it to explain your answers to (i).

⑥ Evaluate the following definite integrals.

(i) $\displaystyle\int_{-1}^2 \left(3x^2 + 2 \right) dx$

(ii) $\displaystyle\int_{-2}^0 \left(5x^4 - 4x \right) dx$

(iii) $\displaystyle\int_{-3}^{-1} (x - 2) \,dx$

⑦ (i) Sketch the curve $y = x^2 + 1$ for values of x between -3 and 3.

(ii) Evaluate the integral $\displaystyle\int_1^2 \left(x^2 + 1 \right) dx$.

(iii) Indicate on your sketch the area corresponding to your answer to (ii).

⑧ (i) Sketch the curve $y = -x^3$ for $-2 \leqslant x \leqslant 2$.

(ii) Shade the area bounded by the curve, the x axis, and the lines $x = -1$ and $x = 0$.

(iii) Calculate the area that you have shaded.

⑨ (i) Shade, on a suitable sketch, the region with an area given by $\displaystyle\int_{-1}^2 (9 - x^2) \,dx$.

(ii) Find the area of the shaded region.

⑩ (i) Sketch the curve $y = 1 - x^3$, and shade the region bounded by the curve and the x and y axes.

(ii) Find the area of the shaded region.

⑪ (i) Factorise $x^3 - 6x^2 + 11x - 6$ and hence sketch the curve with equation $y = x^3 - 6x^2 + 11x - 6$ for $0 \leqslant x \leqslant 2$.

(ii) Shade the regions with areas given by $\displaystyle\int_1^2 \left(x^3 - 6x^2 + 11x - 6 \right) dx$ and $\displaystyle\int_3^4 \left(x^3 - 6x^2 + 11x - 6 \right) dx$.

(iii) Find the values of these two areas.

(iv) Find the value of $\displaystyle\int_1^{1.5} \left(x^3 - 6x + 11x - 6 \right) dx$. What can you deduce about the position of the maximum point between $x = 1$ and $x = 2$?

PS ⑫ Sketch on the same diagram the curve $y = x^3$ and the lines $y = 1$ and $y = 8$. Find the area enclosed by the curve $y = x^3$, the lines $y = 1$ and $y = 8$, and the y axis.

⑬ (i) Evaluate $\displaystyle\int_1^4 6x \,dx$ and $\displaystyle\int_3^{12} 2x \,dx$.

(ii) Does $\displaystyle\int_a^b k\mathrm{f}(x)\,dx = \int_{ka}^{kb} \mathrm{f}(x)\,dx$? Explain your answer.

3 Areas below the _x_ axis

> **ACTIVITY 11.3**
>
> (i) Calculate $\int_{-1}^{1} x^3 \, dx$.
>
> (ii) Sketch the curve $y = x^3$, shading the region bounded by the curve, the _x_ axis, and the lines $x = -1$ and $x = 1$.
>
> (iii) Use your sketch to explain your answer to (i).
>
> (iv) What is the area of the shaded region?

When a curve goes below the _x_ axis, the _y_ value becomes negative. So the value of $\delta A = y \delta x$ is negative.

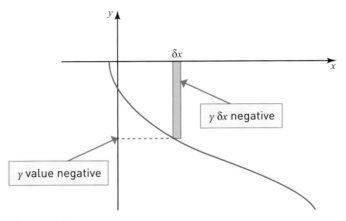

Discussion point

→ When a region is below the _x_ axis, the integral comes out as negative. Can an area be negative?

_y_ δ_x_ negative

_y_ value negative

Figure 11.7

Example 11.6

Find the area bounded by the curve $y = x^2 - 8x + 12$, the _x_ axis, and the lines $x = 3$ and $x = 4$.

Solution

The region in question is shown in Figure 11.8.

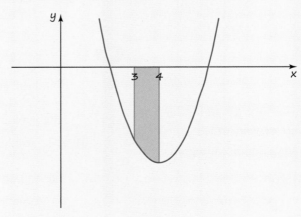

Figure 11.8

Calculate the value of the integral in the normal way.

$$\int_3^4 \left(x^2 - 8x + 12 \right) dx = \left[\frac{x^3}{3} - 4x^2 + 12x \right]_3^4$$

$$= \left(\frac{4^3}{3} - 4(4)^2 + 12(4) \right) - \left(\frac{3^3}{3} - 4(3)^2 + 12(3) \right)$$

$$= \frac{16}{3} - 9$$

$$= -\frac{11}{3}$$

Because the area is below the x axis, the integral comes out as negative.

Therefore the area is $\frac{11}{3}$ square units.

When you have areas that are partly above and partly below the x axis, you must calculate the areas of the regions separately and add them together. It is very important to sketch the curve, to see whether it goes below the x axis.

Example 11.7

Find the area of the region bounded by the curve $y = x^2 + 3x$, the x axis, and the lines $x = -1$ and $x = 2$.

Solution

The region is shown in Figure 11.9.

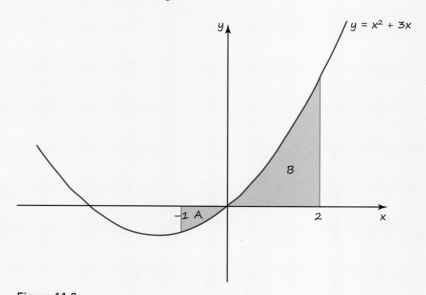

Figure 11.9

You can see that the region is below the x axis for $-1 < x < 0$ (A), and above the x axis for $0 < x < 2$ (B). So we must calculate these areas separately.

$$\int_{-1}^0 \left(x^2 + 3x \right) dx = \left[\frac{x^3}{3} + \frac{3x^2}{2} \right]_{-1}^0$$

Discussion point

➜ What is the value of $\int_{-1}^{2}\left(x^2 + 3x\right)dx$?

➜ Why is this not the same as your answer for the total area?

$$= (0 + 0) - \left(-\frac{1}{3} + \frac{3}{2}\right)$$

$$= -\frac{7}{6}$$ ← This is negative because region A is below the *x* axis.

Area of region A is $\frac{7}{6}$.

$$\int_{0}^{2}\left(x^2 + 3x\right)dx = \left[\frac{x^3}{3} + \frac{3x^2}{2}\right]_{0}^{2}$$

$$= \left(\frac{8}{3} + \frac{3(2)^2}{2}\right) - (0 + 0)$$

$$= \frac{26}{3}$$ ← This is positive because region B is above the *x* axis.

Area of region B is $\frac{26}{3}$.

The total area is $\frac{7}{6} + \frac{26}{3} = \frac{59}{6}$.

Exercise 11.3

① (i) Sketch the curve $y = -x^2$.

   (ii) On your sketch shade the region bounded by the curve, the *x* axis, and the lines $x = 1$ and $x = 2$.

   (iii) Find the area of the shaded region.

② (i) Sketch the curve $y = 9 - x^2$ for $-5 \leqslant x \leqslant 5$.

   (ii) On your sketch shade the region bounded by the curve, the *x* axis, and the lines $x = 3$ and $x = 4$.

   (iii) Find the area of the shaded region.

③ (i) Sketch the curve $y = x^2 - 4$ for $-3 \leqslant x \leqslant 3$.

   (ii) On your sketch shade the region bounded by the curve and the *x* axis.

   (iii) Find the area of the shaded region.

④ (i) Sketch the curve $y = x^3$ for $-3 \leqslant x \leqslant 3$.

   (ii) On your sketch shade the region bounded by the curve, the *x* axis, and the lines $x = -2$ and $x = -1$.

   (iii) Find the area of the shaded region.

⑤ Sketch the curve $y = x^2 - 1$ for $-3 \leqslant x \leqslant 3$, and find the area bounded by the curve, the *x* axis, and the lines $x = 0$ and $x = 2$.

⑥ Sketch the curve $y = x^2 - x - 2$, and find the area bounded by the curve, the *x* axis, and the lines $x = -2$ and $x = 3$.

⑦ Sketch the curve $y = x^2 - 3x + 2$, and shade the regions bounded by the curve, the *x* and *y* axes, and the line $x = 3$. Find the total area of the shaded regions.

⑧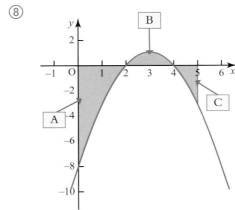

Figure 11.10

240

Figure 11.10 shows the curve $y = f(x)$ where $f(x) = -x^2 + 6x - 8$.

(i) Evaluate $\int_0^5 f(x)\,dx$.

(ii) Find the sum of the areas of the regions A, B and C.

(iii) Explain the difference in your answers to (i) and (ii).

⑨ Figure 11.11 shows part of the curve with equation $y = 5x^4 - x^5$.

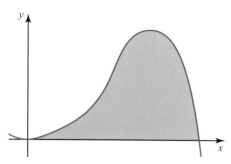

Figure 11.11

(i) Find $\dfrac{dy}{dx}$ and hence calculate the coordinates of the stationary points.

(ii) Calculate the area of the shaded region.

(iii) Calculate $\int_0^6 \left(5x^4 - x^5\right) dx$.
Comment on your result.

⑩ Factorise $x^3 + x^2 - 2x$, and hence sketch the curve $y = x^3 + x^2 - 2x$. Find the area bounded by the curve, the x axis, and the lines $x = -3$ and $x = 2$.

4 Further integration

Prior knowledge

You need to know the meaning of negative, fractional and zero indices. This is covered in Chapter 2.

So far in this chapter, you have only integrated expressions where the powers have been positive integers.

However, the formula $\int kx^n dx = \dfrac{kx^{n+1}}{n+1} + c,\ n \neq -1$ can still be used when n is a negative integer or a fraction.

Example 11.8

Find y for the following gradient functions.

(i) $\dfrac{dy}{dx} = \dfrac{5}{x^2}$ (ii) $\dfrac{dy}{dx} = \sqrt{x}$

Solution

First write $\dfrac{dy}{dx}$ as a power of x.

(i) $\dfrac{dy}{dx} = 5x^{-2}$

Add one to the power and divide by the new power.

$\Rightarrow y = \dfrac{5x^{-2+1}}{-2+1} + c$

Simplify your answer.

$= \dfrac{5x^{-1}}{-1} + c$

$= -\dfrac{5}{x} + c$

(ii) $\dfrac{dy}{dx} = x^{\frac{1}{2}}$

$\Rightarrow y = \dfrac{x^{\frac{3}{2}}}{\frac{3}{2}} + c$ $\dfrac{1}{2} + 1 = \dfrac{3}{2}$

$= \dfrac{2x^{\frac{3}{2}}}{3} + c$

Note

Dividing by a fraction is equivalent to multiplying by its reciprocal so $\div \frac{3}{2} \rightarrow \times \frac{2}{3}$.

Example 11.9

Find the following integrals.

(i) $\displaystyle\int\left(\frac{\sqrt{y}-6y^2+2y}{y}\right)dy$ 　　　(ii) $\displaystyle\int\left(\frac{3}{\sqrt{t}}-\frac{\sqrt{t}}{3}\right)dt$

Note

$\div\frac{1}{2}\to\times 2$

Solution

(i) $\displaystyle\int\left(\frac{\sqrt{y}-6y^2+2y}{y}\right)dy=\int\left(\frac{y^{\frac{1}{2}}}{y}-\frac{6y^2}{y}+\frac{2y}{y}\right)dy$

> Write each term as a power of y.

$$=\int\left(y^{-\frac{1}{2}}-6y+2\right)dy$$

$$=\frac{y^{\frac{1}{2}}}{\frac{1}{2}}-\frac{6y^2}{2}+2y+c$$

$$=2\sqrt{y}-3y^2+2y+c$$

> Simplify.

(ii) $\displaystyle\int\left(\frac{3}{\sqrt{t}}-\frac{\sqrt{t}}{3}\right)dt=\int\left(3t^{-\frac{1}{2}}-\frac{1}{3}t^{\frac{1}{2}}\right)dt$

> Write each term as a power of t.

$$=\frac{3t^{\frac{1}{2}}}{\frac{1}{2}}-\frac{1}{3}\times\frac{t^{\frac{3}{2}}}{\frac{3}{2}}+c$$

$$=6\sqrt{t}-\frac{2}{9}t^{\frac{3}{2}}+c$$

> Simplify.

Example 11.10

Sketch three members of the family of curves given by $y=\displaystyle\int\frac{1}{x^2}\,dx,\ x>0$.

Solution

$y=\displaystyle\int x^{-2}\,dx$

$=\dfrac{x^{-1}}{-1}+c$

$=-\dfrac{1}{x}+c$

The family of curves is the set of curves with equation

$y=-\dfrac{1}{x}+c,\ x>0$

for different values of the arbitrary constant, c. Three members of this family are shown in Figure 11.12.

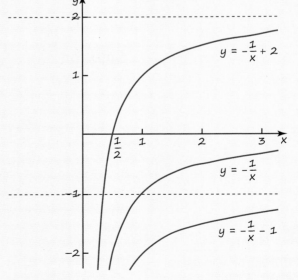

Figure 11.12

Example 11.11

Given that $\dfrac{dy}{dx} = \sqrt{x} + \dfrac{1}{x^2}$

(i) find the equation of the family of solution curves

(ii) find the equation of the particular curve that passes through $(1, 5)$.

Solution

(i) $\dfrac{dy}{dx} = x^{\frac{1}{2}} + x^{-2}$ ⟵ [First write the terms as powers of x.]

$y = \displaystyle\int x^{\frac{1}{2}} + x^{-2}\,dx$ ⟵ [Now integrate.]

$= \dfrac{x^{\frac{3}{2}}}{\frac{3}{2}} + \dfrac{x^{-1}}{-1} + c$

$= \dfrac{2x^{\frac{3}{2}}}{3} - \dfrac{1}{x} + c$

(ii) When $x = 1, y = 5$

$5 = \dfrac{2(1)^{\frac{2}{3}}}{3} - \dfrac{1}{1} + c$ ⟵ [Substitute values into general equation.]

$5 = -\dfrac{1}{3} + c$ ⟵ [Simplify and rearrange to find the value of c.]

$c = 5\tfrac{1}{3}$

Therefore the equation of the curve is $y = \dfrac{2x^{\frac{3}{2}}}{3} - \dfrac{1}{x} + 5\tfrac{1}{3}$.

Exercise 11.4

① Match the equivalent expressions.

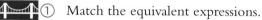

$30x \qquad \dfrac{15x}{2} \qquad \dfrac{6x}{\frac{1}{5}} \qquad \dfrac{10x}{3} \qquad \dfrac{3x}{\frac{5}{2}}$

$\dfrac{5x}{\frac{2}{3}} \qquad \dfrac{6x}{5} \qquad \dfrac{5x}{\frac{3}{2}}$

② The expression $\dfrac{1}{x\sqrt{x}}$ is equal to which

of the following expressions?

A $x^{-\frac{1}{2}}$ B x^{-1} C $x^{-\frac{3}{2}}$ D $x^{-\frac{2}{3}}$

③ Given that $\dfrac{dy}{dx} = \dfrac{1}{2\sqrt{x}},\ x > 0,$

find the general equation of the family of
of solution curves and sketch three
members of this family.

④ Find a general expression for $f(x)$ in
each of these cases.

(i) $f'(x) = x^{-3}$ (ii) $f'(x) = 2x^{-5}$

(iii) $f'(x) = \dfrac{1}{x^4}$ (iv) $f'(x) = \dfrac{4}{x^7}$

⑤ Find the following integrals.

(i) $\displaystyle\int x^{\frac{1}{3}}\,dx$ (ii) $\displaystyle\int 2x^{-\frac{1}{4}}\,dx$

(iii) $\displaystyle\int \sqrt[4]{x}\,dx$ (iv) $\displaystyle\int \dfrac{2}{x\sqrt{x}}\,dx$

⑥ Evaluate the following definite integrals.

(i) $\displaystyle\int_1^9 \dfrac{1}{\sqrt{x}}\,dx$

(ii) $\displaystyle\int_0^4 x\sqrt{x}\,dx$

(iii) $\displaystyle\int_{-1}^2 \dfrac{1}{x^2}\,dx$

⑦ Figure 11.13 shows part of the curves $y = x^2$ and $y = \sqrt{x}$.

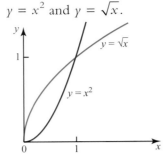

Figure 11.13

(i) Use integration to find the area under the curve $y = x^2$ from $x = 0$ to $x = 1$.

(ii) Use integration to find the area under the curve $y = \sqrt{x}$ from $x = 0$ to $x = 1$.

(iii) Hence write down the area of the region between the two curves.

(iv) Show that this region is bisected by the line $y = x$.

⑧ Find the following integrals.

(i) $\displaystyle\int\left(5x^2 + \frac{1}{3x^{\frac{1}{3}}}\right)\,\mathrm{d}x$

(ii) $\displaystyle\int 6 - \sqrt{x} + \frac{1}{2x^3}\,\mathrm{d}x$

(iii) $\displaystyle\int\left(5x^2 \times \frac{1}{3x^{\frac{1}{3}}}\right)\,\mathrm{d}x$

(iv) $\displaystyle\int\frac{6 - \sqrt{x}}{2x^3}\,\mathrm{d}x$

⑨ Given that $\dfrac{\mathrm{d}y}{\mathrm{d}x} = \sqrt{x}$, $x \geqslant 0$

(i) find the equation of the family of solution curves and sketch three members of this family

(ii) explain why x cannot be negative

(iii) find the equation of the particular curve which passes through $(9, 20)$.

⑩ Figure 11.14 shows part of the curve $y = \dfrac{1}{x^2}$.

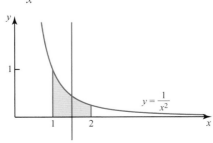

Figure 11.14

The vertical line shown bisects the shaded area. Find the equation of this line.

⑪ The acceleration of a car t seconds after it passes point A is given by $a = \dfrac{2}{\sqrt{t}}$ for $t > 0$ where $a = \dfrac{\mathrm{d}v}{\mathrm{d}t}$, the gradient function of the velocity $v\,\mathrm{m\,s^{-1}}$ of the car. Given that the car has velocity $5\,\mathrm{m\,s^{-1}}$ 1 second after it passes A, find the equation for the velocity at time t seconds.

⑫ Evaluate the following definite integrals.

(i) $\displaystyle\int_1^8\left(\frac{x^2 + 3x + 4}{x^4}\right)\,\mathrm{d}x$

(ii) $\displaystyle\int_1^4\left(3x + 2 - \frac{1}{x^{\frac{5}{2}}}\right)\,\mathrm{d}x$

(iii) $\displaystyle\int_0^9\left(2x - \sqrt{x}\right)^2\,\mathrm{d}x$

⑬ Figure 11.15 shows the line $y = 3$ and the curve $y = 4 - \dfrac{16}{x^2}$.

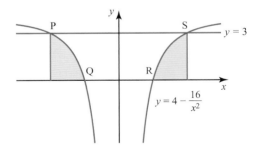

Figure 11.15

(i) Find the coordinates of the points P, Q, R and S.

(ii) Find the area of the shaded region.

⑭ (i) Sketch the curve with equation $y = -\sqrt{x}$ for $x \geqslant 0$.

(ii) Find the equation of the normal to the curve at the point P where $x = 1$.

(iii) The normal to the curve at the point P meets the x axis at the point Q. Find the coordinates of Q.

(iv) Find the area of the region bounded by the curve, the x axis, and the line segment PQ.

⑮ Figure 11.16 shows the curves $y = 1 - \dfrac{3}{x^2}$ and $y = -\dfrac{2}{x^3}$, for $x \leqslant 0$.

The shaded region is bounded by the curves and the line $x = -1$.

Find the area of the shaded region.

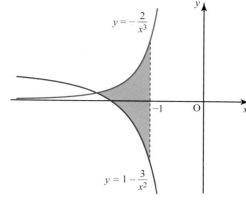

Figure 11.16

LEARNING OUTCOMES

When you have completed this chapter you should be able to:

➤ know and use the Fundamental Theorem of Calculus

➤ integrate x^n (excluding $n = -1$) and related sums, differences and constant multiples

➤ evaluate definite integrals

➤ use a definite integral to find the area under a curve.

KEY POINTS

1 $\dfrac{dy}{dx} = x^n \Rightarrow y = \dfrac{x^{n+1}}{n+1} + c,\ n \neq -1$

2 $\displaystyle\int_a^b x^n\, dx = \left[\dfrac{x^{n+1}}{n+1}\right]_a^b = \dfrac{b^{n+1} - a^{n+1}}{n+1},\ n \neq -1$

3 Area $A = \displaystyle\int_a^b y\, dx = \int_a^b f(x)\, dx$

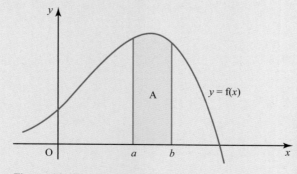

Figure 11.17

4   Area B $= \int_a^b y \, dx = -\int_a^b f(x) \, dx$

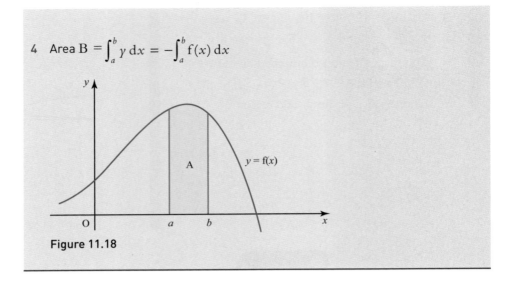

Figure 11.18

FUTURE USES

- In Chapter 21 you will see how integration is used in the study of motion.
- You will learn to integrate a wider variety of functions in the A-level mathematics textbook.

12 Vectors

➜ What information do you need to decide how close the two aircraft which left these vapour trails passed to each other?

1 Vectors

A quantity which has both size (magnitude) and direction is called a **vector**. The velocity of an aircraft through the sky is an example of a vector, as it has size (e.g. 600 mph) and direction (on a course of 254°).

The mass of the aircraft (100 tonnes) is completely described by its size and no direction is associated with it; such a quantity is called a **scalar**.

Vectors are used extensively in mechanics to represent quantities such as force, velocity and momentum, and in geometry to represent displacements.

This chapter focuses on two-dimensional vectors. You will learn about three-dimensional vectors in the A-level mathematics textbook.

Terminology

You can represent a vector by drawing a straight line with an arrowhead. The length of the line represents the **magnitude** (size) of the vector. The **direction** of the vector is indicated by the line and the arrowhead. Direction is usually given as the angle the vector makes with the positive x axis, with the anticlockwise direction taken to be positive.

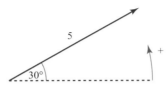

Figure 12.1

The vector in Figure 12.1 has magnitude 5, direction $+30°$. This is written $(5, 30°)$ and said to be in **magnitude–direction form**. This is also called **polar form**. The general form of a vector written in this way is (r, θ) where r is its magnitude and θ its direction.

You can also describe a vector in terms of **components** in given directions. The vector in Figure 12.2 is 4 units in the x direction, and 2 in the y direction, and this is denoted by $\begin{pmatrix} 4 \\ 2 \end{pmatrix}$.

> A unit vector has length 1.

This can also be written as $4\mathbf{i} + 2\mathbf{j}$, where $\mathbf{i}$ is a vector of magnitude 1 (a **unit vector**) in the x direction and $\mathbf{j}$ is a unit vector in the y direction (see Figure 12.3).

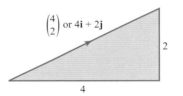

Figure 12.2   Figure 12.3

In a book, a vector may be printed in bold, for example **p** or **OP**, or as a line between two points with an arrow above it to indicate its direction, such as $\overrightarrow{OP}$.

When you write a vector by hand, it is usual to underline it, for example, $\underline{\mathbf{p}}$ or $\underline{OP}$, or to put an arrow above it, as in $\overrightarrow{OP}$.

You can convert a vector from component form to magnitude–direction form (or vice versa) by using trigonometry on a right-angled triangle.

| **Example 12.1** | Write the vector $\mathbf{a} = 4\mathbf{i} + 2\mathbf{j}$ in magnitude–direction form. |

Solution

The magnitude of $\mathbf{a}$ is given by the length a in Figure 12.4.

$a = \sqrt{4^2 + 2^2}$ ← Using Pythagoras' theorem.

   $= 4.47$ to 3 s.f.

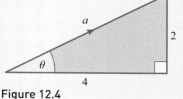

Figure 12.4

The direction is given by the angle θ.

$\tan \theta = \dfrac{2}{4} = 0.5$

$\Rightarrow \theta = 26.6°$ to 3 s.f.

So the vector $\mathbf{a}$ is $(4.47, 26.6°)$. ← Write your answer in the form (r, θ).

The magnitude of a vector is also called its **modulus** and denoted by the symbols $|\ |$. In the example $\mathbf{a} = 4\mathbf{i} + 2\mathbf{j}$, the modulus of $\mathbf{a}$, written $|\mathbf{a}|$, is 4.47.

Say 'mod a'.

Another convention for writing the magnitude of a vector is to use the same letter, but in italics and not bold type; so the magnitude of $\mathbf{a}$ is written a.

| **Example 12.2** | Write the vectors (i) $(5, 60°)$ and (ii) $(5, 300°)$ in component form. |

Prior knowledge

The CAST rule is covered in Chapter 6, p104.

Solution

(i) $(5, 60°)$ means $r = 5$ and $\theta = 60°$.

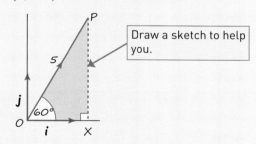

Draw a sketch to help you.

Figure 12.5

In the right-angled triangle OPX

$OX = 5 \cos 60° = 2.5$

$XP = 5 \sin 60° = \dfrac{5\sqrt{3}}{2} = 4.33$ to 2 d.p.

$\overrightarrow{OP} = \begin{pmatrix} 2.5 \\ 4.33 \end{pmatrix}$ or $2.5\mathbf{i} + 4.33\mathbf{j}$.

(ii)

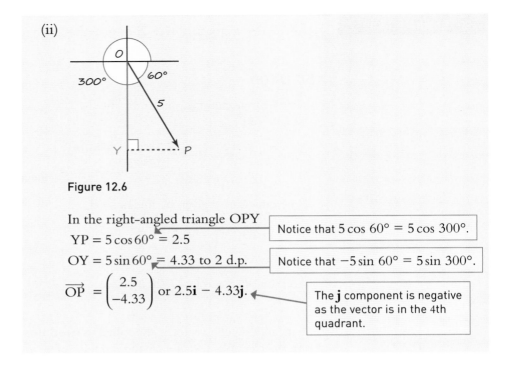

Figure 12.6

In the right-angled triangle OPY

$YP = 5\cos 60° = 2.5$

$OY = 5\sin 60° = 4.33$ to 2 d.p.

$\overrightarrow{OP} = \begin{pmatrix} 2.5 \\ -4.33 \end{pmatrix}$ or $2.5\mathbf{i} - 4.33\mathbf{j}$.

Notice that $5\cos 60° = 5\cos 300°$.

Notice that $-5\sin 60° = 5\sin 300°$.

The **j** component is negative as the vector is in the 4th quadrant.

In general, for all values of θ

$$(r, \theta) \Rightarrow \begin{pmatrix} r\cos\theta \\ r\sin\theta \end{pmatrix} = (r\cos\theta)\mathbf{i} + (r\sin\theta)\mathbf{j}$$

When you convert from component form to magnitude–direction form it is useful to draw a diagram and use it to see which quadrant the angle lies in so that you get the correct signs. This is shown in the next example.

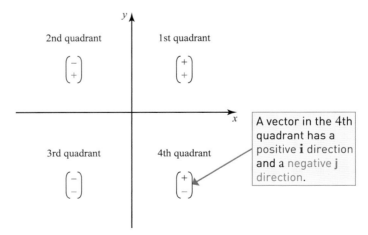

Figure 12.7

Example 12.3

Write $-5\mathbf{i} + 4\mathbf{j}$ in magnitude–direction form.

Solution

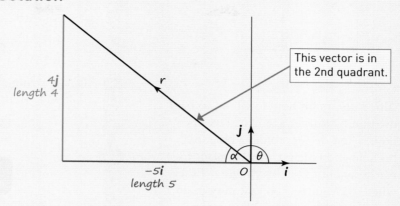

This vector is in the 2nd quadrant.

Figure 12.8

The magnitude $r = \sqrt{5^2 + 4^2} = 6.40$ to 2 d.p.

The direction is given by the angle θ in Figure 12.8.

Remember the direction is the angle the vector makes with the positive x axis.

You need to work out angle α first.

$$\tan \alpha = \frac{4}{5} \rightarrow \alpha = 38.7° \text{ (to nearest 0.1°)}.$$

So $\theta = 180 - \alpha = 141.3°$

The vector is $(6.40, 141.3°)$ in magnitude–direction form.

> ### Note
>
> **Using your calculator**
>
> Many calculators include the facility to convert from polar coordinates (magnitude–direction form) (r, θ) to rectangular coordinates (x, y), and vice versa. This is the same as converting one form of a vector into the other. You may prefer to use your calculator to convert between the different forms.
>
> **Line segments**
>
> A line joining two points is a **line segment**. A line segment is just that particular part of the infinite straight line that passes through those two points.

Exercise 12.1

 ① Which of these vectors are in component form and which in magnitude – direction form?

$(5, 25°)$ $6\mathbf{i} - \mathbf{j}$ $\begin{pmatrix} -2 \\ 4 \end{pmatrix}$

$(0, 0)$ $(0.5, 124°)$ $\begin{pmatrix} 0 \\ 0 \end{pmatrix}$

 ② What is the direction of $\mathbf{i} - \mathbf{j}$?

A 135°   B 45°

C −45°   D 225°

 ③ For each of the following vectors
    (a) express the vector in component form
    (b) find the magnitude of the vector.

(i)

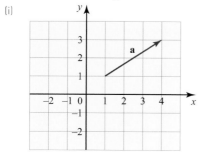

(ii)

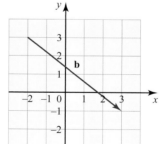

(iii)

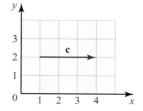

(iv)

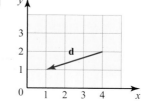

(v)

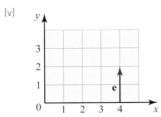

Figure 12.9

④ Alfred knows his home is about to be overrun by thieves. He has a box of money and decides to dig a hole in his field and hide it there. He plans to run away but to tell his daughter Ethel where the money is so that she can go back later and dig it up. How many pieces of information does Alfred need to give Ethel?

⑤ Draw diagrams to show these vectors and then write them in magnitude–direction form.

> Use your calculator to check your answers.

  (i) $3\mathbf{i} + 4\mathbf{j}$

  (ii) $\begin{pmatrix} 4 \\ 3 \end{pmatrix}$

  (iii) $3\mathbf{i} - 4\mathbf{j}$

  (iv) $\begin{pmatrix} -4 \\ 3 \end{pmatrix}$

  (v) $\begin{pmatrix} -3 \\ -4 \end{pmatrix}$

⑥ Draw sketch diagrams to show these vectors and then write them in component form.

  (i) (10, 45°)   (ii) (5, 210°)
  (iii) (5, 360°)   (iv) (10, 270°)
  (v) (10, 135°)   (vi) (5, 330°)

⑦ Write, in component form, the vector $\overrightarrow{AB}$ represented by the line segments joining the following points.

> Remember a line segment is part of a line joining two particular points. Draw a diagram to help you.

  (i) A(2, 3) to B(4, 1)
  (ii) A(4, 1) to B(2, 3)
  (iii) A(−2, −3) to B(4, 1)
  (iv) A(−2, 3) to B(−4, 1)
  (v) A(3, 2) to B(1, 4)
  (vi) A(−3, −2) to B(−1, −4)

⑧ A triangle is formed by the points A(1, 4), B(4, 7) and C(1, 7).

  (i) (a) Find the vectors representing the three medians of the triangle.

  > A median joins one vertex to the midpoint of the opposite side.

    (b) Find the magnitude of each median. Give your answers in the form $a\sqrt{b}$ where a and b are constants.

  The point X(2, 6) is the centroid of the triangle.

  (ii) Find $\left|\overrightarrow{AX}\right|$. Give your answer in the form $a\sqrt{b}$ where a and b are constants.

  (iii) Show that X divides each of the three medians in the ratio 2:1.

⑨ P, Q and R are points on a plane. Investigate whether these situations are possible and, if so, what can be said about the position of R.

  (i) $\left|\overrightarrow{PR}\right| = \left|\overrightarrow{QR}\right|$

  (ii) The direction of $\overrightarrow{PR}$ is the same as the direction of $\overrightarrow{QR}$.

  (iii) Both $\left|\overrightarrow{PR}\right| = \left|\overrightarrow{QR}\right|$ and the direction of $\overrightarrow{PR}$ is the same as the direction of $\overrightarrow{QR}$.

2 Working with vectors

Equal vectors

The statement that two vectors **a** and **b** are equal means two things.

■   The direction of **a** is the same as the direction of **b**.

■   The magnitude of **a** is the same as the magnitude of **b**.

If the vectors are given in component form, each component of **a** equals the corresponding component of **b**.

Position vectors

Saying the vector **a** is given by $4\mathbf{i} + 2\mathbf{j}$ tells you the components of the vector, or equivalently its magnitude and direction. It does not tell you where the vector is situated; indeed it could be anywhere.

All of the lines in Figure 12.10 represent the vector **a**.

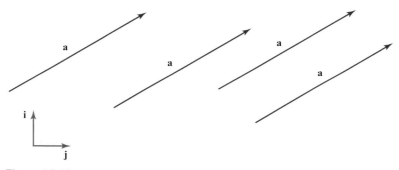

Figure 12.10

A **position vector** is a vector that starts at the origin.

So the line segment joining the origin to the point $(3, 5)$ is the position vector $\begin{pmatrix} 3 \\ 5 \end{pmatrix}$ or $3\mathbf{i} + 5\mathbf{j}$. You can say that

the point $(3, 5)$ has the position vector $\begin{pmatrix} 3 \\ 5 \end{pmatrix}$.

> A position vector does have a fixed location.

Example 12.4

Points L, M and N have coordinates $(4, 3)$, $(-2, -1)$ and $(2, 2)$.

(i)   Write down, in component form, the position vector of L and the vector $\overrightarrow{MN}$.

(ii)   What do your answers to part (i) tell you about the line segments OL and MN?

Solution

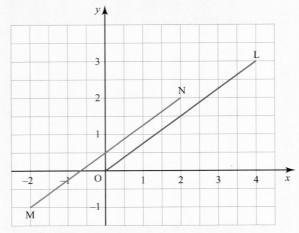

Figure 12.11

(i) The position vector of L is $\overrightarrow{OL} = \begin{pmatrix} 4 \\ 3 \end{pmatrix}$

The vector $\overrightarrow{MN}$ is also $\begin{pmatrix} 4 \\ 3 \end{pmatrix}$. ← Using the diagram

(ii) Since $\overrightarrow{OL} = \overrightarrow{MN}$, the lines OL and MN are parallel and equal in length.

Multiplying a vector by a scalar

When a vector is multiplied by a number (a scalar) its length is altered but its direction remains the same.

The vector **2a** in Figure 12.12 is twice as long as the vector **a** but in the same direction.

Figure 12.12

When the vector is in component form, each component is multiplied by the number. For example:

$$2 \times (3\mathbf{i} - 5\mathbf{j}) = 6\mathbf{i} - 10\mathbf{j}$$

$$2 \times \begin{pmatrix} 3 \\ -5 \end{pmatrix} = \begin{pmatrix} 6 \\ -10 \end{pmatrix}$$ ← Twice '3 right and 5 down' is '6 right and 10 down'.

The negative of a vector

In Figure 12.13 the vector $-\mathbf{a}$ has the same length as the vector **a** but the opposite direction.

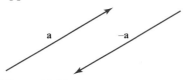

Figure 12.13

When **a** is given in component form, the components of $-$**a** are the same as those for **a** but with their signs reversed.

So $\qquad -\begin{pmatrix} 13 \\ -7 \end{pmatrix} = \begin{pmatrix} -13 \\ 7 \end{pmatrix}$

> This is the same as multiplying by -1.

Adding vectors

When vectors are given in component form, they can be added component by component. This process can be seen geometrically by drawing them on squared paper, as in Example 12.5.

Example 12.5

Add the vectors $2\mathbf{i} - 3\mathbf{j}$ and $3\mathbf{i} + 5\mathbf{j}$.

Solution

$2\mathbf{i} - 3\mathbf{j} + 3\mathbf{i} + 5\mathbf{j} = 5\mathbf{i} + 2\mathbf{j}$

$$\begin{pmatrix} 2 \\ -3 \end{pmatrix} + \begin{pmatrix} 3 \\ 5 \end{pmatrix} = \begin{pmatrix} 5 \\ 2 \end{pmatrix}$$

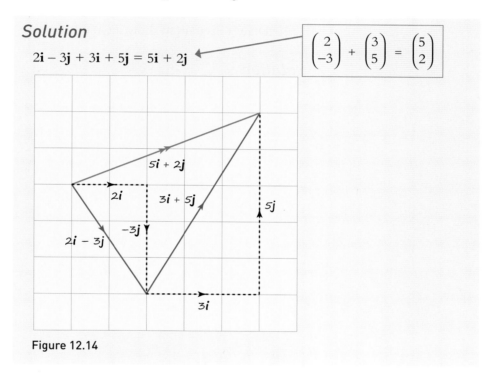

Figure 12.14

The sum of two (or more) vectors is called the **resultant** and is usually indicated by being marked with *two* arrowheads.

Adding vectors is like adding the stages of a journey to find its overall outcome (see Figure 12.15).

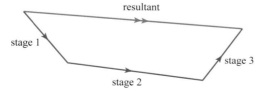

Figure 12.15

In mechanics, vectors are used to represent the forces acting on an object. You can add together all of the forces to find the resultant force acting on an object.

Example 12.6

An object has three forces acting on it as shown in Figure 12.16.

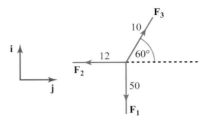

Figure 12.16

(i) Find the resultant force acting on the object.

(ii) Draw a diagram to show the resultant force.

Solution

(i) Write each force in component form.

$$\mathbf{F}_1 = \begin{pmatrix} 0 \\ -50 \end{pmatrix}$$

> Both of these forces are acting in the negative direction.

$$\mathbf{F}_2 = \begin{pmatrix} -12 \\ 0 \end{pmatrix}$$

and $\mathbf{F}_3$ is $(10, 60°) \Rightarrow \mathbf{F}_3 = \begin{pmatrix} 10\cos 60° \\ 10\sin 60° \end{pmatrix} = \begin{pmatrix} 5 \\ 8.66 \end{pmatrix}$

Now add together the three forces:

$$\mathbf{F}_1 + \mathbf{F}_2 + \mathbf{F}_3 = \begin{pmatrix} 0 \\ -50 \end{pmatrix} + \begin{pmatrix} -12 \\ 0 \end{pmatrix} + \begin{pmatrix} 5 \\ 8.66 \end{pmatrix} = \begin{pmatrix} -7 \\ -41.3 \end{pmatrix}$$

(ii)

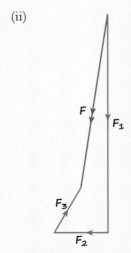

Figure 12.17

Subtracting vectors

Subtracting one vector from another is the same as adding the negative of the vector.

Example 12.7

Two vectors **a** and **b** are given by **a** = 2**i** + 3**j** and **b** = −**i** + 2**j**.

(i) Find **a** − **b**.

(ii) Draw diagrams showing **a**, **b**, **a** − **b**.

Solution

(i)   **a** − **b** = (2**i** + 3**j**) − (−**i** + 2**j**)

$\qquad\quad$ = 3**i** + **j**

(ii)

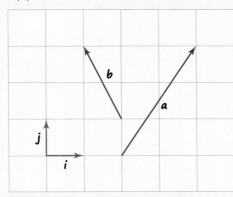
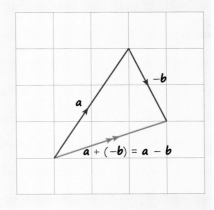

Figure 12.18

Unit vectors

A unit vector is a vector with a magnitude of 1, like **i** and **j**. To find the unit vector in the same direction as a given vector, divide that vector by its magnitude.

So the vector 3**i** + 5**j** (in Figure 12.19) has magnitude $\sqrt{3^2 + 5^2} = \sqrt{34}$, and the vector $\dfrac{3}{\sqrt{34}}\mathbf{i} + \dfrac{5}{\sqrt{34}}\mathbf{j}$ is a unit vector in the same direction.

Its magnitude is 1.

> **Note**
>
> The unit vector in the direction of vector **a** is written as **â** and read as 'a hat'.

> **ACTIVITY 12.1**
>
> Show that $\dfrac{a}{\sqrt{a^2 + b^2}}\mathbf{i}$ $+ \dfrac{b}{\sqrt{a^2 + b^2}}\mathbf{j}$ is the unit vector in the direction of $a\mathbf{i}+b\mathbf{j}$.

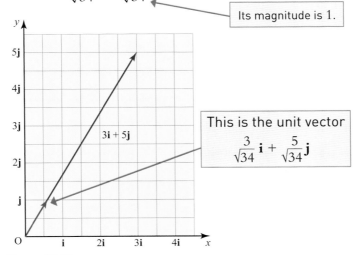

This is the unit vector

$\dfrac{3}{\sqrt{34}}\mathbf{i} + \dfrac{5}{\sqrt{34}}\mathbf{j}$

Figure 12.19

Exercise 12.2

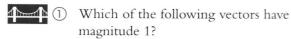

① Which of the following vectors have magnitude 1?

$$\begin{pmatrix} 1 \\ 1 \end{pmatrix} \qquad \begin{pmatrix} \frac{1}{\sqrt{2}} \\ \frac{1}{\sqrt{2}} \end{pmatrix} \qquad \begin{pmatrix} -1 \\ 0 \end{pmatrix} \qquad \begin{pmatrix} -1 \\ 2 \end{pmatrix}$$

② Which of the following is the expression $\begin{pmatrix} 1 \\ -2 \end{pmatrix} + \begin{pmatrix} -2 \\ 4 \end{pmatrix}$ equal to?

A $\begin{pmatrix} 3 \\ 6 \end{pmatrix}$ B $\begin{pmatrix} -1 \\ 2 \end{pmatrix}$ C $\begin{pmatrix} 1 \\ 2 \end{pmatrix}$ D 1

③ Simplify the following.

(i) $\begin{pmatrix} 2 \\ 3 \end{pmatrix} + \begin{pmatrix} 4 \\ 5 \end{pmatrix}$ (ii) $\begin{pmatrix} 2 \\ -3 \end{pmatrix} + \begin{pmatrix} -3 \\ 2 \end{pmatrix}$

(iii) $\begin{pmatrix} 2 \\ 3 \end{pmatrix} + \begin{pmatrix} -2 \\ -3 \end{pmatrix}$ (iv) $4\begin{pmatrix} 2 \\ -3 \end{pmatrix} + 5\begin{pmatrix} -3 \\ 2 \end{pmatrix}$

(v) $2(4\mathbf{i} + 5\mathbf{j}) - 3(2\mathbf{i} + 3\mathbf{j})$

(vi) $4(3\mathbf{i} - 2\mathbf{j}) - 5(2\mathbf{i} - 3\mathbf{j})$

④ Show that the vector $\frac{3}{5}\mathbf{i} - \frac{4}{5}\mathbf{j}$ is a unit vector.

⑤ The vectors $\mathbf{p}$, $\mathbf{q}$ and $\mathbf{r}$ are given by
$\mathbf{p} = 3\mathbf{i} + 2\mathbf{j} \quad \mathbf{q} = 2\mathbf{i} + 2\mathbf{j} \quad \mathbf{r} = -3\mathbf{i} - \mathbf{j}$.
Find, in component form, the following vectors.

(i) $\mathbf{p} + \mathbf{q} + \mathbf{r}$

(ii) $\mathbf{p} - \mathbf{q}$

(iii) $\mathbf{p} + \mathbf{r}$

(iv) $4\mathbf{p} - 3\mathbf{q} + 2\mathbf{r}$

(v) $3\mathbf{p} + 2\mathbf{q} - 2(\mathbf{p} - \mathbf{r})$

(vi) $3(\mathbf{p} - \mathbf{q}) + 2(\mathbf{p} + \mathbf{r})$

⑥ (i) Given that $a\begin{pmatrix} 5 \\ 3 \end{pmatrix} - \begin{pmatrix} 8 \\ b \end{pmatrix} = \begin{pmatrix} 12 \\ 6 \end{pmatrix}$, find the values of a and b.

(ii) Given that $p\begin{pmatrix} 1 \\ 4 \end{pmatrix} + q\begin{pmatrix} 3 \\ 5 \end{pmatrix} = \begin{pmatrix} 5 \\ 6 \end{pmatrix}$, find the values of p and q.

⑦ (i) Given that the vectors $\mathbf{a}$ and $\mathbf{b}$ are parallel, find the value of x when
$\mathbf{a} = \begin{pmatrix} 4 \\ 8 \end{pmatrix}$ and $\mathbf{b} = \begin{pmatrix} 2 \\ x \end{pmatrix}$.

(ii) Given that the vectors $\mathbf{p}$ and $\mathbf{q}$ are parallel, find the value of x when
$\mathbf{p} = \begin{pmatrix} -3 \\ 8 \end{pmatrix}$ and $\mathbf{q} = \begin{pmatrix} 6 \\ x \end{pmatrix}$.

(iii) Given that the vectors $\mathbf{r}$ and $\mathbf{s}$ are parallel, find the value of x when
$\mathbf{r} = \begin{pmatrix} -12 \\ 9 \end{pmatrix}$ and $\mathbf{s} = \begin{pmatrix} 4 \\ x \end{pmatrix}$.

⑧ Find unit vectors in the same directions as the following vectors.

(i) $3\mathbf{i} + 4\mathbf{j}$ (ii) $\begin{pmatrix} -3 \\ 4 \end{pmatrix}$

(iii) $4\mathbf{i} - 3\mathbf{j}$ (iv) $2\mathbf{i} + 3\mathbf{j}$

(v) $\begin{pmatrix} 3 \\ 2 \end{pmatrix}$ (vi) $\begin{pmatrix} -3 \\ -2 \end{pmatrix}$

(vii) $6\mathbf{i}$ (viii) $-6\mathbf{j}$

(ix) $\begin{pmatrix} 6 \\ -6 \end{pmatrix}$

⑨ Each diagram in Figure 12.20 shows the magnitude and direction of the forces acting on an object.

Find the resultant force in component form in each case.

(i)

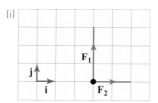

(ii)

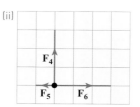

(iii)

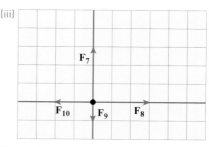

Figure 12.20

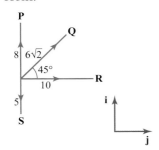

⑩ Two forces **P** and **Q** are given by
$$\mathbf{P} = 5\mathbf{i} - 2\mathbf{j} \qquad \mathbf{Q} = -3\mathbf{i} + 2\mathbf{j}.$$

(i) Find the resultant of **P** and **Q**.

(ii) Draw a diagram showing **P**, **Q** and their resultant.

⑪ The diagram in Figure 12.21 shows the magnitude and direction of the forces acting on an object. The magnitude of each force is shown next to it.

Find the resultant force in component form.

Figure 12.21

⑫ Find unit vectors in the same directions as the following vectors.

(i) $\begin{pmatrix} r\cos\alpha \\ r\sin\alpha \end{pmatrix}$ (ii) $\begin{pmatrix} 1 \\ \tan\beta \end{pmatrix}$

PS ⑬ A and B are points on a plane with position vectors **a** and **b**.

What can you say about the positions of O, A and B when

(i) every point on the plane can be written in the form $p\mathbf{a} + q\mathbf{b}$

(ii) some points on the plane cannot be written in the form $p\mathbf{a} + q\mathbf{b}$?

3 Vector geometry

When you find the vector represented by the line segment joining two points, you are, in effect, subtracting their position vectors.

For example, let P be the point $(2, 1)$ and Q the point $(3, 5)$. Then $\overrightarrow{PQ} = \begin{pmatrix} 1 \\ 4 \end{pmatrix}$ (see Figure 12.22).

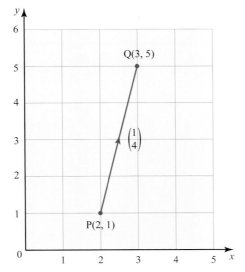

Figure 12.22

You find this by saying

$$\overrightarrow{PQ} = \overrightarrow{PO} + \overrightarrow{OQ} = -\mathbf{p} + \mathbf{q} \qquad \boxed{\text{Look at Figure 12.23.}}$$

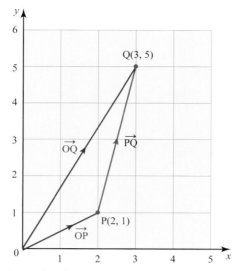

Figure 12.23

In this case, this gives

$$\overrightarrow{PQ} = -\binom{2}{1} + \binom{3}{5} = \binom{1}{4} \text{ as expected.}$$

This is an important result, that

$$\overrightarrow{PQ} = \overrightarrow{OQ} + \overrightarrow{OP} = +\mathbf{q} - \mathbf{p}$$

where $\mathbf{p}$ and $\mathbf{q}$ are the position vectors of P and Q.

The length of PQ is the same as its magnitude.

So $\left|\overrightarrow{PQ}\right| = \sqrt{1^2 + 4^2} = \sqrt{17}$.

Geometrical figures

It is often useful to be able to express lines in a geometrical figure in terms of given vectors, as in the next example.

Example 12.8

Figure 12.24 shows a hexagon ABCDEF.

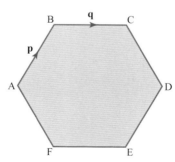

Figure 12.24

The hexagon is regular and consequently $\overrightarrow{AD} = 2\overrightarrow{BC}$.

$\overrightarrow{AB} = \mathbf{p}$ and $\overrightarrow{BC} = \mathbf{q}$. Express the following in terms of $\mathbf{p}$ and $\mathbf{q}$.

(i) $\overrightarrow{AC}$ (ii) $\overrightarrow{AD}$ (iii) $\overrightarrow{CD}$

(iv) $\overrightarrow{DE}$ (v) $\overrightarrow{EF}$ (vi) $\overrightarrow{BE}$

Solution

(i) $\overrightarrow{AC} = \overrightarrow{AB} + \overrightarrow{BC}$

$\qquad = \mathbf{p} + \mathbf{q}$

(ii) $\overrightarrow{AD} = 2\overrightarrow{BC}$

$\qquad = 2\mathbf{q}$

(iii) Since $\overrightarrow{AC} + \overrightarrow{CD} = \overrightarrow{AD}$

$\qquad \mathbf{p} + \mathbf{q} + \overrightarrow{CD} = 2\mathbf{q}$

and so $\qquad \overrightarrow{CD} = \mathbf{q} - \mathbf{p}$

(iv) $\overrightarrow{DE} = -\overrightarrow{AB}$

$\qquad = -\mathbf{p}$

(v) $\overrightarrow{EF} = -\overrightarrow{BC}$

$\qquad = -\mathbf{q}$

(vi) $\overrightarrow{BE} = \overrightarrow{BC} + \overrightarrow{CD} + \overrightarrow{DE}$

$\qquad = \mathbf{q} + (\mathbf{q} - \mathbf{p}) + -\mathbf{p}$

$\qquad = 2\mathbf{q} - 2\mathbf{p}$

So $\overrightarrow{BE}$ and $\overrightarrow{CD}$ are parallel.

Notice that $\overrightarrow{BE} = 2\overrightarrow{CD}$.

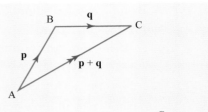
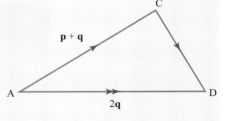
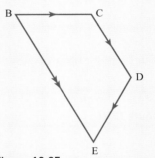

Figure 12.25

Exercise 12.3

 ① Is it always true, sometimes true or never true that $\overrightarrow{AB} = \overrightarrow{OB} - \overrightarrow{OA}$?

 ② Match together the parallel vectors.

| $2\mathbf{i} + 3\mathbf{j}$ | $-4\mathbf{i} + 6\mathbf{j}$ | $2\mathbf{i} - 3\mathbf{j}$ | $6\mathbf{i} + 9\mathbf{j}$ |

| $\dfrac{2}{\sqrt{13}}\mathbf{i} - \dfrac{3}{\sqrt{13}}\mathbf{j}$ | $3\mathbf{i} + 2\mathbf{j}$ |

| $-\dfrac{3}{\sqrt{13}}\mathbf{i} - \dfrac{2}{\sqrt{13}}\mathbf{j}$ | $-12\mathbf{i} - 8\mathbf{j}$ |

③ For each pair of coordinates A and B, find

(a) $\overrightarrow{OA}$ (b) $\overrightarrow{OB}$

(c) $\overrightarrow{AB}$ (d) $|\overrightarrow{AB}|$

(i) A(3, 7) and B(2, 4)

(ii) A(−3, −7) and B(−2, −4)

(iii) A(−3, −7) and B(2, 4)

(iv) A(3, 7) and B(−2, −4)

(v) A(3, −7) and B(−2, 4)

④ The points A, B and C have coordinates (2, 3), (0, 4) and (−2, 1).

(i) Write down the position vectors of A and C.

(ii) Write down the vectors of the line segments joining AB and CB.

(iii) What do your answers to parts (i) and (ii) tell you about

(a) AB and OC

(b) CB and OA?

(iv) Describe the quadrilateral OABC.

⑤ A, B and C are the points (2, 5), (4, 9) and (−3, −5).

(i) Write down the position vector of each point.

(ii) Find the vectors $\overrightarrow{AB}$ and $\overrightarrow{BC}$.

PS (iii) Show that all three points are collinear. ◄─── Collinear means they lie on a straight line.

⑥ In Figure 12.26, PQRS is a parallelogram, and $\overrightarrow{PQ} = \mathbf{a}$ and $\overrightarrow{PS} = \mathbf{b}$.

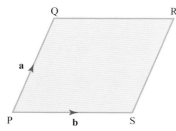

Figure 12.26

(i) Write $\overrightarrow{PR}$ and $\overrightarrow{QS}$ in terms of **a** and **b**.

(ii) The midpoint of PR is M. Find $\overrightarrow{PM}$ and $\overrightarrow{QM}$.

(iii) Hence prove that the diagonals of a parallelogram bisect each other.

⑦ In Figure 12.27, ABCD is a kite. AC and BD meet at M.
$$\overrightarrow{AB} = \mathbf{i} + \mathbf{j} \qquad \text{and} \qquad \overrightarrow{AD} = \mathbf{i} - 2\mathbf{j}$$

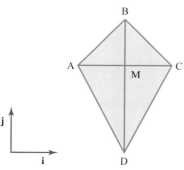

Figure 12.27

(i) Use the properties of a kite to find, in terms of **i** and **j**

(a) $\overrightarrow{AM}$ (b) $\overrightarrow{AC}$

(c) $\overrightarrow{BC}$ (d) $\overrightarrow{CD}$.

(ii) Verify that $\left|\overrightarrow{AB}\right| = \left|\overrightarrow{BC}\right|$ and $\left|\overrightarrow{AD}\right| = \left|\overrightarrow{CD}\right|$.

⑧ $\overrightarrow{OP} = 9\mathbf{p}$ and $\overrightarrow{OQ} = 6\mathbf{q}$.

Given that X lies on $\overrightarrow{PQ}$ and $\overrightarrow{PX} = \frac{1}{2} XQ$, find in terms of **p** and **q**:

(i) $\overrightarrow{PQ}$ (ii) $\overrightarrow{PX}$

(iii) $\overrightarrow{QX}$ (iv) $\overrightarrow{OX}$.

⑨ In Figure 12.28, ABC is a triangle.

L, M and N are the midpoints of the sides BC, CA and AB.

$\overrightarrow{AB} = \mathbf{p}$ and $\overrightarrow{AC} = \mathbf{q}$.

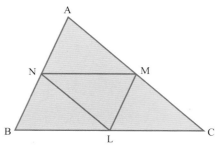

Figure 12.28

(i) Find $\overrightarrow{BC}$ and $\overrightarrow{NM}$ in terms of **p** and **q**.

(ii) What can you conclude about the triangles ABC and LMN? Justify your answer.

⑩ In Figure 12.29, $\overrightarrow{AB} = 3\mathbf{a}$, $\overrightarrow{AD} = \mathbf{b}$ and $\overrightarrow{DC} = 2\mathbf{a}$.

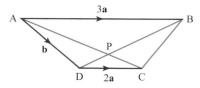

Figure 12.29

(i) Explain how you know that ABCD is a trapezium.

(ii) Find $\overrightarrow{AC}$ and $\overrightarrow{BC}$.

In Figure 12.29, $\overrightarrow{AP} = \mu\overrightarrow{AC}$ and $\overrightarrow{BP} = \lambda\overrightarrow{BD}$.

(iii) Find expressions in terms of μ for $\overrightarrow{AP}$ and $\overrightarrow{BP}$.

(iv) Hence find the value of μ and λ.

LEARNING OUTCOMES

When you have completed this chapter you should be able to:

➤ use vectors in two dimensions
➤ calculate the magnitude and direction of a vector
➤ convert between component form and magnitude/direction form
➤ add vectors diagrammatically
➤ perform the algebraic operations of:
  ○ vector addition
  ○ multiplication by scalars
➤ understand their geometrical interpretations
➤ understand and use position vectors
➤ calculate the distance between two points represented by position vectors
➤ use vectors to solve problems in pure mathematics and in context.

KEY POINTS

1 A vector quantity has magnitude and direction.
2 A scalar quantity has magnitude only.
3 Vectors are typeset in bold, $\mathbf{a}$ or $\mathbf{OA}$, or in the form $\overrightarrow{OA}$.
  They are handwritten either in the underlined form $\underline{a}$, or as $\overrightarrow{OA}$.
4 Unit vectors in the x and y directions are denoted by $\mathbf{i}$ and $\mathbf{j}$, respectively.
5 A vector may be specified in
  ■ magnitude–direction form: (r, θ)
  ■ component form: $x\mathbf{i} + y\mathbf{j}$ or $\begin{pmatrix} x \\ y \end{pmatrix}$.

6 The resultant of two (or more) vectors is found by the sum of the vectors.
  A resultant vector is usually denoted by a double-headed arrow.
7 The position vector $\overrightarrow{OP}$ of a point P is the vector joining the origin to P.
8 The vector $\overrightarrow{AB} = \mathbf{b} - \mathbf{a}$, where $\mathbf{a}$ and $\mathbf{b}$ are the position vectors of A and B.
9 The length (or modulus or magnitude) of the vector $\mathbf{r}$ is written as r or as $|\mathbf{r}|$.

$$\mathbf{r} = a\mathbf{i} + b\mathbf{j} \Rightarrow |\mathbf{r}| = \sqrt{a^2 + b^2}$$

10 A unit vector in the same direction as $\mathbf{r} = a\mathbf{i} + b\mathbf{j}$ is $\dfrac{a}{\sqrt{a^2 + b^2}}\mathbf{i} + \dfrac{b}{\sqrt{a^2 + b^2}}\mathbf{j}$

FUTURE USES

■ You will use vectors to represent forces in Chapter 20.
■ Work on vectors will be extended to three dimensions in the A-level textbook.
■ If you study Further Mathematics, you will learn about how vectors can be used to solve problems involving lines and planes.

13 Exponentials and logarithms

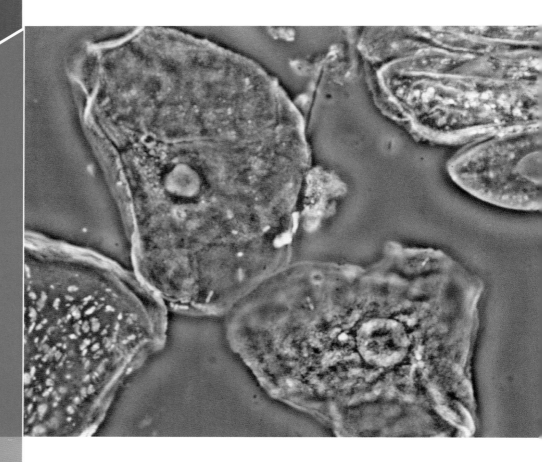

During growth or reproduction in the human body, a cell divides into two new cells roughly every 24 hours.

Assuming that this process takes exactly 1 day, and that none of the cells die off:

➔ Starting with one cell, how many cells will there be after

  (i) 5 days

  (ii) 10 days?

➔ Approximately how many days would it take to create one million cells from a single cell?

To answer the question above, you need to use powers of 2.

An **exponential function** is a function which has the variable as the power, such as 2^x. (An alternative name for power is **exponent**.)

ACTIVITY 13.1

Figure 13.1 shows the curves $y = 2^x$ and $y = x^2$.

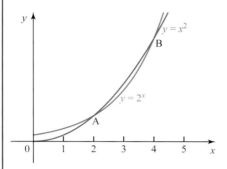

Figure 13.1

(i) (a) Find the coordinates of the points A and B.
  (b) For what values of x is $2^x > x^2$?

(ii) Using any graphing software available to you, draw
  (a) $y = 3^x$ and $y = x^3$
  (b) $y = 4^x$ and $y = x^4$.

(iii) Is it true that for all values of a
  (a) $y = a^x$ and $y = x^a$ intersect when $x = a$
  (b) for large enough values of x, $a^x > x^a$?

1 Exponential functions

From Activity 13.1 you have seen that the graphs of all the exponential functions pass through the point $(0, 1)$ and have a gradient that is increasing.

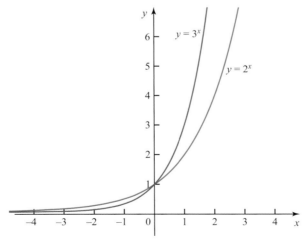

Figure 13.2

Exponential functions model many real-life situations, such as the growth of cells in the question at the beginning of the chapter. Exponential functions increase at an ever-increasing rate. This is described as **exponential growth**.

It is also possible to have exponential functions for which the power is negative, for example $y = 2^{-x}$, $y = 3^{-x}$, etc.

The effect of replacing x by $-x$ for any function of x is to reflect the graph in the y axis.

Discussion points

→ The graphs of $y = a^x$ and $y = a^{-x}$ have a horizontal asymptote at $y = 0$ (the x axis) and go through the point $(0, 1)$.

→ How will these features change if the graphs are

- translated vertically

- translated horizontally

- stretched horizontally

- stretched vertically?

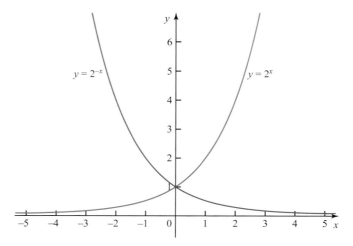

Figure 13.3

Notice that as x becomes very large, the value of 2^{-x} becomes very small. The graph of $y = 2^{-x}$ approaches the x axis ever more slowly as x increases. This is described as **exponential decay**.

Example 13.1

The cost, £C, of a widget t years after initial production is given by $C = 10 + 20 \times 2^{-t}$.

(i)   What is the initial cost of the widget?

(ii) What happens to the cost as t becomes very large?

(iii) Sketch the graph of C against t.

Solution

(i)   When $t = 0$, $C = 10 + 20 \times 2^0 = 10 + 20 = 30$.

So the initial cost is £30. ⎢‾‾‾‾‾‾‾‾‾‾‾ Remember that $a^0 = 1$.

(ii) As t becomes very large, 2^{-t} becomes very small, and so the cost approaches £10.

(iii)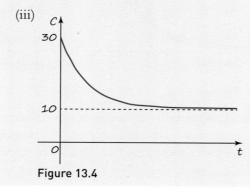

Figure 13.4

Exercise 13.1

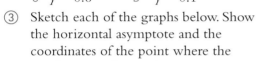

TECHNOLOGY

Graphing software will be useful here to check your answers.

 ① Which of the following functions are increasing functions and which are decreasing functions?

$y = 8^x$ $N = 0.7^t$ $y = 3^{-x}$ $P = 1.5^t$

$y = 0.3^x$ $N = 20^{-t}$ $y = 0.8^x$

 ② Which one of the following functions is decreasing?

A $y = 2^x$ B $y = 1.618^x$

C $y = 0.6^x$ D $y = 0.1^{-x}$

③ Sketch each of the graphs below. Show the horizontal asymptote and the coordinates of the point where the graph crosses the y axis.

(i) $y = 2^x$ (ii) $y = 2^x + 1$

(iii) $y = 2^x - 1$

④ Sketch each of the graphs below. Show the horizontal asymptote and the coordinates of the point where the graph crosses the y axis.

(i) $y = 3^{-x}$ (ii) $y = 3^{-x} + 2$

(iii) $y = 3^{-x} - 1$

⑤ The growth in population P of a certain town after time t years can be modelled by the equation $P = 10\,000 \times 10^{0.1t}$.

(i)   State the initial population of the town and calculate the population after 5 years.

(ii)   Sketch the graph of P against t.

Ⓜ ⑥ The height h m of a species of pine tree t years after planting is modelled by the equation

$$h = 20 - 19 \times 0.9^t$$

(i)   What is the height of the trees when they are planted?

(ii)   Calculate the height of the trees after 2 years.

(iii)   Use trial and improvement to find the number of years for the height to reach 10 metres.

(iv)   What height does the model predict that the trees will eventually reach?

Ⓜ ⑦ A flock of birds was caught in a hurricane and blown far out to sea. Fortunately, they were able to land on a small and very remote island and settle there. There was only a limited sustainable supply of suitable food for them on the island.

Their population size, n birds, at a time t years after they arrived, can be modelled by the equation $n = 200 + 320 \times 2^{-t}$.

(i)   How many birds arrived on the island?

(ii)   How many birds were there after 5 years?

(iii)   Sketch a graph of n against t.

(iv)   What is the long-term size of the population?

⑧ (i)   Show that $5^x > x^5$ when $x = 1$ and that $5^x < x^5$ when $x = 3$.

(ii)   Use a spreadsheet to find, correct to 2 significant figures, the value of x between 1 and 3 for which $5^x = x^5$.

⑨ In music, the notes in the middle octave of the piano and their frequencies in hertz, are as follows. (The frequencies have been rounded to the nearest whole numbers.)

A   220   B   247   C   262

D   294   E   330   F   349

G   392

In the next octave up, the notes are also called A to G and their frequencies are exactly twice those given above; so, for example, the frequency of A in the next octave is

$2 \times 220 = 440$ hertz.

The same pattern of multiplying by 2 continues for higher octaves. Similarly, dividing by 2 gives the frequencies for the notes in lower octaves.

(i)   Find the frequency of B three octaves above the middle.

(ii)   The lowest note on a standard piano has frequency 27.5 hertz. What note is this and how many octaves is it below the middle?

(iii)   Julia's range of hearing goes from 75 up to 9000 hertz. How many notes on the standard scale can she hear?

2 Logarithms

In the questions above, you are trying to solve the equation $2^t = 1000$. When you want to solve an equation like $x^2 = 20$, you need to use the square root function, which is the **inverse** of the square function. So to solve the equation $2^t = 1000$, you need the inverse of the exponential function.

You may have used trial and error to find that $2^{10} = 1024$, so the number of bacteria is greater than 1000 after 10 hours. In this equation, you found that the power is 10. Another name for a power is a logarithm, and the inverse of an exponential function is a logarithm.

You can write the statement $2^{10} = 1024$ as $\log_2 1024 = 10$. You say that 'the logarithm to base 2 of 1024 is 10'.

Similarly, since $81 = 3^4$, the logarithm to the base 3 of 81 is 4. The word logarithm is often abbreviated to log and this statement would be written:

$$\log_3 81 = 4.$$

In general, $a^x = b \Leftrightarrow \log_a b = x$.

Example 13.2

(i) Find the logarithm to the base 2 of each of these numbers.

(a) 64   (b) $\frac{1}{2}$ (c) 1   (d) $\sqrt{2}$.

(ii) Show that $2^{\log_2 64} = 64$.

Solution

(i) (a) $64 = 2^6$ and so $\log_2 64 = 6$.

(b) $\frac{1}{2} = 2^{-1}$ and so $\log_2 \frac{1}{2} = -1$.

(c) $1 = 2^0$ and so $\log_2 1 = 0$.

(d) $\sqrt{2} = 2^{\frac{1}{2}}$ and so $\log_2 \sqrt{2} = \frac{1}{2}$.

(ii) Using (i)(a), $2^{\log_2 64} = 2^6 = 64$ as required.

Logarithms to the base 10

Any positive number can be expressed as a power of 10. Before the days of calculators, logarithms to base 10 were used extensively as an aid to calculation. There is no need for that any more but the logarithmic function remains an

Note

Your calculator will have a button for the logarithm to base 10. Check that you know how to use this. Some calculators will also give logarithms to other bases; however it is important that you understand the relationship between logarithms and indices, and do not just rely on your calculator.

important part of mathematics. Base 10 logarithms are a standard feature on calculators, and they also occur in some specialised contexts such as in the pH value of a liquid.

Since $1000 = 10^3$, $\log_{10} 1000 = 3$

Similarly $\qquad \log_{10} 100 = 2$

$$\log_{10} 10 = 1$$

$$\log_{10} 1 = 0$$

$$\log_{10} \tfrac{1}{10} = \log_{10} 10^{-1} = -1$$

$$\log_{10} \tfrac{1}{100} = \log_{10} 10^{-2} = -2 \qquad \text{and so on.}$$

The laws of logarithms

The laws of logarithms follow from those for indices:

Multiplication

Prior knowledge

You should know the laws of indices, covered in Chapter 2.

$$xy = x \times y \longleftarrow \boxed{xy = a^{\log_a(xy)} \text{ and } y = a^{\log_a(y)}}$$

Writing $xy = x \times y$ in the form of powers (or logarithms) to the base a and using the result that $x = a^{\log_a x}$ gives:

$$a^{\log_a xy} = a^{\log_a x} \times a^{\log_a y}$$

$$\Rightarrow \quad a^{\log_a xy} = a^{(\log_a x) + (\log_a y)}$$

$$\Rightarrow \quad \log_a xy = \log_a x + \log_a y$$

Power zero

Since $a^0 = 1$, $\log_a 1 = 0$

Division

You can prove this in a similar way to the multiplication law.

$$\log_a\left(\frac{x}{y}\right) = \log_a x - \log_a y$$

Substituting $x = 1$ and using the result for power zero gives $\log_a \frac{1}{y} = -\log_a y$.

Indices

Since $\qquad x^k = x \times x \times x \times \cdots \times x \qquad$ (k times)

It follows that $\qquad \log_a(x^k) = \log_a x + \log_a x + \cdots + \log_a x \qquad$ (k times)

$$\Rightarrow \log_a(x^k) = k \log_a x$$

This result is also true for non-integer values of k.

Since $\log_a \sqrt[n]{x} = \log_a x^{\frac{1}{n}}$, using the result for indices gives $\log_a \sqrt[n]{x} = \frac{1}{n} \log_a x$.

Special case – the logarithm of a number to its own base

Since $a^1 = a$ for all values of a, it follows that $\log_a a = 1$ but this result, as with all the results for logarithms, is only true for $a > 0$.

Note

It is more usual to state these laws without reference to the base of the logarithms, except where necessary, since the same base is used throughout each result, and this convention is adopted in the Key points at the end of this chapter.

So the laws of logarithms may be written as follows:

Multiplication	$\log xy = \log x + \log y$
Division	$\log\left(\dfrac{x}{y}\right) = \log x - \log y$
Logarithm of 1	$\log 1 = 0$
Powers	$\log x^n = n \log x$
Reciprocals	$\log\left(\dfrac{1}{y}\right) = -\log y$
Roots	$\log \sqrt[n]{x} = \log x^{\frac{1}{n}} = \dfrac{1}{n}\log x$
Logarithm to its own base	$\log_a a = 1$

Example 13.3

(i) Write $3\log p + n\log q - 4\log r$ as a single logarithm.

(ii) Express $\log pq^2\sqrt{r}$ in terms of $\log p$, $\log q$ and $\log r$.

Solution

(i) $3\log p + n\log q - 4\log r = \log p^3 + \log q^n - \log r^4$

$$= \log \frac{p^3 q^n}{r^4}$$

(ii) $\log pq^2\sqrt{r} = \log p + \log q^2 + \log \sqrt{r}$

$$= \log p + 2\log q + \tfrac{1}{2}\log r$$

Example 13.4

Given that $\log y = \log(x - 2) + 2\log 3$, express y in terms of x.

Solution

The first step is to express the right hand side as a single logarithm:

Using the inverse of the law for indices.

$$\log(x - 2) + 2\log 3 = \log(x - 2) + \log 3^2$$

$$= \log[(x - 2) \times 3^2]$$

Using the inverse of the multiplication law.

Using this, $\log y = \log[9(x - 2)]$

$$\Rightarrow y = 9(x - 2)$$

This step is possible since the logarithm function is single-valued.

Graphs of logarithms

The relationship $y = a^x$ may be written as $x = \log_a y$, and so the graph of $x = \log_a y$ is exactly the same as that of $y = a^x$. Interchanging x and y has the effect of reflecting the graph in the line $y = x$, and changing the relationship into $y = \log_a x$, as shown in Figure 13.5.

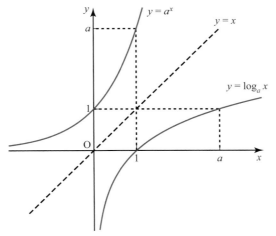

Figure 13.5

ACTIVITY 13.2

Use a graphical calculator to draw the graph of $y = 10^x$ and $y = \log_{10} x$ on the same axes. Describe the relationship between the two graphs.

The logarithmic function $y = \log_a x$ is the inverse of the exponential function $y = a^x$. The effect of applying a function followed by its inverse is to bring you back to where you started.

Thus $\log_a(a^x) = x$ ← | Applying the exponential function to x followed by the logarithmic function brings you back to x.

| Applying the logarithmic function to x followed by the exponential function brings you back to x. | → and $a^{(\log_a x)} = x$

The idea that you can 'undo' an exponential function by applying a logarithmic function is useful in solving equations, as shown in the next example.

Example 13.5

Solve the equation $2^n = 1000$.

Solution

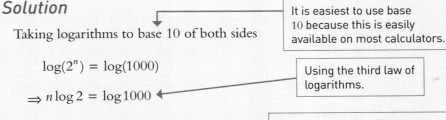

Taking logarithms to base 10 of both sides | It is easiest to use base 10 because this is easily available on most calculators.

$$\log(2^n) = \log(1000)$$

$$\Rightarrow n \log 2 = \log 1000$$ ← | Using the third law of logarithms.

$$\Rightarrow \quad n = \frac{\log 1000}{\log 2} = 9.97 \text{ (3 s.f.)}$$

| You can use your calculator to find the top and bottom lines separately: $\log_{10} 1000 = 3$ and $\log_{10} 2 = 0.301\ldots$ However, you would usually enter it all as a single calculation.

> **Note**
>
> You could solve the above equation more directly by taking logarithms to base 2 of both sides, giving $n = \log_2 1000$. Some calculators have functions to calculate the logarithm to any base, but it is often easier to use logarithms to base 10.

Logarithms

Exercise 13.2

 ① Simplify the following.

$a^{\log_a 5}$ $\qquad \log_a a^{4.2} \qquad 3^{\log_3 7}$

$\log_{10} 10^{3.1} \qquad 10^{\log_{10} x} \qquad \log_c c^x$

 ② Which of the following is **not** equivalent to $\log_2 8$?

A $\log_2 2 + \log_2 4$ $\qquad$ B $3\log_2 2$

C 3 $\qquad\qquad\qquad\qquad$ D 4

 ③ $2^x = 32 \Leftrightarrow x = \log_2 32$

Write similar logarithmic equivalents of these equations. In each case find also the value of x, using your knowledge of indices.

(i) $3^x = 9$ $\qquad$ (ii) $4^x = 64$

(iii) $2^x = \frac{1}{4}$ $\qquad$ (iv) $5^x = \frac{1}{5}$

(v) $7^x = 1$ $\qquad$ (vi) $16^x = 2$

④ Write the equivalent of these equations in exponential form. Find also the value of y in each case.

(i) $y = \log_3 81$ $\qquad$ (ii) $y = \log_5 125$

(iii) $y = \log_4 2$ $\qquad$ (iv) $y = \log_6 1$

(v) $y = \log_5 \frac{1}{125}$

⑤ Write the following expressions in the form $\log x$, where x is a number.

(i) $\log 5 + \log 2$ $\qquad$ (ii) $\log 6 - \log 3$

(iii) $2\log 6$ $\qquad$ (iv) $-\log 7$

(v) $\frac{1}{2}\log 9$

⑥ Express the following in terms of $\log x$.

(i) $\log x^2$ $\qquad$ (ii) $\log x^5 - 2\log x$

(iii) $\log \sqrt{x}$ $\qquad$ (iv) $3\log x + \log x^3$

(v) $\log(\sqrt{x})^5$

⑦ Sketch each of the graphs below. Show the vertical asymptote and the coordinates of the point where the graph crosses the x axis.

(i) $y = \log_{10} x$

(ii) $y = \log_2(x + 1)$

(iii) $y = \log_3(x - 2)$

⑧ Write the following expressions in the form $\log x$, where x is a number.

(i) $\frac{1}{4}\log 16 + \log 2$

(ii) $\log 5 + 3\log 2 - \log 10$

(iii) $\log 12 - 2\log 2 - \log 9$

(iv) $\frac{1}{2}\log \sqrt{16} + 2\log \frac{1}{2}$

(v) $2\log 4 + \log 9 - \frac{1}{2}\log 144$

⑨ (i) Express $2\log x - \log 7$ as a single logarithm.

(ii) Hence solve $2\log x - \log 7 = \log 63$.

⑩ Use logarithms to base 10 to solve the following equations.

(i) $2^x = 1\,000\,000$ (ii) $2^x = 0.001$

(iii) $1.08^x = 2$ $\qquad$ (iv) $1.1^x = 100$

(v) $0.99^x = 0.000\,001$.

PS ⑪ The strength of an earthquake is recorded on the Richter scale. This provides a measure of the energy released. A formula for calculating earthquake strength is

Earthquake strength

$$= \log_{10}\left(\frac{\text{Energy released in joules}}{63\,000}\right)$$

(i) The energy released in an earthquake is estimated to be 2.5×10^{11} joules.

What is its strength on the Richter scale?

(ii) An earthquake is 7.4 on the Richter scale. How much energy is released in it?

(iii) An island had an earthquake measured at 4.2 on the Richter scale. Some years later it had another earthquake and this one was 7.1. How many times more energy was released in the second earthquake than in the first one?

⑫ The loudness of a sound is usually measured in decibels. A decibel is $\frac{1}{10}$ of a bel, and a bel represents an increase in loudness by a factor of 10. So a sound of 40 decibels is 10 times louder than a sound of 30 decibels; similarly a sound of 100 decibels is 10 times louder than one of 90 decibels.

PS

(i) Solve the equation $x^{10} = 10$.

(ii) Sound A is 35 decibels and sound B is 36 decibels. Show that (to the nearest whole number) B is 26% louder than A.

(iii) How many decibels increase are equivalent to a doubling in loudness? Give your answer to the nearest whole number.

(iv) Jamie says 'The percentage increase in loudness from 35 to 37 decibels is given by

$$\frac{37 - 35}{35} \times 100 = 5.7\%'.$$

Explain Jamie's mistake.

⑬ The acidity of a liquid is measured as its pH value. This is defined as

$$pH = \log_{10}\left(\frac{1}{\text{Concentration of } H^+ \text{ ions in moles per litre}}\right).$$

(i) Calculate the pH values of

(a) pure water with an H^+ concentration of 10^{-7} moles per litre

(b) the liquid in the soil in a field with an H^+ concentration of 5×10^{-6} moles per litre

(c) the water from a natural spring with an H^+ concentration of 1.6×10^{-8} moles per litre.

(ii) Figure 13.6 illustrates the colours of indicator paper when placed in liquids of different acidity levels.

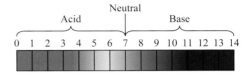

Figure 13.6 The pH scale

You can see that between pH values of about 4.4 and 5.5 it can be described as orange. What are the corresponding concentrations of H^+ ions?

3 The exponential function

Example 13.6

Emma needs to borrow £1000 to go to her friend's hen party. The loan shark says his interest rate will be 100% per annum but he may charge a corresponding smaller amount more often, like 50% every 6 months or 25% every 3 months.

(i) How much does Emma owe at the end of one year if she makes no repayments and the interest is charged

(a) once a year

(b) every 6 months

(c) every 3 months

(d) every month

(e) every day

(f) every hour.

(ii) After one year, Emma's debt to the loan shark has increased by a factor f so that it becomes $f \times £1000$. What is the greatest value that f can take?

Solution

(i) (a) 100% interest rate means that the debt is doubled.
So the debt after a year is £1000 × 2 = £2000.

(b) Every six months means that the interest rate is 50%, so the debt is multiplied by 1.5 twice a year.

So after a year the debt is £1000 × 1.5² = £2250.

(c) This is a rate of 25% four times a year, so the debt is
$$£1000 × 1.25^4 = £2441.41$$

(d) This is a rate of $\frac{100}{12}$% twelve times a year so the debt is
$$£1000\left(1 + \frac{1}{12}\right)^{12} = £2613.04$$

(e) $£1000\left(1 + \frac{1}{365}\right)^{365} = £2714.57$

(f) $£1000\left(1 + \frac{1}{8760}\right)^{8760} = £2718.13$

(ii) If the interest is charged n times per year, the debt at the end of the year is $1000(1 + \frac{1}{n})^n$.

As n becomes very large, $(1 + \frac{1}{n})^n$ approaches the number 2.71828..., which is therefore the greatest value that f can take.

The number found in part (ii) of Example 13.6 is a very important number, and is denoted by the letter e.

You will find a button for e on your calculator. It is an irrational number, like π, and (also like π) appears in many different areas of mathematics and has many interesting properties.

Continuous compound interest

Sometimes, interest is calculated continuously, which means that the interest is added at infinitesimally small intervals. In Example 13.6, if the loan shark used continuous compound interest, Emma's debt after a year would be 1000e, and after t years it would be $1000e^t$. Of course, interest rates are not usually 100%, so if the interest rate is r%, the debt would be $1000e^{\frac{r}{100} \times t}$. If r is expressed as a decimal the formula is $1000e^{rt}$.

In general, continuous compound interest on an investment is given by
$$A = Pe^{\frac{r}{100} \times t}$$

where A = the final amount, P = the amount invested, r% is the annual interest rate and t is the number of years that the money is invested.

Example 13.7

An amount of £5000 is deposited in a bank paying an annual interest rate of 5%, compounded continuously. If no withdrawals are made, how much will the investment be worth in 3 years' time?

Discussion point

➜ How much more is this than the amount that would be received if the interest was added just once per year?

Solution

Using the continuous compound interest formula, $A = Pe^{\frac{r}{100} \times t}$, with $P = 5000$, $r = 5$% and $t = 3$

$$A = Pe^{\frac{r}{100} \times t}$$

| Note that 5% is $\frac{5}{100}$ or 0.05.

$$\Rightarrow A = (5000)e^{(0.05 \times 3)}$$

$$\Rightarrow A = 5809.171......$$

The investment would be worth £5809.17.

ACTIVITY 13.3

Using any graphing software available to you, draw the graph of $y = e^x$ and find the gradient of the graph at several different points. Write down the x and y coordinates of each point, and the gradient at that point. What do you notice?

In Activity 13.3, you probably noticed that for the graph of $y = e^x$, the gradient is equal to the y coordinate at any point. Therefore for $y = e^x$, $\dfrac{dy}{dx} = e^x$.

More generally, if $y = e^{kx}$ then $\dfrac{dy}{dx} = ke^{kx}$. ⟵

So the rate of increase of an exponential quantity at any time is proportional to its value at that time.

> You can check this using graphing software, as in Activity 13.3.

Example 13.8

The number N of bacteria in a culture grows according to the formula $N = 50 + 10e^{0.1t}$, where N is the number of bacteria present and t is the time in days from the start of the experiment.

(i)   How many bacteria are present at the start of the experiment?

(ii) Calculate the number of bacteria after

    (a)   one week   (b)   10 weeks.

(iii) What assumption have you made in part (ii)?

(iv) Calculate the rate of growth after

    (a)   one week   (b)   10 weeks.

Solution

(i)   Initially, $t = 0$ so the number present $= 50 + 10e^0 = 60$.

(ii) (a)   After one week, $t = 7$ so the number present $= 50 + 10e^{(0.1 \times 7)}$

$$= 70.13$$

      So there are 70 bacteria after one week.

    (b)   After 10 weeks, $t = 70$ so the number present $= 50 + 10e^{(0.1 \times 70)}$

$$= 11\,016.33 \ldots$$

      So there are 11 016 bacteria after 10 weeks.

(iii) It has been assumed that none of them die off.

(iv) The rate of growth is given by $\dfrac{dN}{dt}$.

$$N = 50 + 10e^{0.1t} \rightarrow \frac{dN}{dt} = 10 \times (0.1)e^{0.1t}$$

$$= e^{0.1t}$$

> Using the result that if $y = e^{kx}$ then $\dfrac{dy}{dx} = ke^{kx}$.

    (a)   After one week, $\dfrac{dN}{dt} = e^{(0.1 \times 7)} = 2.0137 \ldots$

      The rate of growth after one week is approximately 2 bacteria per day.

    (b)   After 10 weeks, $\dfrac{dN}{dt} = e^{(0.1 \times 70)} = 1096.63 \ldots$

      The rate of growth after 10 weeks is approximately 1096 bacteria per day.

Example 13.9

A radioactive substance of initial mass 200 g is decaying so that after t days the amount remaining is given by $M = 200e^{-0.0015t}$.

(i) Sketch the graph of M against t.

(ii) How much of the substance remains after one year (take 1 year = 365 days)?

(iii) What is the rate of decay of the substance after 200 days?

Solution

(i) The graph of $M = 200e^{-0.0015t}$ has the standard shape for a negative exponential curve and passes through $(0, 200)$.

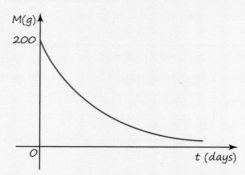

Figure 13.7

(ii) When $t = 365$, $M = 200 \times e^{-(0.0015 \times 365)} = 115.67\ldots$

The amount remaining after 1 year is 116 g.

(iii) To find the rate of decay start by differentiating M with respect to t. This gives the rate at which the mass changes.

$$M = 200e^{-0.0015t}$$

$$\Rightarrow \frac{dM}{dt} = 200 \times (-0.0015)e^{-0.0015t}$$

$$\Rightarrow \frac{dM}{dt} = -0.3e^{-0.0015t}$$

When $t = 200$, $\dfrac{dM}{dt} = -0.3 \times e^{-0.0015 \times 200}$

$$\Rightarrow \frac{dM}{dt} = -0.222245\ldots$$

The rate of decay after 200 days is 0.22 g per day.

> Note that the rate of decay is a positive quantity, since the negative sign is included in the word 'decay'.

Exercise 13.3

 ① Match the transformations of e^x to the graphs in Figure 3.8.

(i) translation of $\begin{pmatrix} 0 \\ 2 \end{pmatrix}$

(ii) stretch of 2 in the y direction

(iii) reflection in the x axis

A

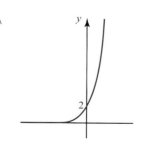

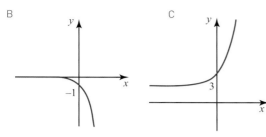

Figure 13.8

 ② $y = e^x$ is stretched by a scale factor of 5 in the x-direction. What is the equation of the new function?

A $y = e^x + 5$ B $y = 5e^x$

C $y = e^{5x}$ D $y = e^{\frac{x}{5}}$

③ Sketch each of the graphs below. Show the horizontal asymptote and the coordinates of the point where the graph crosses the y axis.

(i) $y = e^x$ (ii) $y = e^x + 1$

(iii) $y = e^x - 1$

④ Sketch each of the graphs below. Show the horizontal asymptote and the coordinates of the point where the graph crosses the y axis.

(i) $y = e^{-2x}$ (ii) $y = e^{-2x} + 1$

(iii) $y = e^{-2x} - 1$

⑤ Differentiate the following.

(i) $y = e^{2x}$ (ii) $y = e^{-3x}$

(iii) $y = e^{0.5x}$ (iv) $y = e^{-6x}$

⑥ Sketch the following curves, giving the equations of any asymptotes and the coordinates of the points where they cross the y axis.

(i) $y = 5 + 2e^{-x}$ (ii) $y = 5 - 2e^{-x}$

(iii) $y = 5 + 2e^x$ (iv) $y = 5 - 2e^x$

M ⑦ The value £V of a car after t years is given by the formula $V = 20000e^{-\frac{t}{10}}$.

(i) State the new value of the car.

(ii) Find its value (to the nearest £) after 3 years.

(iii) Find the rate at which its value is falling

(a) when it is new

(b) after 2 years.

(iv) Sketch the graph of V against t.

(v) Comment on how good you think the model is.

⑧ The value £A of an investment after t years varies according to the formula $A = Pe^{\frac{t}{20}}$.

(i) If Marcus invests £2000, how much (to the nearest £) will his investment be worth after 2 years?

(ii) Find how much (to the nearest £) he would need to invest to get £10000 after 5 years.

M ⑨ A colony of humans settles on a previously uninhabited planet. After t years their population, P, is given by $P = 100e^{0.05t}$.

(i) How many settlers land on the planet initially?

(ii) Sketch the graph of P against t.

(iii) Find the rate at which the population is increasing after 10 years.

(iv) What is the population after 50 years?

(v) What is the minimum number needed to settle for the population to be at least 2000 after 50 years?

PS **M** ⑩ The mass m of the radioactive substance fermium-253 is given by the formula $m = m_0e^{-kt}$, where t is measured in minutes and m_0 is the initial mass.

The half-life of a radioactive substance is the time it takes for half of the mass of the substance to decay. Fermium-253 has a half-life of 3 minutes. Find the value of k.

⑪ Samantha deposits £10000 in an investment paying an annual interest rate of 4.7% compounded continuously.

(i) If no withdrawals are made, how much (to the nearest £) will her investment be worth in 5 years' time?

PS (ii) However, after 3 years she needs to withdraw £5000 to pay the deposit on a new car. After this withdrawal, how long will it be until her investment is worth £10000 again, assuming that interest is paid at the same rate?

PS ⑫ (i) Show that the curves $y = e^x$ and $y = x^e$ both pass through the point (e, e^e) and find their gradients at that point.

(ii) Draw a sketch showing the two curves.

(iii) Without using your calculator, say which is greater, π^e or e^π.

4 The natural logarithm function

You have already met $y = \log_{10} x$ as the inverse of $y = 10^x$, and similarly the inverse of e^x is $\log_e x$. This is called the **natural logarithm** and is denoted by $\ln x$.

Most calculators have a button for $\ln x$. ◄————— | ln is short for 'log natural'.

The natural logarithm function obeys all the usual laws of logarithms.

Since $\ln x$ and e^x are inverses of each other, it follows that

$$\ln(e^x) = x$$

and

$$e^{\ln x} = x.$$

Example 13.10

Make x the subject of the formula $a \ln 2x = b$.

Solution

$$a \ln 2x = b$$
$$\Rightarrow \ln 2x = \frac{b}{a}$$
$$\Rightarrow \quad 2x = e^{\frac{b}{a}}$$
$$\Rightarrow \quad x = \tfrac{1}{2} e^{\frac{b}{a}}$$

Applying the exponential function to both sides: $e^{\ln x} = x$ so $e^{\ln 2x} = 2x$.

Example 13.11

A radioactive substance of initial mass $200\,\text{g}$ is decaying so that after t days the amount remaining is given by $M = 200e^{-0.0015t}$.

The half-life of the substance is the time it takes to decay to half of its initial mass. Find the half-life of the substance.

> **Note**
>
> You have already solved equations of the form $a^x = k$ by using logarithms to base 10. If you are using logarithms to solve an equation that involves e then it is convenient to use the natural logarithm function, $\ln$.

Solution

$$M = 200e^{-0.0015t}$$

When $M = 100$, $100 = 200e^{-0.0015t}$

$$\Rightarrow 0.5 = e^{-0.0015t}$$
$$\Rightarrow \ln 0.5 = -0.0015t$$
$$\Rightarrow \quad t = -\frac{\ln 0.5}{0.0015}$$
$$\Rightarrow \quad t = 462.098\ldots \text{ days.}$$

Taking natural logarithms of both sides: $\ln(e^x) = x$ so $\ln(e^{-0.0015t}) = -0.0015t$.

The half-life of the substance is 462 days.

Example 13.12

Find the coordinates of the point on the curve $y = e^{3x}$ where the gradient is 10.

> **Note**
>
> Since
> $$y = e^{3x} \Rightarrow \frac{dy}{dx} = 3e^{3x}$$
> it follows that
> $$\frac{dy}{dx} = 3y.$$
> This provides an alternative way of finding the y coordinate when the gradient is 10.

Solution

$$y = e^{3x} \Rightarrow \frac{dy}{dx} = 3e^{3x}$$

$$\frac{dy}{dx} = 10 \Rightarrow 3e^{3x} = 10$$

$$\Rightarrow \quad e^{3x} = \frac{10}{3}$$

$$\Rightarrow \quad 3x = \ln\left(\frac{10}{3}\right)$$

Taking $\ln$ of both sides to 'undo' the exponential: $\ln e^{3x} = 3x$.

$$\Rightarrow \qquad x = 0.40132\ldots$$
$$y = e^{3x} \Rightarrow \qquad y = e^{(3 \times 0.40132\ldots)} = 3.3333\ldots$$
$$\text{The point is } (0.401, 3.33) \text{ (3 s.f.)}$$

Exercise 13.4

① Simplify:

$e^{\ln 2}$ $\quad e^{\ln x}$ $\quad (e^{\ln 2})^x$ $\quad (e^{\ln 5})^t$

$\ln e^x$ $\quad \ln e^{3.8}$ $\quad \ln e^{4t}$

② Which graph shows $y = \ln x$?

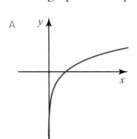
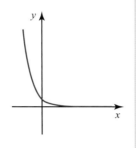
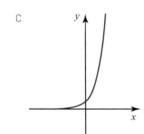
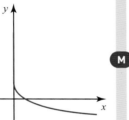

Figure 13.9

③ Sketch the following pairs of graphs on the same axes. Find the exact coordinates of the points where the graphs cut the coordinate axes.

(i) $y = \ln x$ and $y = \ln(x + 1)$

(ii) $y = \ln x$ and $y = \ln 2x$

(iii) $y = \ln x$ and $y = 2 + \ln x$

④ (i) Make x the subject of $\ln x - \ln x_0 = kt$

(ii) Make t the subject of $s = s_0 e^{-kt}$

(iii) Make p the subject of $\ln\left(\dfrac{p}{25}\right) = -0.02t$

(iv) Make x the subject of $y - 5 = (y - y_0)e^x$

⑤ Solve the following equations, giving your answers in exact form.

(i) $e^x = 6$ (ii) $e^{2x} = 6$

(iii) $e^{x-1} = 6$ (iv) $\ln x = 5$

(v) $\ln \dfrac{x}{2} = 5$ (vi) $\ln 2x = 5$

⑥ Find the coordinates of the point on the curve $y = e^{-2x}$ where the gradient is -1.

(M) ⑦ Alex lives 800 metres from school. One morning he set out at 8.00 a.m. and t minutes later the distance $s\,\mathrm{m}$ which he had walked was given by $s = 800(1 - e^{-0.1t})$.

(i) Sketch the graph of s against t.

(ii) How far had he walked by 8.15 a.m.?

(iii) At what time has Alex walked half way to school?

(iv) Registration is at 8.45 a.m. Does Alex get to school in time for this?

(v) When does Alex get to school?

(M) ⑧ A parachutist jumps out of an aircraft and some time later opens the parachute. Her speed at time t seconds from when the parachute opens is $v\,\mathrm{m\,s^{-1}}$, where $v = 8 + 22e^{-0.07t}$.

(i) Sketch the graph of v against t.

(ii) State the speed of the parachutist when the parachute opens, and the final speed that she would attain if she jumped from a very great height.

(iii) Find the value of v as the parachute lands, 60 seconds later.

(iv) Find the value of t when the parachutist is travelling at $20\,\mathrm{m\,s^{-1}}$.

PS ⑨ A curve has equation $y = 8e^x - 4e^{2x}$.

(i) State the exact coordinates of the points at which it intersects the axes.

(ii) Prove that the curve has only one turning point and that it lies on the y axis.

(iii) Sketch the graph.

5 Modelling curves

When you obtain experimental data, you are often hoping to establish a mathematical relationship between the variables in question. Should the data fall on a straight line, you can do this easily because you know that a straight line with gradient m and intercept c has equation $y = mx + c$.

Example 13.13

In an experiment the temperature θ (in °C) was measured at different times t (in seconds), in the early stages of a chemical reaction.

The results are shown in Table 13.1.

Table 13.1

t	20	40	60	80	100	120
θ	16.3	20.4	24.2	28.5	32.0	36.3

(i)   Plot a graph of θ against t.

(ii) What is the relationship between θ and t?

Solution

(i)

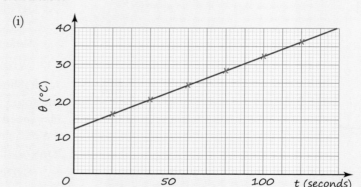

Figure 13.10

(ii) Figure 13.10 shows that the points lie reasonably close to a straight line and so it is possible to estimate its gradient and intercept.

Intercept: $c \approx 12.5$

Gradient: $m \approx \dfrac{23 - 18}{54 - 28}$

$\approx 0.19\ldots$

> To estimate the gradient as accurately as possible, choose two points on the line that lie as close as possible to grid intersections. In this case, two suitable points are $(28, 18)$ and $(54, 23)$.

In this case the equation is not $y = mx + c$ but $\theta = mt + c$, and so is given by

$$\theta = 0.2t + 12.5$$

> Because it is an estimate, it is sensible to give the gradient to just one decimal place in this case.

It is often the case, however, that your results do not end up lying on a straight line but on a curve, so that this straightforward technique cannot be applied. The appropriate use of logarithms can convert some curved graphs into straight lines. This is the case if the relationship has one of two forms, $y = kx^n$ or $y = ka^x$.

The techniques used in these two cases are illustrated in the following examples. In theory, logarithms to any base may be used, but in practice you would only use those available on your calculator: logarithms to the base 10 or natural logarithms.

Relationships of the form $y = kx^n$

Example 13.14

A water pipe is going to be laid between two points and an investigation is carried out as to how, for a given pressure difference, the rate of flow R litres per second varies with the diameter of the pipe d cm. The following data are collected.

Table 13.2

d	1	2	3	5	10
R	0.02	0.32	1.62	12.53	199.80

It is suspected that the relationship between R and d may be of the form

$$R = kd^n$$

where k is a constant.

(i) Explain how a graph of log d against log R tells you whether this is a good model for the relationship.

(ii) Make a table of values of $\log_{10} d$ against $\log_{10} R$ and plot these on a graph.

(iii) If appropriate, use your graph to estimate the values of n and k.

Solution

(i) If the relationship is of the form $R = kd^n$, then taking logarithms gives

$$\log R = \log k + \log d^n$$

or $\log R = n \log d + \log k$.

This equation is of the form $y = mx + c$, with log R replacing y, n replacing m, log d replacing x and log k replacing c. It is thus the equation of a straight line.

Consequently if the graph of log R against log d is a straight line, the model $R = kd^n$ is appropriate for the relationship and n is given by the gradient of the graph. The value of k is found from the intercept, log k, of the graph with the vertical axis.

$$\log_{10} k = \text{intercept} \Rightarrow k = 10^{\text{intercept}}$$

(ii) Working to 2 decimal places (you would find it hard to draw the graph to greater accuracy) the logarithmic data are as follows.

Table 13.3

$\log_{10} d$	0	0.30	0.48	0.70	1.00
$\log_{10} R$	−1.70	−0.49	0.21	1.10	2.30

(iii)

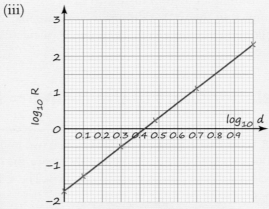

Figure 13.11

In this case the graph in Figure 13.11 is indeed a straight line, with gradient 4 and intercept −1.70,

so $n = 4$ and $c = 10^{-1.70} = 0.020$ (to 2 s.f.).

The proposed equation linking R and d is a good model for their relationship, and may be written as

$$R = 0.02d^4.$$

Note

In Example 13.14, logs to base 10 were used but the same result could have been obtained using logs to base e, or indeed to any base, although bases 10 and e are the only ones in general use.

Exponential relationships

Example 13.15

The temperature, θ in °C, of a cup of coffee at time t minutes after it is made is recorded as follows.

Table 13.4

t	2	4	6	8	10	12
θ	81	70	61	52	45	38

(i)   Plot the graph of θ against t.

(ii) Show how it is possible, by drawing a suitable graph, to test whether the relationship between θ and t is of the form $\theta = ka^t$, where k and a are constants.

(iii) Carry out the procedure.

Solution

(i)

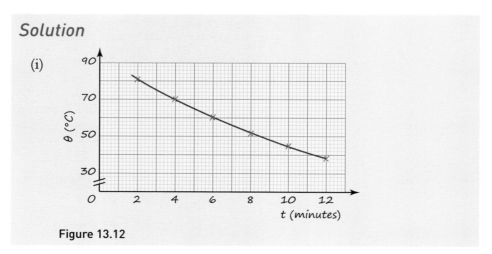

Figure 13.12

(ii) If the relationship is of the form $\theta = ka^t$, taking logarithms of both sides (to base e this time, although the same end result would be obtained using any base) gives

$$\ln \theta = \ln k + \ln a^t$$

$$\Rightarrow \ln \theta = t \ln a + \ln k$$

This is an equation of the form $y = mx + c$ where $y = \ln \theta$, $m = \ln a$, $x = t$ and $c = \ln k$.

This is the equation of a straight line and consequently, if the graph of $\ln \theta$ against t is a straight line, the model $\theta = ka^t$ is appropriate for the relationship.

Measuring the gradient gives the value of $\ln a$, so $a = e^{\text{gradient}}$.

The intercept on the y axis is $\ln k$, so $k = e^{\text{intercept}}$.

(iii) Table 13.5 gives values of $\ln \theta$ for the given values of t.

Table 13.5

t	2	4	6	8	10	12
$\ln \theta$	4.39	4.25	4.11	3.95	3.81	3.64

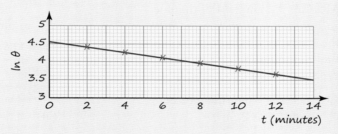

Figure 13.13

The graph is a straight line so the proposed model is appropriate.

The gradient is -0.08 and so $a = e^{-0.08} = 0.92$

The intercept is 4.55 and so $k = e^{4.55} = 94.6$

The relationship between θ and t is $\theta = 94.6 \times 0.92^t$

Exercise 13.5

In this exercise, you will be estimating values as a result of taking measurements from graphs, and these will at best be approximate. It would be misleading to think that you could achieve much better than 2 s.f. accuracy in this type of question.

 ① Match the equivalent models.

$y = ab^x \qquad y = ax^b \qquad y = ba^x \qquad y = bx^a$

$\ln y = \ln b + x \ln a \qquad \ln y = \ln a + b \ln x$

$\ln y = \ln a + x \ln b \qquad \ln y = \ln b + a \ln x$

 ② Which is **not** an appropriate graph to draw for $P = ka^t$?

A $\ln P = \ln k + t \ln a$

B $\ln P = t \ln a + \ln k$

C $\log_{10} P = \log_{10} k + t \log_{10} a$

D $\log_{10} P = a \log_{10} t + \log_{10} k$

③ For each of the following models, where k, a and b are constants, use logs to write the relationship in the form $y = mx + c$. Write down expressions for m and c in each case.

(i) $A = kt^b$ (ii) $N = kt^a$

(iii) $P = kV^n$

④ The area $A \, \text{cm}^2$ occupied by a patch of mould is measured each day.

It is believed that A may be modelled by a relationship of the form $A = kb^t$, where t is the time in days.

(i) By taking logarithms to base 10 of each side, show that the model may be written as $\log A = \log k + t \log b$.

(ii) Figure 13.14 shows $\log A$ plotted on the vertical axis against t on the horizontal axis.

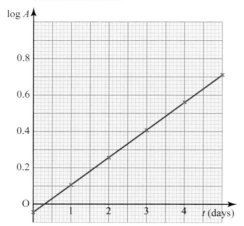

Figure 13.14

Measure the gradient of the line and use it to calculate b and use the intercept on the vertical axis to find the value of k.

(iii) Using these values, estimate

(a) the time when the area of mould was $2\,\text{cm}^2$.

(b) the area of the mould after 3.5 days.

⑤ The time after a train leaves a station is recorded in minutes as t and the distance that it has travelled in metres as s. It is suggested that the relationship between s and t is of the form $s = kt^n$ where k and n are constants.

(i) By taking logarithms to base e of each side, show that the model may be written as $\ln s = \ln k + n \ln t$.

(ii) Figure 13.15 shows $\ln s$ plotted on the vertical axis against $\ln t$ on the horizontal axis.

Measure the gradient of the line and use it to calculate n and use the intercept on the vertical axis to find the value of k.

(iii) Estimate how far the train travelled in the first 100 seconds.

(iv) Explain why you would be wrong to use your results to estimate the distance the train has travelled after 10 minutes.

⑥ The inhabitants of an island are worried about the rate of deforestation taking place. A research worker uses records from the last 200 years to estimate the number of trees at different dates. Her results are as follows:

Table 13.6

Year	1800	1860	1910	1930	1950	1970	1990	2010
Trees	3000	2900	3200	2450	1800	1340	1000	740

NB The number of trees is measured in thousands, so in 1800 there were 3 million trees.

From the table it appears that deforestation may have begun in 1910 when there were 3 200 000 trees. It is suggested that the number of trees has been decreasing exponentially since 1910, so that the number of trees N may be modelled by the equation

$$N = ka^t$$

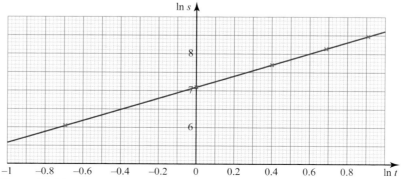

Figure 13.15

where t is the number of years since 1910.

(i) By taking logarithms to base e of both sides show that the model can be written as

$$\ln N = \ln k + t \ln a$$

(ii) Make a table of values for t and $\ln N$, starting from the year 1910 (when $t = 0$).

(iii) The equation in (i) may be written as $(\ln N) = (\ln a)t + (\ln k)$, and in this form it compares directly with the equation $y = mx + c$ for a straight line. Draw a graph to represent this relationship, plotting the values of t on the horizontal axis and the values of $\ln N$ on the vertical axis.

(iv) Measure the gradient of the line and use it to calculate a, and use the intercept on the vertical axis to calculate k.

(v) If this continues to be an appropriate model for the deforestation, in what year can the number of trees be expected to fall below $500\,000$?

⑦ The planet Saturn has many moons. The table below gives the mean radius of orbit and the time taken to complete one orbit, for five of the best known of them.

Table 13.7

Moon	Tethys	Dione	Rhea	Titan	Iapetus
Radius R ($\times 10^5$ km)	2.9	3.8	5.3	12.2	35.6
Period T (days)	1.9	2.7	5.4	15.9	79.3

It is believed that the relationship between R and T is of the form $R = kT^n$.

(i) By taking logarithms to base 10 of both sides show that the model can be written as $\log R = \log k + n \log T$.

(ii) Add the values of $\log R$ and $\log T$ to the table above.

(iii) The equation in (i) may be written as $(\log R) = n(\log T) + (\log k)$, and in this form it compares directly with the equation $y = mx + c$ for a straight line. Draw a graph to represent this relationship, plotting the values of $\log T$ on the horizontal axis and the values of $\log R$ on the vertical axis.

(iv) Measure the gradient of the line and use it to calculate n, and use the intercept on the vertical axis to calculate k. Write down the equation in the form $R = kT^n$.

In 1980 a Voyager spacecraft photographed several previously unknown moons of Saturn.

One of these, named 1980 S.27, has a mean orbital radius of 1.4×10^5 km.

(v) Estimate how many days it takes this moon to orbit Saturn.

⑧ (i) After the introduction of a vaccine, the number of new cases of a virus infection each week is given in the table below.

Table 13.8

Week number, t	1	2	3	4	5
No. of new cases, y	240	150	95	58	38

Plot the points (t, y) on graph paper and join them with a smooth curve.

(ii) Decide on a suitable model to relate t and y, and plot an appropriate graph to find this model. (You may wish to use a spreadsheet or graphing software.)

(iii) Find the value of y when $t = 15$, and explain what you find.

⑨ A local newspaper wrote an article about the number of residents using a new online shopping site based in the town. Based on their own surveys, they gave the following estimates.

Table 13.9

Time in weeks t	2	3	4	5	6
Number of people P	4600	5000	5300	5500	5700

(i) Investigate possible models for the number of people, P, using the site at time t weeks after the site began, and determine which one matches the data most closely.

(ii) Use your model to estimate how many people signed on in the first week that the site was open.

(iii) Discuss the long term validity of the model.

LEARNING OUTCOMES

When you have completed this chapter you should be able to:

➤ know and use the function a^x and its graph, where a is positive

➤ know and use the function e^x and its graph

➤ know that the gradient of e^{kx} is equal to ke^{kx} and hence understand why the exponential model is suitable in many applications

➤ know and use the definition of $\log_a x$ as the inverse of a^x, where a is positive and $x \geq 0$

➤ know and use the function $\ln x$ and its graph

➤ know and use $\ln x$ as the inverse function of e^x

➤ understand and use the laws of logarithms:

  ○ $\log_a x + \log_a y = \log_a xy$

  ○ $\log_a x - \log_a y = \log_a \frac{x}{y}$

  ○ $k \log_a x = \log_a x^k$ including $k = -1$ or $-\frac{1}{2}$

➤ solve equations of the form $a^x = b$

➤ use logarithmic graphs to estimate parameters in relationships of the form $y = ax^n$ and $y = kb^x$, given data for x and y

➤ understand and use exponential growth and decay

➤ use in modelling (examples may include the use of e in continuous compound interest, radioactive decay, drug concentration decay, exponential growth as a model for population growth)

➤ consider the limitations and refinements of exponential models.

KEY POINTS

1 A function of the form a^x is described as exponential.

2 $y = e^{kx} \Rightarrow \dfrac{dy}{dx} = ke^{kx}$

3 $y = \log_a x \Leftrightarrow a^y = x$

4 $\log_e x$ is called the natural logarithm of x and is denoted by $\ln x$.

5 The rules for logarithms to any base are:

- Multiplication $\qquad\qquad\qquad \log xy = \log x + \log y$

- Division $\qquad\qquad\qquad\qquad \log\left(\dfrac{x}{y}\right) = \log x - \log y$

- Logarithm of 1 $\qquad\qquad\quad \log 1 = 0$

- Powers $\qquad\qquad\qquad\qquad \log x^n = n \log x$

- Reciprocals $\qquad\qquad\qquad \log\left(\dfrac{1}{y}\right) = -\log y$

- Roots $\qquad\qquad\qquad\qquad \log \sqrt[n]{x} = \log x^{\frac{1}{n}} = \dfrac{1}{n}\log x$

- Logarithm to its own base $\quad \log_a a = 1$

6 Logarithms may be used to discover the relationship between variables in two types of situation:

(i) $y = kx^n \Leftrightarrow \log y = n \log x + \log k$

Plotting $\log y$ against $\log x$ gives a straight line where n is the gradient and $\log k$ is the y axis intercept.

(ii) $y = ka^x \Leftrightarrow \log y = (\log a)x + \log k$

Plotting $\log y$ against x gives a straight line where $\log a$ is the gradient and $\log k$ is the y axis intercept.

FUTURE USES

- In the A-level book you will learn more about differentiating and integrating exponential and logarithmic functions.

- If you study Further Mathematics, exponential functions are important in many topics, such as solving differential equations.

PRACTICE QUESTIONS: PURE MATHEMATICS 3

MP ①

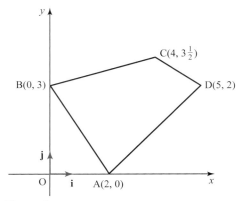

Figure 1

Figure 1 shows the coordinates of the vertices of a quadrilateral ABCD.

Use vectors to prove that ABCD is a trapezium. [3 marks]

MP ② The population P of a town at time t years is modelled by the equation $P = ae^{kt}$.

When $t = 0$, the population is 52 300, and after 5 years, it is 58 500.

(i) Find the values of a and k. [4 marks]

(ii) Using the model, which of the following is an estimate of the population, to the nearest 100, after 8 years:

62654, 427900, 62600, 427878 [1 mark]

MP ③ (i) Find $\int_{-1}^{1}\left(3x - x^3\right)\mathrm{d}x$. [3 marks]

(ii) Given that $y = 3x - x^3$, find $\frac{\mathrm{d}y}{\mathrm{d}x}$.

Hence verify that one of the turning points of the curve $y = 3x - x^3$ is (1, 2) and find the other one. [5 marks]

(iii) Sketch the curve $y = 3x - x^3$. [2 marks]

(iv) Using your sketch, explain the result of your calculation in part (i). [1 mark]

T ④ Figure 2 is a copy of a spreadsheet to calculate the gradient of a chord PQ of the curve $y = 3x^2$, where P has coordinates $(x(\mathrm{P}), y(\mathrm{P}))$, Q has coordinates $(x(\mathrm{Q}), y(\mathrm{Q}))$, and $x(\mathrm{Q}) = x(\mathrm{P}) + h$.

Home Insert Page Layout Formulas Data Review

	A	B	C	D	E	F
1	curve	$y = 3x^2$				
2	x(P)	y(P)	h	x(Q)	y(Q)	gradient of PQ
3	2	12	0.1	2.1	13.23	12.3
4	2	12	0.01	2.01	12.1203	12.03
5	2	12	0.001			
6						

Figure 2

(i)   Write down the calculation for the value 12.3 in cell F3.   [2 marks]

(ii)   Complete the values in D5, E5 and F5.   [3 marks]

(iii) Write down the limit of the gradient PQ as h tends to zero.   [1 mark]

(iv)   Differentiate $3x^2$. Hence find the gradient of the curve at (2, 12).   [2 marks]

(PS) ⑤ A cardboard matchbox is designed as an open box with a sleeve. The box is of length 5 cm, width x cm and height y cm. The box has a sleeve which fits round it (ignoring the thickness of the cardboard). The box and sleeve, together with their nets, are shown in Figure 3.

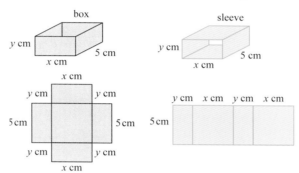

Figure 3

(i)   Find, in terms of x and y, the total area A cm² of cardboard in one box and its sleeve.   [3 marks]

To hold the required number of matches, the volume of the box has to be 60 cm³.

(ii)   Show that $A = 24 + 15x + \dfrac{240}{x}$.   [2 marks]

(iii) Find the dimensions of the box that make the area of cardboard used in a box as small as possible.   [4 marks]

(MP) ⑥

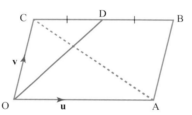

Figure 4

In Figure 4, OABC is a parallelogram with $\overrightarrow{OA} = \mathbf{u}$ and $\overrightarrow{OC} = \mathbf{v}$.

D is the midpoint of CB and E divides the line OD in the ratio 2:1.

(i)   Find $\overrightarrow{OD}$ in terms of $\mathbf{u}$ and $\mathbf{v}$, and show that
$$\overrightarrow{OE} = \frac{1}{3}\mathbf{u} + \frac{2}{3}\mathbf{v}.$$ [2 marks]

F divides AC in the ratio 2:1.

(ii)   Prove that E and F are the same point.   [4 marks]

(M) ⑦ Table 1 shows the world records for men's middle and long distance running events at the start of the 2016 Olympics (*source:* Wikipedia).

Table 1

Distance d m	800	1500	3000	5000	10 000
Time t s	100.91	206.00	440.67	757.35	1577.53

Figure 5 shows a graph of the points (ln d, ln t) for these values.

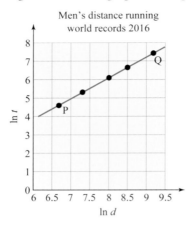

Men's distance running world records 2016

Figure 5

(i) Find the coordinates, correct to 3 decimal places, of the points labelled P and Q. [2 marks]

These data are modelled by the equation ln $t = a + b$ ln d, where a and b are constants.

(ii) Use the points P and Q to determine the values of a and b correct to 2 decimal places. [4 marks]

(iii) Using these values for a and b, find the equation for t in terms of d. [2 marks]

(iv) At the start of the 2016 Olympics, the world record for the men's 100 metres was 9.58 s.

Investigate how this compares with the record predicted by model in part (iii) and comment on it. [3 marks]

PS ⑧ The Annual Percentage Rate (APR) of a compound interest payment is the equivalent annual interest payable on an investment of £100.

If compound interest of p% is paid monthly on a loan of £100, the total amount owed after a year (assuming no repayments are made) is

$$£100\left(1 + \frac{p}{100}\right)^{12}.$$

(i) Show that the APR of a loan with compound interest of 2% per month is 26.82%. [2 marks]

(ii) Find, correct to 1 decimal place, the monthly compound percentage interest payable on a loan of £100 which gives an APR of 50%. [5 marks]

14 Data collection

THE AVONFORD STAR

Another cyclist seriously hurt.

Will you be next?

On her way back home from school on Wednesday afternoon, little Denise Cropper was knocked off her bicycle and taken to hospital with suspected concussion.

Denise was struck by a Ford transit van, only 50 metres from her own house.

Denise is the fourth child from the Springfields estate to be involved in a serious cycling accident this year.

The busy road where Denise Cropper was knocked off her bicycle yesterday

After this report, the editor of the *Avonford Star* asks Robin, one of the paper's journalists, to investigate the situation and write an in-depth analysis and a leading article for the paper on it. She explains to Robin that there is growing concern locally about cycling accidents involving children. She emphasises the need for good quality data to support whatever Robin writes.

1 Using statistics to solve problems

Before starting this assignment, Robin thinks about a number of questions. They come under four headings.

Problem specification and analysis

- What problem am I going to address?
- What data do I need to collect?

Information collection

- How will I collect the data?

Processing and representation

- How am I going to process and represent the data?

Interpretation

- How will I be able to interpret the data and explain my conclusions?

Problem specification and analysis

Robin needs to understand the problem before actually collecting any data. He decides it can be written in the form of two questions.

- How dangerous is it to be a cyclist in Avonford?
- Are any age groups particularly at risk?

To answer these questions he needs to collect data on recent cycling accidents. After a lot of thought he decides the data should cover the following fields: name; sex; age; distance from home; cause; injuries; whether the cyclist was wearing a helmet; nights in hospital; time, day of the week and month of the accident and the reporting police officer.

Information collection

Robin obtains information from the police and the local hospital. Both are willing to help him and he is able to collect a list of 93 people involved in accidents reported over the last year. Figure 14.1 shows part of it. The full table is given at the back of this book and is available on a spreadsheet.

first name	last name	sex	age	distance from home	cause	injuries	wearing a helmet?	nights in hospital	time of accident	day of accident	month of accident	officer reporting
Farhan	Ali	M	13	250 m	hit lamp-post	compression		1	9.15 am	Saturday	October	39014
Martin	Anderson	M	31		drunk	abrasions	n	1	23.30	Friday	February	78264
Marcus	Appleton	M	64	2 miles	car pulled out	concussion	n	2	8.25 am	Monday	August	97655
Lucy	Avon	F	52	500 m	lorry turning	abrasions		0	7.50 am	Thursday		39014
Thomas	Bailey	M	10	500 m	hit friend	suspected concussion	y	0	4 pm	Sunday	June	78264

Figure 14.1 Part of the list of cycling accidents

Here are the ages of 92 of the people involved in the accidents. The age of the other person was not recorded.

13	31	64	52	10	18	18	16	32	44
61	60	18	7	46	20	62	34	10	9
9	6	14	44	67	13	7	28	63	66
26	19	22	29	22	39	14	8	7	9
37	62	5	61	59	16	21	15	55	23
26	10	6.5	28	22	61	16	20	36	20
36	25	42	21	21	67	17	16	34	66
17	11	11	138	60	13	9	15	12	45
88	37	46	18	9	9	12	50	64	52
35	8.5								

Processing and representation

The figures above for people's ages are **raw data**. The first thing Robin has to do is to **clean** these data. This involves dealing with **outliers, missing data** and **errors**.

Outliers are extreme values that are far away from the bulk of the data. You need to decide whether to include or to exclude them. In this case there are two: 88 and 138. Robin investigates both of them. He finds that the 88 is genuine; it is the age of Millie Smith who is a familiar sight cycling to the shops. Needless to say the 138 is not genuine; it was the written response of a man who was insulted at being asked his age. Since no other information was available, Robin decides that the information from this man is unreliable and so should be excluded from the whole data set reducing its size from 93 to 92.

> Always try to understand an outlier before deciding whether to include or exclude it. It may be giving you valuable information.

Discussion point

→ Do you agree with Robin's decision?

There was also one missing data item. In this case the age of someone involved in a minor accident was not recorded. Robin excludes that from the list of ages, reducing it to 91, but not from the rest of the data set which stays at 92.

> In this case the missing item is not particularly important. That is not always the case and sometimes it is best to estimate it or to give it an average value if this can be justified by other available information.

Robin looks through the remaining age data carefully and decides there are no obvious errors, but two ages are recorded as 6.5 and 8.5. This is not the style of the other entries for ages which are whole numbers, so he decides to change them to 6 and 8 years respectively. Age is normally given as whole years so it makes sense to round down in this context.

> A different style for recording data may cause problems with presentation and processing.

Note

Stem-and-leaf diagrams are explained here for illustrative purposes only. They do not feature in the AQA specification and you will not see them in your exam.

Having cleaned the data, he then wishes to display them and decides to use this ordered stem-and-leaf diagram, see Figure 14.2.

The column of figures on the left, going from 0 down to 8, corresponds to the tens digits of the ages. This is called the **stem**. In this example there are 9 **branches**, going across the page. The numbers on the branches are called the **leaves**. Each leaf represents the units digit of a data value.

Key   0 | 5   means 5 years old

Stem-and-leaf diagrams are easy to interpret and work out summary values from.

0	5	6	6	7	7	7	8	8	9	9	9	9	9	9											
1	0	0	0	1	1	2	2	3	3	3	4	4	5	5	6	6	6	6	7	7	8	8	8	8	9
2	0	0	0	1	1	1	2	2	2	3	5	6	6	8	8	9									
3	1	2	4	4	5	6	6	7	7	9															
4	2	4	4	5	6	6																			
5	0	2	2	5	9																				
6	0	0	1	1	1	2	2	3	4	4	6	6	7	7											
7																									
8	8																								

Figure 14.2

If the values on the basic stem-and-leaf diagram are too cramped (that is, if there are so many leaves on a line that the diagram is not clear) you may stretch it. To do this you put values 0, 1, 2, 3, 4 on one line and 5, 6, 7, 8, 9 on another.

This diagram gives Robin a lot of information about those involved in accidents at a glance.

- The youngest child to be involved in an accident was 5 years old.
- The most common age is 9.
- 14 of those involved in accidents were under 10 and 39 under 20.
- There were also a lot of people (16 people) in their 20s.
- Another common age group (14 people) were those in their 60s.

Interpretation

Robin now has to write two pieces for the *Avonford Star*, an editorial and a detailed article to back it up.

This is the editorial.

THE AVONFORD STAR

A council that does not care

The level of civilisation of any society can be measure by how much it cares for its most vulnerable members.

On that basis our town council rates somewhere between savages and barbarians. Every day they sit back complacently while those least able to defend themselves, the very old and the very young, run the gauntlet of our treacherous streets.

I refer of course to the lack of adequate safety measures for our cyclists, 60% of whom are children or senior citizens. Statistics show that they only have to put one wheel outside their front doors to be in mortal danger. 80% of cycling accidents happen within 1500 metres of home.

Last week Denise Cropper became the latest unwitting addition to these statistics. Luckily she is now on the road to recovery but that is no thanks to the members of our unfeeling council who set people on the road to death and injury without a second thought.

What, this paper asks our councillors, are you doing about providing safe cycle tracks from our housing estates to our schools and shopping centres? And what are you doing to promote safety awareness among our cyclists, young and old?

Answer: Nothing.

Discussion point

→ Is this editorial fair? Is there evidence to back it up?

Figure 14.3

The problem solving cycle

In the example you have just read, Robin was presented with a problem which required him to use statistics. The steps he took fit into the **problem solving cycle**. This is a general process used for investigation and problem solving.

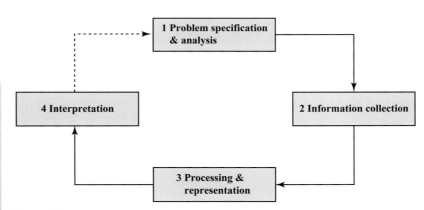

Prior knowledge

You will need to know the following summary measures: mean, mode, median. You will also need to be able to draw and interpret bar charts and pie charts.

Figure 14.4

Stage 1 **Problem specification and analysis** covers the preliminary work at the start, beginning with an analysis of the problem to be solved or the question to be investigated. So the first step is to be clear exactly what you want to achieve. Then you have to plan how you are going to go about it. A critical question is what data you need but you also need to be clear how you are going to collect them and what you expect to do with the data once you have them. The ideas involved are introduced in this chapter.

Stage 2 **Information collection** often involves taking a sample from all the possible data (the parent population). However, sometimes you are able to collect the whole population; such a 100% sample is called a **census**. There are many different ways of collecting samples and they are described later in this chapter.

Stage 3 Several tasks come under the heading **processing and representation**. The first of these is cleaning the data. This is usually followed by presenting the data in a diagram that shows their main features. At this stage it is also common to calculate summary measures, typically covering central tendency (for example the mean) and spread (for example the range and standard deviation). Most of this work is described in Chapter 15.

Stage 4 The final stage involves **interpretation**. This is where you draw conclusions about the situation you are investigating. You will typically have carried out some calculations already; this is where you explain what they tell you. If you are carrying out a hypothesis test, as for example in Chapter 18, this is where it fits in. The investigation will not, however, be complete until you have reported what you have found in words. Finally you should ask yourself if your findings are realistic and if they answer the question you were addressing. If not you will have to do further work; this may involve collecting more data, and so repeating stages 2, 3 and 4, or collecting different data, and so repeating the whole cycle again. Most of the exercise questions in the next chapter are about interpretation.

Exercise 14.1

This exercise is based on the data that Robin, a reporter from the Avonford Star, *collected in order to investigate cycling accidents. He collected data for 93 accidents covering 12 fields. The full data set is available as a spreadsheet on* www.hoddereducation.co.uk/ AQAMathsYear1

You are expected to use a spreadsheet or statistical software in order to answer most of the questions.

 ① What is the difference between an outlier and an error?

② Which of the following entries for age of the cyclist would you consider an error?

(A) $8\frac{5}{7}$ (B) 6   (C) 95   (D) 3

③ Check the whole data set for possible outliers, anomalies and incorrect entries. Decide how you want to deal with these and clean the data set, saving it under a different file name. This new data set is the one you are going to work with.

④ Use the spreadsheet to answer the following questions.

(i) What was the average age of the cyclists?

> You may want to use the 'filter' function to focus on these data for some of the individual question parts.

(ii) How many children were involved in accidents?

(iii) What was the average distance from home at which accidents occurred?

(iv) What percentage of the accidents were no more than 1 km from home?

(v) What percentage of accidents involved a motor vehicle?

(vi) What percentage of accidents resulted in an overnight stay in hospital?

(vii) Among those involved in accidents, what percentages were wearing helmets among

(a)   the under 18 age group

(b)   those aged at least 60?

(viii) Did all the police officers report whether helmets were being worn?

⑤   Robin's assistant says, 'To the nearest whole number, the average police officer is number 67591'. Is the assistant right?

T ⑥   Use suitable software to produce diagrams to illustrate the following information:

(i)   the ages of those involved in the accidents

(ii)   the number of nights they spent in hospital

(iii) the days of the week when accidents occurred

(iv)   the months when the accidents occurred

(v)   the time of day of the accidents.

PS ⑦   For each part of question 6, state at least one important feature of the data that is highlighted by the diagram.

⑧   The data set includes information on the causes of the accidents and on the types of injuries. Suggest ways in which Robin could classify these fields so that they could be displayed to show any important information they contain.

PS ⑨   Robin was asked to look at the dangers of being a cyclist in Avonford, particularly for children. Choose some calculations and diagrams which could help to answer these questions.

⑩   What should Robin write in his supporting article?

⑪   What other data and information would it have been sensible for Robin to collect?

2 Sampling

THE AVONFORD STAR
Student set to become Local MP

Next week's Avonford by-election looks set to produce the youngest Member of Parliament, according to an opinion poll conducted by the Star.

When 30 potential voters were asked who they thought would make the best MP, 15 opted for 20-year-old art student Meena Mehta. None of the other four candidates got more than 5 votes.

Meena Mehta

Assuming the figures quoted in the article are true, does this really mean that 20-year-old Meena Mehta will be elected to parliament next week? Only time will tell, but meanwhile the newspaper report raises a number of questions that should make you suspicious of its conclusion.

> In general the larger the sample size the more accurate will be the information it gives you.

Discussion points

How would you answer the following questions?

→ How was the sample selected? Was it representative of the whole population?

→ Was the sample large enough?

→ Those interviewed were asked who they thought would make the best MP not who they would vote for. Does this matter?

Terminology and notation

A **sample** typically provides a set of data values of a **random variable**, drawn from all such possible values.

This set is called the **parent population** (often just called the **population**). The parent population can be finite, such as all professional golf players, or infinite, for example the points where a dart can land on a dart board. A sample is intended to give information about the parent population and so it must be representative of it.

A parent population is usually described in terms of its **parameters**, such as its mean μ. By convention, Greek letters are used to denote these parameters and Roman letters are used to denote the equivalent sample values.

A list (or other representation) of the items available to be sampled is called the **sampling frame**. This could, for example, be a list of the sheep in a flock, a map marked with a grid or an electoral register. In many situations no sampling frame exists, nor is it possible to devise one, for example for the cod in the North Atlantic. The proportion of the available items that are actually sampled is called the **sampling fraction**.

Discussion points

Both of the situations below involve a *population* and a *sample*. In each case identify both, briefly but precisely, and also any likely source of bias.

➜ An MP is interested in whether her constituents support proposed legislation to restore capital punishment for murder. Her staff report that letters on the proposed legislation have been received from 361 constituents of whom 309 support it.

➜ A flour company wants to know what proportion of Manchester households bake some or all of their own bread. A sample of 500 residential addresses in Manchester is taken and interviewers are sent to these addresses. The interviewers are employed during regular working hours on weekdays and interview only during these hours.

Points to consider

There are several reasons why you might wish to take a sample, including the following.

- To obtain information as part of a pilot study to inform a proposed investigation.
- To estimate the values of the parameters of the parent population.
- To conduct a hypothesis test.

There are different points to think about. First you should consider how the sample data will be collected and what steps you can take to ensure their quality.

An estimate of a parameter derived from sample data will in general differ from its true value. The difference is called the **sampling error**. To reduce the sampling error, you want your sample to be as representative of the parent population as you can make it. This, however, may be easier said than done.

Here are a number of questions that you should ask yourself when about to take a sample.

1 Are the data relevant?

It is a common mistake to replace what you need to measure by something else for which data are more easily obtained.

You must ensure that your data are relevant, giving values of whatever it is that you really want to measure. This was clearly not the case in the example of the *Avonford Star*, where the question people were asked, 'Who would make the best MP?', was not the one whose answer was required. The questions should have been 'Which candidate do you intend to vote for?'.

2 Are the data likely to be biased?

Bias is a systematic error. If, for example, you wish to estimate the mean time of young women running 100 metres and did so by timing the women members of a hockey team over that distance, your result would be biased. The women from the hockey team would be fitter and more athletic than the general population of young women and so your estimate for the time would be too low.

You must try to avoid bias in the selection of your sample.

3 Does the method of collection distort the data?

The process of collecting data must not interfere with the data. It is, for example, very easy when designing a questionnaire to frame questions in such a way as to lead people into making certain responses. 'Are you a law-abiding citizen?' and 'Do you consider your driving to be above average?' are both questions inviting the answer 'Yes'.

In the case of collecting information on voting intentions another problem arises. Where people put the cross on their ballot paper is secret and so people are being asked to give away private information. There may well be those who find this offensive and react by deliberately giving false answers.

People may give the answer they think the questioner wants to receive.

4 Is the right person collecting the data?

Bias can be introduced by the choice of those taking the sample. For example, a school's authorities want to estimate the proportion of the students who smoke, which is against the school rules. Each class teacher is told to ask five students whether they smoke. Almost certainly some smokers will say 'No' to their teacher for fear of getting into trouble, even though they might say 'Yes' to a different person.

5 Is the sample large enough?

The sample must be sufficiently large for the results to have some meaning. In the case of the *Avonford Star*, the intention was to look for differences of support between the five candidates and for that a sample of 30 is totally inadequate. For opinion polls, a sample size of over 1000 is common.

If a sample has been well chosen, it will be the case that the larger it is the more accurate will be the information it gives about the population, and so the smaller the sampling error. The sample size you choose will depend on the precision you need. For example, in opinion polls for elections, a much larger sample is required if you want the estimate to be reliable to within 1% than for 5% accuracy.

6 Is the sampling procedure appropriate in the circumstances?

The method of choosing the sample must be appropriate. Suppose, for example, that you were carrying out the survey for the *Avonford Star* of people's voting intentions in the forthcoming by-election. How would you select the sample of people you are going to ask?

If you stood in the town's high street in the middle of one morning and asked passers-by you would probably get an unduly high proportion of those who, for one reason or another, were not employed. It is quite possible that this group has different voting intentions from those in work.

If you selected names from the telephone directory, you would automatically exclude various groups such as people who are ex-directory and those who use a mobile phone instead of a land line.

It is actually very difficult to come up with a plan which will yield a fair sample and that is not biased in some way or another. There are, however, a number of established sampling techniques.

Sampling techniques

Simple random sampling

In a **simple random sampling** procedure, every possible sample of a given size is equally likely to be selected. It follows that in such a procedure every member of the parent population is equally likely to be selected. However, the converse is not true. It is possible to devise a sample procedure in which every member is equally likely to be selected but some samples are not permissible.

Simple random sampling is fine when you can do it, but you must have a sampling frame. The selection of items within the frame is often done using random numbers; these can be generated using a calculator or computer, or you can use tables of random numbers.

Stratified sampling

You have already thought about the difficulty of conducting a survey of people's voting intentions in a particular area before an election. In that situation it is possible to identify a number of different sub-groups which you might expect to have different voting patterns: low, medium and high income groups; urban, suburban and rural dwellers; young, middle-aged and elderly voters; men and women; and so on. The sub-groups are called **strata**. In **stratified sampling**, you would ensure that all strata were sampled. You would need to sample from high income, suburban, elderly women; medium income, rural young men; etc. In this example, 54 strata ($3 \times 3 \times 3 \times 2$) have been identified. If the numbers sampled in the various strata are proportional to the size of their populations, the procedure is called **proportional stratified sampling**.

The selection of the items to be sampled within each stratum is usually done by simple random sampling. Stratified sampling will usually lead to more accurate results about the entire population, and will also give useful information about the individual strata.

> ### Discussion points
>
> → A school has 20 classes, each with 30 students. One student is chosen at random from each class, giving a sample size of 20. Why is this not a simple random sampling procedure?
>
> → If you write the name of each student in the school on a slip of paper, put all the slips in a box, shake it well and then take out 20, would this be a simple random sample?

Cluster sampling

Cluster sampling also starts with sub-groups, or strata, of the population, but in this case the items are chosen from one or several of the sub-groups. The sub-groups are now called clusters. It is important that each cluster should be reasonably representative of the entire population. If, for example, you were asked to investigate the incidence of a particular parasite in the puffin population of Northern Europe, it would be impossible to use simple random sampling. Rather you would select a number of sites and then catch some puffins at each place. This is cluster sampling. Instead of selection from the whole population you are choosing from a limited number of clusters.

Systematic sampling

Systematic sampling is a method of choosing individuals from a sampling frame. If you were surveying telephone subscribers, you might select a number at random, say 66, and then sample the 66th name on every page of the directory. If the items in the sampling frame are numbered $1, 2, 3, \ldots$, you might choose a random starting point such as 38 and then sample numbers 38, 138, 238 and so on.

Systematic sampling is particularly useful if you want to sample from items listed in a spreadsheet.

When using systematic sampling you have to beware of any cyclic patterns within the frame. For example, a school list which is made up class by class, each of exactly 25 children, in order of merit, means that numbers $1, 26, 51, 76, 101, \ldots$, in the frame are those at the top of their classes. If you sample every 50th child starting with number 26, you will conclude that the children in the school are very bright.

Quota sampling

Quota sampling is a method often used by companies employing people to carry out opinion polls. An interviewer's quota is usually specified in stratified terms: for example, how many males and how many females, etc. The choice of who is sampled is then left up to the interviewer and so is definitely non-random.

Opportunity sampling

As its name suggests, **opportunity sampling** is used when circumstances make a sample readily available. As an example, the delegates at a conference of hospital doctors are used as a sample to investigate the opinions of hospital doctors in general on a particular issue. This can obviously bias the results and opportunity sampling is often viewed as the weakest form of sampling. However it can be useful for social scientists who want to study behaviours of particular groups of people, such as criminals, where research will lead to individual case studies rather than results which are applied to the whole population. Opportunity sampling can also be useful for an initial pilot study before a wider investigation is carried out.

Self-selecting sample

A sample is **self-selected** when those involved volunteer to take part, or are given the choice to participate or decline. Examples are an online survey or when a researcher on the street asks people if they want to take part in a survey, they can agree to answer questions or say no. If enough people say no then this may affect the results of the survey. If someone who agrees to take part in the survey then refuses to answer some of the questions, this is also a form of self-selection and can bias the results.

It is also possible to have a self-selecting sample when people volunteer to answer questions, such as a text-in survey on a radio station to identify the 'best' song; however, as this can be easily identified as a self-selecting survey, it is not usually used in serious research.

A broader perspective

This is by no means a complete list of sampling techniques and you may come across others that are used in particular subjects, for example **snowball sampling** in sociology.

Sampling fits into the information collection stage of the problem solving cycle. It is rarely used for its own sake but to provide the information to shed light on a particular problem. The terms **survey design** and **experimental design** are used to describe the formulation of the most appropriate sampling procedure in a given situation and this is a major topic within statistics.

Exercise 14.2

① What are the advantages of an opportunity sample over a simple random sample?

② What do you need in order to select a simple random sample?

[A] a cluster   [B] a sampling frame
[C] clean data   [D] a calculator

③ An accountant is sampling from a spreadsheet. The first number is selected randomly and is item 47; the rest of the sample is selected automatically and comprises items 97, 147, 197, 247, 297, ...

What type of sampling procedure is being used?

④ Pritam is a student at Avonford High School. He is given a copy of the list of all students in the school. The list numbers the students from 1 to 2500. Pritam generates a four-digit random number between 0 and 1 on his calculator, for example 0.4325. He multiplies the random number by 2500 and notes the integer part. For example, 0.4325 × 2500 results in 1081 so Pritam chooses the student listed as 1081. He repeats the process until he has a sample of 100 names.

(i) What type of sampling procedure is Pritam carrying out?

(ii) What is the sampling fraction in this case?

⑤ Mr Jones wishes to find out if a mobile grocery service would be popular in Avonford. He chooses four streets at random in the town and calls at 15 randomly selected houses in each of the streets to seek the residents' views.

(i) What type of sampling procedure is he using?

(ii) Does the procedure give a simple random sample?

 ⑥ Teegan is trying to encourage people to shop at her boutique. She has produced a short questionnaire and has employed four college students to administer it.

The questionnaire asks people about their fashion preferences. Each student is told to question 20 women and 20 men and then to stop.

(i) What type of sampling procedure is Teegan using?

(ii) Does the procedure give a simple random sample?

(iii) Comment on the number of people that are surveyed.

⑦ Identify the sampling procedures that would be appropriate in the following situations.

(i) A local education officer wishes to estimate the mean number of children per family on a large housing estate.

(ii) A consumer protection body wishes to estimate the proportion of trains that are running late.

(iii) A marketing consultant wishes to investigate the proportion of households in a town with two or more cars.

(iv) A parliamentary candidate wishes to carry out a survey into people's views on solar farms within the constituency.

(v) A health inspector wishes to investigate what proportion of people wear spectacles.

(vi) Ministry officials wish to estimate the proportion of cars with illegally worn tyres.

(vii) A company wishes to estimate the proportion of dog owners who do not have their dogs micro-chipped.

(viii) The police want to find the average speed at which cars travel in the outside lane of a motorway.

(ix) A sociologist wants to know how many friends, on average, 18-year-old boys have.

(x) The headteacher of a large school wishes to estimate the average number of hours of homework done per week by the students.

⑧ There are five year groups in the school Jane attends. She wishes to survey opinion about what to do with an unused section of field next to the playground. Because of a limited budget she has produced only 60 questionnaires.

There are 140 students in each of Years 1 and 2.

There are 100 students in each of Years 3 and 4.

There are 120 students in Year 5.

Jane plans to use a stratified sampling procedure.

(i) What is the sampling fraction?

(ii) How many students from each year should Jane survey?

⑨ A factory safety inspector wishes to inspect a sample of vehicles to check for faulty tyres. The factory has 280 light vans, 21 company cars and 5 large-load vehicles.

He has decided that a sampling fraction of $\frac{1}{10}$ should be used and that each type of vehicle should be represented proportionately in the sample.

(i) How many of each type of vehicle should be inspected?

(ii) How should the inspector choose the sample? What is the sampling procedure called?

⑩ A small village has a population of 640. The population is classified by age as shown in the table below.

Age (years)	Number of people
0–5	38
6–12	82
13–21	108
22–35	204
36–50	180
51+	28

A survey of the inhabitants of the village is intended. A sample of size 80 is proposed.

(i) What is the overall sampling fraction?

(ii) A stratified sample is planned. Calculate the approximate number that should be sampled from each age group.

PS (11) You have been given the job of refurnishing the canteen of a sixth form college. You wish to survey student opinion on this. You are considering a number of sampling methods. In each case name the sampling method and list the advantages and disadvantages.

(i) Select every 25th student from the college's alphabetical listing of students.

(ii) Ask for 50 volunteers to complete a questionnaire.

(iii) Select students as they enter the canteen.

(iv) Select students at random from first and second year-group listings and in proportion to the number on each list.

(12) A beauty company wants to know whether a new moisturiser is better than the current product. After safety testing, they put a special label on the current product, inviting users to contact the company, receive free samples of the new product and give them feedback via their website. Just over 5000 people contact the company to receive free samples. The company receives feedback from 1235 people; 1003 say that the product is much better, 150 say it is about the same and the rest say it is not as good. The company releases a press statement saying that over 90% of users prefer the new product. How do they justify this statement? What are the issues around this survey?

PS (13) A factory produces computer chips. It has five production lines. Each production line produces, on average, 100 000 chips per week. One week the quality control manager says that from each machine the first chip to be produced each hour should be used as a sample.

The machines work 24 hours per day, 7 days per week.

(i) What is the sampling fraction?

(ii) What sampling method is she using? State two advantages of this method in this situation.

In the following questions you are required to devise, name and describe a suitable sampling strategy. Each answer is expected to take several lines.

PS (14) A company producing strip lighting wishes to find an estimate of the life expectancy of a typical strip light. Suggest how they might obtain a suitable sample.

PS (15) A tree surgeon wishes to estimate the number of damaged trees in a large forest. He has available a map of the forest. Suggest how he might select a sample.

PS (16) Avonford Sixth Form College is anxious to monitor the use of its staff car park, which has parking spaces for 100 cars. It is aware that the number of staff employed by the College is greater than this but also that a proportion of staff use public transport or car share sometimes. It is considering giving staff a choice of a free parking permit or paying for them to use public transport.

How would you survey staff views on these proposals?

PS (17) Some of your fellow students have shown concern about the lack of available space to do private study. You have been asked to represent them in approaching the Principal in order to press for some improvement in appropriate study space. Before you do this you want to be sure that you are representing a majority view, not just the feelings of a few individuals with strong opinions.

Describe how you would survey the students to gain the required information.

LEARNING OUTCOMES

When you have completed this chapter you should be able to:

➤ understand and use the terms 'population' and 'sample'

➤ use samples to make informal inferences about the population

➤ understand and use sampling techniques, including simple random sampling and opportunity sampling

➤ select or critique sampling techniques in the context of solving a statistical problem, including understanding that different samples can lead to different conclusions about the population

➤ be able to clean data, including dealing with missing data, errors and outliers.

KEY POINTS

1 Statistics are often used to help solve problems and the stages of the problem solving cycle are used:

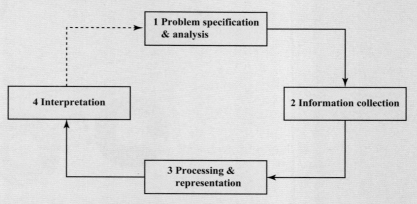

2 Information collection often requires you to take a sample. Here is a check list. When taking a sample you should ensure that:

- the data are relevant
- the data are unbiased
- the data are not distorted by the act of collection
- a suitable person is collecting the data
- the sample is of a suitable size
- a suitable sampling procedure is being followed.

Some sampling procedures are:

- simple random sampling
- stratified sampling
- cluster sampling
- systematic sampling
- quota sampling
- opportunity sampling
- self-selecting sample.

3 Processing and representation involves:

- cleaning the data
- calculating summary measures
- presenting the data using suitable diagrams.

You will usually use statistics software or a spreadsheet for this.

FUTURE USES

- The problem solving cycle pervades all the statistics you will meet.

- Data processing and representation is the focus of the next chapter, Chapter 15.

- Interpretation is continued in the next and subsequent chapters; hypothesis testing is covered in Chapter 18.

15

Data processing, presentation and interpretation

→ All the items in the collage above were taken from national newspapers on one day. How many different types of data can you see? How many ways of displaying data?

Once you have collected and cleaned your data, you need to organise them in ways that make it easy to see their main features. This involves using suitable data displays and summary measures. These are covered in this chapter.

You have already met some of the terms and notation used to describe data, words such as **outlier**, **sample** and **population**. Some more are given overleaf and others are described and defined later on, as they arise.

Important vocabulary

Variable, random variable

The score you get when you throw two dice is one of the values 2, 3, . . . up to 12. Rather than repeatedly using the phrase 'The score you get when you throw two dice', it is usual to use an upper case letter, like X, to denote it. Because its value varies, X is called a **variable**. Because it varies at random it is called a **random variable**.

Particular values of the variable X are often denoted by the equivalent lower case letter, x. So in this case x can be 2, 3, . . . up to 12.

Frequency

The number of times a particular value of a variable occurs in a data set is called its **frequency**.

Categorical (or qualitative) data

Categorical (or qualitative) data come in classes or categories, like types of bird or makes of car. Categorical data are also called **qualitative**, particularly if they can be described without using numbers.

Numerical (or quantitative) data

Numerical (or quantitative) data are defined in some way by numbers. Examples include the times people take to run a race, the numbers of trucks in freight trains and the birth weights of babies.

Ranked data

Ranked data are given by their position within a group rather than by measurements or score. For example the competitors in a competition could be given their positions as 1st, 2nd, 3rd . . . .

Discrete variables

The number of children in families (0, 1, 2, 3), the number of goals a football team scores in a match

(0, 1, 2, 3, . . .) and shoe sizes in the UK (1, 1½, 2, 2½, . . .) are all examples of **discrete variables**. They can take certain particular values but not those in between.

Continuous variables

Distance, mass, temperature and speed are all **continuous variables**. A continuous variable, if measured accurately enough, can take any appropriate value. You cannot list all the possible values.

Distribution, unimodal and bimodal distributions, positive and negative skew

The pattern in which the values of a variable occur is called its **distribution**. This is often displayed in a diagram with the variable on the horizontal scale and the frequency (or probability) on the vertical scale. If the diagram has one peak the distribution is **unimodal**; if there are two distinct peaks it is **bimodal**. If the peak is to the left of the middle the distribution has **positive skew** and if it is to the right there is **negative skew**.

Grouped data

When there are many possible values of the variable, it is convenient to allocate the data to groups. An example of the use of **grouped data** is the way people are allocated to age groups.

Bivariate and multivariate data

In **bivariate data** two variables are assigned to each item, for example the age and mileage of second-hand cars. The data are described as **multivariate** when more than two variables are involved. Multivariate data may be stored on a spreadsheet with different fields used for the different variables associated with each item.

1 Presenting different types of data

Before you display your data, or calculate summary measures, it is essential to ask yourself what type of data you have. Different types of data often require different techniques.

Categorical data

Common ways of displaying categorical data are **pictograms**, **bar charts**, **dot plots** and **pie charts**. These are illustrated for the following data collected from a survey of bats at a particular place one evening.

Table 15.1

Species	Frequency
Brown long-eared	4
Noctule	5
Pipistrelle	20
Serotine	5
Others	6

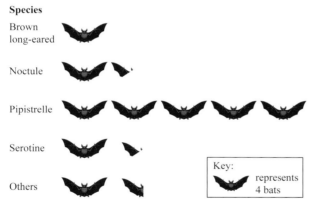

Figure 15.1 Pictogram

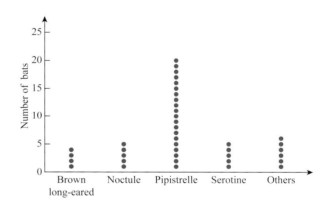

Figure 15.2 Dot plot

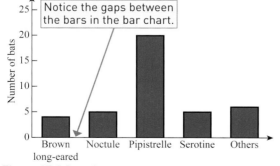

Figure 15.3 Bar chart

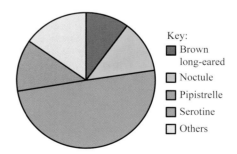

Figure 15.4 Pie chart

Discussion point

→ Always ask yourself what the total of the frequencies represented by the sectors of a pie chart actually means. If it is not something sensible, you should not be using a pie chart.

TECHNOLOGY

Explore the statistical graphs available in your spreadsheet and graphing software. Check they do not give a misleading graph under default settings.

These are simple versions of these types of display diagrams. If you want to use one of them for a particular purpose, for example to illustrate a report you are writing, you can find out about variants of them online. They are all quite straightforward and a few of them are used in Exercise 15.1.

The most commonly used summary measure for categorical data is the **modal class**. This is the class with the highest frequency. In the example above it is the Pipistrelle.

Notice the term **modal class** is used here and not **mode**. The mode is a number and so it is not right to use it to describe a category.

Exercise 15.1

① (i) What do the **mode** and the **modal class** have in common?

(ii) What is the difference between **mode** and **modal** class?

② What type of data are GCSE marks?
 A categorical B ranked
 C continuous D discrete

③ The compound bar chart in Figure 15.5 shows smoking- and non-smoking-attributable deaths. H and CD is the abbreviation for Heart and Coronary Disease, and COPD is Chronic Obstructive Pulmonary Disease.

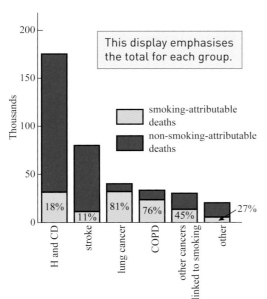

Figure 15.5

(i) What does this chart show the greatest cause of death to be?

(ii) The smoking-attributable deaths for 'stroke' and 'other cancers' seem to be about the same size, but have different percentages marked in the bars. Explain what these percentages mean.

(iii) What does the chart suggest to be the greatest causes of deaths which are
 (a) not smoking-attributable
 (b) smoking-attributable?

④ The multiple bar chart in Figure 15.6 compares the level of sales of three products of a company over four years.

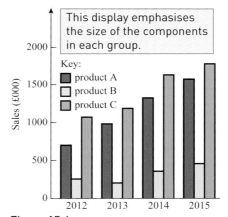

Figure 15.6

(i) Which product has shown the greatest increase in sales over the four years?

(ii) Describe the sales pattern for product B.

⑤ The education budget of a developing country is divided as shown in the multiple bar chart in Figure 15.7. Primary education has been universal since 1950.

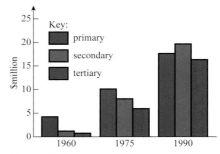

Figure 15.7

(i) In which sector was there the largest proportional change between 1960 and 1990?

(ii) In which sector was there the largest actual growth?

(iii) How do you think the country's education provision has changed over this time? Give your reasons.

⑥ Figures 15.8 and 15.9 illustrate the number of young people aged 16–24 who were Not in Education, Employment or Training (NEETs) during 2013–2015. The figures in the charts are in thousands.

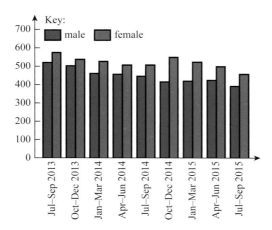

Figure 15.8

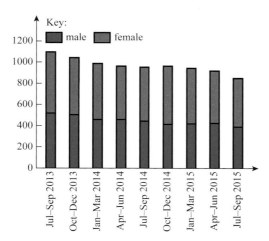

Figure 15.9

(i) What do the graphs tell you about the number of NEETs during the period July 2013–September 2015?

(ii) State one advantage for each of the methods of display used in Figures 15.8 and 15.9.

(iii) Discuss whether it would be appropriate to use a pie chart to show the data for male NEETs for the period July 2013–September 2015.

⑦
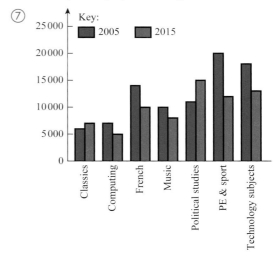

Figure 15.10

Figure 15.10 shows the UK entry numbers, rounded to the nearest thousand, for seven A-level subjects in the years 2005 and 2015.

(i) What is the name of the graph being used to represent these data?

(ii) In which subjects were the greatest and smallest percentage change in entries from 2005 and 2015? Estimate the percentage in each case.

The UK entry data (rounded to the nearest thousand) for 2005 and 2015 in four other A-level subjects is shown in Table 15.2.

Table 15.2

Subject	2005	2015
Biology	54000	63000
Chemistry	39000	53000
Mathematics	53000	93000
Physics	28000	36000

(iii) What potential issues might there be in adding these new data to the existing chart?

⑧ The pie charts in Figures 15.11, 15.12 and 15.13 summarise three aspects of a research study about the population of the UK in 2012.

Key:
- ■ single, less than 65 years old
- ■ single, 65 and older
- ■ single with children living at home
- ■ two adults less than 65, no child living at home
- ■ two adults at least one 65 or older, no child living at home
- ■ two adults with less than 3 children living at home
- □ two adults with at least 3 children living at home
- □ other

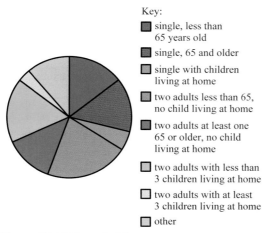

Figure 15.11 Household types

Key:
- ■ rental
- ■ owner-occupied with mortgage
- □ owner-occupied without mortgage

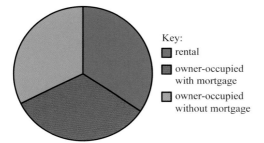

Figure 15.12 Housing status

Key:
- ■ male
- ■ female

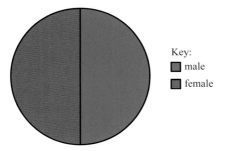

Figure 15.13 Gender

There were estimated to be approximately 62 million people, living in 26.5 million individual households.

Five statements are given below. Decide whether each of them is

- definitely a TRUE conclusion from the pie charts
- definitely a FALSE conclusion
- UNCERTAIN because it is impossible to say from the information given.

In each case explain your decision.

A The majority of households in the UK include two adults.

B The population in the UK is roughly equally split between male and female.

C Most people own their houses outright.

D There are about 2 million men living on their own over the age of 65.

E About a third of households with two adults and some children living at home own their houses and do not have a mortgage.

Working with a large data set

⑨ Using the cyclists' data set create several different charts to illustrate the time of day in which most accidents occurred. You will need to choose how to group your data. Then decide which of them are most helpful to you in analysing the data.

You should include a bar chart and a pie chart and explore the many other charts available in your software package. How does changing the colour or shading in your graph affect the impression it gives? Does changing to three-dimensional affect the impression the chart gives? Which chart(s) best identify the most 'dangerous' times for cyclists? Could any chart(s) be used to mislead people about this?

⑩ Using a large data set of your choice, create at least six different charts to display your data. How does changing the colour or shading in your graph affect the impression it gives? Does changing to three-dimensional affect the impression the chart gives?

Which is the most helpful to you in analysing the data and so answering your question? Can you find a chart which will suggest a different answer to your question (a misleading graph)?

Compare your answers with those from other students in your group. Which outputs from your software are the most useful and which are the most misleading?

2 Ranked data

In statistics, data are sometimes ranked in order of size and the ranks are used in preference to the original values. Table 15.3 gives the times in minutes and the ranks of 15 athletes doing a half-marathon.

Table 15.3

Name	Time	Rank	Name	Time	Rank	Name	Time	Rank
Alice	80	5	Bronwyn	105	9	Harriet	119	12
Alison	81	6	Carol	76	3	Lucy	125	13
Akosua	74	1	Cheryl	107	10	Ping	75	2
Ama	90	7	Fatima	77	4	Sally	182	15
Bella	111	11	Fauzia	133	14	Susmita	101	8

Ranked data gives rise to some useful summary measures.

The **median** is the value of the middle item. If there are n items, the median is the one with rank $\frac{n+1}{2}$.

In the example of the 15 athletes $n = 15$, so the median is the time of number $\frac{15+1}{2} = 8$.

The athlete ranked number 8 is Susmita and her time is 101 minutes, so that is the median.

You have to be aware, when working out the median, whether n is odd or even. If it is odd, as in the example above, $\frac{n+1}{2}$ works out to be a whole number but if n is even that is not the case.

For example if $n = 20$, $\frac{n+1}{2} = 10\frac{1}{2}$. In that case the data set does not have a single middle value; those ranked 10 and 11 are equally spaced either side of the middle and so the median is half way between their values.

The median is a typical middle value and so is sometimes called an **average**. More formally it is a **measure of central tendency**. It usually provides a good representative value.

The median is easy to work out if the data are stored on a spreadsheet since that will do the ranking for you. Notice that extreme values have little, if any, effect on the median. It is described as resistant to outliers. It is often useful when some data values are missing but can be estimated.

The median divides the data into two groups, those with high ranks and those with low ranks. The **lower quartile** and the **upper quartile** are the middle values for these two groups so between them the two quartiles and the median divide the data into four equal sized groups according to their ranks. These three measures are sometimes denoted by Q_1, Q_2 and Q_3.

> Q_2 is the median.

> There is the possibility of confusion because Q1, Q2, Q3 and Q4 are sometimes used to mean the four quarters of the year rather than quartiles.

Quartiles are used mainly with large data sets and their values found by looking at the $\frac{1}{4}$, $\frac{1}{2}$ and $\frac{3}{4}$ points. So for a data set of, say, 1000 you would often take Q_1 to be the value of the 250th data item, Q_2 that of the 500th item and Q_3 the 750th item.

There is no standard method or formula for finding Q_1 and Q_3 and you may meet different strategies. The method used here is consistent with the output from some calculators; it displays the quartiles of a data set and depends on whether the number of items, n, is even or odd.

If n is *even* then there will be an equal number of items in the lower half and upper half of the data set. To calculate the lower quartile, Q_1, find the median of the lower half of the data set. To calculate the upper quartile, Q_3, find the median of the upper half of the data set.

For example, for the data set $\{1, 3, 6, 10, 15, 21, 28, 36, 45, 55\}$ the median, Q_2, is $\frac{15 + 21}{2} = 18$.

The lower quartile, Q_1, is the median of $\{1, 3, 6, 10, 15\}$, i.e. 6.

The upper quartile, Q_3, is the median of $\{21, 28, 36, 45, 55\}$, i.e. 36.

If n is *odd* then define the 'lower half' to be all data items *below* the median. Similarly define the 'upper half' to be all data items *above* the median. Then proceed as if n were even.

For example, for the data set $\{1, 3, 6, 10, 15, 21, 28, 36, 45\}$ the median, Q_2, is 15.

The lower quartile, Q_1, is the median of $\{1, 3, 6, 10\}$, i.e. $\frac{3 + 6}{2} = 4.5$.

The upper quartile, Q_3, is the median of $\{21, 28, 36, 45\}$, i.e. $\frac{28 + 36}{2} = 32$.

Example 15.1

Catherine is a junior reporter at the *Avonford Star*. As part of an investigation into consumer affairs she purchases 0.5 kg of lean mince from 12 shops and supermarkets in the town. The resulting data, with the prices in rank order, are as follows:

£1.39, £1.39, £1.46, £1.48, £1.48, £1.50, £1.52, £1.54, £1.60, £1.66, £1.68, £1.72

Find Q_1, Q_2 and Q_3.

Solution

Table 15.4

Price in £	1.39	1.39	1.46	1.48	1.48	1.50	1.52	1.54	1.60	1.66	1.68	1.72
Rank order	1	2	3	4	5	6	7	8	9	10	11	12

$$Q_3 = £1.47 \qquad Q_3 = £1.51 \qquad Q_3 = £1.63$$

Discussion points

→ What are the range and the interquartile range for the times of the half-marathon athletes in the earlier example?

→ Which is the more representative measure of spread?

The simplest measure of spread for ranked data is the **range**, the difference between the values of the highest and lowest ranked items. The range is often not a very good measure as it is unduly affected by extreme values or outliers. A better measure is the **interquartile range** (also sometimes called the quartile spread), which is the difference between the two quartiles. This tells you the difference between typically high and typically low values.

Half the interquartile range is comparable to **standard deviation** which you will meet later in the chapter.

Interquartile range is sometimes used to identify possible outliers. A standard procedure is to investigate items that are at least 1.5 × interquartile range above or below the nearer quartile.

The median and quartiles are not the only common divisions used with ranked data. For example, the **percentiles** lie at the percentage points and are widely used with large data sets.

A **box-and-whisker plot**, or **boxplot**, is often used to display ranked data. This one (Figure 15.14) shows the times in minutes of the half-marathon athletes.

Figure 15.14 Box-and-whisker plot of half-marathon times

Box-and-whisker plots are sometimes drawn horizontally, like the one above, and sometimes vertically. They give an easy-to-read representation of the location and spread of a distribution. The box represents the middle 50% of the distribution and the whiskers stretch out to the extreme values.

In this case the interquartile range is $108 - 77 = 31$ and $1.5 \times 31 = 47.5$. At the lower end, 74 is only 3 below 77 so nowhere near being an outlier. However at the upper end 182 is 74 above 108 and so needs to be investigated as an outlier. The question 'why did Sally take so long?' clearly needs to be looked into. Some software identifies outliers for you.

Ranked data can also be displayed on **cumulative frequency curves**. However, it is usual to use these curves for grouped data, often when the individual values are not known. Cumulative frequency curves are therefore covered later in this chapter.

Exercise 15.2

 ① Which of the following are measures of spread and which are measures of central tendency?

median   range   mode
interquartile range   mean

 ② Which of these measures is not resistant to extreme values?

A median   B interquartile range
C mode   D mean

 ③ Karolina is investigating life expectancy in different parts of London. She starts with this sample of 11 boroughs. The figures are the average life expectancies in the year 2000.

Table 15.5

| | Life expectancy for people born in different London boroughs | | Ranking | |
	Female	Male	Female	Male
Barking and Dagenham	79.7	74.7		
Barnet	81.8	77.6		2.5
Camden	80.5	74.3		
Croydon	80.5	76.7		
Greenwich	80.1	74.2		
Islington	79.1	73.5		
Kensington & Chelsea	84.0	78.8	1	1
Kingston upon Thames	81.3	77.6		2.5
Redbridge	81.5	76.7		
Tower Hamlets	78.9	72.7		
Westminster	82.7	77.0		4

Barnet and Kingston upon Thames have the same life expectancy so they are both given the mean of the relevant rankings.

(i) Copy the table and complete the rankings for the different London boroughs. What are the rankings for Croydon?

(ii) Fill in the missing figures for Table 15.6.

Table 15.6

	Female	Male
Minimum	78.9	
Q_1	79.7	
Q_2	80.5	
Q_3		
Maximum	84.0	

(iii) Using the same scale, draw box-and-whisker plots for the male and female data sets.

(iv) Using your graphs, compare the life expectancy of females and males born in the different London boroughs.

(v) Suggest one reason each for the differences between

(a) the different boroughs

(b) men and women.

 ④ The Food Standards agency uses a traffic light system for food labelling to help shoppers to make healthy choices when buying food. The content of sugars, fat, saturates and salt are measured in 100 g of the food (Figure 15.15).

	High	Medium	Low
	Unhealthy	Eat with caution	Healthy
Sugars	Over 15 g	5 g to 15 g	5 g and below
Fat	Over 20 g	3 g to 20 g	3 g and below
Saturates	Over 5 g	1.5 g to 5 g	1.5 g and below
Salt	Over 1.5 g	0.3 g to 1.5 g	0.3 g and below

Figure 15.15

Note

A stem-and-leaf diagram is used to order the data.

Stem-and-leaf diagrams are used here to support understanding. They do not feature in the AQA specification and you will not see them in your exam.

Figure 15.16 shows a stem–and–leaf diagram containing information about the salt content, in grams per 100 g, of 45 different breakfast cereals.

Key 0.3 | 7 = 0.37 g

Stem	Leaf
0.2	3 8
0.3	
0.4	2
0.5	
0.6	5 5 9
0.7	0 0 0 0 0 0 0 0 0 0 3 3 3 3 4 4 4 5 5 5 6
0.8	0 0 0 0 0 0 0 9
0.9	0 0
1.0	0
1.1	5 5
1.2	0 0 4 5
1.3	0

Figure 15.16

(i) Find the median of the data.

(ii) Find the interquartile range of the data.

(iii) Comment on whether the cereals tested have a healthy salt content, using the data to justify your answer.

⑤ Female athletes from England, Ireland, Scotland and Wales take part in a team competition to run 10 km. There are three people in each team, making 12 in all. Table 15.7 shows their times in minutes and seconds.

Table 15.7

England	Ireland	Scotland	Wales
31.47	31.23	31.08	32.48
32.03	31.46	31.24	32.49
32.04	33.30	42.56	32.50

Here are four methods of deciding the order of the teams.

A Rank the athletes and add the three ranks for each team. The lowest total rank comes 1st, and so on.

B Add the times of the three athletes in each team. The lowest total time comes 1st, and so on.

C Use the median rank for each team to decide the order.

D Use the median time for each team to decide the order.

(i) Work out which teams came 1st, 2nd, 3rd and 4th under each method.

(ii) Explain why the orders for C and D were the same.

(iii) Say which method you consider the fairest. Give your reasons.

⑥ The stem-and-leaf diagram in Figure 15.17 covers 48 different European countries and 32 countries from North America, Central America and the Caribbean. The figures show the percentage of each country's population that have access to the internet.

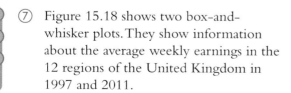

Key 3 | 8 = 38%	Europe		America & Caribbean
		0	
		1	2 2 4 7 9
		2	3 6 7 8 9
	8 0	3	3 8
	6 6 5 3 0	4	1 5 6 8
	9 9 8 7 6 2 2 1 0	5	0 1 6 9
	9 9 7 7 7 5 2 0	6	4 9
	9 7 6 5 4 1	7	4 6 8 8 8
	9 9 9 8 7 7 5 4 3 3 0	8	6 7 9
	7 6 5 4 3 3 2	9	2 7

Figure 15.17

(i) Find the median and interquartile range for both sets of data.

(ii) Comment on what the data show you about the percentages of people with internet access in both regions.

Note

Stem-and-leaf diagrams are used here for illustrative purposes only. They do not feature in the AQA specification and you will not see them in your exam.

⑦ Figure 15.18 shows two box-and-whisker plots. They show information about the average weekly earnings in the 12 regions of the United Kingdom in 1997 and 2011.

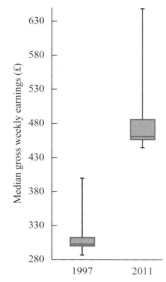

Figure 15.18 Average (median) gross weekly earnings in 1997 and 2011

Five statements are given below. Decide whether each of them is

- definitely a TRUE conclusion from the diagram

- definitely a FALSE conclusion from the diagram

- UNCERTAIN because it is impossible to say from the information given.

In each case explain your decision.

A The average wages in all 12 regions have all increased.

B The median across the regions of the average wage increased from £303 to £460 over the time period covered.

C The spread of the data, as shown by both the interquartile range and the range, has decreased from 1997 to 2011.

D The wages are evenly spread through the range.

E The vertical scale has been deliberately truncated to give a false impression.

Working with a large data set

Median and quartiles

(T) ⑧ Take a sample of 15 cyclists from the cyclists' data set. Use your software package to find the median and quartiles of the ages of cyclists involved in accidents. Compare this with the results you get when you calculate them by hand and when you use a graphical calculator. Find out the method the spreadsheet uses to determine the position of the lower and upper quartiles. How could different methods affect results from a large data set?

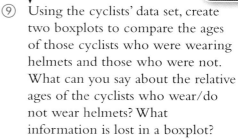

Boxplots

(T) ⑨ Using the cyclists' data set, create two boxplots to compare the ages of those cyclists who were wearing helmets and those who were not. What can you say about the relative ages of the cyclists who wear/do not wear helmets? What information is lost in a boxplot?

(T) ⑩ Using a large data set of your choice, use the median and quartiles to make a comparison between two subsets of the data (for example male and female). What can you say about the differences and similarities?

3 Discrete numerical data

Prior knowledge

Sorting data: you will already be familiar with methods to sort data and some of the calculations covered here. What matters now is to select the appropriate method for your data and to interpret it.

You will sometimes collect or find yourself working with **discrete** numerical data. Your first step will often be to sort the data in a frequency table. A **tally** can be helpful when you are doing this.

Here, for example, are the scores of the teams in the various matches in the 2014 Football World Cup.

Table 15.8

Group A	Group B	Group C	Group D	Group E	Group F	Group G	Group H
3−1	1−5	3−0	1−3	2−1	2−1	4−0	2−1
1−0	3−1	2−1	1−2	3−0	0−0	1−2	1−1
0−0	2−3	2−1	2−1	2−5	1−0	2−2	1−0
0−4	0−2	0−0	0−1	1−2	1−0	2−2	2−4
1−4	0−3	2−1	0−0	0−0	3−1	2−1	1−1
1−3	2−0	1−4	0−1	0−3	2−3	0−1	0−1
Knock-out stage							
1−1	2−0	2−1	1−1	2−0	2−1	1−0	2−1
Quarter finals		0−1	2−1	1−0	0−0		
Semi finals		1−7	0−4				
Final		1−0		Play-off		0−3	

15

Chapter 15 Data processing, presentation and interpretation
317

ACTIVITY 15.1

Copy and complete this tally and then the frequency column, from the data in Table 15.8.

	Numbers of goals scored by teams	Goals	Frequency
0	JHT JHT JHT JHT JHT JHT JHT I	0	36
1	JHT JHT JHT JHT JHT JHT JHT JHT IIII	1	44
2		2	
3		3	
4		4	
5		5	
6		6	
7		7	

Figure 15.19

Several measures of central tendency are commonly used for discrete numerical data: the **mode**, the **mean** and the **median**.

The mode is the value that occurs most frequently. If two non-adjacent values occur more frequently than the rest, the distribution is said to be **bimodal**, even if the frequencies are not the same for both modes.

Bimodal data may indicate that the sample has been taken from two populations. For example, the heights of a sample of students (male and female) would probably be bimodal reflecting gender differences in heights.

For a small set of discrete data the mode can often be misleading, especially if there are many values that the data can take. Several items of data can happen to fall on a particular value. The mode is used when the most probable or most frequently occurring value is of interest. For example a dress shop manager who is considering stocking a new style might first buy dresses of the new style in the modal size as she would be most likely to sell those.

The **mean** is found by adding the individual values together and dividing by the number of individual values. This is often just called the **average** although strictly it is the **arithmetic mean**.

In Activity 15.1 you completed the frequency table and that makes finding the mean much easier, as you can see from the table below.

Table 15.9

x	f	$x \times f$
0	36	0
1	44	44
2	27	54
3	12	36
4	6	24
5	2	10
6	0	0
7	1	7
Σ	128	175

Discussion point

➜ Give an example of situations where the weighted mean is better than an ordinary arithmetic mean.

Notice that in this calculation $\overline{x} = \dfrac{\Sigma xf}{n}$, so that this is actually a form of **weighted mean**. The answer is $\overline{x} = \dfrac{175}{128} = 1.37$ (to 2 d.p.). It is not the same as taking the mean of 0, 1, 2, 3, 4, 5, 6 and 7 and coming out with an answer of $\dfrac{28}{8} = 3.5$

Notice the notation that has been used.

- The value of the variable, in this case the number of goals, is denoted by x.
- The frequency is denoted by f.
- Summation is indicated by the use of the symbols Σ (sigma).
- The total number of data items is n so $\Sigma f = n$.
- The mean value of x is denoted by $\overline{x}$ and $\overline{x} = \dfrac{\Sigma xf}{n}$.

So the mean of the number of goals per team per match in the 2014 World Cup finals was 1.37.

Another measure of central tendency is the **median**. As you have already seen, this is used for ranked data.

The football scores are not ranked but it is easy enough to use the frequency table to locate the median. There are 128 scores and since $128 \div 2 = 64$, the median is half way between number 64 and number 65. Table 15.10 gives the ranks of the various scores; it has been turned upside down so that the highest score comes first at the top of the table.

Table 15.10

Goals	f	Ranks
7	1	1
6	0	–
5	2	2–3
4	6	4–9
3	12	10–21
2	27	22–48
1	44	49–92
0	36	93–128

Clearly numbers 64 and 65 are both in the row representing 1 goal, so the median is 1.

The simplest measure of **spread** for this type of data is the **range**, the difference between the highest value and the lowest. If you rank the data you can find the quartiles and so the **interquartile range**. If the data are in a frequency table you can find the quartiles using a similar method to that used above for the median. However, the most widely used measure of spread is **standard deviation** and this is covered later in the chapter.

> **Discussion point**
>
> ➜ Did the table need to be turned upside down? Did it matter whether you gave the highest number of goals or the lowest the rank of 1?

In the case of the goals scored by teams in the World Cup final the variable has only a small number of possible values and they are close together. In a case like this, the most appropriate diagram for displaying the data is a **vertical line chart**, for example Figure 15.20.

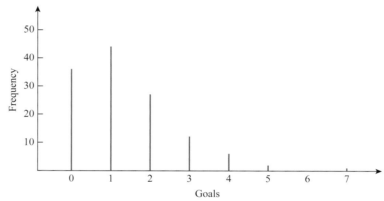

Figure 15.20 Vertical line chart showing the goals scored in the group stage of the World Cup

Sometimes, however, your data will be spread out over too wide a range for a vertical line chart to be a helpful way of displaying the information. In that case a stem-and-leaf diagram may be better. Figure 15.21 shows the results for 80 people taking an aptitude test for becoming long-distance astronauts; the maximum possible score is 100.

> **Note**
>
> Stem-and-leaf diagrams are used here for illustrative purposes only. They do not feature in the AQA specification and you will not see them in your exam.

0	8
1	2 7
2	4 5 5 8
3	3 5 5 5 6 8 9
4	0 1 1 2 4 4 4 5 7 8 8 9 9 9
5	0 0 0 1 4 4 5 5 5 6 6 8 8 9 9
6	0 2 2 2 2 2 2 3 3 3 6 7 7 8 8 8 9
7	0 1 1 2 4 4 4 4 5 5 7 8 9
8	0 0 2 5 7
9	1 3

Key: 0 | 8 mean 8

Figure 15.21

A stem-and-leaf diagram provides a form of grouping for discrete data. The data in Figure 15.21 could be presented in the **grouped frequency table** given in Table 15.11.

Table 15.11

Scores	Frequency
0 to 9	1
10 to 19	2
20 to 29	4
30 to 39	7
40 to 49	14
50 to 59	15
60 to 69	17
70 to 79	13
80 to 89	5
90 to 99	2

Discussion point

→ How would you estimate the mean of the data as given in the grouped frequency table (Table 15.11)?

Grouping means putting the data into a number of classes. The number of data items falling into any class is called the **frequency** for that class. When numerical data are grouped, each item of data falls within a **class interval** lying between **class boundaries** (Figure 15.22).

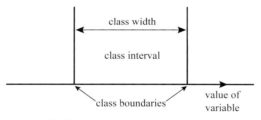

Figure 15.22

Every item of data must belong to exactly one class. There must be no ambiguity. Always check your class intervals to make sure that no classes overlap and that there are no gaps in between them.

The main advantage of grouping is that it makes it easier to display the data and to estimate some of the summary measures. However, this comes at a cost; you have lost information, in the form of the individual values and so any measures you work out can only be estimates.

The easiest way to display grouped discrete data is to use a bar chart with the different classes as categories and gaps between them, as shown below.

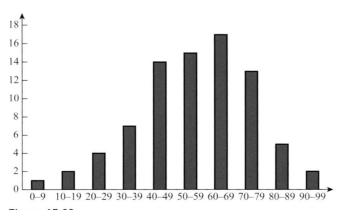

Figure 15.23

321

Discussion points

→ When would you want to group discrete data?

→ Discuss whether you would group your data in these cases.

■ You have conducted a survey of the number of passengers in cars using a busy road at commuter time.

■ You have carried out a survey of the salaries of 200 people five years after leaving university.

■ It is July and you have trapped and weighed a sample of 50 small mammals, recording their masses to the neared 10 grams.

Exercise 15.3

 ① What are the benefits of keeping data ungrouped?

 ② Which of these can be displayed using a vertical line chart?

A categorical data   B continuous data

C discrete data   D grouped data

③ The number of goals scored by a hockey team in its matches one season are illustrated on the vertical line chart in Figure 15.24.

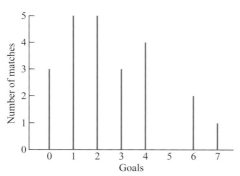

Figure 15.24

(i) Draw a box-and-whisker plot to illustrate the same data.

(ii) State, with reasons, which you think is the better method of display in this case.

④ Rachel plays cricket for the Avonford Cricket Club. She is a fast bowler. At the end of her first season she draws this diagram to show the frequency with which she took different numbers of wickets in matches.

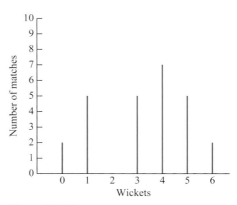

Figure 15.25

(i) Construct a frequency table showing the information in the diagram.

(ii) Find the mode and mean of the number of wickets Rachel took per match.

(iii) Which of these measures do you think she will quote when she is talking to her friends?

⑤ A biologist is concerned about the possible decline in numbers of a type of bird in a wood. The bird eats insects and the biologist thinks that the use of insecticides on nearby fields may be one of the causes of the possible decline.

The biologist observes 12 female birds during the breeding season and counts the number of fledglings (young that leave their nests). The results are shown in Figure 15.26.

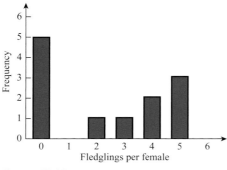

Figure 15.26

She applies the following modelling assumptions to this situation.

- Half of the fledglings are male and half are female.

- 75% of the fledglings will be taken by predators before they are old enough to breed.

- A female breeds three times in her lifetime.

- A female breeds once a year.

(i) Show that these observations and modelling assumptions lead to the conclusion that the birds will reduce in numbers.

(ii) Is there good evidence for the biologist to ask the farmer to stop using insecticides on the fields near the wood? Give three comments.

⑥ (i) A shoe shop has to make sure that it has enough shoes in each size. The manager asks Zahra to work out the average shoe size so they can order more in this size. Which average should Zahra use and why?

(ii) Robert is asked to investigate the heights of students at his sixth form college. Suggest how he might do this and design a collection sheet he could use. Identify any issues he may have.

⑦ Table 15.12 gives the mean GDP per person (to the nearest 100 US$) and the populations for the five North African countries (to the nearest 1000).

Table 15.12

Country	GDP (US$)	Population
Algeria	7 500	38 814 000
Egypt	6 600	86 895 000
Libya	11 300	6 244 000
Morocco	5 500	32 987 000
Tunisia	9 900	10 938 000

(i) Find

(a) the arithmetic mean of the five GDP figures,

(b) their weighted mean taking population into account.

(ii) Which figure is more representative of a typical person's GDP?

(PS) ⑧ The manager of a fast food restaurant wants to attract new workers. He advertises the position saying that the average wages in the restaurant are nearly £11 per hour.

When Anna asks about this she is told that she will be paid £7.50 per hour.

She asks about other rates of pay and the manager tells her the rates are:

   1 manager: £27 per hour

   2 deputy managers:

    £19.50 per hour each

   5 team leaders:

    £14.50 per hour each

   15 restaurant workers:

    £7.50 per hour each

Find the weighted mean of the rates of pay.

Is the manager justified in stating that the average wages are nearly £11 per hour? What might be a fairer way of describing the wages?

(PS) ⑨ The data in Table 15.13 were published in 1898. They give the numbers of men dying as a result of being kicked by a horse per year in a number of Prussian cavalry corps for the 20-year period 1875 to 1894.

Table 15.13

Deaths per year	0	1	2	3	4	5+
Frequency	144	91	32	11	2	0

(i) How many cavalry corps do the data cover?

(ii) How many men's deaths are covered by the data?

(iii) What was the mean number of deaths per cavalry corps per year?

(iv) Which out of the mode, median and mean is the most representative measure? Explain your answer briefly.

⑩ Thomas keeps a record over one year, starting on 1 December 2014, of how many eggs his hens lay each day. He has eight hens so the greatest possible number of eggs in any day is eight.

The data are summarised in Table 15.14.

Table 15.14

| Number of eggs | Frequency | | | |
	Winter Dec–Feb	Spring Mar–May	Summer June–Aug	Autumn Sep–Nov
0	20	0	0	6
1	31	6	8	27
2	22	12	14	21
3	15	10	16	15
4	2	24	19	11
5	0	19	18	6
6	0	16	13	4
7	0	5	4	1
8	0	0	0	0

Five statements are given below. Decide whether each of them is
- definitely a TRUE conclusion from the data
- definitely a FALSE conclusion from the data
- UNCERTAIN because it is impossible to say from the information given.

In each case explain your decision.

A One of the hens did not lay any eggs.

B Some days are missing from Thomas's records.

C The mean number of eggs per day was 2.97 (to 2 d.p.).

D All the data in the table could be illustrated on a pie chart with one sector for each season.

Working with a large data set

⑪ Using the cyclists' data set, find the mean, median and mode of the ages of the cyclists involved in accidents. Which measure best describes the data set?

⑫ Using a large data set of your choice, calculate the mean, median, and mode for different variables. Which measure of central tendency is the best for each variable? Is there a measure which is always useful? Is there a measure which is never useful? Are there variables for which different measure of central tendency give different interpretations?

Measures of central tendency

4 Continuous numerical data

When your data are continuous you will almost always need to group them. This includes two special cases.

- The variable is actually discrete but the intervals between values are very small. For example, cost in £ is a discrete variable with steps of £0.01 (that is, one penny) but this is so small that the variable may be regarded as continuous.

- The underlying variable is continuous but the measurements of it are rounded (for example, to the nearest mm), making your data discrete. All measurements of continuous variables are rounded and providing the rounding is not too coarse, the data should normally be treated as continuous. A particular case of rounding occurs with people's age; this is a continuous variable but is usually rounded down to the nearest completed year.

Displaying continuous grouped data

There are two ways of displaying continuous grouped data using vertical bars. They are **frequency charts** and **histograms**. In both cases there are no gaps between the bars.

> You will sometimes meet equivalent diagrams but with the bars drawn horizontally.

Frequency chart

A frequency chart is used to display data that are grouped into classes of equal width. Figure 15.27 illustrates the lengths of the reigns of English kings and queens from 827 to 1952 in completed years.

Years	$0 \leq y < 10$	$10 \leq y < 20$	$20 \leq y < 30$	$30 \leq y < 40$	$40 \leq y < 50$	$50 \leq y < 60$	$60 \leq y < 70$
Frequency	22	16	11	7	1	3	1

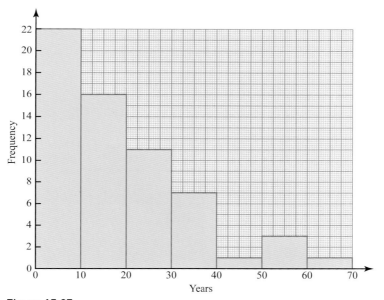

Figure 15.27

As you can see, the vertical axis represents the frequency. The horizontal scale is marked 0, 10, 20, . . . , 70 and labelled 'Years'. There are no gaps between the bars.

You can only use a frequency chart if all the classes are of equal width. If the classes are not of equal width you must use a histogram.

Histograms

> A histogram may also be used for classes of equal width.

There are two key differences between histograms and frequency charts.

- In a histogram, frequency is represented by the *area* of a bar and not by its *height*.

- The vertical scale of a histogram represents **frequency density** not **frequency**.

You can see these points in the following example.

Example 15.2

The heights of 80 broad bean plants were measured, correct to the nearest centimetre, 10 weeks after planting.

Table 15.15

Height, h cm	Frequency, f
$7.5 \leqslant h < 11.5$	1
$11.5 \leqslant h < 13.5$	3
$13.5 \leqslant h < 15.5$	7
$15.5 \leqslant h < 17.5$	11
$17.5 \leqslant h < 19.5$	19
$19.5 \leqslant h < 21.5$	14
$21.5 \leqslant h < 23.5$	13
$23.5 \leqslant h < 25.5$	9
$25.5 \leqslant h < 28.5$	3

Draw a histogram to display these data.

Solution

The first step is to add two more columns to the frequency table, one giving the class width and the other giving the frequency density.

Table 15.16

Height, h cm	Frequency, f	Class width, w cm	Frequency density, $\frac{f}{w}$
$7.5 \leqslant h < 11.5$	1	4	0.25
$11.5 \leqslant h < 13.5$	3	2	1.5
$13.5 \leqslant h < 15.5$	7	2	3.5
$15.5 \leqslant h < 17.5$	11	2	5.5
$17.5 \leqslant h < 19.5$	19	2	9.5
$19.5 \leqslant h < 21.5$	14	2	7.0
$21.5 \leqslant h < 23.5$	13	2	6.5
$23.5 \leqslant h < 25.5$	9	2	4.5
$25.5 \leqslant h < 28.5$	3	3	1.0

Use the left hand column to find the class width, for example $11.5 - 7.5 = 4.0$.

Use the middle two columns to find the class frequency density, for example $1 \div 4 = 0.25$.

Now draw the histogram with height on the horizontal axis and frequency density on the vertical axis.

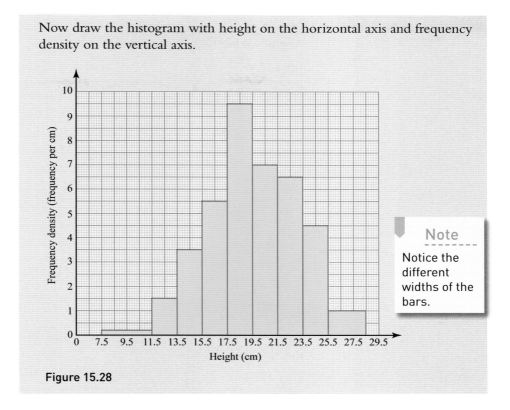

Figure 15.28

There are two things to notice about this histogram.

- As an example, look at the bar for $15.5 \leqslant h < 17.5$. Its width is 2 and its height is 5.5 so its *area* is $2 \times 5.5 = 11$ and this is the *frequency* it represents. You might like to check the other bars for yourself.

- The horizontal scale is in centimetres so the vertical scale is frequency per centimetre. The horizontal and vertical scales of a histogram are linked in such a way that multiplying them gives frequency.

In this case

$$\text{centimetres} \times \frac{\text{frequency}}{\text{centimetres}} = \text{frequency}.$$

Estimating summary measures from grouped continuous data

The first example shows how to estimate the **mean** and **modal class** of a grouped continuous data set.

Example 15.3

As part of an on-going dispute with her parents, Emily times all calls during August on her mobile phone.

Table 15.17

Time in seconds (s)	Frequency, f
0–	10
30–	21
60–	23
120–	12
180–	8
300–	4
500–1000	1
Σ	79

(i)   Estimate the mean length of her calls.

(ii) Find the modal class.

Solution

To complete both parts of this question, four more columns need to be added (Table 15.18).

Table 15.18

Time (s)	Frequency, f	Mid-value t (s)	$t \times f$	Class width, w (s)	Freq density, $f \div w$
0–	10	15	150	30	0.33
30–	21	45	945	30	0.70
60–	23	90	2070	60	0.38
120–	12	150	1800	60	0.20
180–	8	240	1920	120	0.13
300–	4	400	1600	200	0.07
500–1000	1	750	750	500	0.02
Σ	79		9235		

(i) To complete the estimation of the mean you need the new third and fourth columns, mid-value and $t \times f$.

$$\text{Mean} = \frac{\sum tf}{\sum f} = \frac{9235}{79} = 117 \text{ seconds (to 3 s.f.)}$$

(ii) To find the modal class you need the histogram in Figure 15.29. It is the class with the highest bar.

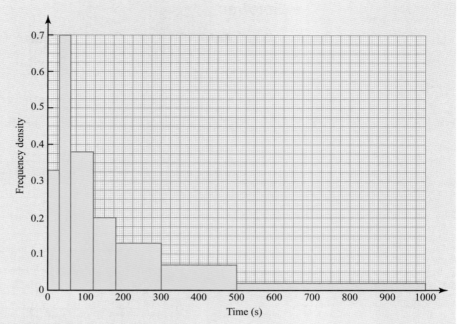

Figure 15.29

The highest bar is the second one. So the modal class is from 30 to 60 seconds. Notice that the class with the highest frequency is 60 to 120 but that has a greater class width.

Discussion point

→ Did you actually need to draw the histogram?

! When you are estimating summary measures of grouped continuous data you need to look very carefully at how the class intervals are defined. This is especially so if, as in the next example, the data have been rounded coarsely in comparison with the class width.

Here are two further points to notice from Example 15.3.

■ In this example the intervals are written 0–, 60–, 120– and so on. You will sometimes see intervals written like this and it is clear that the meaning is

$$0 \leqslant \text{time} < 60, \qquad 60 \leqslant \text{time} < 120 \qquad \text{and so on.}$$

■ There is no indication of the level of precision of the recorded data but it is reasonable to assume stop-watch accuracy from Emily's mobile phone so better than 1 second.

Example 15.4

Robert is a market gardener and sells his produce to a supermarket. He collects sample data about the weight of the tomatoes he plans to sell. He wants to know the mean value, but he also needs to know about the distribution as the supermarket will not accept any that are too small or too large.

(i) Robert's first thought is to describe the intervals, in grams, as 57–59, 59–61, and so on. What is wrong with these intervals?

(ii) Robert then says, 'I am going to record my measurements to the nearest 1 g and then denote them by m g. My classes will be $57 < m \leqslant 59$, $59 < m \leqslant 61$, and so on'.

What is the mid-value, x g, of the first interval?

(iii) Robert has six intervals. The last one is $67 < m \leqslant 69$.

Their frequencies, in order, are 4, 11, 19, 8, 5, 1.

Estimate the mean weight of the tomatoes.

Solution

(i) It is not clear which group some tomatoes should be in, for example one of mass 59 g, as the intervals overlap.

(ii) The lower bound of the weight for the group $57 < m \leqslant 59$ is 57.5 g.

The upper bound of the weight for the group $57 < m \leqslant 59$ is 59.5 g.

So the mid-value of the first interval is 58.5 g.

(iii) **Table 15.19**

Mass, m (g)	Mid–value x (g)	Frequency, f	xf
$57 < m \leqslant 59$	58.5	4	234.0
$59 < m \leqslant 61$	60.5	11	665.5
$61 < m \leqslant 63$	62.5	19	1187.5
$63 < m \leqslant 65$	64.5	8	516.0
$65 < m \leqslant 67$	66.5	5	332.5
$67 < m \leqslant 69$	68.5	3	205.5
	Σ	50	3141.0

Look carefully at this answer. It is not entirely intuitive. You might expect the mid-value to be 58. If the data were discrete then 58 would be the mid-value of a class consisting of 57, 58 and 59, but with continuous data that have been rounded before grouping that is not the case. You have to consider the upper and lower bounds of the group.

Estimated mean $= \frac{3141}{50} = 62.82$, so 63 g to the nearest whole number.

Cumulative frequency curves

To estimate the **median** and **quartiles** and related measures it can be helpful to use a **cumulative frequency curve**.

Cumulative frequency curves may also be used with other types of data but they are at their most useful with grouped continuous data.

THE AVONFORD STAR

Letters to the editor

Dear Sir,

I am a student trying to live on a government loan. I'm trying my best to allow myself a sensible monthly budget but my lecturers have given me a long list of textbooks to buy. If I just buy half of them I will have nothing left to live on this month. The majority of books on my list are over £16.

I want to do well at my studies but won't do well without books and I won't do well if I am ill through not eating properly.

Please tell me what to do, and don't say 'go to the library' because the books I need are never there.

Yours faithfully
Sheuli Roberts

After receiving this letter the editor wondered if there was a story in it. She asked a student reporter to carry out a survey of the prices of textbooks in a big bookshop. The student reporter took a large sample of 470 textbooks and the results are summarised in Table 15.20.

Table 15.20

Cost, C (£)	Frequency (no. of books)
$0 \leqslant C < 10$	13
$10 \leqslant C < 15$	53
$15 \leqslant C < 20$	97
$20 \leqslant C < 25$	145
$25 \leqslant C < 30$	81
$30 \leqslant C < 35$	40
$35 \leqslant C < 40$	23
$40 \leqslant C < 45$	12
$45 \leqslant C < 50$	6

If the student reporter had kept a record of the actual prices, he could have given accurate values of the median and quartiles rather than estimates.

The student reporter decided to estimate the median, the upper quartile and the lower quartile of the prices.

The first steps were to make a cumulative frequncy table and to use it to draw a cumulative frequency curve.

Table 15.21

Cost, C (£)	Frequency (no. of books)	Cost	Cumulative frequency
$0 \leqslant C < 10$	13	$C < 10$	13
$10 \leqslant C < 15$	53	$C < 15$	66
$15 \leqslant C < 20$	97	$C < 20$	163
$20 \leqslant C < 25$	145	$C < 25$	308
$25 \leqslant C < 30$	81	$C < 30$	389
$30 \leqslant C < 35$	40	$C < 35$	429
$35 \leqslant C < 40$	23	$C < 40$	452
$40 \leqslant C < 45$	12	$C < 45$	464
$45 \leqslant C < 50$	6	$C < 50$	470

Notice that the interval $C < 15$ means $0 \leqslant C < 15$ and so includes the 13 books in the interval $0 \leqslant C < 10$ and the 53 books in the interval $10 \leqslant C < 15$, giving 66 books in total.

Similarly, to find the total for the interval $C < 20$ you must add the number of books in the interval $15 \leqslant C < 20$ to your previous total, giving you $66 + 97 = 163$.

To draw the cumulative frequency curve you plot the cumulative frequency (vertical axis) against the upper boundary of each class interval (horizontal axis).

Then you join the points with a smooth curve (Figure 15.30).

It is usual in a case like this, based on grouped data, to estimate the median as the value of the $\frac{n}{2}$th term and the quartiles as the values of terms $\frac{n}{4}$ and $\frac{3n}{4}$.

In this case there are 470 items, so the median is at number 235 and the quartiles at 117.5 and 352.5.

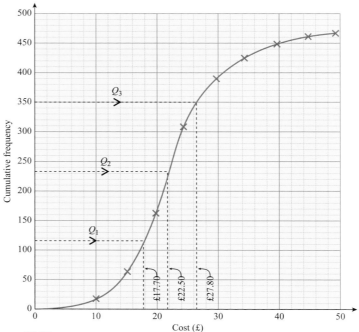

Figure 15.30

As you can see from Figure 15.30, these give the following values.

Lower quartile, Q_1 £17.70

Median, Q_2 £22.50

Upper quartile, Q_3 £27.80

> The cumulative frequency scale on the graph goes from 0 to 17, the middle is at $\frac{n}{2}$, so it is this value that is used to estimate the median.

Exercise 15.4

① Explain how to estimate the values of the greatest and least items of data from the cumulative frequency graph.

② Which of these measures do you not use to construct a boxplot?

A median   B greatest value

C mode   D lower quartile

③ The cumulative frequency graphs below show the weights in kg of 3152 male patients (the red curve) and 3690 female patients (the blue curve).

(i) On the same set of axes construct boxplots for both sets of data.

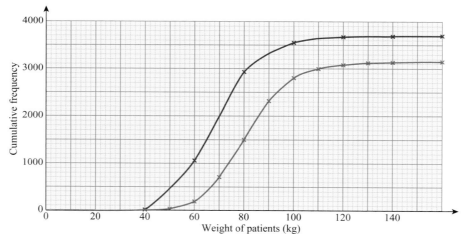

Figure 15.31

(ii) Use your boxplots to compare the two data sets.

(iii) Which is easier to interpret, the cumulative frequency curves or the boxplots? Explain your answer.

PS ④ As part of a quality control check at a supermarket, a case of oranges was opened and each orange was weighed. The masses, m g, are given in Table 15.22.

Table 15.22

Mass, m (g)	Frequency
$60 < m \leqslant 100$	20
$100 < m \leqslant 120$	60
$120 < m \leqslant 140$	80
$140 < m \leqslant 160$	87
$160 < m \leqslant 220$	23

(i) Estimate the mean mass of the oranges from the crate.

(ii) David drew the cumulative frequency curve in Figure 15.32 to illustrate the data.

(iii) Use David's graph to estimate the median, quartiles and interquartile range.

(iv) David has plotted one point incorrectly. Which point is it?

(v) Which of the answers to part (ii) are affected by David's error? Estimate their correct values.

(vi) The manager tells David this is not a helpful way of displaying the data. Do you agree with the manager? Give your reasons.

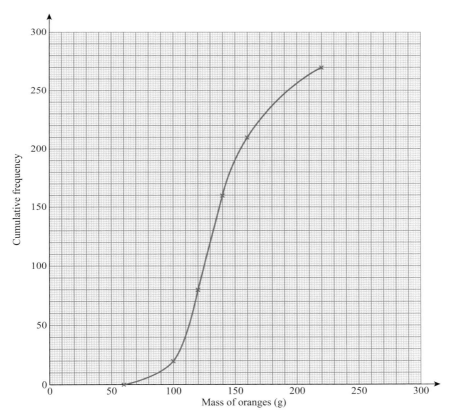

Figure 15.32

⑤ The doctors in a hospital in an African country are concerned about the health of pregnant women in the region. They decide to record and monitor the weights of newborn babies.

(i) A junior office clerk suggests recording the babies' weights in this frequency table (Table 15.23).

Table 15.23

Weight (kg)	2.0–2.5	2.5–3.0	3.0–3.5	3.5–4.0
Frequency	Doctors	to	fill	in

Give three criticisms of the table.

(ii) After some months the doctors have enough data to draw this histogram.

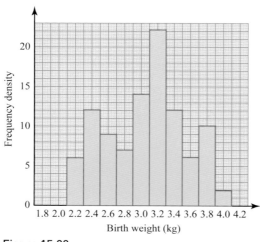

Figure 15.33

They then consider a number of possible courses of action.

- Carry on as at present.
- Induce any babies that are more than a week overdue.
- Set up a special clinic for 'at risk' pregnant women.

Comment on what features of the histogram give rise to each of these suggestions.

(iii) The doctors decide that the best option is to set up a special clinic. Suggest what other data they will need in order to make a strong case for funding.

⑥ The dot plot in Figure 15.34 shows the average number of mobile phone accounts per 100 people in 189 different countries in 2011.

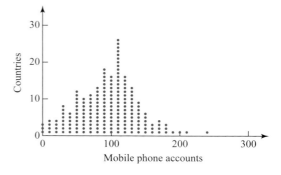

Figure 15.34

Six statements are given below. Decide whether each of them is

- definitely a TRUE conclusion from the diagram
- definitely a FALSE conclusion from the diagram
- UNCERTAIN because it is impossible to say from the information given.

In each case explain your decision.

A Each figure has been rounded to the nearest 10 before being displayed on the dot plot.

B The mode number of mobile phone accounts is 110.

C The median of the data is at least 100.

D The large number of people with more than one mobile phone account is partly a result of parents holding accounts for their children.

E Each dot on the dot plot represents the same number of people.

F The data could have been represented by a frequency chart or by a histogram.

Working with a large data set

⑦ Using the cyclists' data set and your software package, draw a number of histograms to illustrate the times of accidents. You will need to try out different ways to divide the time into classes. Decide which are most helpful in determining the times of day when accidents are most likely.

Does it matter if you look at weekdays and weekends separately?

Do school holidays affect your results?

Histograms

⑧ Using a large data set of your choice and a charting package, create histograms with different class widths. What effect does changing the class widths have on your charts? Is it possible to have too many classes? Is it possible to have too few classes?

Compare your results with those from other students in your group; can you agree on a good number of classes to have? Does it depend on the spread of the data?

5 Bivariate data

Table 15.24 shows the final results for the Premiership football teams in the 2014–15 season.

Table 15.24

Place	Team	P	W	D	L	GF	GA	PTS
1	Manchester City	38	27	5	6	102	37	86
2	Liverpool	38	26	6	6	101	50	84
3	Chelsea	38	25	7	6	71	27	82
4	Arsenal	38	24	7	7	68	41	79
5	Everton	38	21	9	8	61	39	72
6	Tottenham Hotspur	38	21	6	11	55	51	69
7	Manchester United	38	19	7	12	64	43	64
8	Southampton	38	15	11	12	54	46	56
9	Stoke City	38	13	11	14	45	52	50
10	Newcastle United	38	15	4	19	43	59	49
11	Crystal Palace	38	13	6	19	33	48	45
12	Swansea City	38	11	9	18	54	54	42
13	West Ham	38	11	7	20	40	51	40
14	Sunderland	38	10	8	20	41	60	38
15	Aston Villa	38	10	8	20	39	61	38
16	Hull City	38	10	7	21	38	53	37
17	West Bromwich Albion	38	7	15	16	43	59	36
18	Norwich City	38	8	9	21	28	62	33
19	Fulham	38	9	5	24	40	85	32
20	Cardiff City	38	7	9	22	32	74	30

Discussion point

➜ What are the types of the nine variables?

The data in Table 15.24 are **multivariate**. There are nine columns; each row represents one data item covering nine variables.

Tom is investigating the relationship between, Goals For (GF) and Points (PTS). So his data items can be written (102, 86), (101, 84) and so on. These items cover just **two** variables and so the data are described as **bivariate**.

Displaying bivariate data

Discussion point

→ Does this scatter diagram show any of the teams to be outliers?

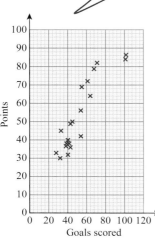

Figure 15.35

Bivariate data are often displayed on a scatter diagram like Figure 15.35.

Looking at the spread of the data points, it is clear that on the whole teams scoring many goals tend to have high points totals. There is a high level of **association** between the two variables. If, as in this case, high values of both variables occur together, and the same for low values, the association is **positive**. If, on the other hand, high values of one variable are associated with low values of the other, the association is **negative**.

A **line of best fit** is often drawn through the points on a scatter diagram. If the points lie close to a straight line the association is described as **correlation**. So correlation is linear association.

In this case consideration of the context suggests there may be a causal link. Many goals scored makes a win more likely so more points are gained. In general, you cannot assume a causal link simply because there is correlation. Correlation does not imply causation. For example, there is positive correlation between average teachers' salaries and average house prices, but a change in one does not directly cause the other to change. They both rise as a consequence of inflation.

Dependent and independent variables

The scatter diagram (Figure 15.35) was drawn with the goals scored on the horizontal axis and the points total on the vertical axis. It was done that way to emphasise that the number of points is dependent on the number of goals scored. (A team gains points as a result of scoring goals. It does not score goals as a result of gaining points.) It is normal practice to plot the **dependent variable** on the vertical (y) axis and the **independent variable** on the horizontal (x) axis. Table 15.25 shows some more examples of dependent and independent variables.

Table 15.25

Independent variable	Dependent variable
Number of people in a lift	Total weight of passengers
The amount of rain falling on a field while the crop is growing	The weight of the crop yielded
The number of people visiting a bar in an evening	The volume of beer sold

Random and non-random variables

In the examples above, both the variables have unpredictable values and so are **random**. The same is true for the example about the goals scored and the points totals in the premier league. Both variables are random variables, free to assume any of a particular set of values in a given range. All that follows about correlation will assume that both variables are random.

Sometimes one or both variables is or are **controlled**, so that the variable only takes a set of predetermined values; for example, the times at which temperature measurements are taken at a meteorological station. Controlled variables are independent and so are usually plotted on the horizontal axis of scatter diagrams.

Interpreting scatter diagrams

Assuming both variables are random, you can often judge whether they are correlated by looking at the scatter diagram. This will also show up some common situations where you might think there is correlation but would be incorrect.

When correlation is present you expect most of the points on the scatter diagram to lie roughly within an ellipse. You can think of drawing ellipses round the first two of the diagrams below.

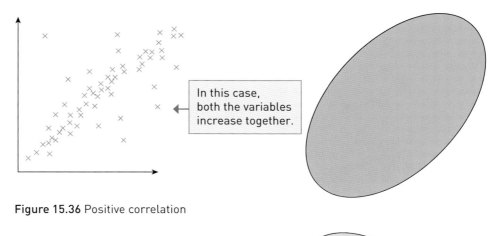

Figure 15.36 Positive correlation

> In this case, both the variables increase together.

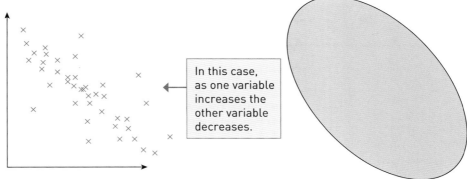

Figure 15.37 Negative correlation

> In this case, as one variable increases the other variable decreases.

> ! You should be aware of scatter diagrams that at first sight appear to indicate correlation but in fact do not.

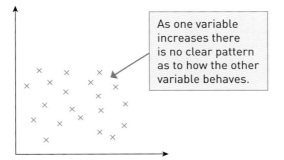

Figure 15.38 No correlation

> As one variable increases there is no clear pattern as to how the other variable behaves.

The scatter diagram is probably showing two quite different groups, neither of them having any correlation.

This is a small data set with no correlation. However the two outliers give the impression there is positive correlation.

> **Note**
>
> In these three false cases, notice that the distribution is not even approximately elliptical.

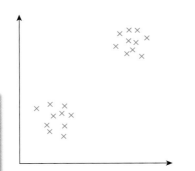

Figure 15.39 Two islands

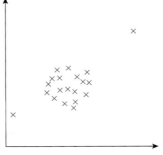

Figure 15.40 Outliers

The bulk of these data have no correlation but a few items give the impression that there is correlation.

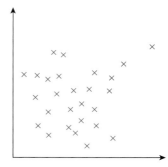

Figure 15.41 A funnel-shaped distribution

Summary measures for bivariate data

Association and correlation

There are two summary measures that you may meet when using statistical software or in other subjects.

Spearman's rank correlation coefficient is a measure of association that is used when both variables are ranked. Its interpretation depends on the sample size. It is used as a test statistic.

Pearson's product moment correlation coefficient is a measure of correlation. It is given by most statistical software. It is used as a measure in its own right and as a test statistic. In both situations its interpretation depends on the sample size.

Lines of best fit

When bivariate data are displayed on a scatter diagram you often need to draw a line of best fit through the points. You can do this roughly by eye, trying to ensure that about the same numbers of points lie above and below the line. If you know the mean values of the two variables, you know the coordinates of the mean point and should draw the line of best fit so that it passes through it. However, drawing a line of best fit by eye is a somewhat haphazard process and in slightly more advanced use it is common to calculate its equation using the **least squares regression line**. This is a standard output from statistical software and can also be found from many spreadsheets.

 Always be careful when using a line of best fit or regression line; be certain that it makes sense, given the data you are dealing with.

Exercise 15.5

 ① Research suggests that there is a positive correlation between success in school and playing video games. Why do you think that is? Does playing video games improve your achievement in school?

 ②
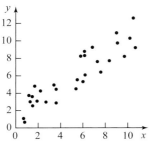
Figure 15.42

Describe the correlation in this scatter graph.

A positive   B strong negative

C none   D weak negative

③ For each of these scatter diagrams comment on whether there appears to be any correlation.

A   The maths and physics test results of 14 students.

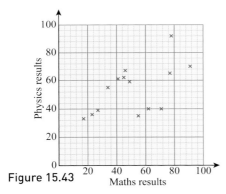
Figure 15.43

B   The value of a used car and the mileage the car has completed.

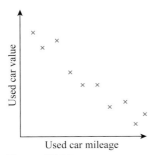
Figure 15.44

C   The number of hours of sunshine and the monthly rainfall in centimetres in an eight-month period.

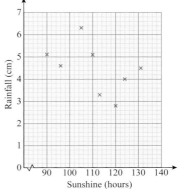
Figure 15.45

D   The mean temperature in degrees Celsius and the amount of icecream sold in a supermarket in hundreds of litres.

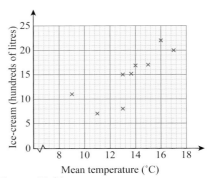
Figure 15.46

④ Data taken from Blackbirds data set at www.hoddereducation.co.uk/AQAMathsYear1

PS

The scatter diagram shows the wingspan (in millimetres) and the mass (in grams) of a sample of 100 blackbirds.

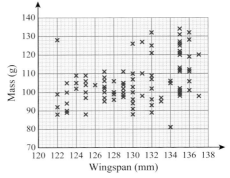
Figure 15.47

(i) Describe the correlation between wingspan and mass.

(ii) Identify and comment on two outliers in this data set.

(iii) A blackbird is brought into a rescue centre. Its wingspan is 126 mm and its mass is 72 g.

What can you say about this bird?

(iv) It is suggested that a regression line with equation $y = x - 27$ provides a good model for predicting the mass, y g, of a blackbird with wingspan x mm.

> **Note**
>
> A regression line is a calculated line of best fit.

(a) A blackbird has wingspan 125 mm. Use this model to predict its mass.

Is your answer consistent with the data shown on the scatter diagram?

(b) Suggest and justify a realistic margin of error to accompany this model.

PS ⑤ The scatter diagram in Figure 15.48 shows the maximum monthly temperature, C, in °C, and the total rainfall, T mm, for June in Whitby in Yorkshire. The data cover the 25 summers included in the period 1991 to 2015.

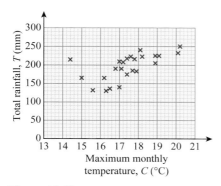

Figure 15.48

(i) Explain why a scatter diagram is an appropriate way to display this type of data.

(ii) Describe the relationship between the maximum monthly temperature and the total monthly rainfall in June in Whitby over this 25-year period.

(iii) Which point(s) might be described as outlier(s)? Comment on your answer.

A regression line for these data is calculated to be $T = 16C - 83$.

(iv) Use this line to estimate the maximum monthly temperature in June 1990, given that the total monthly rainfall for that month was 166.1 mm.

(v) A pensioner had his honeymoon in Whitby in June 1975. He knows the maximum temperature that month was 22 °C and uses the regression line to estimate the rainfall.

Calculate the value he finds and comment on whether it is likely to be a good estimate.

PS ⑥ The scatter diagram in Figure 15.49 shows the length, l metres, and the weight, w kg, of 15 adult elephant seals on a beach.

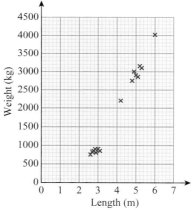

Figure 15.49

(i) By referring to Figure 15.49 and the context of the question:

(a) suggest a reason why the scatter diagram appears to show two distinct sections in the sample

(b) identify and comment on the two outliers on the scatter diagram.

(ii) Explain why it would not be appropriate to produce a regression line based on this sample of data.

A sample of 200 elephant seals from one distinct section of the population is taken.

This sample is used to calculate the regression line $w = 210l + 255$.

(iii) With reference to your answer to part (i)(a), suggest which section of the population was sampled and use the regression line to predict the length of an elephant seal of weight 775 kg.

(PS) ⑦ Abbi is a sociology student. She is investigating the pattern of crime over time. Tables 15.26 and 15.27 give the numbers of homicides (including murder, manslaughter and infanticide) in England and Wales at the start and end of the 20th century.

Table 15.26

Year	Homicides	Year	Homicides
1900	312	1990	669
1901	341	1991	725
1902	330	1992	687
1903	312	1993	670
1904	316	1994	726
1905	287	1995	745
1906	263	1996	679
1907	273	1997	739
1908	321	1998	748
1909	300	1999	750
1910	288	2000	766

The population of England and Wales for certain years is given below in thousands.

Table 15.27

Year	Population	Year	Population
1901	32 528	1991	50 954
1911	36 073	2001	52 042

Five statements are given below. Decide whether each of them is

■ definitely a TRUE conclusion from the data

■ definitely a FALSE conclusion from the data

■ UNCERTAIN because it is impossible to say from the information given.

In each case explain your decision.

A The probability of a randomly selected person being the victim of homicide in 1901 was 1.05×10^{-5} and by 1991 it had risen to 1.42×10^{-5}.

B The data in Table 15.26 are bivariate and both variables are random.

C The data in Table 15.27 could have been taken from the National Census.

D The figures show that in the second half of the 21st century there will be over 1000 murders per year in England and Wales.

E Abbi has drawn this scatter diagram of the data in Table 15.26 and the red line of best fit on it (see Figure 15.50). This shows there is strong positive correlation.

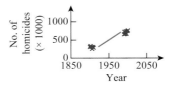

Figure 15.50

Working with a large data set

Scatter diagrams

(T) ⑧ In Exercise 15.4 you looked at what time of day the cyclists' accidents took place and also investigated whether looking at different subsets of the data (mainly based on age) changed the results of that analysis.

Now draw scatter diagrams using the fields 'age' and 'time of accident'.

Is there a relationship between the age of the cyclist and the time of the accident? Notice that you may need to clean the original data.

(T) ⑨ Using a large data set of your choice produce scatter diagrams and find a correlation coefficient for different pairs of variables if appropriate. Which show correlation?

6 Standard deviation

You have already met range and interquartile range as measures of the spread of a data set. Sometimes neither of these will quite meet your requirements.

- The range does not use all the available information, only the extreme values which may well be outliers. In most situations this is unsatisfactory, although in quality control it can be an advantage as it is very sensitive to something going wrong, for example on a production line.

- The interquartile range requires that either the data are ranked, or else that an estimate is made, for example using a cumulative frequency curve.

Very often you want to know how much a typical value is above or below a central value such as the mean.

A new measure of spread

Kim and Jo play as strikers for local hockey teams. The selectors for the county team want to choose one of them.

These are their goal-scoring records over the last 10 matches.

| Kim | 0 | 0 | 0 | 3 | 0 | 1 | 0 | 0 | 1 | 5 |
| Jo | 2 | 1 | 1 | 0 | 0 | 2 | 1 | 0 | 1 | 2 |

Who should be selected?

The means of their numbers of goals per match should give an indication of their overall performance.

They have both scored 10 goals so have the same mean of 1.0 goals per match. That does not help the selectors.

The spread of their numbers of goals could give an indication of their reliability.

Start by finding how far each score is from the mean. This is the **deviation**, $(x - \bar{x})$. Table 15.28 is for Kim's data.

Table 15.28

	Kim's goals										Total, Σ
Number of goals scored, x	0	0	0	3	0	1	0	0	1	5	10
Deviation $(x - \bar{x})$	-1	-1	-1	$+2$	-1	0	-1	-1	0	$+4$	0

The total of the deviations is zero.

> The reason the total is zero is because of the definition of the mean. The sum of the values above it must be equal to the sum of those below it.

Instead you want the absolute value of the deviation.

Whether it is positive or negative, it is counted as positive.

It is denoted by $|x - \bar{x}|$.

You can see this in Table 15.29. Remember that in this case $\bar{x} = 1$.

Table 15.29

	Kim's goals										Total, Σ
Number of goals scored, x	0	0	0	3	0	1	0	0	1	5	10
Absolute deviation, $|x - \bar{x}|$	1	1	1	2	1	0	1	1	0	4	12

The next step is to find the means of their absolute deviations.

For Kim it is $\frac{12}{10} = 1.2$. For Jo the mean absolute deviation works out to be $\frac{6}{10} = 0.6$.

> Check Jo's mean absolute deviation for yourself.

So there is a greater spread in the numbers of goals Kim scored.

Which of the two players should be picked for the county team?

If the selectors want a player with a steady reliable performance, they probably should select Jo; on the other hand if they want someone who has good days and bad days but is very good at her best, then it should be Kim.

Mean absolute deviation is given by $\dfrac{\sum |x - \bar{x}|}{n}$. It is an acceptable measure of spread but it is not widely used because it is difficult to work with. Instead the thinking behind it is taken further with **standard deviation** which is more important mathematically and consequently is very widely used.

To work out the mean absolute deviation, you had to treat all deviations as if they were positive. A different way to get rid of the unwanted negative signs is to square the deviations. For Kim's data this is shown in Table 15.30.

Table 15.30

	Kim's goals										Total, Σ
Number of goals scored, x	0	0	0	3	0	1	0	0	1	5	10
Squared deviation $(x - \bar{x})^2$	+1	+1	+1	+4	+1	0	+1	+1	0	+16	26

So the mean of the squared deviations is $\frac{26}{10} = 2.6$.

This is not a particularly easy measure to interpret, but if you take its square root and so get the **root mean squared deviation**, this can then be compared with the actual data values.

> Find the root mean squared deviation for Jo's data for yourself.

In this example the root mean square deviation for Kim's data is $\sqrt{2.6} = 1.61\ldots$.

It is a measure of how much the value of a typical item of data item might be above or below the mean.

It is good practice to use standard notation to explain your working, and this is given below.

Discussion points

➔ How can you show that
$$\sum (x - \bar{x})^2 = \sum x^2 - n\bar{x}^2,$$
in other words that the two formulae for S_{xx} are equivalent?

➔ Are there any situations where one method might be preferred to the other?

Notation

The sum of the squared deviations is denoted by S_{xx}.

So $S_{xx} = \displaystyle\sum_{i=1}^{i=n} (x_i - \bar{x})^2$.

This is often written more simply as $S_{xx} = \sum (x - \bar{x})^2$.

An equivalent form is $S_{xx} = \sum x^2 - n\bar{x}^2$.

Mean squared deviation is given by $\dfrac{S_{xx}}{n}$

and root mean squared deviation is $\sqrt{\dfrac{S_{xx}}{n}}$.

If you are working by hand and the mean, $\bar{x}$, has several decimal places, working out the individual deviations can be tedious. So in such cases it is easier to use the $\sum x^2 - n\bar{x}^2$ form. However, it is good practice to get into the habit of using the statistical functions on your calculator or spreadsheet.

You are now one step away from finding **variance** and **standard deviation**.

> **Notation**
> ----------
> Variance $s^2 = \dfrac{S_{xx}}{n-1}$
>
> Standard deviation
>
> $s = \sqrt{\dfrac{S_{xx}}{n-1}}$

The calculation of the mean squared deviation and the root mean squared deviation involves dividing the sum of squares, S_{xx}, by n. If, instead, S_{xx} is divided by $(n-1)$ you obtain the **variance** and the **standard deviation**. For sample data, they are denoted by s^2 and s.

The reason for dividing by $(n-1)$ rather than by n is that when you come to work out the n deviations of a data set from the mean there are only $(n-1)$ independent variables.

Take, for example, the simple data set 2, 3, 7. The mean is $\dfrac{2+3+7}{3} = 4$.

The deviation of 2 is $2 - 4 = -2$ and the deviation of 3 is $3 - 4 = -1$.

Since the sum of the deviations is 0, it follows that the deviation of the third data item must be $+3$, and you do not need to work out $7 - 4 = 3$. So only two of the deviations are independent: the third is dependent on them.

The same argument can be applied to the deviations for a set with n data items. Only $(n-1)$ of them are independent; the last deviation is not independent as it can be calculated from all the rest.

> **ACTIVITY 15.2**
>
> What are the variance and standard deviation for Jo's data?

For Kim's data, $S_{xx} = 26$ and $n = 10$.

So the variance, $s^2 = \dfrac{26}{10-1} = 2.888\ldots$ and the standard deviation

$s = \sqrt{2.888\ldots} = 1.699\ldots$ (or 1.700 to 3 d.p.).

Standard deviation is by far the most important measure of spread in statistics. It is used for both discrete and continuous data and with ungrouped and grouped data.

Example 15.5 illustrates the calculation of the standard deviation of a continuous (and so grouped) data set.

Example 15.5

As part of her job as a quality control inspector Stella commissioned these data relating to the lifetime, h hours, of 80 light bulbs produced by her company.

Table 15.31

Lifetime	$h < 200$	$200 \leqslant h < 300$	$300 \leqslant h < 400$	$400 \leqslant h < 500$	$500 \leqslant h < 600$	$600 \leqslant h < 700$	$700 \leqslant h < 800$	$h \geqslant 800$
Frequency	0	22	12	8	6	21	11	0

(i) Estimate the mean and standard deviation of the light bulb lifetimes. ← It is easiest to do a calculation like this in a table.

(ii) What do you think Stella should tell her company?

Solution

(i) **Table 15.32**

Interval	f	Mid-point, x	fx	fx^2
$200 \leqslant h < 300$	22	250	5 500	1 375 000
$300 \leqslant h < 400$	12	350	4 200	1 470 000
$400 \leqslant h < 500$	8	450	3 600	1 620 000
$500 \leqslant h < 600$	6	550	3 300	1 815 000
$600 \leqslant h < 700$	21	650	13 650	8 872 500
$700 \leqslant h < 800$	11	750	8 250	6 187 500
Σ	80		38 500	21 340 000

$$\text{Mean} = \frac{\Sigma fx}{\Sigma f} = \frac{38\,500}{80} = 481.25$$

$$S_{xx} = \sum fx^2 - n\bar{x}^2 = 21\,340\,000 - 80 \times 481.25^2 = 2\,811\,875$$

$$\text{Variance } s^2 = \frac{S_{xx}}{(n-1)} = \frac{2\,811\,875}{79} = 35\,593.3\ldots$$

$$\text{Standard deviation} = \sqrt{\text{Variance}} = 188.6\ldots$$

To the nearest whole numbers of hours, the mean is 481 and the standard deviation is 189.

(ii) The distribution is bimodal. This is shown both by the pattern of the data and the high standard deviation. The data suggest that something is probably going wrong in the production process but they cannot say what; that requires an investigation of the process itself.

Discussion point

➜ Can standard deviation have a negative value?

Identifying outliers

You have already seen how interquartile range can be used to identify possible outliers in a data set. Standard deviation can be used in a similar way.

A commonly used check is to investigate all items which lie more than 2 standard deviations from the mean and to decide whether they should be included in your analysis or not. The distributions of many populations are approximately normal and in such cases this test can be expected to highlight the most extreme 5% of data values.

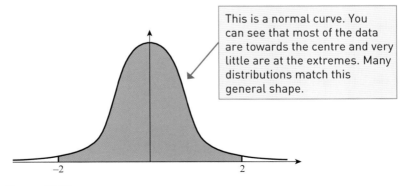

This is a normal curve. You can see that most of the data are towards the centre and very little are at the extremes. Many distributions match this general shape.

Figure 15.51

Exercise 15.6

(1) You are able to read off values for the root mean squared deviation, $\sqrt{\dfrac{S_{xx}}{n}}$, and the standard deviation, $\sqrt{\dfrac{S_{xx}}{n-1}}$, from your calculator. Which is larger? How do you know?

(2) Which of these expressions does **not** calculate S_{xx}?

A $\displaystyle\sum (x - \bar{x})^2$ B $\displaystyle\sum x^2 - n\bar{x}^2$

C $\displaystyle\sum x^2 - \dfrac{\left(\sum x\right)^2}{n}$ D $\displaystyle\sum x^2 - \bar{x}^2$

PS (3) An engineer is commissioning a new machine to cut metal wire to specific lengths. The machine is set to 1 metre (1000 mm). 100 lengths are cut and then measured. The errors in mm above or below the set length are recorded. For example a wire which is 2 mm below the set 1 m length is recorded as −2. The results of the trial are shown in the frequency chart in Figure 15.52.

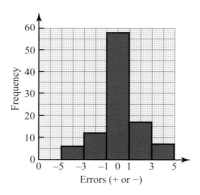

Figure 15.52

(i) Estimate the mean error.
(ii) Estimate the standard deviation of the error.
(iii) The sample mean of the raw data is 0.09 mm, and the standard deviation is 1.99. Explain why these are slightly different from your estimates in parts (i) and (ii).

PS (4) Alisa and Bjorn gather data about the heights of plants for a biology project. They measure the same plants. Alisa records the data in mm whilst Bjorn measures to the same accuracy, but uses cm.

Here are their data.

Table 15.33

	Alisa	Bjorn
Plant	Plant height (mm)	Plant height (cm)
1	135.0	13.5
2	124.0	12.4
3	142.0	14.2
4	113.0	11.3
5	127.0	12.7
6	121.0	12.1
7	132.0	13.2
8	129.0	12.9
9	128.0	12.8
10	130.0	13.0
Mean	128.1	12.8
SD	7.46	0.746

Bjorn remarks, 'That's strange, we measured the same plants, but your data have a larger standard deviation which means they are more variable than mine'.

Are Alisa's data more variable? Explain your answer.

PS (5) In an A-level German class the examination marks at the end of the year are shown below.

35   52   55   61   96   63
50   58   58   49   61

(i) Show these marks on a number line.

(ii) Find their mean and standard deviation.

(iii) (a) Show that one of the marks is more than 2 standard deviations from the mean.

(b) How would you describe that mark?

(iv) Suggest three possible explanations for the mark and in each case decide whether it should be accepted or rejected.

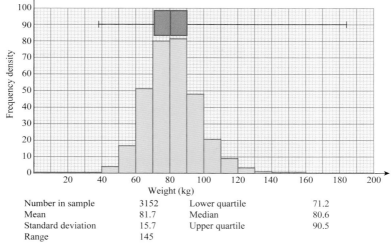

Number in sample	3152	Lower quartile	71.2
Mean	81.7	Median	80.6
Standard deviation	15.7	Upper quartile	90.5
Range	145		

Figure 15.53

⑥ The histogram, boxplot and data summary in Figure 15.53 shows the weights of 3152 male patients, aged 14 and over, at a hospital over a period of one year.

(i) State two things that the boxplot shows for these data.

(ii) Calculate the number of the patients with weights between 50 and 80 kg.

(iii) State the weights that are 2 standard deviations above and below the mean.

(iv) Hence estimate the number of patients with weights that can be considered outliers.

(v) Without doing any further calculations, say whether the outliers are about equally split between very heavy people and very light people.

Say what evidence you have used in answering this question.

⑦ As part of a biology experiment Andrew caught and weighed 121 newts. He used his calculator to find the mean and standard deviation of their weights.

Mean = 16.231 g

Standard deviation = 4.023 g

(i) Find the total weight, $\sum x$, of Andrew's 121 newts.

(ii) Use the formula

$$s = \sqrt{\frac{\sum x^2 - n\bar{x}^2}{n - 1}}$$ to find $\sum x^2$

for Andrew's newts.

Another member of the class, Sara, did the same experiment with newts caught from a different pond. Her results are summarised by:

$n = 81$

$\bar{x} = 15.214$

$s = 3.841$

Their supervisor suggests they should combine their results into a single set but by then they have both thrown away their measurements and the newts are back in their ponds.

(iii) Find n, $\sum x$ and $\sum x^2$ for the combined data set.

(iv) Find the mean and the standard deviation for the combined data set.

⑧ Sam finds this definition of a Normal distribution in a textbook.

For a Normal distribution, approximately

■ 68% of the data lie within 1 standard deviation of the mean

■ 95% lie within 2 standard deviations of the mean

■ 99.75% lie within 3 standard deviations of the mean.

Investigate whether this is the case for the data about male hospital patients in question 6.

Working with a large data set

⑨ For the cyclists' data set, find the mean and standard deviation of the ages of those who had accidents. Compare these with the median and the semi-interquartile range.

⑩ For a large data set of your choice, select a variable that seems to have a distribution that is roughly symmetrical about the mean. Find the mean and standard deviation for this variable and check whether the figures given in question 8 for a Normal distribution are about right.

LEARNING OUTCOMES

When you have completed this chapter you should be able to:

➤ interpret diagrams for single-variable data, including understanding that area in a histogram represents frequency

➤ interpret scatter diagrams and regression lines for bivariate data:
  ○ recognition of scatter diagrams which include distinct sections of the population (calculations involving regression lines are excluded)

➤ understand informal interpretation of correlation

➤ understand that correlation does not imply causation

➤ interpret measures of central tendency and variation, extending to standard deviation

➤ be able to calculate standard deviation, including from summary statistics

➤ recognise and interpret possible outliers in data sets and statistical diagrams

➤ select or critique data presentation techniques in the context of a statistical problem.

15

KEY POINTS

Summary of types of data

Type of data	Typical display	Summary measures Central tendency	Summary measures Spread
Categorical	Bar chart Pie chart Dot plot	Modal class	
Ranked	(Stem-and-leaf diagram) Boxplot	Median	Range Interquartile range
Discrete numerical (ungrouped)	(Stem-and-leaf diagram) Vertical line chart	Mean Weighted mean Mode (Median)	Range Standard deviation
Discrete numerical (grouped)	Bar chart (using groups as categories)	Mean Weighted mean Modal class	Range Standard deviation
Continuous	Frequency chart Histogram	Mean Weighted mean Median Modal class	Range Standard deviation
	Cumulative frequency curve	Median	Interquartile range
Bivariate	Scatter diagram Line of best fit	Mean point	Correlation coefficient (e.g. Spearman's, Pearson's)

FUTURE USES

- The material in this and the previous chapter covers the basic techniques you need when using the Statistics cycle. Consequently it pervades much of the future use of statistics.
- Ranked data are widely used in non-parametric statistics, including Spearman's rank correlation coefficient.

16 Probability

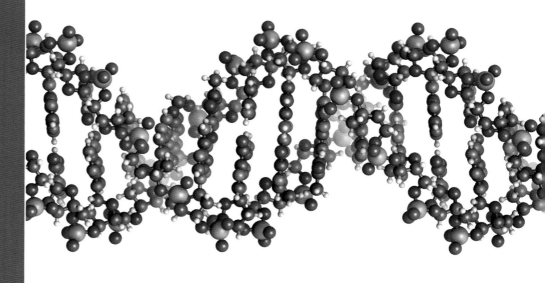

THE AVONFORD STAR

A library without books

If you plan to pop into the Avonford library and pick up the latest bestseller, then forget it. All the best books 'disappear' practically as soon as they are put on the shelves.

I talked about the problem with the local senior librarian, Gina Clarke.

'We have a real problem with unauthorised loans at the moment,' Gina told me. 'Out of our total stock of, say, 80 000 books, something like 44 000 are out on loan at any one time. About 20 000 are on the shelves and I'm afraid the rest are unaccounted for.'

Librarian Gina Clarke is worried about the problem of 'disappearing books'

That means that the probability of finding the particular book you want is exactly $\frac{1}{4}$. With odds like that, don't bet on being lucky next time you visit your library.

➔ How do you think the figure of $\frac{1}{4}$ at the end of the article was arrived at?

➔ Do you agree that the probability is *exactly* $\frac{1}{4}$?

The information about the different categories of book is summarised in Table 16.1.

Table 16.1

Category of book	Typical numbers
On the shelves	20 000
Out on loan	44 000
Unauthorised loan	16 000
Total stock	80 000

On the basis of these figures it is possible to estimate the probability of finding the book you want. Of the total stock of 80 000 books bought by the library, you might expect to find about 20 000 on the shelves at any one time. As a fraction this is $\frac{20}{80}$ or $\frac{1}{4}$ of the total. So, as a rough estimate, the probability of you finding a particular book is 0.25 or 25%.

Similarly, 16 000 out of the total of 80 000 books are on unauthorised loan (a euphemism for stolen) and this is 20%, or $\frac{1}{5}$.

An important assumption underlying these calculations is that all the books are equally likely to be unavailable, which is not very realistic since popular books are more likely to be stolen. Also, the numbers given are only rough approximations, so it is definitely incorrect to say that the probability is *exactly* $\frac{1}{4}$.

1 Working with probability

Measuring probability

Probability (or chance) is a way of describing the likelihood of different possible **outcomes** occurring as a result of some **experiment**.

In the example of the library books, the experiment is looking in the library for a particular book. Let us assume that you already know that the book you want is in the library's stocks. The three possible outcomes are that the book is on the shelves, out on loan or missing.

It is important in probability to distinguish experiments from the outcomes which they may generate. A few examples are given in Table 16.2.

Table 16.2

Experiments	Possible outcomes
Guessing the answer to a four-option multiple choice question	A B C D
Predicting the stamp on the next letter I receive	first class second class foreign other
Tossing a coin	heads tails

Another word for experiment is **trial**. This is used to describe the binomial situation where there are just two possible outcomes.

Another word you should know is **event**. This often describes several outcomes put together. For example, when rolling a die, an event could be 'the die shows an even number'. This event corresponds to three different outcomes from the trial, the die showing 2, 4 or 6. However, the term event is also often used to describe a single outcome.

Estimating probability

Probability is a number which measures likelihood. It may be estimated experimentally or theoretically.

Experimental estimation of probability

In many situations probabilities are estimated on the basis of data collected experimentally, as in the following example.

Of 30 drawing pins tossed in the air, 21 of them were found to have landed with their pins pointing up. From this you would estimate that the probability that the next pin tossed in the air will land with its pin pointing up is $\frac{21}{30}$ or 0.7.

You can describe this in more formal notation.

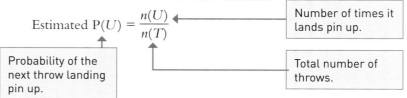

$$\text{Estimated } P(U) = \frac{n(U)}{n(T)}$$

Probability of the next throw landing pin up.

Total number of throws.

Number of times it lands pin up.

Theoretical estimation of probability

There are, however, some situations where you do not need to collect data to make an estimate of probability.

For example, when tossing a coin, common sense tells you that there are only two realistic outcomes and, given the symmetry of the coin, you would expect them to be equally likely. So the probability, $P(H)$, that the next coin will produce the outcome heads can be written as follows.

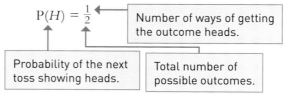

$$P(H) = \frac{1}{2}$$

Probability of the next toss showing heads.

Total number of possible outcomes.

Number of ways of getting the outcome heads.

Example 16.1

Using the notation described above, write down the probability that the correct answer for the next four-option multiple choice question will be answer A. What assumptions are you making?

Solution

Assuming that the test-setter has used each letter equally often, the probability, P(A), that the next question will have answer A can be written as follows:

$$P(A) = \frac{1}{4} \longleftarrow \boxed{\text{Answer A}}$$

$$\boxed{\text{Answers A, B, C and D}}$$

Equiprobability is an important assumption underlying most work on probability.

Notice that we have assumed that the four options are equally likely.

For equiprobable events (or outcomes) the probability, P(A), of event occurring can be expressed formally as:

$$P(A) = \frac{n(A)}{n(\varepsilon)}$$

$\boxed{\text{Number of ways that event A can occur.}}$

$\boxed{\text{Probability of event } A \text{ occurring.}}$

$\boxed{\text{Total number of ways that the possible events can occur. Notice the use of the symbol } \varepsilon, \text{ the universal set of all the ways that the possible events can occur.}}$

Probabilities of 0 and 1

The two extremes of probability are **certainty** at one end of the scale and **impossibility** at the other.

Table 16.3 shows some examples of certain and impossible events.

Table 16.3

Experiments	Certain events	Impossible events
Rolling two dice	The total is in the range 2 to 12 inclusive	The total is 13
Tossing a coin	Getting either heads or tails	Getting neither a head nor a tail

Certainty

As you can see from Table 16.3, for events that are certain, the number of ways that the event can occur, $n(A)$ in the formula, is equal to the total number of possible events, $n(\varepsilon)$.

$$\frac{n(A)}{n(\varepsilon)} = 1$$

So the probability of an event which is certain is one.

Impossibility

The number of ways that an impossible event, A, can occur, $n(A)$, is zero.

$$\frac{n(A)}{n(\varepsilon)} = \frac{0}{n(\varepsilon)} = 0$$

So the probability of an event which is impossible is zero.

Typical values of probabilities lie between 0 and 1 and so might be something like 0.3 or 0.9. If you arrive at probability values of, for example, -0.4 or 1.7, you will know that you have made a mistake since these are meaningless.

$$0 \leqslant P(A) \leqslant 1 \quad \boxed{\text{Certain event.}}$$

$$\boxed{\text{Impossible event.}}$$

The complement of an event

The complement of an event A, denoted by A', is the event not-A, that is the event 'A does not happen'.

Example 16.2

It was found that, out of a box of 50 matches, 45 lit but the others did not. What was the probability that a randomly selected match would not have lit?

Solution

The probability that a randomly selected match lit was

$$P(A) = \frac{45}{50} = 0.9$$

The probability that a randomly selected match did not light was

$$P(A') = \frac{50 - 45}{50} = \frac{5}{50} = 0.1 \quad \boxed{\text{Notice that } 0.1 = 1 - 0.9}$$

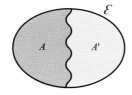

Figure 16.1 Venn diagram showing events A and A'

From this example you can see that

$$P(A') = 1 - P(A)$$

$\boxed{\text{The probability of } A \text{ not occurring.}}$

This is illustrated in Figure 16.1.

$\boxed{\text{The probability of } A \text{ occurring.}}$

Expectation

THE AVONFORD STAR

Avonford braced for flu epidemic

Local health services are poised for their biggest challenge in years. The virulent strain of flu, currently sweeping across Europe, is expected to hit Avonford any day.

With a chance of one in three of any individual contracting the disease, and 120 000 people within the Health Area, surgeries and hospitals are expecting to be swamped with patients.

Local doctor Aloke Ghosh says 'Immunisation seems to be ineffective against this strain'.

How many people can the Health Area expect to contract flu? The answer is easily seen to be $120\,000 \times \frac{1}{3} = 40\,000$. This is called the **expectation** or **expected frequency** and is given in this case by np where n is the population size and p the probability.

Expectation is a technical term and need not be a whole number. Thus the expectation of the number of heads when a coin is tossed 5 times is $5 \times \frac{1}{2} = 2.5$. You would be wrong to go on to say 'That means either 2 or 3' or to qualify your answer as 'about $2\frac{1}{2}$'. The expectation is 2.5.

The probability of either one event or another

So far you have looked at just one event at a time. However, it is often useful to bracket two or more of the events together and calculate their combined probability.

Example 16.3

This example is based on the data at the beginning of this chapter and shows the probability of the next book requested falling into each of the three categories listed, assuming that each book is equally likely to be requested.

Table 16.4

Category of book	Typical numbers	Probability
On the shelves (S)	20 000	0.25
Out on loan (L)	44 000	0.55
Unauthorised loan (U)	16 000	0.20
Total ($S + L + U$)	80 000	1.00

What is the probability that a randomly requested book is *either* out on loan *or* on unauthorised loan (i.e. that it is not available)?

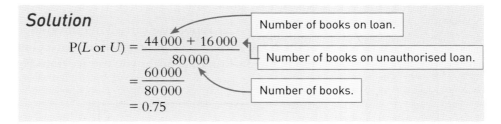

Solution

$$P(L \text{ or } U) = \frac{44\,000 + 16\,000}{80\,000}$$
$$= \frac{60\,000}{80\,000}$$
$$= 0.75$$

Number of books on loan.
Number of books on unauthorised loan.
Number of books.

This can be written in more formal notation as

$$P(L \cup U) = \frac{n(L \cup U)}{n(\varepsilon)}$$
$$= \frac{n(L)}{n(\varepsilon)} + \frac{n(U)}{n(\varepsilon)}$$
$$P(L \cup U) = P(L) + P(U)$$

Notice the use of the **union** symbol, $\cup$, to mean **or**. This is illustrated in Figure 16.2.

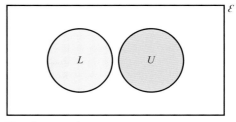

Key: L = out on loan
U = out on unauthorized loan

Figure 16.2 Venn diagram showing events L and U. It is not possible for both to occur

In this example you could add the probabilities of the two events to get the combined probability of either one or the other event occurring. However, you have to be very careful adding probabilities as you will see in the next example.

Example 16.4

Table 16.5 shows further details of the categories of books in the library.

Table 16.5

Category of book	Number of books
On the shelves	20 000
Out on loan	44 000
Adult fiction	22 000
Adult non-fiction	40 000
Junior	18 000
Unauthorised loan	16 000
Total stock	80 000

Assuming all the books in the library are equally likely to be requested, find the probability that the next book requested will be either out on loan or a book of adult non-fiction.

> ⚠ **Incorrect solution**
>
>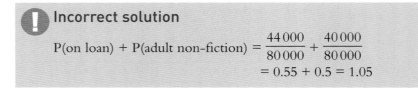
> $$P(\text{on loan}) + P(\text{adult non-fiction}) = \frac{44\,000}{80\,000} + \frac{40\,000}{80\,000}$$
> $$= 0.55 + 0.5 = 1.05$$

This is clearly nonsense as you cannot have a probability greater than 1.

So what has gone wrong?

The way this calculation was carried out involved some double counting. Some of the books classed as adult non-fiction were counted twice because they were also in the on-loan category, as you can see from Figure 16.3.

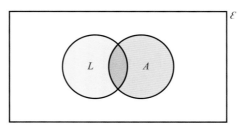

Key: L = out on loan
A = adult non-fiction

Figure 16.3 Venn diagram showing events L and A. It is possible for both to occur

If you add all six of the book categories together, you find that they add up to 160 000, which represents twice the total number of books owned by the library.

Solution

A more useful representation of these data is given in the two–way table below.

Table 16.6

	Adult fiction	Adult non-fiction	Junior	Total
On the shelves	4 000	12 000	4 000	20 000
Out on loan	14 000	20 000	10 000	44 000
Unauthorised loan	4 000	8 000	4 000	16 000
Total	22 000	40 000	18 000	80 000

If you simply add 44 000 and 40 000, you *double count* the 20 000 books which fall into both categories. So you need to subtract the 20 000 to ensure that it is only counted once. Thus:

Number either out on loan or adult non-fiction

$$= 44\,000 + 40\,000 - 20\,000$$

$$= 64\,000 \text{ books}$$

So the required probability $= \dfrac{64\,000}{80\,000} = 0.8$

Mutually exclusive events

The problem of double counting does not occur when adding two rows in Table 16.6. Two rows cannot overlap, or **intersect**, which means that those categories are **mutually exclusive** (i.e. the one excludes the other). The same is true for two columns within the table.

Where two events, A and B, are mutually exclusive, the probability that either A or B occurs is equal to the sum of the separate probabilities of A and B occurring (Figure 16.4a).

Where two events, A and B, are *not* mutually exclusive, the probability that either A or B occurs is equal to the sum of the separate probabilities of A and B occurring minus the probability of A and B occurring together (Figure 16.4b).

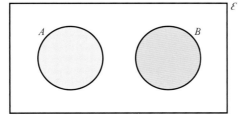

Figure 16.4 (a) Mutually exclusive events
$\mathrm{P}(A \cup B) = \mathrm{P}(A) + \mathrm{P}(B)$

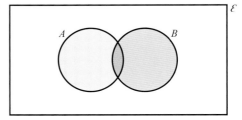

(b) Not mutually exclusive events
$\mathrm{P}(A \cup B) = \mathrm{P}(A) + \mathrm{P}(B) - \mathrm{P}(A \cap B)$

Notice the use of the intersection sign, $\cap$, to mean *both . . . and . . .*

Example 16.5

A card is selected at random from a normal pack of 52 playing cards.

Event A is 'The card is a heart'.

Event B is 'The card is a King'.

(i) Draw a Venn diagram showing events A and B.

(ii) Find the probability that the card is

    (a) a heart

    (b) a King

    (c) the King of hearts

    (d) a King or a heart that is not the King of hearts.

(iii) Verify that in this case $P(A \cup B) = P(A) + P(B) - P(A \cap B)$.

Solution

(i)

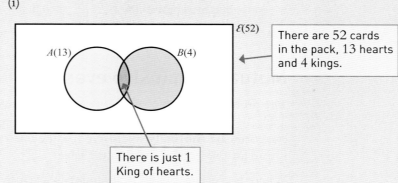

There are 52 cards in the pack, 13 hearts and 4 kings.

There is just 1 King of hearts.

Figure 16.5

(ii) (a) $P(A) = \frac{13}{52} = \frac{1}{4}$

    (b) $P(B) = \frac{4}{52} = \frac{1}{13}$

    (c) $P(A \cap B) = \frac{1}{52}$ There is just 1 King of hearts.

    (d) There are 4 Kings.

       Apart from the King of hearts there are 12 other hearts.

       So $P(A \cup B) = \frac{4 + 12}{52} = \frac{16}{52}$ $13 - 1 = 12$

(iii) Substituting in

    $P(A \cup B)$ and $P(A) + P(B) - P(A \cap B)$

        $\frac{16}{52} \quad = \quad \frac{13}{52} + \frac{4}{52} - \frac{1}{52} = \frac{16}{52}$ ✓

The probability of two events happening

Sometimes two events can happen at the same time, even though they are independent. For example, the selection of a playing card and the throw of a coin do not affect each other, but both can be done simultaneously.

Example 16.6

If you select a card at random from a standard pack and toss a coin, what is the probability of picking a spade and throwing a tail?

Solution

You could list all of the possible outcomes:

> spade head, heart head, ...

> A benefit of using a sample space diagram is that it makes it difficult to miss out possible outcomes.

but you can also show the possible outcomes in a **sample space diagram**.

		Outcomes for the card			
		Spade ♠	Heart ♥	Diamond ♦	Club ♣
Outcomes for the coin	Head	♠H	♥H	♦H	♣H
	Tail	♠T	♥T	♦T	♣T

From this sample space diagram you can see easily that there are eight possible outcomes, only one of which is both a spade and a tail. They are all equally likely. This gives a probability of $\frac{1}{8}$.

For some situations it is not feasible to list all of the outcomes, but you know the probabilities of the individual outcomes. In such cases you can calculate the probability of the combined event. This is usually quicker anyway.

In this example, the probability of getting a spade, $P(♠) = \frac{1}{4}$ and the probability of getting a tail, $P(T) = \frac{1}{2}$.

If you multiply these you get

$$P(♠\ T) = \frac{1}{4} \times \frac{1}{2} = \frac{1}{8}$$

which of course is the same answer as before.

Another way of showing probabilities is on a **tree diagram**.

Example 16.7

On my way to college I drive through two sets of traffic lights; the probability that I am stopped at the first set is 0.7 and at the second set is 0.4. The timings of the two sets of lights are independent of each other.

(i)   What is the probability that I am stopped at both sets?

(ii) What is the probability that I am stopped at least once?

Solution

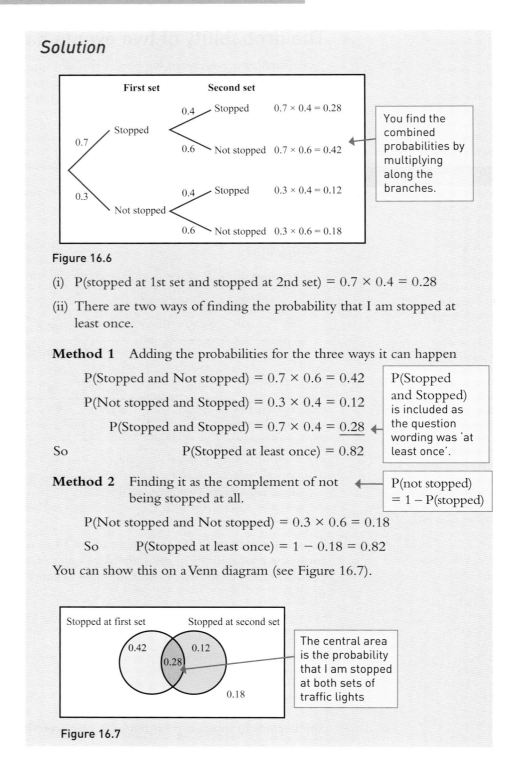

Figure 16.6

(i) P(stopped at 1st set and stopped at 2nd set) = $0.7 \times 0.4 = 0.28$

(ii) There are two ways of finding the probability that I am stopped at least once.

Method 1   Adding the probabilities for the three ways it can happen

P(Stopped and Not stopped) = $0.7 \times 0.6 = 0.42$

P(Not stopped and Stopped) = $0.3 \times 0.4 = 0.12$

P(Stopped and Stopped) = $0.7 \times 0.4 = \underline{0.28}$

So   P(Stopped at least once) = 0.82

> P(Stopped and Stopped) is included as the question wording was 'at least once'.

Method 2   Finding it as the complement of not being stopped at all.

> P(not stopped) = 1 − P(stopped)

P(Not stopped and Not stopped) = $0.3 \times 0.6 = 0.18$

So   P(Stopped at least once) = $1 - 0.18 = 0.82$

You can show this on a Venn diagram (see Figure 16.7).

Figure 16.7

Probability distributions

Table 16.15 shows the probabilities of different possible scores when two dice are thrown. It shows the **probability distribution**. The random variable 'the total of the two dice' is denoted by X and the possible values it can take by x.

Table 16.7

Score, x	2	3	4	5	6	7	8	9	10	11	12
$P(X = x)$	$\frac{1}{36}$	$\frac{2}{36}$	$\frac{3}{36}$	$\frac{4}{36}$	$\frac{5}{36}$	$\frac{6}{36}$	$\frac{5}{36}$	$\frac{4}{36}$	$\frac{3}{36}$	$\frac{2}{36}$	$\frac{1}{36}$

Notice that the total of all the probabilities is 1.

$$\frac{1}{36} + \frac{2}{36} + \frac{3}{36} + \frac{4}{36} + \frac{5}{36} + \frac{6}{36} + \frac{5}{36} + \frac{4}{36} + \frac{3}{36} + \frac{2}{36} + \frac{1}{36} = \frac{36}{36} = 1$$

Table 16.7 covers all the possible outcomes.

You can use the figures in the table to find the mean, or expectation, of the random variable, as in the next example.

> This can be expressed formally as
> $$E(x) = \sum_{\text{All } x} P(X = x) = 1$$

Example 16.8

Find the expectation of the total score when two dice are thrown.

> This can be expressed formally as $E(x) = \sum_{\text{All } x} x \times P(X = x)$

Solution

The expectation is given by $\sum_{\text{All } x} x \times P(X = x)$.

$$\frac{1}{36} \times 2 + \frac{2}{36} \times 3 + \frac{3}{36} \times 4 + \frac{4}{36} \times 5 + \frac{5}{36} \times 6 + \frac{6}{36} \times 7 + \frac{5}{36} \times 8$$

$$+ \frac{4}{36} \times 9 + \frac{3}{36} \times 10 + \frac{2}{36} \times 11 + \frac{1}{36} \times 12 = \frac{252}{36} = 7$$

So the expectation is 7.

> Notice that you can see from symmetry that the mean value is 7. Expectation and mean are the same thing.

Discussion point

→ What situation could be represented by this particular uniform distribution?

A particular example of a probability distribution is the discrete uniform distribution in which all the possible values of the variable have the same probability, as in Table 16.8.

Table 16.8

x	1	2	3	4	5	6
$P(X = x)$	$\frac{1}{6}$	$\frac{1}{6}$	$\frac{1}{6}$	$\frac{1}{6}$	$\frac{1}{6}$	$\frac{1}{6}$

Sometimes it is possible to express a probability distribution algebraically, as in the following example.

Example 16.9

A probability distribution is given

$$P(X = r) = kr^2 \quad \text{for } r = 2, 3 \text{ and } 5$$

$$P(X = r) = 0 \quad \text{otherwise}$$

(i)   Represent the probability distribution as a table.

(ii) Find the value of k.

(iii) Find $P(X < 5)$.

Solution

(i) Table 16.9

r	2	3	5
$P(X = r)$	$4k$	$9k$	$25k$

(ii) $4k + 9k + 25k = 1$ ← Using the fact that the sum of the probabilities must be 1.

$38k = 1$

$k = \frac{1}{38}$

(iii) $P(X < 5) = P(X = 2 \text{ or } 3)$

$= \frac{4}{38} + \frac{9}{38} = \frac{13}{38}$

Risk

In everyday language, probability is often described as **risk**, particularly when it is associated with something we don't want to happen.

The risk of a particular place flooding is often described as a 'once in 100 years event' (or some other number of years such as 200, 500, 1000, etc.). Sometimes this is given as a ratio, for example, 'a 1 : 100 year flood event', meaning that the probability of a flood at that place in any year is $\frac{1}{100}$.

The flooding map for London in Figure 16.8 was published by the Environment Agency. Without flood defences, the probability of flooding, p, in the dark blue areas would be given by $p \geq 0.01$ and in the light blue areas $0.01 < p \leq 0.001$.

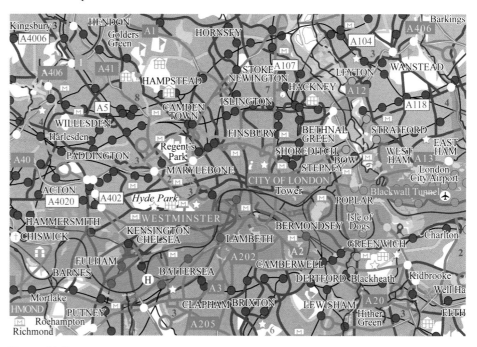

Figure 16.8

Many organisations maintain risk registers. The risks to their smooth running are identified, together with the impact they would have. These are then combined to give a measure of the seriousness of the various possibilities. Sometimes the risks are given as numerical probabilities, like 0.25, at other times they are just described in words, for example as 'high', 'medium' and 'low'.

Example 16.8

The risk of a street being flooded in any year is 1 in 30. What is the risk of it being flooded in at least one of the next 2 years?

Solution

In any year, the probability the street is flooded is $\frac{1}{30}$ so the probability it is not flooded is $1 - \frac{1}{30} = \frac{29}{30}$.

For 2 years

P(The street is flooded in at least one year) = 1 − P(it is flooded in neither year)

> You can draw a tree diagram to illustrate the situation.

$$= 1 - \frac{29}{30} \times \frac{29}{30} = \frac{59}{900}$$

$$\frac{59}{900} = \frac{1}{15.25...} \text{ so the risk is about 1 in 15.}$$

Discussion point

➜ How would you work out the risk of the street being flooded at least once in 5 years?

Exercise 16.1

① This Venn diagram illustrates the occurrence of two events F and G.

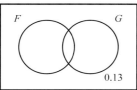

Figure 16.9

P(G') = 0.42
P(F) = 0.6
Work out P($F \cap G$) and P($G \cap F'$).

② Given that P(A) = 0.12, what is the value of P(A')?
A 0.144   B 0.78   C 0.88   D 0.98

> All subsequent questions involve modelling.

PS

③ Three separate electrical components, switch, bulb and contact point, are used together in the construction of a pocket torch. Of 534 defective torches examined to identify the cause of failure, 468 are found to have a defective bulb. For a given failure of the torch, what is the probability that either the switch or the contact point is responsible for the failure? State clearly any assumptions that you have made in making this calculation.

④ One year a team of biologists set up cameras to observe the nests of 30 female birds of a particular species. As a result of their study they make the following estimates.

- The total number of eggs is 120.
- The number of eggs hatching is 90.
- The number of young birds fledging (flying away from the nest) is 80.
- Half of the fledglings are female and half are male.

Say whether each of the following five statements is definitely TRUE, FALSE or UNCERTAIN because it is impossible to say from the information given.

In each case explain your decision.

A The probability of any egg hatching is about $\frac{3}{4}$.

B Each female lays exactly 4 eggs.

C The expected number of fledglings per female is less than 3.

D This year's 30 female birds will be replaced by more than 30 next year.

E The probability of any egg becoming a male fledgling is $\frac{1}{3}$.

⑤ A card is selected at random from a standard pack of 52 playing cards. Find the probability that it is

(i)   a 4

(ii) a number between 2 and 10 inclusive

(iii) not an Ace, King, Queen or Jack

(iv) an Ace or a spade

⑥ A bag containing Scrabble® tiles has the following letter distribution.

A	B	C	D	E	F	G
9	2	2	4	12	2	3

H	I	J	K	L	M	N
2	9	1	1	4	2	6

O	P	Q	R	S	T	U
8	2	1	6	4	6	4

V	W	X	Y	Z	Blanks
2	2	1	2	1	2

The first tile is chosen at random from the bag.

Find the probability that it is

(i) an E

(ii) in the first half of the alphabet

(iii) in the second half of the alphabet

(iv) a vowel

(v) a consonant

(vi) the only one of its kind.

⑦ Marla is a playing a game of chance in which she scores points. She predicts that her number of points on any game will have a uniform distribution from 0 to a maximum possible of 4.

(i) According to this distribution, what is the probability that Marla scores at least 3 points on any game?

Marla then finds that the probability distribution is actually quite different. After playing the game for some time she makes this table showing a revised probability distribution but one of the figures is missing.

Table 16.10

Number of points	0	1	2	3	4
Frequency	0.25	0.1	0.15		0.3

(ii) What is the probability that Marla scores at least 3 points according to her new distribution?

(iii) Find the mean number of points Marla would get according to each of the distributions.

⑧ A reading from an instrument on a machine is denoted by the random variable X. The reading can take the values 0, 1, 2, 3 or 4. Its probability distribution is given in Table 16.11 but the entry for $P(X = 4)$ is missing.

Table 16.11

Reading r	0	1	2	3	4
$P(X = r)$	0.1	0.4	0.1	0.2	

(i) Find $P(X = 4)$.

(ii) A reading is taken. What is the probability that it is less than 3?

(iii) Two further readings are taken. What is the probability that they are the same?

What assumption have you made in answering part (ii)?

⑨ Sally often drives to see her parents. She always takes the same route but her journey time, t minutes, varies considerably. She keeps the following record.

Table 16.12

Journey time	Frequency
$60 < t \leqslant 70$	2
$70 < t \leqslant 80$	6
$80 < t \leqslant 90$	19
$90 < t \leqslant 100$	22
$100 < t \leqslant 110$	8
$110 < t \leqslant 120$	3

(i) Illustrate these data on a cumulative frequency graph.

(ii) Use your graph to estimate the probability that on any occasion the journey will take more than 95 minutes.

(iii) Hence estimate the probability that Sally will take more than 95 minutes on each of two consecutive journeys.

(iv) State an assumption you made in answering part (iii) and identify circumstances under which it may not be valid.

⑩ A lottery offers five prizes of £100 each and a total of 2000 lottery tickets are sold. You buy a single ticket for 20p.

(i) What is the probability that you will win a prize?

(ii) What is the probability that you will not win a prize?

(iii) How much money do the lottery organisers expect to make or lose?

(iv) How much money should the lottery organisers charge for a single ticket in order to break even?

(v) If they continue to charge 20p per ticket, how many tickets would they need to sell in order to break even?

PS ⑪ A sporting chance

(i) Two players, A and B, play tennis. On the basis of their previous results when playing each other, the probability of A winning, P(A), is calculated to be 0.65. What is P(B), the probability of B winning?

(ii) Two hockey teams, A and B, play a game. On the basis of their previous results, the probability of A winning, P(A), is calculated to be 0.65. Why is it not possible to calculate directly P(B), the probability of team B winning, without further information?

(iii) In a tennis tournament, player A, the favourite, is estimated to have a 0.3 chance of winning the competition. Player B is estimated to have a 0.15 chance. Find the probability that either A or B will win the competition.

(iv) In the Six Nations Rugby Championship, France and England are each given a 25%

chance of winning. It is also estimated that there is a 5% chance that they will share the cup. Estimate the probability that either England or France will win or share the cup.

⑫ The probability of a discrete random variable X is given by

$$P(X = r) = \frac{kr}{8} \qquad \text{for } r = 2, 4, 6, 8$$
$$P(X = r) = 0 \qquad \text{otherwise.}$$

(i) Find the value of k and tabulate the probability distribution.

(ii) If two successive values of X are generated independently find the probability that

(a) the two values are equal

(b) the first value is greater than the second value.

⑬ Two dice are thrown. The scores on the dice are added.

(i) Copy and complete Table 16.13 showing all possible outcomes.

Table 16.13

		First die					
+		1	2	3	4	5	6
Second die	1						
	2						
	3						
	4						10
	5						11
	6	7	8	9	10	11	12

(ii) What is the probability of a score of 4?

(iii) What is the most likely outcome?

(iv) Criticise this argument: 'There are 11 possible outcomes: 2, 3, 4, up to 12. Therefore each of them has a probability of $\frac{1}{11}$.'

PS ⑭ Around 0.8% of men are red–green colour-blind (the figure is slightly different for women) and roughly 1 in 5 men is left-handed.

Assuming these characteristics occur independently, calculate with the aid of a tree diagram the probability that a man chosen at random will be

(i) both colour-blind and left-handed

(ii) colour-blind and not left-handed

(iii) colour-blind or left-handed

(iv) neither colour-blind nor left-handed.

⑮ Spiro is unwell. His doctor prescribes three types of medicine: A, B and C. Each of them has a risk of side effects, as shown in Figure 16.10.

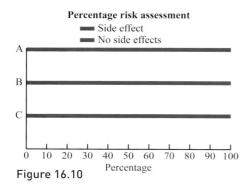

Figure 16.10

Find the risk of Spiro suffering at least one side effect, giving your answer in the form 1 in n where n is a number to be determined to 1 decimal place.

PS ⑯ Every Friday and Saturday Jake drives to his local pub and drinks many pints of beer, even though he has to drive home. He says 'I will be all right. The risk of my being caught by the police is only 1 in 100.' If Jake is caught he will be prosecuted immediately and disqualified from driving for at least 5 years, as well as being given a heavy fine.

Using Jake's figure of 1 in 100, estimate the probability that Jake will not be driving

(i) 1 week from now

(ii) 1 year from now

(iii) 5 years from now.

Give your answers to 2 significant figures.

⑰ The town of Solice is subject to extremes of temperature with very cold winters and very hot summers. The frequency chart (Figure 16.11) shows the estimated average number of days per year with given midday temperatures. (A figure which is on a class boundary is included in the interval above it.)

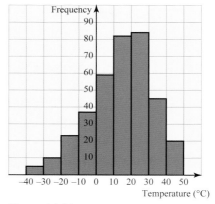

Figure 16.11

The extremes of temperature present health risks for the elderly. Doctors have quantified these risks for those over 70 years old. They say that on a day when the temperature is

■ below −20 °C the probability of death from hypothermia is $\frac{1}{2000}$

■ at least 40 °C the probability of death from heat stroke is $\frac{1}{1000}$.

(i) Estimate the probability that on a randomly selected day the midday temperature is

(a) below −20 °C

(b) at least 40 °C

(c) between −20° and 40 °C.

(ii) Jeremiah is a 75-year-old resident of Solice. Estimate the risk of his dying as a result of the temperature on a randomly selected day.

(iii) Which presents the greater risk for someone over 70 in Solice, the cold or the heat?

⑱ An ornithologist carries out a study of the number of eggs laid per pair of a species of rare birds in its annual breeding season. He concludes that it may be considered as a discrete random variable X with probability distribution given by

$P(X = r) = 0.2$

$P(X = r) = k(4r - r^2)$ for $r = 1, 2, 3, 4$

$P(X = r) = 0$ otherwise.

(i) Find the value of k and write the probability distribution as a table.

The ornithologist observes that the probability of survival (that is of the egg hatching and of the chick living to the stage of leaving the nest) is dependent on the number of eggs in the nest. He estimates the probabilities to be as follows.

Table 16.14

r	Probability of survival
1	0.8
2	0.6
3	0.4

(ii) Find, in the form of a table, the probability distribution of the number of chicks surviving per pair of adults.

Problem solving

Alphabet puzzle

21	2	24	17	4	18	13	6	21
9		13		3		16		8 Q
4	1	21		21		15	22	23
16			5	20	22			13
23	3	11	22		4	10	21	5
19			22	26	26			4
4	18	25		13		5	4	10
5		13		14		21		19
21	7	5	12 R	21	2	21	19	15 Y

A B C D E F G H I J K L M N O P ~~Q~~ ~~R~~ S T U V W X ~~Y~~ Z

1	2	3	4	5	6	7	8	9	10	11	12	13	14	15	16	17	18	19	20	21	22	23	24	25	26
							Q				R			Y											

Figure 16.12

This is an example of a popular type of puzzle. Each of the 26 numbers corresponds to a different letter of the alphabet. Three of the letters have been given to you. You have to decide on the rest and so find all the words.

Devise a strategy for solving problems like this.

1 **Problem specification and analysis**

There are two related problems here. The main one is to devise a strategy for all such puzzles. Then you should show how it works in this particular case.

The obvious first step is to use the information you are given, that: 8 is Q, 12 is R and 15 is Y. Fill in all the squares numbered 8, 12 and 15.

You can then try to spot some words but you are unlikely to get very far without further information. So you will need to use the fact that in English some letters are more common than others.

2 **Information collection**

A randomly selected square is more likely to contain some letters than others. To make use of this fact, you need two types of information.

What information do you need? How will you obtain it?

3 **Processing and representation**

Now that you have collected the information, there are several things for you to think about.

How will you display the information in a way that will be helpful to you?

How exactly you are going to use the information?

An important question to think about is whether at this stage the puzzle is effectively solved.

Is the letter corresponding to each number definitely known? Or is there still an element of trial and error? And if so why?

Now complete the solution of the puzzle.

4 **Interpretation**

To complete this problem you should provide two things.

(i) A step-by-step strategy for solving this type of puzzle. This should include an explanation of how it makes use of probability and statistics, and also your knowledge of the English language.

(ii) You should then relate your strategy to how you solved this particular puzzle.

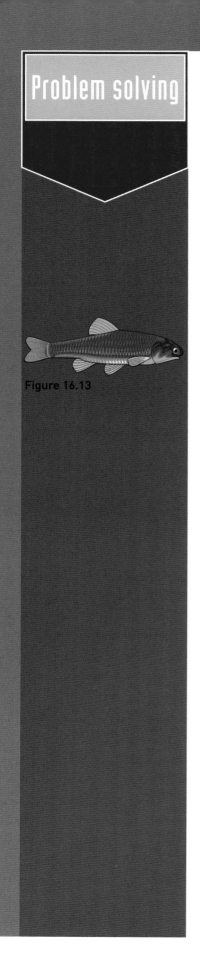

Figure 16.13

Estimating minnows

A building company is proposing to fill in a pond which is the home for many different species of wildlife. The local council commission a naturalist called Henry to do a survey of what wildlife is dependent on the pond. As part of this exercise Henry decides to estimate how many minnows are living in the pond. The minnows cannot all be seen at the same time so it is not possible just to count them.

This task involves the various stages of the problem solving cycle.

1 **Problem specification and analysis**

Henry starts by planning how he will go about the task.

The procedure he chooses is based on a method called **capture–recapture**. To carry it out he needs a minnow trap and some suitable markers for minnows.

2 **Information collection**

Henry sets the trap once a day for a week, starting on Sunday. Each time that he opens the trap he counts how many minnows he has caught and how many of them are already marked. He then marks those that are not already marked and returns the minnows to the pond.

Table 16.15 shows his results for the week.

Table 16.15

Day	Sun	Mon	Tues	Wed	Thus	Fri	Sat
Number caught	10	12	8	15	6	12	16
Number already marked	–	1	1	3	2	4	6

3 **Processing and representation**

(i) After Henry has returned the minnows to the pond on Monday, he estimates that there are 120 minnows in the pond. After Tuesday's catch he makes a new estimate of 168 minnows. Show how he calculates these figures.

(ii) Use the figures for the subsequent days' catches to make four more estimates of the number of minnows in the pond.

(iii) Draw a suitable diagram to illustrate the estimates.

4 **Interpretation**

(i) Henry has to write a report to the council. It will include a short section about the minnows. Comment briefly what it might say about the following:

■ the best estimate of the number of minnows

■ any assumptions required for the calculations and if they are reasonable

■ how accurate the estimate is likely to be.

(ii) Suggest a possible improvement to Henry's method for data collection.

LEARNING OUTCOMES

When you have completed this chapter you should be able to:

➤ understand and use mutually exclusive and independent events when calculating probabilities.

➤ Interpret diagrams for single-variable data in connection with probability distributions.

➤ Link probability to discrete and continuous distributions.

➤ Understand and use simple, discrete probability distributions (excluding calculating mean and variance of discrete random variables).

KEY POINTS

1 The probability of an event A,

$$P(A) = \frac{n(A)}{n(\varepsilon)}$$

where $n(A)$ is the number of ways that A can occur and $n(\varepsilon)$ is the total number of ways that all possible events can occur, all of which are equally likely.

2 For any two events, A and B, of the same experiment,

$$P(A \cup B) = P(A) + P(B) - P(A \cap B)$$

Where the events are mutually exclusive (i.e. where the events do not overlap) the rule still holds but, since $P(A \cap B)$ is now equal to zero, the equation simplifies to:

$$P(A \cup B) = P(A) + P(B).$$

3 Where an experiment produces two or more mutually exclusive events, the probabilities of the separate events sum to 1:

$$P(A) + P(A') = 1$$

FUTURE USES

■ The basic probability in this chapter underpins almost all of statistics from now on, particularly distributions (for example, the binomial distribution in Chapter 17) and inference (for example, hypothesis testing in Chapter 18).

17

The binomial distribution

THE AVONFORD STAR

Samantha's great invention

Mother of three, Samantha Weeks, has done more than her bit to help improve people's health. She has invented the first cheap health monitors that can be worn as wristbands.

Now Samantha is out to prove that she is not only a clever scientist but a smart business woman as well. For Samantha is setting up her own factory to make and sell her health-monitoring wristbands.

Samantha admits there are still some technical problems . . .

Samantha Weeks hopes to make a big success of her health-monitoring wristbands

Samantha's production process is indeed not very good and there is a probability of 0.1 that any health-monitoring wristband will be substandard and so not function as accurately as it should.

She decides to sell her health-monitoring wristbands in packs of three and believes that if one in a pack is substandard the customers will not complain but if two or

⑪ There are 15 children in a class.

(i) What is the probability that
(a) 0 (b) 1 (c) 2 (d) at least 3
were born in January?

(ii) What assumption have you made in answering this question? How valid is this assumption in your view?

⑫ Criticise this argument.

If you toss two coins they can come down three ways: two heads, one head and one tail, or two tails. There are three outcomes and so each of them must have probability one-third.

Example 17.2

2 Using the binomial distribution

The expectation of $B(n, p)$

The number of good wristbands in a packet of three is modelled by the random variable X.

$$X \sim B(3, 0.9).$$

(i) Find the expected frequencies of obtaining $0, 1, 2$ and 3 good wristbands in 2000 packets.

(ii) Find the mean number of good wristbands per packet.

Solution

(i) $P(X = 0) = 0.001$, so the expected frequency of packets with no good wristbands is $2000 \times 0.001 = 2$ ← ⎯ Check: $2 + 54 + 486 + 1458 = 2000$ ✓

Similarly, the other expected frequencies are:

For 1 good wristband: $2000 \times 0.027 = 54$

For 2 good wristbands: $2000 \times 0.243 = 486$

For 3 good wristbands: $2000 \times 0.729 = 1458$

(ii) The expected total of good wristbands in 2000 packets is

$$0 \times 2 + 1 \times 54 + 2 \times 486 + 3 \times 1458 = 5400$$

Therefore the mean number of good wristbands per packet is

$$\frac{5400}{2000} = 2.7 ← \boxed{\text{This is also called the } \textbf{expectation}.}$$

Notice the steps in this example to calculate the mean.

- Multiply each probability by 2000 to get the frequency.
- Multiply each frequency by the number of good wristbands.
- Add these numbers together.
- Finally divide by 2000.

Of course you could have obtained the mean with less calculation by just multiplying each number of good wristbands by its probability and then summing, i.e. by finding

$$\sum_{r=0}^{3} rP(X = r).$$

This is the standard method for finding an expectation.

Notice also that the mean or expectation of X is $2.7 = 3 \times 0.9 = np$.

The result for the general binomial distribution is the same.

> If $X \sim B(n, p)$, then the expectation of X, written $E(X)$, is given by $E(X) = np$.

This seems obvious. If the probability of success in each single trial is p, then the expected numbers of successes in n independent trials is np. However, since what seems obvious is not always true, this result needs to be justified.

Take the case when $n = 5$. The expectation of X is

$$E(X) = 0 \times q^5 + 1 \times 5q^4 p^1 + 2 \times 10q^3 p^2 + 3 \times 10q^2 p^3 + 4 \times 5q^1 p^4 + 5 \times p^5$$

$$= 5q^4 p + 20q^3 p^2 + 30q^2 p^3 + 20qp^4 + 5p^5$$

$$= 5p\left(q^4 + 4q^3 p + 6q^2 p^2 + 4qp^3 + p^4\right) \quad \leftarrow \boxed{\text{Take out the common factor } 5p.}$$

$$= 5p(q + p)^4$$

$$= 5p \quad \leftarrow \boxed{\text{Since } p + q = 1.}$$

The proof in the general case follows the same pattern: the common factor is now np, and the expectation simplifies to $np(q + p)^{n-1} = np$. The details are more fiddly because of the manipulations of the binomial coefficients.

Using the binomial distribution

Example 17.3

A spinner has numbers 1 to 6. They are all equally likely to come up. Which is more likely: that you get at least one 6 when you spin it six times, or that you get at least two 6s when you spin it twelve times?

Solution

On a single spin, the probability of getting a 6 is $\frac{1}{6}$ and that of not getting a 6 is $\frac{5}{6}$.

So the probability distributions for the two situations in this problem are $B(6, \frac{1}{6})$ and $B(12, \frac{1}{6})$ giving probabilities of:

$$1 - \binom{6}{0}\left(\frac{5}{6}\right)^6 = 1 - 0.335 = 0.665 \text{ (at least one 6 in six spins)}$$

and

$$1 - \left[\binom{12}{0}\left(\frac{5}{6}\right)^{12} + \binom{12}{1}\left(\frac{5}{6}\right)^{11}\left(\frac{1}{6}\right)\right] = 1 - (0.112 + 0.269)$$

$$= 0.619 \text{ (at least two 6s in 12 spins)}$$

So at least one 6 in six spins is slightly more likely.

Example 17.4

Extensive research has shown that 1 person out of every 4 is allergic to a particular grass pollen.

A group of 20 university students volunteer to try out a new treatment.

(i)   What is the expectation of the number of allergic people in the group?

(ii) What is the probability that

(a) exactly two   (b)   no more than two of the group are allergic?

(iii) How large a sample would be needed for the probability of it containing at least one allergic person to be greater than 99.9%?

(iv) What assumptions have you made in your answer?

Solution

This situation is modelled by the binomial distribution with $n = 20$, $p = 0.25$ and $q = 0.75$.

The number of allergic people is denoted by X.

(i)   Expectation $= np = 20 \times 0.25 = 5$ people.

(ii) $X \sim B(20, 0.25)$

(a) $P(X = 2) = \binom{20}{2} (0.75)^{18} (0.25)^2 = 0.067$

(b) $P(X \leqslant 2) = P(X = 0) + P(X = 1) + P(X = 2)$

$$= (0.75)^{20} + \binom{20}{1} (0.75)^{19} (0.25) + \binom{20}{2} (0.75)^{18} (0.25)^2$$

$$= 0.003 + 0.021 + 0.067$$

$$= 0.091$$

(iii) Let the sample size be n (people), so that $X \sim B(n, 0.25)$.

The probability that none of them is allergic is

$$P(X = 0) = (0.75)^n$$

and so the probability that at least one is allergic is

$$P(X \geqslant 1) = 1 - P(X = 0)$$

$$= 1 - (0.75)^n$$

So we need $1 - (0.75)^n > 0.999$

$$(0.75)^n < 0.001$$

Solving $(0.75)^n = 0.001$

gives $n \log 0.75 = \log 0.001$

$$n = \log 0.001 \div \log 0.75$$

$$= 24.01$$

So 25 people are required.

> You can also use a spreadsheet to find n or trial and improvement on your calculator.

> Although 24.01 is very close to 24 it would be incorrect to round down. $1 - (0.75)^{24} = 0.998\,996\,6$ which is just less than 99.9%.

(iv) The assumptions made are:

(a)   That the sample is random. This is almost certainly untrue. Most university students are in the 18–25 age range and so a sample of them cannot be a random sample of the whole population. They may well also be unrepresentative of the whole population in other ways. Volunteers are seldom truly random.

(b)   That the outcome for one person is independent of that for another. This is probably true unless they are a group of friends from, say, an athletics team, where those with allergies may be less likely to be members.

Does the binomial distribution really work?

INVESTIGATION

This is a true story. During voting at a by-election, an exit poll of 1700 voters indicated that 50% of people had voted for the Labour Party candidate. When the real votes were counted it was found that he had in fact received 57% support.

Carry out a computer simulation of the situation and use it to decide whether the difference was likely to have occurred because of the random nature of the sample, faulty sampling techniques or other possible reasons you can think of.

EXPERIMENT

Imagine that you throw six dice and record how many 6s you get. The result could be 0, 1, 2, 3, 4, 5 or 6. If you did this repeatedly you should generate the probability distribution below.

$X \sim B(6, \frac{1}{6})$ and this gives the probabilities in Table 17.7.

Table 17.7

Number of 6s	Probability
0	0.335
1	0.402
2	0.201
3	0.054
4	0.008
5	0.001
6	0.000

So if you carry out the experiment of throwing six dice 1000 times and record the number of 6s each time, you should get none about 335 times, one about 402 times and so on. What does 'about' mean? How close an agreement can you expect between experimental and theoretical results?

You could carry out the experiment with dice, but it would be very tedious even if several people shared the work. Alternatively you could simulate the experiment on a spreadsheet using a random number generator.

Exercise 17.2

> ### Note
> All questions involve modelling.

 ① What is the difference between the score you 'expect to get' and the 'expectation' of your score?

 ② The probability of winning a game is $\frac{1}{20}$. How many times would you expect to win if you played 100 games?

A 5   B 20   C 10   D 2

③ In a game five dice are rolled together.

(i) What is the probability that

(a) all five show 1

(b) exactly three show 1

(c) none of them shows 1?

(ii) What is the most likely number of times for 6 to show?

④ There are eight colours of Smarties™ which normally occur in equal proportions: red, orange, yellow, green, blue, purple, pink and brown. Daisy's mother gives each of her children 16 smarties. Daisy says that the blue ones are much nicer than the rest and is very upset when she received less than her fair share of them.

(i) How many blue smarties did Daisy expect to get?

(ii) What was the probability that she would receive fewer blue ones than she expected?

(iii) What was the probability that she would receive more blue ones than she expected?

⑤ In a particular area 30% of men and 20% of women are underweight. Four men and three women work in an office.

(i) Find the probability that there are

    (a)   0   (b)   1

    (c)   2 underweight men;

    (d)   0   (e)   1

    (f)   2 underweight women;

    (g)   2 underweight people in the office.

(ii) What assumption have you made in answering this question?

⑥ On her drive to work Stella has to go through four sets of traffic lights. She estimates that, for each set, the probability of her finding them red is $\frac{2}{3}$ and green $\frac{1}{3}$. (She ignores the possibility of them being amber.) Stella also estimates that when a set of lights is red she is delayed by one minute.

(i) Find the probability of

    (a)   0   (b)   1   (c)   2   (d)   3

    sets of lights being against her.

(ii) Find the expected extra journey time due to waiting at lights.

PS ⑦ A man steps out of a shop into a narrow alley which runs from west to east. At each step he chooses at random whether to go east or west. After 12 steps he stops for a rest.

(i) Why is it impossible for him then to be 1 step away from the shop?

(ii) What is the probability that he is then 10 steps east of the shop?

(iii) What is his most likely position?

(iv) What is the probability that he is 4 steps away from the shop, either to the east or west?

(v) What is the expectation of his number of steps away from the shop, either to the east or to the west?

PS ⑧ Pepper moths are found in two varieties, light and dark. The proportion of dark moths increases with certain types of atmospheric pollution. At the time of the question 30% of the moths in a particular town are dark. A research student sets a moth trap and catches nine moths, four light and five dark.

(i) What is the probability of that result for a sample of nine moths?

(ii) What is the expected number of dark moths in a sample of nine?

The next night the student's trap catches ten pepper moths.

(iii) What is the probability that the number of dark moths in this sample is the same as the expected number?

⑨ A company uses machines to manufacture wine glasses. Because of imperfections in the glass it is normal for 10% of the glasses to leave the machine cracked. The company takes regular samples of 10 glasses from each machine. If more than 2 glasses in a sample are cracked, they stop the machine and check that it is set correctly.

(i) What is the probability that a sample of 10 glasses contains

    (a)   0   (b)   1   (c)   2

    faulty glasses, when the machine is set correctly?

(ii) What is the probability that as a result of taking a sample a machine is stopped when it is correctly set?

(iii) A machine is in fact incorrectly set and 20% of the glasses from it are cracked. What is the probability that this is undetected by a particular sample?

LEARNING OUTCOMES

When you have completed this chapter you should be able to:

➤ understand and use simple, discrete probability distributions (calculation of mean and variance of discrete random variables is excluded)

➤ use the binomial distribution as a model

➤ calculate probabilities using the binomial distribution

➤ understand the link with the binomial expansion.

KEY POINTS

1 The binomial distribution, $\mathrm{B}(n, p)$, may be used to model situations in which:
   - you are conducting trials on random samples of a certain size, n
   - there are two possible outcomes, often referred to as success and failure
   - both outcomes have fixed probabilities, p and q, and $p + q = 1$
   - the probability of success in any trial is independent of the outcomes of previous trials.

2 The probability that the number of successes, X, has the value r, is given by
$$P(X = r) = \binom{n}{r} q^{n-r} p^r$$

3 For $\mathrm{B}(n, p)$ the expectation of the number of successes is np.

To be and not to be, that is the answer.

Piet Hein
(1905–1996)

FUTURE USES

   - The binomial distribution will be used for binomial hypothesis testing in the next chapter.

18

Statistical hypothesis testing using the binomial distribution

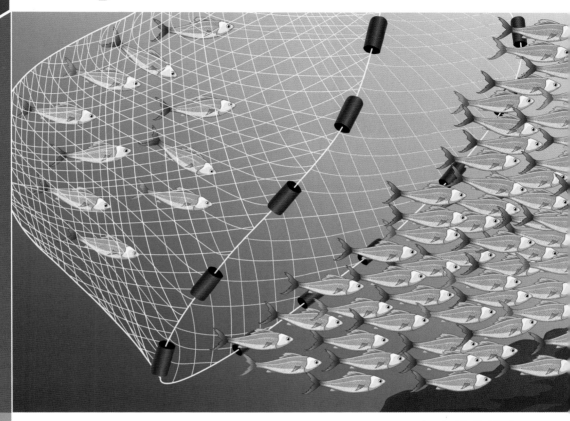

THE AVONFORD STAR

Machoman Stan

Beer drinking Stan Edmonds has just become a father for the eighth time and they are ALL BOYS.

Find Stan at his local and he will tell you that having no girls among the eight shows that he is a real man.

'When I was just a lad at school I said I had macho chromosomes,' Stan claims.

'Now I have been proved right.'

(*Posed by model*)
Cheers! Stan Edmonds reckons his all-boy offspring prove he's a real man

What do you think?

There are two quite different points here.

The first is that you have probably decided that Stan is a male chauvinist, preferring boys to girls. However, you should not let your views on that influence your judgement on the second point, his claim to be biologically different from other people, with special chromosomes.

There are two ways this claim could be investigated, to look at his chromosomes under a high magnification microscope or to consider the statistical evidence.

Since you have neither Stan Edmonds nor a suitable microscope to hand, you must consider the statistical evidence.

If you had eight children, you would expect them to be divided about evenly between the sexes, 4–4, 5–3 or perhaps 6–2. When you realised that another baby was on its way you would think it equally likely to be a boy or a girl until it was born or a scan was carried out, when you would know for certain.

In other words, you would say that the probability of its being a boy was 0.5 and that of its being a girl 0.5. So you can model the number of boys among eight children by the binomial distribution B(8, 0.5).

This gives these probabilities, shown in Table 18.1 and in Figure 18.1.

Table 18.1

Boys	Girls	Probability
0	8	$\frac{1}{256}$
1	7	$\frac{8}{256}$
2	6	$\frac{28}{256}$
3	5	$\frac{56}{256}$
4	4	$\frac{70}{256}$
5	3	$\frac{56}{256}$
6	2	$\frac{28}{256}$
7	1	$\frac{8}{256}$
8	0	$\frac{1}{256}$

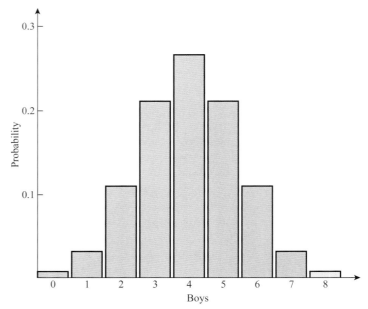

Figure 18.1

So you can say that, if a biologically normal man fathers eight children, the probability that they will all be boys is $\frac{1}{256}$ (shaded green in Figure 18.1).

This is unlikely but by no means impossible.

> **Note**
>
> The probability of a baby being a boy is not in fact 0.5 but about 0.503. Boys are less tough than girls and so more likely to die in infancy and this seems to be nature's way of compensating. In most societies men have a markedly lower life expectancy as well.

Discussion point

→ You may think Stan Edmonds is a thoroughly objectionable character but in some countries large sections of society value boy children more highly than girls. Some parents choose to terminate a pregnancy when a scan reveals an unborn baby is a girl. What would be the effect of this on a country's population if, say, half the parents decided to have only boys and the other half to let nature take its course?

(This is a real problem. The social consequences can be devastating.)

1 The principles and language of hypothesis testing

Defining terms

In the opening example you investigated Stan Edmonds' claim by comparing it to the usual situation, the unexceptional. If you use p for the probability that a child is a boy then the normal state of affairs can be stated as

$p = 0.5$

This is called the **null hypothesis**, denoted by H_0.

> The null hypothesis is the default position which will only be rejected if the evidence is strong enough.

Stan's claim (made, he says, before he had any children) was that

$p > 0.5$

and this is called the **alternative hypothesis**, H_1.

The word hypothesis (plural **hypotheses**) means a theory which is put forward either for the sake of an argument or because it is believed or suspected to be true. An investigation like this is usually conducted in the form of a test, called a **hypothesis test**. There are many different sorts of hypothesis test used in statistics; in this chapter you meet only one of them.

It is never possible to prove something statistically in the sense that, for example, you can prove that the angle sum of a triangle is 180°. Even if you tossed a coin a million times and it came down heads every single time, it is still possible that the coin is unbiased and just happened to land that way. What you can say is that it is very unlikely; the probability of it happening that way is $(0.5)^{1\,000\,000}$ which is a decimal that starts with over 300 000 zeros. This is so tiny that you would feel quite confident in declaring the coin biased.

There comes a point when the probability is so small that you say, 'That's good enough for me. I am satisfied that it hasn't happened that way by chance.'

The probability at which you make that decision is called the **significance level** of the test. Significance levels are usually given as percentages; 0.05 is written as 5%, 0.01 as 1% and so on.

So in the case of Stan Edmonds, the question could have been worded:

Test, at the 1% significance level, Stan Edmonds' boyhood claim that his children are more likely to be boys than girls.

The answer would then look like this:

Null hypothesis, H_0: $p = 0.5$ (boys and girls are equally likely)

Alternative hypothesis, H_1: $p > 0.5$ (boys are more likely)

Significance level: 1%

Probability of 8 boys from 8 children $= \frac{1}{256} = 0.0039 = 0.39\%$

This probability is called the p-value for the test. Since the p-value of 0.39% is less than the significance level of 1% the null hypothesis is rejected in favour of the alternative hypothesis. The evidence supports Stan Edmonds' claim (as now worded).

> ❗ The letter p is used with two different meanings in a binomial hypothesis test like this.
>
> ■ The value of p is the probability of success in a binomial trial.
>
> ■ The *p-value* is the probability of the observed outcome (or a more extreme outcome).
>
> These are both standard ways of using the letter p and they just happen to come together in this test. You need to be very careful; muddling them up is a mistake waiting to be made!

Hypothesis testing questions

The example of Stan Edmonds illustrates three questions you should consider when carrying out a hypothesis test.

1 Was the test set up before or after the data were known?

The test consists of a null hypothesis, an alternative hypothesis and a significance level.

In this case, the null hypothesis is the natural state of affairs and so does not really need to be stated in advance. Stan's claim 'When I was just a lad at school I said I had macho chromosomes' could be interpreted as the alternative hypothesis, $p > 0.5$.

The problem is that one suspects that whatever children Stan had he would find an excuse to boast. If they had all been girls, he might have been talking about 'my irresistible attraction for the opposite sex' and if they have been a mixture of girls and boys he would have been claiming 'super-virility' just because he had eight children.

Any test carried out retrospectively must be treated with suspicion.

> One way of cheating is to set up a test to fit already existing data.

2 Was the sample involved chosen at random and are the data independent?

The sample was not random and that may have been inevitable. If Stan has lots of children around the country with different mothers, a random sample of eight could have been selected. However, we have no knowledge that this is the case.

The data are the sexes of Stan's children. If there are no multiple births (for example, identical twins), then they are independent.

3 Is the statistical procedure actually testing the original claim?

Stan Edmonds claims to have 'macho chromosomes', whereas the statistical test is of the alternative hypothesis that $p > 0.5$. The two are not necessarily the same. Even if this alternative hypothesis is true, it does not necessarily follow that Stan has macho chromosomes, whatever that means.

The ideal hypothesis test

In the ideal hypothesis test you take the following steps, in this order:

1 Establish the null and alternative hypotheses.
2 Decide on the significance level.
3 Collect suitable data using a random sampling procedure that ensures the items are independent.
4 Conduct the test, doing the necessary calculations.
5 Interpret the result in terms of the original claim, conjecture or problem.

There are times, however, when you need to carry out a test but it is just not possible to do so as rigorously as this.

In the case of Stan Edmonds, you would require him to go away and father eight more children, this time with randomly selected mothers, which is clearly impossible. Had Stan been a laboratory rat and not a human, however, you probably could have organised it.

Choosing the significance level

> **Note**
>
> Another way to think of the significance level is that it is the probability of incorrectly rejecting the null hypothesis.

If, instead of 1%, you had set the significance level at 0.1%, then you would have rejected Stan's claim, since 0.39% > 0.1%. The lower the percentage in the significance level, the more stringent is the test.

The significance level you choose for a test involves a balanced judgement.

Imagine that you are testing the rivets on an aeroplane's wing to see if they have lost their strength. Setting a small significance level, say 0.1%, means that you will only declare rivets weak if you are confident of your finding. The trouble with requiring such a high level of evidence is that even when they are weak you may well fail to register the fact, with possible consequences that the aeroplane crashes. On the other hand if you set a high significance level, such as 10%, you run the risk of declaring the rivets faulty when they are all right, involving the company in expensive and unnecessary maintenance work.

Example 18.1

One evening Leonora lost a lot of money at a casino. She complained to the management that one of the dice was biased, with a tendency to show the number 1.

The management agreed to test it at the 5% significance level, throwing it 20 times. If the test supported Leonora's view she would get her money refunded, otherwise she would be asked to leave the premises and never return.

The results were are follows.

1	6	6	5	5
1	2	3	2	3
4	4	4	1	4
1	1	4	1	3

What happened to Leonora?

Solution

Let p be the probability of getting 1 on any roll.

Null hypothesis, H_0: $p = \frac{1}{6}$ (it is unbiased)

Alternative hypothesis, H_1: $p > \frac{1}{6}$ (it is biased towards 1)

Significance level: 5%

The results may be summarised as follows.

Score	1	2	3	4	5	6
Frequency	6	2	3	5	2	2

Under the null hypothesis, the number of 1s obtained is modelled by the binomial distribution, $B\left(20, \frac{1}{6}\right)$ which gives the probabilities in Table 18.2. The two columns to the right give the cumulative probabilities.

Table 18.2

Number of 1s	Expression	Probability	Number of 1s	Cumulative probability
0	$\left(\frac{5}{6}\right)^{20}$	0.0261	≤ 0	0.0261
1	$\binom{20}{1}\left(\frac{5}{6}\right)^{19}\left(\frac{1}{6}\right)$	0.1043	≤ 1	0.1304
2	$\binom{20}{2}\left(\frac{5}{6}\right)^{18}\left(\frac{1}{6}\right)^{2}$	0.1982	≤ 2	0.3286
3	$\binom{20}{3}\left(\frac{5}{6}\right)^{17}\left(\frac{1}{6}\right)^{3}$	0.2379	≤ 3	0.5665
4	$\binom{20}{4}\left(\frac{5}{6}\right)^{16}\left(\frac{1}{6}\right)^{4}$	0.2022	≤ 4	0.7687
5	$\binom{20}{5}\left(\frac{5}{6}\right)^{15}\left(\frac{1}{6}\right)^{5}$	0.1294	≤ 5	0.8981
6	$\binom{20}{6}\left(\frac{5}{6}\right)^{14}\left(\frac{1}{6}\right)^{6}$	0.0647	≤ 6	0.9628
7	$\binom{20}{7}\left(\frac{5}{6}\right)^{13}\left(\frac{1}{6}\right)^{7}$	0.0259	≤ 7	0.9887
8	$\binom{20}{8}\left(\frac{5}{6}\right)^{12}\left(\frac{5}{6}\right)^{8}$	0.0084	≤ 8	0.9971
...	...	...	...	
20	$\left(\frac{1}{6}\right)^{20}$	0.0000	≤ 20	1.000

> The probability of 1 coming up between 0 and 5 times is found by adding these probabilities to get 0.8981.

> If you worked out all these and added them you would get the probability that the number of 1s is 6 or more (up to a possible 20). It is much quicker, however, to find this as $1 - 0.8981$ (the answers above) $= 0.1019$.

> How can you get the numbers in this column on your calculator?

Calling X the number of 1s occurring when it is rolled 20 times, assuming it is unbiased, the probability of six or more 1s is given by

$$P(X \geq 6) = 1 - (X \leq 5) = 1 - 0.8981 = 0.1019$$

So the p-value is 10.19%.

Since 10.19% > 5%, the null hypothesis (it is unbiased) is accepted.

The probability of a result at least as extreme as that observed is greater than the 5% cut off that was set in advance, that is, greater than the chosen significance level.

Even though the number 1 did come up more often than the other numbers, the evidence is not strong enough to reject the null hypothesis in favour of the alternative hypothesis.

Discussion point

→ Do you think that it is fair that Leonora should be made to leave the casino?

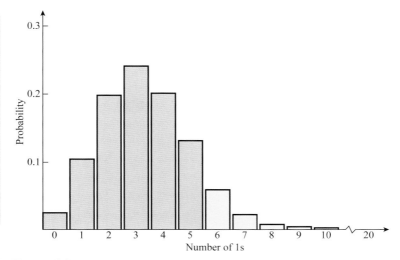

Figure 18.2

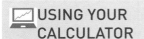
Another method

In the solution to Example 18.1, the probabilities of possible numbers of 1s were given in Table 18.2. These were then added together to find the cumulative probabilities. This was rather a time consuming method. There are two quicker methods:

Using cumulative binomial probability tables

These give $P(X \leqslant x)$ when $X \sim B(n, p)$ for $x = 0, 1, 2, \ldots, n$ and values of p from 0.05 to 0.95 at intervals of 0.05 plus $\frac{1}{6}, \frac{1}{3}, \frac{2}{3}, \frac{5}{6}$. There is a separate table for each value of n from 1 to 20. Look up under $n = 20$.

In this case $p = \frac{1}{6}$ so the probability of up to five scores of 1 is 0.8982, the same result as before apart from the last figure where there is a difference of 1 from rounding.

n	$\dfrac{P}{x}$	0.050	0.100	0.150	$\dfrac{1}{6}$	0.200	0.250	0.300	$\dfrac{1}{3}$	0.350
20	0	0.3585	0.1216	0.0388	0.0261	0.0115	0.0032	0.0008	0.0003	0.0002
	1	0.7358	0.3917	0.1756	0.1304	0.0692	0.0243	0.0076	0.0033	0.0021
	2	0.9245	0.6769	0.4049	0.3287	0.2061	0.0913	0.0355	0.0176	0.0121
	3	0.9841	0.8670	0.6477	0.5665	0.4114	0.2252	0.1071	0.0604	0.0444
	4	0.9974	0.9568	0.8298	0.7687	0.6296	0.4148	0.2375	0.1515	0.1182
	5	0.9997	0.9887	0.9327	0.8982	0.8042	0.6172	0.4164	0.2972	0.2454
	6	1.0000	0.9976	0.9781	0.9629	0.9133	0.7858	0.6080	0.4973	0.4166
	7		0.9996	0.9941	0.9887	0.9679	0.8982	0.7723	0.6615	0.6010
	8		0.9999	0.9987	0.9972	0.9900	0.9591	0.8867	0.8095	0.7624
	9		1.0000	0.9998	0.9994	0.9974	0.9861	0.9520	0.9081	0.8782
	10			1.0000	0.9999	0.9994	0.9961	0.9829	0.9624	0.9468
	11				1.0000	0.9999	0.9991	0.9949	0.9870	0.9804
	12					1.0000	0.9998	0.9987	0.9963	0.9940
	13						1.0000	0.9997	0.9991	0.9985
	14							1.0000	0.9998	0.9997
	15								1.0000	1.0000

Figure 18.3 A section of the cumulative binomial probability tables

Exercise 18.1

 ① What is the difference between a **p-value** and the **value of p**?

 ② It is claimed that a die is biased as it is showing 4 too many times. Which of the following are the correct null and alternative hypotheses?

A $H_0 : p < \dfrac{1}{6}$ $H_1 : p \geqslant \dfrac{1}{6}$

B $H_0 : p = \dfrac{1}{6}$ $H_1 : p > \dfrac{1}{6}$

C $H_0 : p = \dfrac{1}{6}$ $H_1 : p \leqslant \dfrac{1}{6}$

D $H_0 : p = \dfrac{1}{6}$ $H_1 : p < \dfrac{1}{6}$

③ A spinner has 4 sectors. They are meant to be equally likely to come up. Rob wants to test whether the spinner is fair or biased towards the number 4.

He spins it 20 times and obtains the following results.

Number	1	2	3	4
Frequency	3	4	3	10

He then says, 'If the spinner is fair, the probability of getting 10 or more 4s is 0.0139. This is less than 0.05. So it looks as though it is biased.'

(i) State the null and alternative hypotheses that Rob has used.

(ii) What was the *p*-value?

(iii) What significance level has he applied?

(iv) Was the test significant?

> **Note**
> ---------------------------------
> All subsequent questions involve modelling using the binomial distribution.

PS ④ Mrs da Silva is running for President. She claims to have 60% of the population supporting her.

She is suspected of overestimating her support and a random sample of 20 people are asked whom they support. Only nine say Mrs da Silva.

Test, at the 5% significance level, the hypothesis that she has overestimated her support.

PS ⑤ A company developed synthetic coffee and claim that coffee drinkers could not distinguish it from the real product. A number of coffee drinkers challenged the company's claim, saying that the synthetic coffee tasted synthetic. In a test, carried out by an independent consumer protection body, 20 people were given a mug of coffee. Ten had the synthetic brand and ten the natural, but they were not told which they had been given.

Out of the ten given the synthetic brand, eight said it was synthetic and two said it was natural. Use this information to test the coffee drinkers' claim (as against the null hypothesis of the company's claim), at the 5% significance level.

PS ⑥ A group of 18 students decides to investigate the truth of the saying that if you drop a piece of toast it is more likely to land butter-side down. They each take one piece of toast, butter it on one side and throw it in the air. Eleven land butter-side down, the rest butter-side up. Use their results to carry out a hypothesis test at the 10% significance level, stating clearly your null and alternative hypotheses.

PS ⑦ On average 70% of people pass their driving test first time. There are complaints that one examiner, Mr McTaggart, is too harsh and so, unknown to himself, his work is monitored. It is found that he fails 10 out of 20 candidates. Are the complaints justified at the 5% significance level?

PS ⑧ A machine makes bottles. In normal running 5% of the bottles are expected to be cracked, but if the machine needs servicing this proportion will increase. As part of a routine check, 50 bottles are inspected and 5 are found to be unsatisfactory. Does this provide evidence, at the 5% significance level, that the machine needs servicing?

PS ⑨ An annual mathematics contest contains 15 questions, 5 short and 10 long. The probability that I get a short question right is 0.9. The probability that I get a long question right is 0.5. My performances on questions are independent of each other. Find the probability of the following:

(i)   I get all the 5 short questions right

(ii)   I get exactly 8 out of 10 long questions right

(iii)   I get exactly 3 of the short questions and all of the long questions right

(iv)   I get exactly 13 of the 15 questions right.

After some practice, I hope that my performance on the long questions will improve this year.

I intend to carry out an appropriate hypothesis test.

(v)   State suitable null and alternative hypotheses for the test.

In this year's contest I get exactly 8 out of 10 long questions right.

(vi)   Is there sufficient evidence, at the 5% significance level, that my performance on long questions has improved?

2 Extending the language of hypothesis testing

Critical values and critical regions

In Example 18.1 the number 1 came up six times and this was not often enough to get Leonora a refund. What was the least number of times 1 would have had to come up for the test to give Leonora a refund?

Again, use X to denote the number of times 1 comes up in the 20 throws, and so $X = 6$ means that the number 1 comes up 6 times.

You know from your earlier work that the probability that $X \leqslant 5$ is 0.8982 and you can use the binomial distribution to work out the probabilities that $X = 6$, $X = 7$, etc.

$$P(X = 6) = \binom{20}{6} \left(\tfrac{5}{6}\right)^{14}\left(\tfrac{1}{6}\right)^{6} = 0.0647$$

$$P(X = 7) = \binom{20}{7} \left(\tfrac{5}{6}\right)^{13}\left(\tfrac{1}{6}\right)^{7} = 0.0259$$

You have already seen that $P(X \geqslant 6) = 1 - P(X \leqslant 5) = 1 - 0.8982 = 0.1018$.

0.1018 is a little over 10% and so greater than the significance level of 5%.

There is no reason to reject H_0.

What about the case when the number 1 comes up seven times, that is $X = 7$?

Since $P(X \leqslant 6) = P(X \leqslant 5) + P(X = 6)$

$P(X \leqslant 6) = 0.8982 + 0.0647 = 0.9629$

So $P(X \geqslant 7) = 1 - P(X \leqslant 6)$

$= 1 - 0.9629 = 0.0371 = 3.71\%$

Since 3.7% < 5%, H_0 is now rejected in favour of H_1.

You can see that Leonora needed the 1 to come up seven or more times if her case was to be upheld. She missed by just one. You might think Leonora's 'all or nothing' test was a bit harsh. Sometimes tests are designed so that if the result falls within a certain region further trials are recommended.

In this example the number 7 is the **critical** value (at the 5% significance level), the value at which you change from accepting the null hypothesis to rejecting it.

Test procedure

Take 20 pistons

If 3 or more are faulty
REJECT
the batch

Figure 18.4

The range of values for which you reject the null hypothesis, in this case $X \geqslant 7$, is called the **critical region**, or the **rejection region**. The range of values for which you accept the null hypothesis is called the acceptance region.

In this example, 'success' was getting the number 1. Any of the other numbers were 'failure'. The number of successes is an example of a **test statistic**. It allows the observed outcome to be compared with what would be expected under the null hypothesis.

You have now met two approaches to hypothesis testing.

■ Finding the critical region for the given significance level and seeing if the test statistic lies in it.

■ Working out the probability of obtaining a value at least as extreme as the test statistic (the p-value) and comparing it to the significance level.

According to the situation you should be prepared to use either approach. Sometimes one is easier, sometimes the other. Most statistics packages give a p-value and so provide the result of the test. That is fine if you have the right test and a suitable significance level. The critical value and critical region are very helpful in situations like taking regular samples for quality control, as in the following case.

The quality control department of a factory tests a random sample of 20 items from each batch produced. A batch is rejected (or perhaps subject to further tests) if the number of faulty items in the sample, X, is more than 2.

This means that the critical region is $X \geqslant 3$.

It is much simpler for the operator carrying out the test to be told the critical region (determined in advance by the person designing the procedure) than to have to work out a probability for each test result.

Example 18.2

Worldwide 25% of men are colour-blind but it is believed that the condition is less widespread among a group of remote hill tribes. An anthropologist plans to test this by sending field workers to visit villages in that area. In each village 30 men are to be tested for colour-blindness. Find the critical region for the test at the 5% level of significance.

Solution

Let p be the probability that a man in that area is colour-blind.

Null hypothesis, H_0: $\qquad\qquad$ $p = 0.25$

Alternative hypothesis, H_1: $\quad$ $p < 0.25$ $\qquad$ (less colour–blindness in this area)

Significance level: $\qquad\qquad$ 5%

With the hypothesis H_0, if the number of colour-blind men in a sample of 30 is X, then $X \sim B(30, 0.25)$.

The critical region is the region $X \leqslant k$, where

$\qquad$ $P(X \leqslant k) \leqslant 0.05$ $\qquad$ and $\qquad$ $P(X \leqslant k + 1) > 0.05$

Calculating the probabilities gives:

$\qquad$ $P(X = 0) = (0.75)^{30} = 0.00018$

$\qquad$ $P(X = 1) = 30(0.75)^{29}(0.25) = 0.00179$

💻 TECHNOLOGY

Table mode can be used to calculate these probabilities using

$f(X) = {}_{30}C_X \times 0.25^X$

$\times 0.75^{(30-X)}$

$$P(X = 2) = \binom{30}{2}(0.75)^{28}(0.25)^2 = 0.00863$$

$$P(X = 3) = \binom{30}{3}(0.75)^{27}(0.25)^3 = 0.02685$$

$$P(X = 4) = \binom{30}{4}(0.75)^{26}(0.25)^4 = 0.06042$$

So $P(X \leq 3) = 0.00018 + 0.00179 + 0.00863 + 0.02685 \approx 0.0375 \leq 0.05$

but $P(X \leq 4) \approx 0.0929 > 0.05$

Therefore the critical region is $X \leq 3$.

> **Discussion point**
>
> → What is the critical region at the 10% significance level?

1-tail and 2-tail tests

Think back to the two examples earlier in the chapter.

What would Stan have said if his eight children had all been girls?

What would Leonora have said if the number 1 had not come up at all?

In both the examples the claim was not only that something unusual had happened but that it was so in a particular direction. So you looked at only one side of the distributions when working out the probabilities, as you can see in Figure 18.1 on page 384 and Figure 18.2 on page 389. In both cases 1-tailed tests were applied. (The word 'tail' refers to the shaded part at the end of the distribution.)

If Stan had just claimed that there was something odd about his chromosomes, then you would have had to work out the probability of a result as extreme on either side of the distribution, in this case either eight girls or eight boys, and you would then apply a 2-tail test.

Here is an example of a 2-tail test.

Example 18.3

The producer of a television programme claims that it is politically unbiased.

'If you take somebody off the street it is 50:50 whether he or she will say the programme favours the government or the opposition', she says.

However, when ten people, selected at random, are asked the question 'Does the programme support the government or the opposition?', nine say it supports the government.

Does this constitute evidence, at the 5% significance level, that the producer's claim is inaccurate?

Solution

Read the last sentence carefully and you will see that it does not say in which direction the bias must be. It does not ask if the programme is favouring the government or the opposition, only if the producer's claim is inaccurate. So you must consider both ends of the distribution, working out the probability of such an extreme result either way; 9 or 10 saying it favours the government, or 9 or 10 the opposition. This is a 2-tail test.

If p is the probability that somebody believes the programme supports the government, you have

Null hypothesis, H_0: $\qquad$ $p = 0.5$ (claim accurate)

Alternative hypothesis, H_1: $\quad$ $p \neq 0.5$ (claim inaccurate)

Significance level: $\qquad\quad$ 5%

$\qquad\qquad\qquad\qquad\qquad$ 2-tail test

The situation is modelled by the binomial distribution $B(10, 0.5)$ and is shown in Figure 18.5.

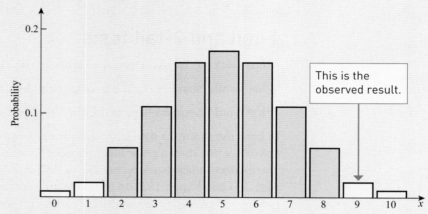

This is the observed result.

Figure 18.5

This gives

$$P(X = 0) = \tfrac{1}{1024}$$

$$P(X = 1) = \tfrac{10}{1024}$$

$$P(X = 10) = \tfrac{1}{1024}$$

$$P(X = 9) = \tfrac{10}{1024}$$

where X is the number of people saying the programme favours the government. Thus the total probability for the two tails is $\tfrac{22}{1024}$ or 2.15%.

Since 2.15% < 5%, the null hypothesis is rejected in favour of the alternative, *that the producer's claim is inaccurate.*

> **Note**
>
> You have to look carefully at the way a test is worded to decide if it should be 1-tail or 2-tail.
>
> Stan Edmonds claimed his chromosomes made him more likely to father boys than girls. That requires a 1-tail test.
>
> Leonora claimed there was bias in the direction of too many 1s; again a 1-tail test.
>
> The test of the television producer's claim was for inaccuracy in either direction and so a 2-tail test was needed.

Asymmetrical cases

In Example 18.3 the distribution was symmetrical and so the 2-tail test was quite simple to apply. In the next case, the distribution is not symmetrical and the test has to be carried out by finding out the critical regions at each tail.

Example 18.4

Peppered moths occur in two varieties, light and dark. The proportion of dark moths increases with certain types of atmospheric pollution.

In a particular village, 25% of the moths are dark, the rest light. A biologist wants to use them as a pollution indicator. She traps samples of 15 moths and counts how many of them are dark.

Note

Another way to think of the significance level is that it is the probability of incorrectly rejecting the null hypothesis.

For what numbers of dark moths among the 15 can she say, at the 10% significance level, that the pollution is changing?

Solution

In this question you are asked to find the critical region for the test:

H_0: $p = 0.25$ (the proportion of dark moths is 25%)

H_1: $p \neq 0.25$ (the proportion is no longer 25%)

Significance level: 10%

2-tail test

where p is the probability that a moth selected at random is dark.

You want to find each tail to be as nearly as possible 5% but both must be less than 5%.

There are two ways you can do this, by using cumulative binomial tables, or by using your calculator.

Using tables

Look under $n = 15$, for $p = 0.25$.

n	x \ p	0.050	0.100	0.150	$\frac{1}{6}$	0.200	0.250	0.300	$\frac{1}{3}$	0.350	0.400
15	0	0.4633	0.2059	0.0874	0.0649	0.0352	0.0134	0.0047	0.0023	0.0016	0.0005
	1	0.8290	0.5490	0.3186	0.2596	0.1671	0.0802	0.0353	0.0194	0.0142	0.0052
	2	0.9638	0.8159	0.6042	0.5322	0.3980	0.2361	0.1268	0.0794	0.0617	0.0271
	3	0.9945	0.9444	0.8227	0.7685	0.6482	0.4613	0.2969	0.2092	0.1727	0.0905
	4	0.9994	0.9873	0.9383	0.9102	0.8358	0.6865	0.5155	0.4041	0.3519	0.2173
	5	0.9999	0.9978	0.9832	0.9726	0.9389	0.8516	0.7216	0.6184	0.5643	0.4032
	6	1.0000	0.9997	0.9964	0.9934	0.9819	0.9434	0.8689	0.7970	0.7548	0.6098
	7		1.0000	0.9994	0.9987	0.9958	0.9827	0.9500	0.9118	0.8868	0.7869
	8			0.9999	0.9998	0.9992	0.9958	0.9848	0.9692	0.9578	0.9050
	9			1.0000	1.0000	0.9999	0.9992	0.9963	0.9915	0.9876	0.9662
	10					1.0000	0.9999	0.9993	0.9982	0.9972	0.9907
	11						1.0000	0.9999	0.9997	0.9965	0.9981
	12							1.0000	1.0000	0.9999	0.9997
	13									1.0000	1.0000
	14										
	15										

Figure 18.6

TECHNOLOGY

Your work on page 389 should mean that you know how to use your calculator to find the cumulative binomial probability for a 1-tail test. Now make sure that you know the steps to use it for a 2-tail test as well.

From this you can see that the left-hand tail includes 0 but not 1 or more; the right-hand tail is 8 or above but not 7.

So the critical regions are less than 1 and more than 7 dark moths in the 15.

For these values she would claim the pollution levels are changing.

> **Note**
>
> This is really quite a crude test. The left-hand tail is 1.34%, the right-hand $(1 - 0.9827)$ or 1.73%. Neither is close to 5%. The situation would be improved if you were to increase the sample size; 15 is a small number of moths on which to base your findings.

Exercise 18.2

 ① What is the difference between a **critical value** and a **critical region**?

 ② A coin is tossed 20 times and comes up heads 13 times. Boris claims it is biased.

The hypothesis test has:

A $\quad$ $H_0 : p = \dfrac{1}{2}$ $\qquad$ $H_1 : p \leq \dfrac{1}{2}$

B $\quad$ $H_0 : p = \dfrac{1}{2}$ $\qquad$ $H_1 : p \geq \dfrac{1}{2}$

C $\quad$ $H_0 : p = \dfrac{1}{2}$ $\qquad$ $H_1 : p > \dfrac{1}{2}$

D $\quad$ $H_0 : p = \dfrac{1}{2}$ $\qquad$ $H_1 : p \neq \dfrac{1}{2}$

③ A particular gold coin is proposed for the toss in the final of a big sporting event. Henry and Mandy are both investigating the coin to see if it is biased. They decide that a 10% significance level is sufficient. They are going to throw the coin 50 times. Henry decides to find out if tails is more likely; Mandy decides to test whether tails is either less likely or more likely. Before they start they want to know the critical values.

(i) What sort of test should Henry carry out? What sort of test should Mandy carry out?

(ii) Calculate the critical values for their tests.

(iii) What are their acceptable regions?

(iv) For what values will Henry say the coin is biased?

(v) For what values will Mandy say the coin is biased?

(vi) Who do you think has designed the better test?

④ To test the claim that a coin is biased, it is tossed 20 times. It comes down heads 7 times. Test at the 10% significance level whether this claim is justified.

> **Note**
>
> All subsequent questions, except Question 8, involve modelling using the binomial distribution.

PS ⑤ A biologist discovers a colony of a previously unknown type of bird nesting in a cave. Out of the 16 chicks which hatch during his period of investigation, 13 are female. Test at the 5% significance level whether this supports the view that the sex ratio for the chicks differs from 1:1.

⑥ People entering an exhibition have to choose whether to turn left or right. Out of the first twelve people, nine turn left and three right. Test at the 5% significance level whether people are more likely to turn one way than another.

⑦ Weather records for a certain seaside resort show that on average one day in four in April is wet, but local people write to their newspaper complaining that the climate is changing.

PS A reporter on the paper records the weather for the next 20 days in April and finds that 10 of them are wet.

Do you think the complaint is justified? (Assume a 10% significance level.)

PS ⑧ In a fruit machine there are five drums which rotate independently to show one out of six types of fruit each (lemon, apple, orange, melon, banana and pear). You win a prize if all five stop showing the same fruit. A customer claims that the machine is fixed; the lemon in the first place is not showing the right number of times. The manager runs the machine 20 times and the lemon shows 6 times in the first place. Is the customer's complaint justified at the 10% significance level?

⑨ A boy is losing in a game of cards and claims that his opponent is cheating.

Out of the last 18 times he shuffled and dealt the cards, the first card to be laid down was a spade on only one occasion. Can he justify his claim at

(i) the 10% significance level

(ii) the 5% significance level?

⑩ A small colony of 20 individuals of a previously unknown animal is discovered. It is believed that it might be the same species as one described by early explorers who said that one-quarter of them were male, the rest female.

What numbers of males and females would lead you to believe, at the 5% significance level, that they are not the same species?

PS ⑪ A multiple choice test has 20 questions, with the answer for each allowing four options, A, B, C, and D. All the students in a class tell their teacher that they guessed all 20 answers. The teacher does not believe them. Devise a 2-tail test at the 10% significance level to apply to a student's mark to test the hypothesis that the answers were not selected at random.

PS ⑫ When a certain language is written down, 15% of the letters are Z. Use this information to devise a test at the 10% significance level which somebody who does not know the language could apply to a short passage, 50 letters long, to determine whether it is written in the same language.

PS ⑬ A seed firm states on the packets of bean seeds that the germination rate is 80%. Each packet contains 25 seeds.

(i) How many seeds would you expect to germinate out of one packet?

(ii) What is the probability of exactly 17 germinating?

A man buys a packet and only 12 seeds germinate.

(iii) Is he justified in complaining?

LEARNING OUTCOMES

When you have completed this chapter you should be able to:

➤ understand the link with histograms

➤ understand and apply the language of statistical hypothesis testing, developed through a binomial model:

  ○ null hypothesis

  ○ alternative hypothesis

  ○ significance level

  ○ test statistic

  ○ 1-tail test

  ○ 2-tail test

  ○ critical value

  ○ critical region

  ○ acceptance region

  ○ *p*-value

➤ conduct a statistical hypothesis test for the proportion in the binomial distribution and interpret the results in context

➤ understand that a sample is being used to make an inference about the population and appreciate that the significance level is the probability of incorrectly rejecting the null hypothesis.

KEY POINTS

Steps for conducting a hypothesis test

1 Establish the null and alternative hypotheses:

  ■ The null hypothesis, H_0, is the default position that nothing special has occurred.

  ■ The alternative hypothesis, H_1, is that there has been a change from the position described by the null hypothesis.

  ■ The alternative hypothesis may be 1-tail or 2-tail according to whether the direction of change is or is not specified.

2 Decide on the significance level. This is the level at which you say there is enough evidence to reject the null hypothesis in favour of an alternative hypothesis.

3 Collect suitable data using a random sampling procedure that ensures the items are independent.

4 Use the data to determine the test statistic. This is the measure that will be used to decide whether the test is significant. In the case of a binomial test it is the number of successes.

5 Conduct the test doing the necessary calculations. Then choose one of the options:

  ■ Work out the p-value, the probability of a result at least as extreme as that obtained and compare it with the significance level.

  ■ Work out the critical value for the particular significance level and compare it with the test statistic.

  ■ Work out the critical (or rejection) region for the particular significance level and determine whether the test statistic lies within it or outside it in the acceptance region. (The terms rejection and acceptance refer to the null hypothesis.)

Typical ways of describing the outcome of a hypothesis test

1 Either:

  ■ The evidence is not strong enough to reject the null hypothesis in favour of the alternative hypothesis.

  ■ There is not enough evidence to conclude that the probability/proportion has changed/increased/decreased.

2 Or:

  ■ There is sufficient evidence to reject the null hypothesis in favour of the alternative hypothesis.

  ■ There is sufficient evidence to conclude that the probability/proportion has changed/increased/decreased.

3 You should then add a comment relating to the situation you are investigating.

FUTURE USES

There are many other hypothesis tests based on the value of a parameter of the parent population, for example

■ the mean

■ correlation or association (for a bivariate distribution)

■ proportions in different categories.

Questions 5 and 6 are based on large data sets published by the government and located at https://www.gov.uk/government/statistical-data-sets/family-food-datasets

You are advised to visit this site and familiarise yourself with the data sets before attempting these questions.

① An online bookseller claims that 75% of books ordered will be delivered 1 working day later, 20% will be delivered 2 working days later and the remainder will be delivered 3 working days later.

Assuming these figures are correct, find the probability that a random sample of 3 books will all be delivered

(i) 1 working day later,   [2 marks]

(ii) on the same working day.   [2 marks]

② In 2015, the BBC Sport website carried the following graphic about the leading goal scorers in the England football team.

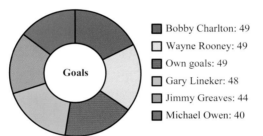

Bobby Charlton: 49
Wayne Rooney: 49
Own goals: 49
Gary Lineker: 48
Jimmy Greaves: 44
Michael Owen: 40

Figure 1 England's all-time leading goal-scorers

(i) Give two reasons why this graphic is unsatisfactory.   [2 marks]

(ii) Suggest a better way of presenting this information.
     Justify your choice.   [2 marks]

PS ③ A large manufacturer of dairy products sold in supermarkets wants to know what its customers think of the health and nutrition information printed on the packaging of its products. It decides to interview a sample of 200 customers in supermarkets.

(i) Explain what simple random sampling is and why it would
     not be possible in this case.   [2 marks]

(ii) Describe how opportunity sampling and quota sampling
     would work in this case, making clear how the two methods
     differ and which of them, if either, would be preferable.   [4 marks]

PS ④ 'Old Faithful' is a geyser in Yellowstone National Park, Wyoming. The geyser erupts frequently, and eruptions last for different lengths of time. The scatter diagram shows *Length of eruption* and *Time to next eruption* for a random sample of 100 occasions.

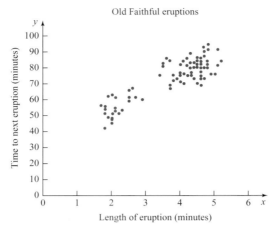

Old Faithful eruptions

Figure 2

(i)   Use the scatter diagram to describe the patterns in eruptions of Old Faithful that a visitor to Yellowstone could expect to see.   [3 marks]

(ii) Ann looks at the data and says that longer eruptions tend to be followed by longer times to the next eruption.

Ben says that her description is too simple.

Explain why Ben might take this view.   [1 marks]

(iii) Suppose that a histogram was drawn of the variable *Length of eruption*.

What distinctive feature would this histogram show?

Would a histogram of the variable *Time to next eruption* show the same feature?   [2 marks]

(iv) The mean value of the variable *Length of eruption* for these data is 3.5 minutes. Discuss briefly whether or not this is a useful measure.   [2 marks]

⑤ Merula is looking at some data on 'Purchased quantities of household food and drink by Government Office Region and Country 2001–2014' from the Living costs and food survey published in December 2015 and she constructs the following spreadsheet.

	A	B
1	**Average per person per week of cheese (g)**	
2	**Year**	**England**
3	2001–02	114
4	2002–03	114
5	2003–04	115
6	2004–05	112
7	2005–06	117
8	2006	118
9	2007	120
10	2008	111
11	2009	117
12	2010	119
13	2011	121
14	2012	116
15	2013	120
16	2014	113

Table 1 Purchased quantities of cheese per week in grammes in England.

(i) Merula decides to discard the data in rows 3 to 7 of the spreadsheet. Explain why that is a sensible thing to do. [2 marks]

Merula uses her spreadsheet to draw a graph of consumption of cheese per week for households in England, as shown in Figure 18.9.

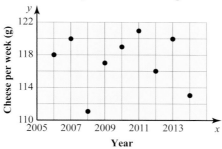

Consumption of cheese in England

Figure 3

(ii) Write a short account of what the graph shows about the variation in cheese consumption in England during this time. [3 marks]

(iii) Merula is puzzled that the data point for 2008 is much lower than the rest. How would you interpret this data point? [2 marks]

(iv) The spreadsheet Merula is using has a facility to model a set of data by fitting straight lines or curves to the data points. Discuss briefly whether or not it would be appropriate to model these data using that facility. [3 marks]

⑥ Consider the data on the amount of sugar, in grammes, eaten per person, per week from the Living costs and food survey published in December 2015. This amount is denoted by x in this question.

For the North West $\Sigma x = 1303$ and $\Sigma x^2 = 123\,557$.

There are 14 years for which the data is given.

(i) Use your calculator to find the mean and standard deviation of x. [3 marks]

(ii) The comparable data for the South East is

110	101	88	100	93	92	87
89	89	92	91	91	93	74

Use your calculator to find the mean and standard deviation of this data. [3 marks]

(iii) Compare the two data sets, with reference to the context. [2 marks]

It is often quoted in the media that people in some areas of the country eat more healthily than people in other areas.

(iv) Does the data support this? Give reasons for your answer. [2 marks]

⑦ In a game, player A throws two ordinary unbiased six-faced dice and the score is given by whichever of the dice shows the higher number.

(i) Show that there are exactly 3 ways in which player A can obtain a score of 2. Hence deduce the probability of player A scoring 2. [2 marks]

Part of the probability distribution for the score, X, obtained by player A is given in the following.

r		1	2	3	4	5	6
P(X = r)		k	3k	5k	7k	9k	

(ii) State the value of k. Complete the probability distribution table. [2 marks]

Player B now throws the dice and obtains a score, Y, in the same way.

(iii) Prove that $P(X = Y) = 286k^2$. Hence find, correct to 3 decimal places, the probability that B beats A in the game. [4 marks]

 ⑧ The traffic police in one area of the country are concerned about the number of cars that have tyres which are unsafe. Over a period of several months, all cars stopped by the police for any reason have their tyres checked. As a result, the police adopt a probability model in which 15% of cars have at least one unsafe tyre.

In one 8-hour shift a police patrol stops 25 cars. Use the police model to:

(i) find the expected number of these cars with at least one unsafe tyre [1 marks]

(ii) find the probability that the number of cars with at least one unsafe tyre is less than the expected number. [2 marks]

The traffic police decide to have a campaign in which publicity is given to the problem of cars having unsafe tyres. Motorists are encouraged to have their tyres checked and replaced if necessary. They are warned that the police will be stopping more cars and issuing penalty notices when tyres are unsafe.

After the campaign has been running for a month, the police carry out a check to see if there is evidence that the proportion of cars with unsafe tyres has reduced. Suppose that n cars are stopped at random and that k of them have unsafe tyres.

(iii) Given that $n = 50$ and $k = 5$, carry out an appropriate hypothesis test at the 5% level of significance and show that there is insufficient evidence to suggest that the proportion has reduced. [4 marks]

(iv) Given instead that $n = 100$, determine the possible values of k for which there is evidence, at the 1% level of significance, that the proportion has reduced. [3 marks]

19 Kinematics

Throw a small object such as a marble straight up in the air and think about the words you could use to describe its motion from the instant just after it leaves your hand to the instant just before it hits the floor. Some of your words might involve the idea of direction. Other words might be to do with the position of the marble, its speed or whether it is slowing down or speeding up. Underlying many of these is time.

1 The language of motion

Direction

The marble that you have thrown straight up in the air moves as it does because of the gravitational pull of the earth. We understand directional words such as up and down because we experience this pull towards the centre of the Earth all the time. The *vertical* direction is along the line towards or away from the centre of the Earth.

In mathematics a quantity which has only size, or magnitude is called a **scalar**. One which has both magnitude and direction is called a **vector**.

Distance, position and displacement

The total **distance** travelled by the marble at any time does not depend on its direction. It is a scalar quantity.

Position and displacement are two vectors related to distance: they have direction as well as magnitude. Here their direction is up or down and you decide which of these is positive. When up is taken as positive, down is negative.

The **position** of the marble is then its distance above a fixed origin, for example the distance above the place where it first left your hand.

When it reaches the top, the marble might have travelled a distance of 1.25 m. Relative to your hand its position is then 1.25 m upwards or +1.25 m.

At the instant it returns to the same level as your hand it will have travelled a total distance of 2.5 m. Its *position,* however, is zero upwards.

A position is always referred to a fixed origin but a **displacement** can be measured from any point. When the marble returns to the level of your hand, its displacement is zero relative to your hand but −1.25 m relative to the highest point the marble reached.

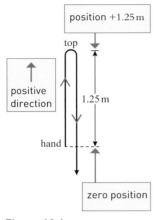

Figure 19.1

Discussion point

→ What are the positions of the particles A, B and C in the diagram below?

Figure 19.2

→ What is the displacement of B   (i)   relative to A   (ii)   relative to C?

Diagrams and graphs

In mathematics, it is important to use words precisely, even though they might be used more loosely in everyday life. In addition, a picture in the form of a diagram or graph can often be used to show the information more clearly.

Figure 19.3 is a **diagram** showing the direction of motion of the marble and relevant distances. The direction of motion is indicated by an arrow. Figure 19.4 is a **graph** showing the position above the level of your hand against the time.

Note

When drawing a graph it is very important to specify your axes carefully. Graphs showing motion usually have time along the horizontal axis. Then you have to decide where the origin is and which direction is positive on the vertical axis. In this graph the origin is at hand level and upwards is positive. The time is measured from the instant the marble leaves your hand.

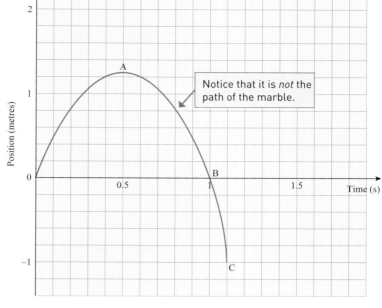

Notice that it is *not* the path of the marble.

→ The graph in Figure 19.4 shows that the position is negative after one second (point B). What does this negative position mean?

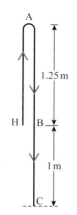

Figure 19.3   **Figure 19.4**

Notation and units

As with most mathematics, you will see that certain letters are commonly used to denote certain quantities. This makes things easier to follow. Here the letters used are:

- s, h, x, y and z for position
- t for time measured from a starting instant
- u and v for velocity
- a for acceleration.

The S.I. (Système International d'Unités) unit for **distance** is the metre (m), that for **time** is the second (s) and that for **mass** is the kilogram (kg). Other units follow from these so that speed, for example, is measured in metres per second, written ms^{-1}.

Exercise 19.1

 ① What is the difference between position and displacement?

 ② A stone is dropped from a height of 2.3 m relative to an origin directly beneath it on the ground. What is its position when it hits the ground?

 A 2.3 m   B −2.3 m

 C 0 m   D 4.6 m

 ③ A marble is thrown up in the air. It leaves your hand, which is 1 m above ground level, and travels a distance of 1.25 m before returning to your hand. The origin for the motion is taken at ground level. What is the position of the marble

(i) when it leaves your hand?

(ii) at its highest point?

④ A boy throws a ball vertically upwards so that its position y m at time t is as shown in Figure 19.5.

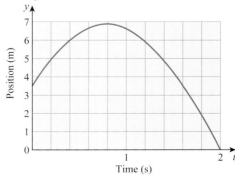

Figure 19.5

(i)   Write down the position of the ball at times $t = 0, 0.4, 0.8, 1.2, 1.6$ and 2.

(ii) Calculate the displacement of the ball relative to its starting position at these times.

(iii) What is the total distance travelled

(a) during the first $0.8\,\text{s}$

(b) during the $2\,\text{s}$ of the motion?

> **Note**
>
> A particle is considered to have no size but it does have mass. It is a simplifying assumption to avoid having to consider rotation or air resistance.

⑤   The position of a particle moving along a straight horizontal groove is given by $x = 2 + t(t - 3)$ for $0 \leqslant t \leqslant 5$ where x is measured in metres and t in seconds.

(i)   What is the position of the particle at times $t = 0, 1, 1.5, 2, 3, 4$ and 5?

(ii) Draw a diagram to show the path of the particle, marking its position at these times.

(iii) Find the displacement of the particle relative to its initial position at these times.

(iv) Calculate the total distance travelled during the motion.

⑥   A particle moves so that its position x metres at time t seconds is given by $x = 2t^3 - 18t$.

(i)   Calculate the position of the particle at times $t = 0, 1, 2, 3$ and 4.

(ii) Draw a diagram showing the position of the particle at these times.

(iii) Sketch a graph of the position against time.

(iv) State the times when the particle is at the origin and describe the direction in which it is moving at those times.

> **Note**
>
> All subsequent questions involve modelling.

> 💻 **TECHNOLOGY**
>
> Table mode is useful here, if you have it.

PS ⑦   For each of the following situations sketch a graph of position against time. Show clearly the origin and the positive direction.

(i)   A stone is dropped from a bridge which is $40\,\text{m}$ above a river.

(ii) A parachutist jumps from a helicopter which is hovering at $2000\,\text{m}$. She opens her parachute after $10\,\text{s}$ of free fall.

(iii) A bungee jumper on the end of an elastic string jumps from a high bridge.

PS ⑧   Figure 19.6 is a sketch of the position–time graph for a fairground ride.

(i)   Describe the motion, stating in particular what happens at O, A, B, C and D.

(ii) What type of ride is this?

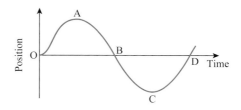

Figure 19.6

2 Speed and velocity

Speed is a scalar quantity and does not involve direction. **Velocity** is the vector related to speed; its magnitude is the speed but it also has a direction. When an object is moving in the negative direction, its velocity is negative.

Amy has a letter to post on her way to college. The postbox is $500\,\text{m}$ east of her house and the college is $2.5\,\text{km}$ to the west. Amy cycles at a steady speed of $10\,\text{m}\,\text{s}^{-1}$ and takes $10\,\text{s}$ at the postbox to find the letter and post it.

Figure 19.7

Figure 19.8 shows Amy's journey using east as the positive direction. The distance of 2.5 km has been changed to metres so that the units are consistent.

Figure 19.8

After she leaves the post box Amy is travelling west so her velocity is negative. It is $-10\,\text{m}\,\text{s}^{-1}$.

The distances and times for the three parts of Amy's journey are shown in Table 19.1.

Table 19.1

	Distance	Time
Home to postbox	500 m	$\frac{500}{10} = 50\,\text{s}$
At postbox	0 m	10 s
Postbox to college	3000 m	$\frac{3000}{10} = 300\,\text{s}$

Prior knowledge

You have met gradients in Chapter 5 on coordinate geometry.

Discussion point

→ Calculate the gradients of the three portions of this graph. What conclusions can you draw?

These can be used to draw the position–time graph using home as the origin, as in Figure 19.9.

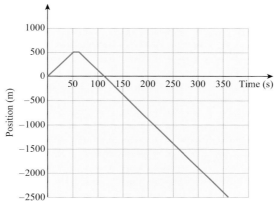

Figure 19.9

The velocity is the rate at which the position changes.

Velocity is represented by the gradient of the position–time graph.

Note

By drawing the graphs below each other with the same horizontal scales, you can see how they correspond to each other.

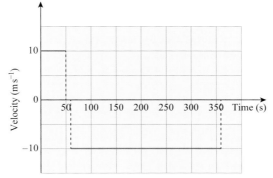

Figure 19.10

Figure 19.10 is the velocity–time graph.

Distance–time graphs

Figure 19.11 is the distance–time graph of Amy's journey. It differs from the position–time graph because it shows how far she travels irrespective of her direction. There are no negative values. The gradient of this graph represents Amy's speed rather than her velocity.

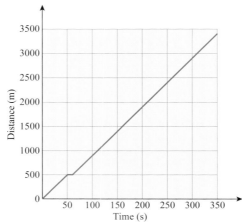

Figure 19.11

Average speed and average velocity

You can find Amy's average speed on her way to college by using the definition

$$\text{average speed} = \frac{\text{total distance travelled}}{\text{total time taken}}$$

When the distance is in metres and the time in seconds, speed is found by dividing metres by seconds and is written as m s^{-1}. So Amy's average speed is

$$\frac{3500\,\text{m}}{360\,\text{s}} = 9.72\,\text{m s}^{-1}$$

Amy's average velocity is different. Her displacement from start to finish is $-2500\,\text{m}$ as the college is in the negative direction.

$$\text{average velocity} = \frac{\text{displacement}}{\text{time taken}}$$

So Amy's average velocity is $-\frac{2500}{360} = -6.94\,\text{m s}^{-1}$

If Amy had taken the same time to go straight from home to college at a steady speed, this steady speed would have been $6.94\,\text{m s}^{-1}$.

Velocity at an instant

The position–time graph for a marble thrown straight up into the air at $5\,\text{m s}^{-1}$ is curved because the velocity is continually changing, see Figure 19.12

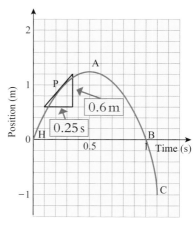

Figure 19.12

The velocity is represented by the gradient of the position–time graph. When a position–time graph is curved like this you can find the **velocity at an instant** of time by drawing a tangent as in Figure 19.12.

The velocity at P is approximately

$$\frac{0.6}{0.25} = 2.4\,\text{m s}^{-1}$$

The velocity–time graph is shown in Figure 19.13.

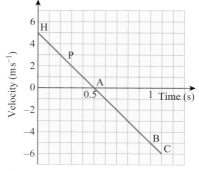

Figure 19.13

At the point A, the velocity and gradient of the position–time graph are zero. We say the marble is **instantaneously at rest**. The velocity at H is positive because the marble is moving in the positive direction (upwards). The velocity at B and at C is negative because the marble is moving in the negative direction (downwards).

Exercise 19.2

> **Note**
>
> Questions 2, 3, 4, 7 and 8 involve modelling.

 ① What are the differences between a speed–time graph and a velocity–time graph?

 ② Mark cycles north at 6 m s^{-1} for 25 s, then at 10 m s^{-1} for 15 s. He rests for 20 s then cycles south, back to his starting point, at 15 m s^{-1}.

Which one of the graphs in Figure 19.14 shows his journey?

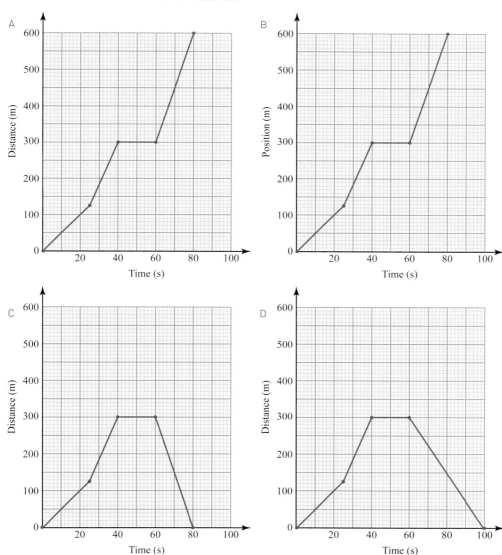

Figure 19.14

③ Amy cycles $500\,\mathrm{m}$ east at a steady speed of $10\,\mathrm{m\,s^{-1}}$, stops for $10\,\mathrm{s}$ and then cycles $3000\,\mathrm{m}$ west at a steady speed of $10\,\mathrm{m\,s^{-1}}$. Draw a speed–time graph for Amy's journey.

④ The distance–time graph in Figure 19.15 shows the relationship between distance travelled and time for a person who leaves home at 9.00 a.m., walks to a bus stop and catches a bus into town.

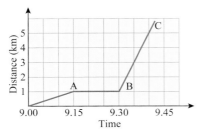

Figure 19.15

(i) Describe what is happening during the time from A to B.

(ii) The section BC is much steeper than OA; what does this tell you about the motion?

(iii) Draw the speed–time graph for the person.

(iv) What simplifications have been made in drawing these graphs?

⑤ For each of the following journeys find

(a) the initial and final positions

(b) the total displacement

(c) the total distance travelled

(d) the velocity and speed for each part of the journey

(e) the average velocity for the whole journey

(f) the average speed for the whole journey.

(i)

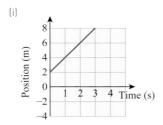

(ii)

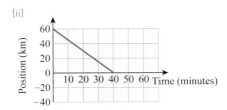

(iii)

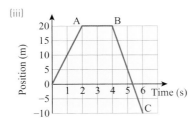

(iv)

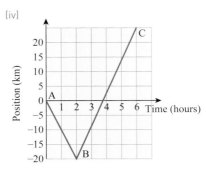

Figure 19.16

⑥ The current world record for the $100\,\mathrm{m}$ sprint is $9.58\,\mathrm{s}$. Find the average speed in $\mathrm{m\,s^{-1}}$ and in $\mathrm{km\,h^{-1}}$.

⑦ The current world record for the men's marathon (a distance of $42.195\,\mathrm{km}$) is 2 hours 2 minutes and 57 seconds. Find the average speed.

 ⑧ An aeroplane flies from London to Toronto, a distance of $5700\,\mathrm{km}$, at an average speed of $1280\,\mathrm{km\,h^{-1}}$. It returns at an average speed of $1200\,\mathrm{km\,h^{-1}}$. Find the average speed for the round trip.

 ⑨ A car travels $50\,\mathrm{km}$ from A to B at an average speed of $100\,\mathrm{km\,h^{-1}}$. It stops at B for 45 minutes and then returns to A. The average speed for the whole journey is $40\,\mathrm{km\,h^{-1}}$. Find the average speed from B to A.

⑩ A train takes 45 minutes to complete its 24 kilometre trip. It stops for 1 minute at each of 7 stations during the trip.

(i) Calculate the average speed of the train.

(ii) What could be the average speed if the stop at each station was reduced to 20 seconds?

PS ⑪ When Louise is planning car journeys she reckons that she can cover distances along main roads at roughly $100\,\text{km}\,\text{h}^{-1}$ and those in town at $30\,\text{km}\,\text{h}^{-1}$.

(i) Find her average speed for each of the following journeys.

(a) 20 km on main roads and then 10 km in a town.

(b) 150 km on main roads and then 2 km in a town.

(c) 20 km on main roads and then 20 km in a town.

(ii) In what circumstances would her average speed be $65\,\text{km}\,\text{h}^{-1}$?

PS ⑫ The triathlon is a combined event which includes a 1500 m swim, followed by a 40 km bicycle ride and finally a 10 km run.

Competitor A completed the swim in 20 minutes and 5 seconds, the ride in 1 hour 5 minutes and 20 seconds and the run in 32 minutes and 45 seconds.

(i) Find the average speed for each of the three legs and for the combined event.

Competitor B swam at an average speed of $1.25\,\text{m}\,\text{s}^{-1}$, cycled at an average speed of $38.2\,\text{km}\,\text{h}^{-1}$ and ran at an average speed of $18.25\,\text{km}\,\text{h}^{-1}$.

(ii) Who out of A or B finishes first?

3 Acceleration

In everyday language, the word 'accelerate' is usually used when an object speeds up and 'decelerate' when it slows down. The idea of deceleration is sometimes used in a similar way by mathematicians but in mathematics the word acceleration is used when there is a change in velocity, whether an object is speeding up, slowing down or changing direction. Acceleration is *the rate at which the velocity changes.*

Over a period of time

$$\text{average acceleration} = \frac{\text{change in velocity}}{\text{time taken}}$$

Acceleration is represented by the gradient of a velocity–time graph. It is a vector and can take different signs in a similar way to velocity. This is illustrated by Tom's cycle journey which is shown in Figure 19.17.

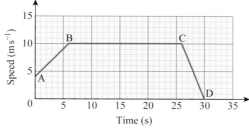

Figure 19.17

Tom turns onto the main road at $4\,\text{m}\,\text{s}^{-1}$, accelerates uniformly, maintains a constant speed and then slows down uniformly to stop when he reaches home.

Between A and B, Tom's velocity increases by $(10 - 4) = 6\,\mathrm{m\,s^{-1}}$ in 6 seconds, that is 1 metre per second every second.

This acceleration is written as $1\,\mathrm{m\,s^{-2}}$ (one metre per second squared) and is the gradient of AB in Figure 19.17.

From B to C, acceleration $= 0\,\mathrm{m\,s^{-2}}$ ◄————— There is no change in velocity.

From C to D, acceleration $= \dfrac{0 - 10}{(30 - 26)} = -2.5\,\mathrm{m\,s^{-2}}$

From C to D Tom is slowing down while still moving in the positive direction towards home, so his acceleration, the gradient of the graph, is negative.

The sign of acceleration

Think again about the marble thrown up into the air with a speed of $5\,\mathrm{m\,s^{-1}}$.

Figure 19.18 represents the velocity when *upwards* is taken as the positive direction and shows that the velocity *decreases* from $+5\,\mathrm{m\,s^{-1}}$ to $-5\,\mathrm{m\,s^{-1}}$ in one second.

This means that the gradient, and hence the acceleration, is *negative*. It is $-10\,\mathrm{m\,s^{-2}}$. (You might recognise the number 10 as an approximation to *g*. See page 424.)

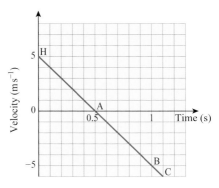

Figure 19.18

Discussion point

→ A car accelerates away from a set of traffic lights. It accelerates to a maximum speed and at that instant starts to slow down to stop at a second set of traffic lights. Which of the graphs in Figure 19.19 could represent

(i) the distance–time graph

(ii) the velocity–time graph

(iii) the acceleration–time graph?

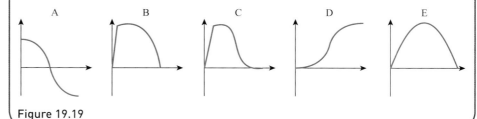

Figure 19.19

Exercise 19.3

> **Note**
> Questions 6 and 8 involve modelling.

① What does a negative acceleration mean?

② The velocity–time graph in Figure 19.20 shows a journey.

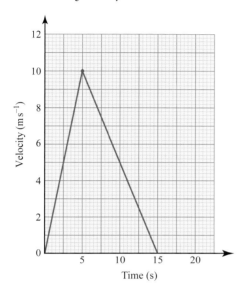

Figure 19.20

Which of the graphs in Figure 19.21 is the acceleration–time graph for this journey?

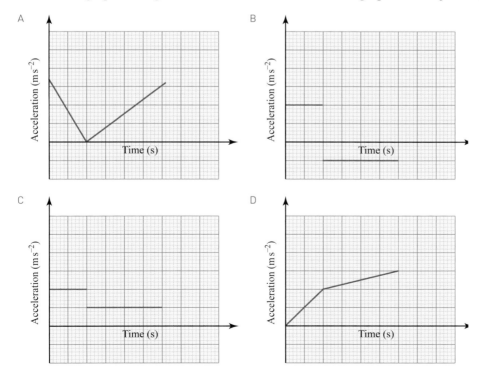

Figure 19.21

③ (i) Calculate the acceleration for each part of the following journey.

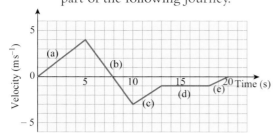

Figure 19.22

(ii) Use your results to sketch an acceleration–time graph.

④ A particle is moving in a straight line with acceleration

$$a = 5, \qquad 0 \leqslant t \leqslant 3$$
$$a = -10, \quad 3 < t \leqslant 10$$

(i) At time $t = 0$ the particle has a velocity of $5\,\mathrm{m\,s^{-1}}$ in the positive direction. Find the velocity of the particle when $t = 3$.

(ii) At what time is the particle travelling in the negative direction with a speed of $25\,\mathrm{m\,s^{-1}}$?

⑤ A particle travels along a straight line. Its acceleration in the interval $0 \leqslant t \leqslant 6$ is shown in the acceleration–time graph.

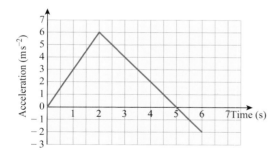

Figure 19.23

(i) Write down the acceleration at time $t = 2$.

(ii) Given that the particle starts from rest at $t = 0$, find the speed at $t = 2$.

(iii) Find an expression for the acceleration as a function of t in the interval $0 \leqslant t \leqslant 2$.

(iv) At what time is the speed greatest?

PS ⑥ A lift travels up and down between the ground floor (G) and the roof garden (R) of a hotel. It starts from rest, takes 5 s to increase its speed uniformly to $2\,\mathrm{m\,s^{-1}}$, maintains this speed for 5 s and then slows down uniformly to rest in another 5 s. In the following questions use upwards as positive.

(i) Sketch a velocity–time graph for the journey from G to R.

On one occasion the lift stops for 5 s at R before returning to G.

(ii) Sketch a velocity–time graph for this journey from G to R and back.

(iii) Calculate the acceleration for each 5 s interval. Take care with the signs.

(iv) Sketch an acceleration–time graph for this journey.

⑦ A particle starts from rest at time $t = 0$ and moves in a straight line, accelerating as follows:

$$a = 2, \qquad 0 \leqslant t \leqslant 10$$
$$a = 0.5, \quad 10 \leqslant t \leqslant 50$$
$$a = -3, \quad 50 < t \leqslant 60$$

where a is the acceleration in $\mathrm{m\,s^{-2}}$ and t is the time in seconds.

(i) Find the speed of the particle when $t = 10, 50$ and 60.

(ii) Sketch a speed–time graph for the particle in the interval $0 \leqslant t \leqslant 60$.

(iii) Find the total distance travelled by the particle in the interval $0 \leqslant t \leqslant 60$.

PS ⑧ A film of a dragster doing a 400 m run from a standing start yields the following positions at 1 second intervals.

Figure 19.24

(i) Draw a displacement–time graph of its motion.

(ii) Use your graph to help you to sketch

(a) the velocity–time graph

(b) the acceleration–time graph.

4 Using areas to find distances and displacements

Discussion points

→ Calculate the area between the speed–time graph and the time axis from

(i) $t = 0$ to 1

(ii) $t = 0$ to 2

(iii) $t = 0$ to 3

→ Compare your answers with the distance that the stone has fallen, shown on the distance–time graph at $t = 1, 2$ and 3.

→ What conclusion do you reach?

The graphs in Figures 19.25 and 19.26 model the motion of a stone falling from rest.

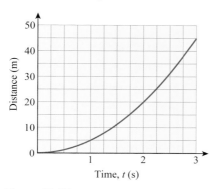

Figure 19.25

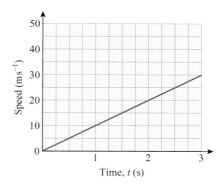

Figure 19.26

The area between a speed–time graph and the time axis represents the distance travelled.

There is further evidence for this if you consider the units on the graphs.

Multiplying metres per second by seconds gives metres. A full justification relies on the calculus methods explained in Chapter 21.

Finding the area under speed–time graphs

Many of these graphs consist of straight-line sections. The area is easily found by splitting it up into triangles, rectangles or trapezia (Figure 19.27).

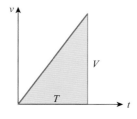

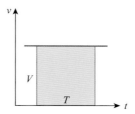

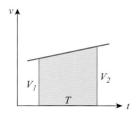

Triangle: area $= \frac{1}{2} VT$

Rectangle: area $= VT$

Trapezium: area $= \frac{1}{2} (V_1 + V_2) T$

Figure 19.27

Example 19.1

The graph shows Tom's journey from the time he turns on to the main road until he arrives home. How far does Tom cycle?

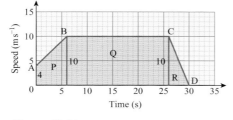

Figure 19.28

Discussion point

→ What is the meaning of the area between a velocity–time graph and the time axis?

Solution

Split the area under the speed–time graph into three regions.

P	trapezium:	area $= \frac{1}{2}(4 + 10) \times 6 = 42\,\text{m}$
Q	rectangle:	area $= 10 \times 20$ $\quad = 200\,\text{m}$
R	triangle:	area $= \frac{1}{2} \times 10 \times 4$ $\quad = 20\,\text{m}$
total area		$= 262\,\text{m}$

Tom cycles 262 m.

The area between a velocity–time graph and the time axis

Example 19.2

David walks east for 6 s at $2\,\text{ms}^{-1}$ then west for 2 s at $1\,\text{ms}^{-1}$. Draw

(i) a diagram for the journey

(ii) the speed–time graph

(iii) the velocity–time graph.

Interpret the area under each graph.

Solution

(i) David's journey is shown in Figure 19.29.

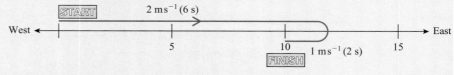

Figure 19.29

(ii) Speed–time graph

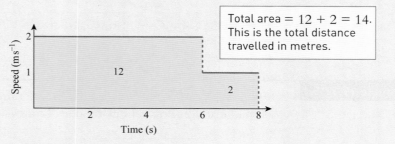

Total area $= 12 + 2 = 14$. This is the total distance travelled in metres.

Figure 19.30

(iii) Velocity–time graph

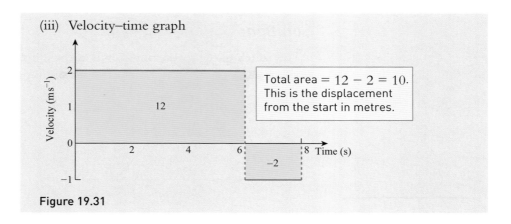

Total area = 12 − 2 = 10. This is the displacement from the start in metres.

Figure 19.31

The area between a velocity–time graph and the time axis represents the change in position, that is the displacement.

When the velocity is negative, the area is below the time axis and represents a displacement in the negative direction, west in this case.

Estimating areas

Sometimes the velocity–time graph does not consist of straight lines so you have to make the best estimate you can. One way is to replace the curve by a number of straight line segments and then find the area underneath them. Sometimes you will count the squares.

Discussion point

→ This speed–time graph shows the motion of a dog over a 60 s period.

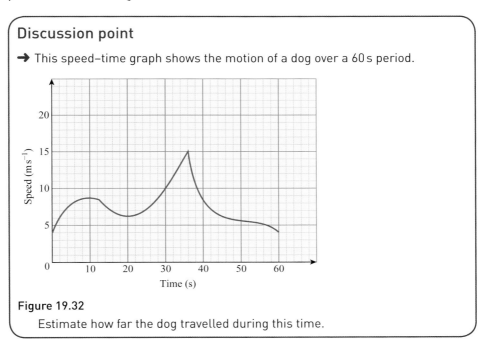

Figure 19.32

Estimate how far the dog travelled during this time.

Example 19.3

On the London Underground, Oxford Circus and Piccadilly Circus are 0.8 km apart. A train accelerates uniformly to a maximum speed when leaving Oxford Circus and maintains this speed for 90 s before decelerating uniformly to stop at Piccadilly Circus. The whole journey takes 2 minutes. Find the maximum speed.

Solution

The sketch of the speed–time graph of the journey shows the given information, with suitable units. The maximum speed is $v\,\text{ms}^{-1}$.

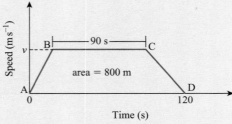

Figure 19.33

The area is $\dfrac{1}{2}(120 + 90) \times v = 800$

$$v = \frac{800}{105} = 7.619$$

The maximum speed of the train is $7.6\,\text{ms}^{-1}$ (to 2 s.f.).

Discussion point

→ Does it matter how long the train takes to speed up and slow down?

Exercise 19.4

 ① What is the difference between what is represented by the area under a speed–time graph and what is represented by the area under a velocity–time graph?

② The graph in Figure 19.34 shows the velocity–time graph for a journey.

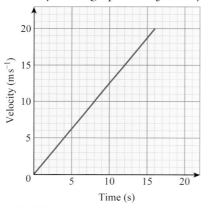

Figure 19.34

Which of the following is **not** true?

A The displacement is 160 m

B The acceleration is 1.25 m s^{-2}

C The distance travelled is 160 m

D The acceleration is 0.8 m s^{-2}

> **Note**
>
> All subsequent questions involve modelling.

 ③ Figure 19.35 shows the speeds of two cars travelling along a street.

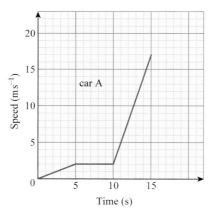

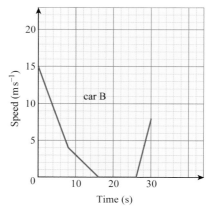

Figure 19.35

For each car find

(i) the acceleration for each part of its motion

(ii) the total distance it travels in the given time

(iii) its average speed.

④ The graph in Figure 19.36 shows the speed of a lorry when it enters a very busy road.

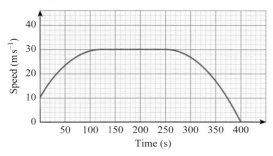

Figure 19.36

(i) Describe the journey over this time.

(ii) Use a ruler to make a tangent to the graph and hence estimate the acceleration at the beginning and end of the period.

(iii) Estimate the distance travelled and the average speed.

⑤ A train leaves a station where it has been at rest and picks up speed at a constant rate for 60 s. It then remains at a constant speed of $17\,\mathrm{m\,s^{-1}}$ for 60 s before it begins to slow down uniformly as it approaches a set of signals. After 45 s it is travelling at $10\,\mathrm{m\,s^{-1}}$ and the signal changes. The train again increases speed uniformly for 75 s until it reaches a speed of $20\,\mathrm{m\,s^{-1}}$. A second set of signals then orders the train to stop, which it does after slowing down uniformly for 30 s.

(i) Draw a speed–time graph for the train.

(ii) Use your graph to find the distance that it has travelled from the station.

⑥ When a parachutist jumps from a helicopter hovering above an airfield her speed increases at a constant rate to $28\,\mathrm{m\,s^{-1}}$ in the first three seconds of her fall. It then decreases uniformly to $8\,\mathrm{m\,s^{-1}}$ in a further 6 s, and then remains constant until she reaches the ground.

(i) Sketch a speed–time graph for the parachutist.

(ii) Find the height of the plane when the parachutist jumps out if the complete jump takes 1 minute.

⑦ A car is moving at $20\,\mathrm{m\,s^{-1}}$ when it begins to increase speed. Every 10 s it gains $5\,\mathrm{m\,s^{-1}}$ until it reaches its maximum speed of $50\,\mathrm{m\,s^{-1}}$ which it retains.

(i) Draw the speed–time graph for the car.

(ii) When does the car reach its maximum speed of $50\,\mathrm{m\,s^{-1}}$?

(iii) Find the distance travelled by the car after 150 s.

(iv) Write down expressions for the speed of the car t seconds after it begins to speed up.

⑧ A train takes 10 minutes to travel between two stations. The train accelerates at a rate of $0.5\,\mathrm{m\,s^{-2}}$ for 30 s. It then travels at a constant speed and is finally brought to rest in 15 s with a constant deceleration.

(i) Sketch a velocity–time graph for the journey.

(ii) Find the steady speed, the rate of deceleration and the distance between the two stations.

⑨ A car is travelling along a straight horizontal road. Initially the car has velocity $20\,\mathrm{m\,s^{-1}}$. It decelerates for 10 seconds at $1\,\mathrm{m\,s^{-2}}$, after which it travels at constant speed for 20 seconds and then accelerates uniformly for 20 seconds until it reaches its initial speed of $20\,\mathrm{m\,s^{-1}}$.

(i) Sketch the speed–time graph for the car.

(ii) Find the average speed for the car in the 50 seconds of its motion.

⑩ A car travels along a straight road. The car starts at A from rest and accelerates for 30 s at a constant rate until it reaches a speed of $25\,\mathrm{m\,s^{-1}}$. The car continues at $25\,\mathrm{m\,s^{-1}}$ for T s, after which it decelerates for 10 s until it reaches a speed of $15\,\text{m s}^{-1}$ as it passes B. The distance AB is 8 km.

(i) Sketch the velocity–time graph for the journey between A and B.

(ii) Find the total time of the journey from A to B.

PS ⑪ A train was scheduled to travel at $50\,\text{m s}^{-1}$ for 15 minutes on part of its journey. The velocity–time graph in Figure 19.37 illustrates the actual progress of the train which was forced to stop because of signals.

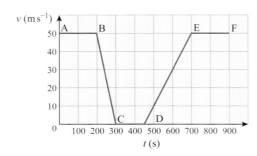

Figure 19.37

(i) Without carrying out any calculations, describe what was happening to the train in each of the stages BC, CD and DE.

(ii) Find the deceleration of the train while it was slowing down and the distance travelled during this stage.

(iii) Find the acceleration of the train when it starts off again and the distance travelled during this stage.

(iv) Calculate by how long the stop will have delayed the train.

(v) Sketch the distance–time graph for the journey between A and F, marking the points A, B, C, D, E and F.

PS ⑫ A train has a maximum allowed speed of $30\,\text{m s}^{-1}$. With its brakes fully applied, the train has a deceleration of $0.6\,\text{m s}^{-2}$. It can accelerate at a constant rate of $0.4\,\text{m s}^{-2}$. Find the shortest time in which it can travel from rest at one station to rest at the next station 12 km away.

PS ⑬ Dan is driving at a constant speed of $12.5\,\text{m s}^{-1}$ on a straight horizontal road, when he sees a red traffic light 50 m

away and, putting on the brakes, comes to rest at the light. He stops for 30 seconds and then resumes his journey, attaining his original speed of $12.5\,\text{m s}^{-1}$ in a distance of 75 m. Assuming that the acceleration and the deceleration are uniform, find how much time has been lost due to the stoppage.

PS ⑭ A car is travelling at $36\,\text{km h}^{-1}$ when the driver has to perform an emergency stop. During the time the driver takes to appreciate the situation and apply the brakes the car has travelled 7 m ('thinking distance'). It then pulls up with constant deceleration in a further 8 m ('braking distance') giving a total stopping distance of 15 m.

(i) Find the initial speed of the car in metres per second and the time that the driver takes to react.

(ii) Sketch the velocity–time graph for the car.

(iii) Calculate the deceleration once the car starts braking.

(iv) What is the stopping distance for a car travelling at $60\,\text{km h}^{-1}$ if the reaction time and the deceleration are the same as before?

PS ⑮ A particle P moves along the straight line AB (Figure 19.38). The midpoint of AB is X. The velocity–time graph for P is shown in Figure 19.39.

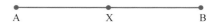

Figure 19.38

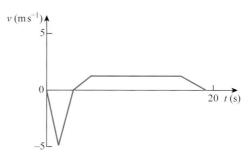

Figure 19.39

The particle starts at rest from a point X halfway between A and B and moves towards A. It comes to rest at A when $t = 3\,\text{s}$.

(i)   Find the distance AX.

After being instantaneously at rest when $t = 3$, the particle starts to move towards B. The particle takes 15 seconds to travel from A to B and comes to rest at B. For the first two seconds of this stage, P is accelerating at $0.6\,\mathrm{m\,s}^{-2}$, reaching a velocity V which it keeps for T seconds after which it decelerates to rest at B.

(ii)   Find V.

(iii)   Find T.

(iv)   Find the deceleration.

⑯   Cyclists are entered for a time trial event. They follow the same course, but start at different times, usually 1 minute apart and race against the clock. They are following a 10 km course.

Starting from rest, cyclist A accelerates to a maximum speed of $10\,\mathrm{m\,s}^{-1}$ which is then maintained till the end of the course. Cyclist B starts 1 minute later at the same point and accelerates uniformly to a maximum speed of $10.5\,\mathrm{m\,s}^{-1}$, which he subsequently maintains. The accelerations of the two cyclists are $0.1\,\mathrm{m\,s}^{-2}$ for A and $0.15\,\mathrm{m\,s}^{-2}$ for B and they are travelling along the same road towards the same destination, 10 km away from the start.

Use a speed–time graph to determine

(i)   the finishing times for both cyclists, and

(ii)   the time at which B overtakes A on the road and the distance from the starting point at that time.

⑰   The maximum acceleration for a lift is $1\,\mathrm{m\,s}^{-2}$ and the maximum deceleration is $4\,\mathrm{m\,s}^{-2}$. Use a speed–time graph to find the minimum time taken for the lift to go from ground level to the top of a skyscraper 40 m high:

(i)   if the maximum speed is $5\,\mathrm{m\,s}^{-1}$

(ii)   if there is no restriction on the speed of the lift.

5 The constant acceleration formulae

A particle is moving in a straight line with a constant acceleration a. It has an initial velocity u and a final velocity v after an interval of time t. Its displacement at time t is s. This is illustrated in the velocity–time graph shown in Figure 19.40.

The increase in velocity is $(v - u)\,\mathrm{m\,s}^{-1}$ and the constant acceleration $a\,\mathrm{m\,s}^{-2}$ is given by

$$a = \frac{v - u}{t}$$

So $v - u = at$

$$v = u + at \qquad \text{①}$$

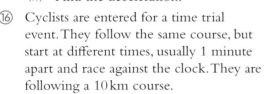

Figure 19.40

The area under the graph in Figure 19.41 represents the displacement s metres and is the area of the trapezium.

$$s = \frac{(u + v)}{2} \times t \qquad \text{②}$$

There are other useful formulae as well. For example you might want to find the displacement, s, without involving v in your calculations. This can be done by looking at the area under the velocity–time graph in a different way, using the rectangle R and the triangle T in Figure 19.42.

Figure 19.41

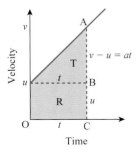

Figure 19.42

$AC = v$ and $BC = u$

So $AB = v - u = at$ from equation ①

total area = area of R + area of T

so $s = ut + \frac{1}{2} \times t \times at$

giving $s = ut + \frac{1}{2} at^2$ ③

Alternatively you could write the area as the difference between the larger rectangle of area $v \times t$ and the triangle giving rise to

$s = vt - \frac{1}{2} at^2$ ④

To find a formula which does not involve t, you need to eliminate t. One way to do this is first to rewrite equations ① and ② as

$$v - u = at \quad \text{and} \quad v + u = \frac{2s}{t}$$

and then multiplying these to give

$$(v - u)(v + u) = at \times \frac{2s}{t}$$
$$v^2 - u^2 = 2as$$
$$v^2 = u^2 + 2as \quad ⑤$$

Discussion point

→ Derive equation ③ algebraically by substituting for v from equation ① into equation ②.

Equations ① − ⑤ are sometimes called the **suvat** equations and can be used for any motion in a straight line with **constant acceleration**. There are five equations involving five variables (s, u, v, a and t). Each one involves four variables with the fifth one missing. So that equation ① has s missing, ② has a missing, ③ has v missing, ④ has u missing and ⑤ has t missing. In any problem involving linear motion with constant acceleration, if you know three of the variables, then you can obtain the other two.

Example 19.4

A bus leaving a bus stop accelerates at $0.8\,\text{m s}^{-2}$ for 5 s and then travels at a constant speed for 2 minutes before slowing down uniformly at $4\,\text{m s}^{-2}$ to come to rest at the next bus stop. Calculate

(i) the constant speed

(ii) the distance travelled while the bus is accelerating

(iii) the total distance travelled by the bus.

Solution

(i) The diagram shows the information for the first part of the motion.

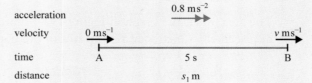

Figure 19.43

Let the constant speed be $v\,\text{m s}^{-1}$.

$a = 0.8, u = 0, t = 5$ and v is to be found

so $\qquad v = u + at \qquad$ is used

$$v = 0 + 0.8 \times 5 = 4$$

The constant speed is $4\,\mathrm{m\,s}^{-1}$.

(ii) Let the distance travelled be $s_1\,\mathrm{m}$.

$u = 0$, $a = 0.8$, $t = 5$ and s is to be found,

so $\qquad s = ut + \frac{1}{2}at^2 \qquad$ is used

| Use the suffix as there are three distances to be found in this question.

$$s_1 = 0 + \frac{1}{2} \times 0.8 \times 5^2$$
$$= 10$$

The bus accelerates over $10\,\mathrm{m}$.

> As $v = 4$ was found in part (i), any of the four remaining equations could have been used to obtain s_1:
>
> $s_1 = \frac{1}{2}(0 + 4) \times 5 = 10$ or $s_1 = 4 \times 5 - \frac{1}{2} \times 0.8 \times 5^2 = 10$ or $s_1 = \dfrac{4^2 - 0^2}{2 \times 0.8} = 10$.
>
> $s = ut + \frac{1}{2}at^2$ was used as it was the equation that only involved the given variables (a, u and t) and not the calculated v.

(iii) Figure 19.44 gives all the information for the rest of the journey.

Velocity decreases so acceleration is negative.

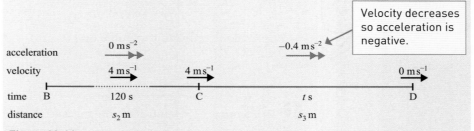

Figure 19.44

Between B and C the velocity is constant so the distance travelled is $4 \times 120 = 480\,\mathrm{m}$.

Let the distance between C and D be $s_3\,\mathrm{m}$.

$$u = 4, a = -0.4, v = 0$$

so the equation involving u, v, a and s is used.

$$v^2 = u^2 + 2as$$

$$0 = 16 + 2(-0.4)s_3$$

$$0.8s_3 = 16$$

$$s_3 = 20$$

Distance taken to slow down $= 20\,\mathrm{m}$

The total distance travelled is $(10 + 480 + 20)\,\mathrm{m} = 510\,\mathrm{m}$.

Units in the *suvat* equations

Constant acceleration usually takes place over short periods of time so it is best to use $\mathrm{m\,s}^{-2}$ for this. When you do not need to use a value for the acceleration you can, if you wish, use the *suvat* equations with other units provided they are consistent. This is shown in the next example.

Example 19.5

When leaving a town, a car accelerates from 30 mph to 60 mph in 5 s. Assuming the acceleration is constant, find the distance travelled in this time.

Solution

To make the units compatible, change 5 s to hours.

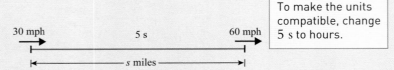

Figure 19.45

Changing 5 s into hours gives $\frac{5}{3600}$ hours.

Let the distance travelled be s miles. You want s and are given $u = 30$, $v = 60$ and $t = \frac{5}{3600}$.

So an equation involving u, v, t and s is used.

$$s = \frac{(u + v)}{2} \times t$$

$$s = \frac{(30 + 60)}{2} \times \frac{5}{3600} = \frac{1}{16}$$

The distance travelled is $\frac{1}{16}$ mile or 110 yards. (One mile is 1760 yards.)

Discussion point

→ Write all the measurements in Example 19.5 in terms of metres and seconds (using the conversion factor of 5 miles to every 8 km) and then find the resulting distance.

❗ In Examples 19.4 and 19.5, the bus and the car are always travelling in the positive direction so it is safe to use s for distance. Remember that s is not the same as the distance travelled if the direction changes during the motion.

The acceleration due to gravity

When a model ignoring air resistance is used, all objects falling under gravity fall with the same constant acceleration, $g\,\text{m s}^{-2}$. The value of g varies over the surface of the Earth. This is due to the fact that g is inversely proportional to the square of the distance (R) to the centre of the Earth. As the Earth is not a perfect sphere (bulging at the equator and flattening at the poles), this gives rise to larger values of R and consequently smaller values of g at the equator. For the same reasons g also depends on height above sea level giving rise to smaller values at altitude. g varies between $9.76\,\text{m s}^{-2}$ and $9.83\,\text{m s}^{-2}$. In this book it is assumed that all situations occur in a place where $g = 9.8\,\text{m s}^{-2}$ except in some cases where g is taken to be $10\,\text{m s}^{-2}$ as an approximation.

Example 19.6

A coin is dropped from rest at the top of a building of height 12.25 m and travels in a straight line with constant acceleration.

Find the time it takes to reach the ground and the speed of impact.

Solution

Suppose the time taken to reach the ground is t seconds.

Using S.I. units, $u = 0$, $a = 9.8$ and $s = 12.25$ when the coin hits the ground, so a formula involving u, a, s and t is required.

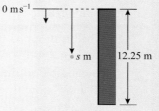

Figure 19.46

$$s = ut + \frac{1}{2}at^2$$

$$12.25 = 0 + \frac{1}{2} \times 9.8 \times t^2$$

$$t^2 = 2.5$$

$$t = 1.58 \text{ (to 3 s.f.)}$$

Down is positive so $a = +9.8$

To find the velocity, v, a formula involving s, u, a and t is required.

$$v^2 = u^2 + 2as$$

$$v^2 = 0 + 2 \times 9.8 \times 12.25$$

$$v^2 = 240.1$$

$$v = 15.5 \text{ (to 3 s.f.)}$$

The coin takes 1.58 s to hit the ground and has speed 15.5 m s^{-1} on impact.

Summary

The equations for motion with **constant acceleration** are

When using these equations make sure that the units you use are consistent. For example, when the time is t seconds and the distance s metres, any speed involved is in m s^{-1}.

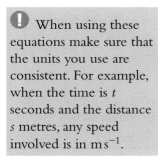

$$v = u + at \qquad ①$$

$$s = ut + \frac{1}{2}at^2 \qquad ③$$

$$v^2 = u^2 + 2as \qquad ⑤$$

$$s = \frac{(u+v)}{2} \times t \qquad ②$$

$$s = vt - \frac{1}{2}at^2 \qquad ④$$

Exercise 19.5

 ① You are given values for u, t and a. You need to work out s.
Which *suvat* equation should you use to do it in one step?
How did you decide?

 ② A particle accelerates uniformly 2 m s^{-1} to 10 m s^{-1} in 4 s.

Which equation should you use to work out the distance travelled in one step?

A $s = ut + \frac{1}{2}at^2$ B $s = vt - \frac{1}{2}at^2$

C $s = \frac{1}{2}(u + v)t$ D $v^2 = u^2 + 2as$

 ③ (i) Find v when $u = 10$, $a = 6$ and $t = 2$.

(ii) Find s when $v = 20$, $u = 4$ and $t = 10$.

(iii) Find s when $v = 10$, $a = 2$ and $t = 10$.

(iv) Find a when $v = 2$, $u = 12$ and $s = 7$.

④ Decide which equation to use in each of these situations.

(i) Given u, s, a, find v.

(ii) Given a, u, t, find v.

(iii) Given u, a, t, find s.

(iv) Given u, v, s, find t.

(v) Given u, s, v, find a.

(vi) Given u, s, t, find a.

(vii) Given u, a, v, find s.

(viii) Given a, s, t, find v.

> **Note**
>
> All subsequent questions, except Question 10, involve modelling.

⑤ A ball is dropped from a window, so when $t = 0$, $v = 0$. Air resistance is negligible so the acceleration is 9.8 ms^{-2} downwards. Find
  (i) its speed after 1 s and after 10 s
  (ii) how far it has fallen after 1 s and after 10 s
  (iii) how long it takes to fall 19.6 m.
  Which of these answers are likely to need adjusting to take account of air resistance? Would you expect your answer to be an over- or underestimate?

⑥ A car starting from rest at traffic lights reaches a speed of 90 km h^{-1} in 12 s. Find the acceleration of the car (in m s^{-2}) and the distance travelled. Write down any assumptions that you have made.

⑦ A top sprinter accelerates from rest to 9 m s^{-1} in 2 s. Calculate his acceleration, assumed constant, during this period and the distance travelled.

PS ⑧ A van skids to a halt from an initial speed of 24 m s^{-1} covering a distance of 36 m. Find the acceleration of the van (assumed constant) and the time it takes to stop.

PS ⑨ A car approaches a bend at a speed of 80 km h^{-1} and has to reduce its speed to 30 km h^{-1} in a distance of 250 m in order to take the bend. Find the required deceleration. After the bend is taken the car regains its original speed in 25 seconds. Find the distance it travels in doing so.

PS ⑩ A firework is projected vertically upwards at 20 m s^{-1}. For what length of time is it at least 5 m above the point of projection?

PS ⑪ An object moves in a straight line with acceleration -8 m s^{-2}. It starts its motion at the origin with velocity 16 m s^{-1}.
  (i) Write down equations for its position and velocity at time t s.
  (ii) Find the smallest non-zero time when
    (a) the velocity is zero
    (b) the object is at the origin.
  (iii) Sketch the position–time, velocity–time and speed–time graphs for $0 \leqslant t \leqslant 4$.

6 Further examples

The next two examples illustrate ways of dealing with more complex problems. In Example 19.7, none of the possible equations has only one unknown and there are also two situations, so simultaneous equations are used.

Example 19.7

Harvey practises using the stopwatch facility on his new watch by measuring the time between lamp posts on a car journey. As the car speeds up, two consecutive times are 1.2 s and 1 s.

Later he finds out that the lamp posts are 30 m apart.

(i) Calculate the acceleration of the car (assumed constant) and its speed at the first lamp post.

(ii) Assuming the same acceleration, find the time the car took to travel the 30 m before the first lamp post.

Solution

(i) Figure 19.47 shows all the information, assuming the acceleration is a m s^{-2} and the velocity at A is u m s^{-1}.

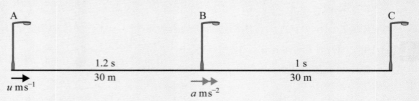

Figure 19.47

For AB, $s = 30$ and $t = 1.2$. You are looking for u and a so you use

$$s = ut + \frac{1}{2}at^2$$

$$30 = 1.2u + \frac{1}{2}a \times 1.2^2$$

$$30 = 1.2u + 0.72a \qquad \textcircled{1}$$

To use the same equation for the part BC you would need the velocity at B and this brings in another unknown. It is much better to go back to the beginning and consider the whole of AC with $s = 60$ and $t = 2.2$. Then using again

$$s = ut + \frac{1}{2}at^2$$

$$60 = 2.2u + \frac{1}{2}a \times 2.2^2$$

$$60 = 2.2u + 2.42a \qquad \textcircled{2}$$

These two simultaneous equations in two unknowns can be solved more easily if they are simplified. First make the coefficients of u integers.

$$\textcircled{1} \quad \times 10 \div 12 \qquad 25 = u + 0.6a \qquad \textcircled{3}$$

$$\textcircled{2} \quad \times 5 \qquad 300 = 11u + 12.1a \qquad \textcircled{4}$$

then $\quad \textcircled{3} \quad \times 11 \qquad 275 = 11u + 6.6a \qquad \textcircled{5}$

Subtracting $\textcircled{5}$ from $\textcircled{4}$ gives

$$25 = 0 + 5.5a$$

$$a = 4.545$$

Now substitute 4.545 for a in $\textcircled{3}$ to find

$$u = 25 - 0.6 \times 4.545 = 22.273$$

The acceleration of the car is $4.55\,\mathrm{m\,s^{-2}}$ and the initial speed is $22.3\,\mathrm{m\,s^{-1}}$ (to 3 s.f.).

Discussion point

➜ Calculate u when $t = 8.19$, $v = 22.3$ and $a = 4.55$. Is $t = 8.19$ a possible answer in Example 19.7 (ii)?

(ii)

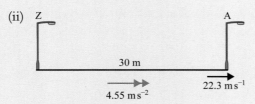

Figure 19.48

For this part you know that $s = 30$, $v = 22.3$ and $a = 4.55$ and you want t so using

$$s = vt - \frac{1}{2}at^2$$

$$30 = 22.3 \times t - \frac{1}{2} \times 4.55 \times t^2$$

$$\Rightarrow 2.275t^2 - 22.3 \times t + 30 = 0$$

Solving this using the quadratic formula gives $t = 1.61$ and $t = 8.19$.

The most sensible answer to this particular problem is $1.61\,\mathrm{s}$.

Using a non-zero initial displacement

What, in the constant acceleration formulae, are v and s when $t = 0$?

Putting $t = 0$ in the *suvat* formulae gives the **initial values**, u for the velocity and $s = 0$ for the position.

Sometimes, however, it is convenient to use an origin which gives a non-zero value for s when $t = 0$.

For example, when you model the motion of an eraser thrown vertically upwards you might decide to find its height above the ground rather than above the point from which it was thrown.

What is the effect on the various *suvat* formulae if the initial position is s_0 rather than 0?

If the height of the eraser above the ground is s at time t and s_0 when $t = 0$ the displacement over time t is $s - s_0$. You then need to replace equation ③ with

$$s - s_0 = ut + \frac{1}{2}at^2$$

Example 19.8

A stone is projected vertically upwards from a point O, with initial speed $8\,\text{m s}^{-1}$. At the same instant another stone is dropped from a point 5 m vertically above O. Find the height above O at which the stones collide.

Solution

The displacements of the stones above O are $s_1\,\text{m}$ and $s_2\,\text{m}$ at time t. The initial positions of the stones are 0 m and 5 m and the initial velocities are $8\,\text{m s}^{-1}$ and $0\,\text{m s}^{-1}$.

For each stone, you know u and a and you want to involve s and t so you use

$$s = s_0 + ut + \frac{1}{2}at^2.$$

For the first stone:

$$s_1 = 0 + 8 \times t + \frac{1}{2} \times (-9.8) \times t^2$$

$$s_1 = 8t - 4.9t^2 \qquad\qquad ①$$

For the second stone:

$$s_2 = 5 + 0 \times t + \frac{1}{2}(-9.8)t^2$$

$$s_2 = 5 - 4.9t^2 \qquad\qquad ②$$

When the stones collide, $s_1 = s_2$, so that

$$8t - 4.9t^2 = 5 - 4.9t^2$$

$$8t = 5$$

$$t = 0.625$$

Substituting $t = 0.625$ in ① or ② gives

$$s_1 = s_2 = 5 - 4.9 \times 0.625^2 = 3.0859$$

The stones collide after 0.625 s, at a height of 3.09 m above O (to 3.s.f.).

> **Note**
> ------------------------------
> All questions involve modelling.

Use g $= 9.8\,m\,s^{-2}$ *in this exercise unless otherwise specified.*

 ① A car accelerates from rest and travels 50 m in T s. It continues to accelerate at the same rate and travels the next section in the same time finishing with a speed of 40 m s^{-1}.

What is that time and what is the acceleration?

 ② A car accelerates, with acceleration a, from 5 m s^{-1} to 10 m s^{-1} in time T, covering a distance x. Which of these equations does **not** describe the motion?

A $\;x = 5T + \frac{1}{2}aT^2$ B $\;x = 7.5T$

C $\;75 = 2ax$ D $\;15 = aT$

 ③ A car is travelling along a straight road. It accelerates uniformly from rest to a speed of $15\,m\,s^{-1}$ and maintains this speed for 10 minutes. It then decelerates uniformly to rest. The acceleration is $5\,m\,s^{-2}$ and the deceleration is $8\,m\,s^{-2}$. Find the total journey time and the total distance travelled during the journey.

 ④ A skier pushes off at the top of a slope with an initial speed of $2\,m\,s^{-1}$. She gains speed at a constant rate throughout her run. After 10s she is moving at $6\,m\,s^{-1}$.

(i) Find an expression for her speed t seconds after she pushes off.

(ii) Find an expression for the distance she has travelled at time t seconds.

(iii) The length of the ski slope is 400 m. What is her speed at the bottom of the slope?

PS ⑤ Towards the end of a half-marathon Sabina is 100 m from the finish line and is running at a constant speed of $5\,m\,s^{-1}$. Dan, who is 140 m from the finish and is running at $4\,m\,s^{-1}$, decides to accelerate to try to beat Sabina. If he accelerates uniformly at $0.25\,m\,s^{-2}$ does he succeed?

PS ⑥ Rupal throws a ball upwards at $2\,m\,s^{-1}$ from a window which is 4 m above ground level.

(i) Write down an equation for the height h m of the ball above the ground after t s (while it is still in the air).

(ii) Use your answer to part (i) to find the time the ball hits the ground.

(iii) How fast is the ball moving just before it hits the ground?

(iv) In what way would you expect your answers to parts (ii) and (iii) to change if you were able to take air resistance into account?

PS ⑦ Nathan hits a tennis ball straight up into the air from a height of 1.25 m above the ground. The ball hits the ground after 2.5 seconds. Air resistance is neglected.

(i) Find the speed with which Nathan hits the ball.

(ii) Find the greatest height above ground reached by the ball.

(iii) Find the speed of the ball when it hits the ground.

(iv) The ball loses 20% of its speed on hitting the ground. How high does it bounce?

(v) Is your answer to part (i) likely to be an over- or underestimate given that you have ignored air resistance?

PS ⑧ A ball is dropped from a building of height 30 m and at the same instant a stone is thrown vertically upwards from the ground so that it hits the ball. In modelling the motion of the ball and stone it is assumed that each object moves in a straight line with a constant downward acceleration of magnitude $9.8\,m\,s^{-2}$. The stone is thrown with initial speed of $15\,m\,s^{-1}$ and is h_s metres above the ground t seconds later.

(i) Draw a diagram of the ball and stone before they collide, marking their positions.

(ii) Write down an expression for h_s at time t.

(iii) Write down an expression for h_b of the ball at time t.

(iv) When do the ball and stone collide?

(v) How high above the ground do the ball and stone collide?

⑨ When Kim rows her boat, the two oars are both in the water for 3s and then both out of the water for 2s. This 5s cycle is then repeated. When the oars are in the water the boat accelerates at a constant $1.8\,\text{ms}^{-2}$ and when they are not in the water it decelerates at a constant $2.2\,\text{ms}^{-2}$.

  (i) Find the change in speed that takes place in each 3s period of acceleration.

  (ii) Find the change in speed that takes place in each 2s period of deceleration.

  (iii) Calculate the change in the boat's speed for each 5s cycle.

  (iv) A race takes Kim 45s to complete. If she starts from rest what is her speed as she crosses the finishing line?

  (v) Discuss whether this is a realistic speed for a rowing boat.

PS ⑩ A ball is dropped from a tall building. The distance between floors in the block is constant. The ball takes 0.5s to fall from the 14th to the 13th floor and 0.3s to fall from the 13th floor to the 12th. What is the distance between floors?

PS ⑪ Two clay pigeons are launched vertically upwards from exactly the same spot at 1s intervals. Each clay pigeon has initial speed $30\,\text{ms}^{-1}$ and acceleration $10\,\text{ms}^{-2}$ downwards. How high above the ground do they collide?

PS ⑫ A train accelerates along a straight, horizontal section of track. The driver notes that he reaches a bridge 120m from the station in 8s and that he crosses the bridge, which is 31.5m long, in a further 2s.

The motion of the train is modelled by assuming constant acceleration. Take

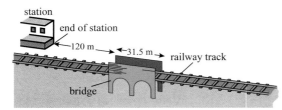

Figure 19.49

PS the speed of the train when leaving the station to be $u\,\text{ms}^{-1}$ and the acceleration to have the value $a\,\text{ms}^{-2}$.

  (i) By considering the part of the journey from the station to the bridge, show that $u + 4a = 15$.

  (ii) Find a second equation involving u and a.

  (iii) Solve the two equations for u and a to show that a is 0.15 and find the value of u.

  (iv) If the driver also notes that he travels 167m in the 10s after he crosses the bridge, have you any evidence to reject the modelling assumption that the acceleration is constant?

PS ⑬ A car travelling with constant acceleration is timed to take 15s over 200m and 10s over the next 200m. What is the speed of the car at the end of the observed motion?

PS ⑭ A train is decelerating uniformly. It passes three posts spaced at intervals of 100m. It takes 8 seconds between the first and second posts and 12 seconds between the second and third. Find the deceleration and the distance between the third post and the point where the train comes to rest.

PS ⑮ A stone is thrown vertically upwards with a velocity of $35\,\text{ms}^{-1}$. Find the distance it travels in the fourth second of its motion.

⑯ A man sees a bus, accelerating uniformly from rest at a bus stop, 50m away. He then runs at constant speed and just catches the bus in 30s. Determine the speed of the man and the acceleration of the bus. If the man's speed is $3\,\text{ms}^{-1}$, how close to the bus can he get?

PS ⑰ In order to determine the depth of a well, a stone is dropped into the well and the time taken for the stone to drop is measured. It is found that the sound of the stone hitting the water arrives 5 seconds after the stone is dropped. If the speed of sound is taken as $340\,\text{ms}^{-1}$ find the depth of the well.

LEARNING OUTCOMES

When you have completed this chapter you should be able to:

➤ understand and use fundamental quantities and units in the S.I. system:
- ○ length
- ○ time

➤ understand and use derived quantities and units:
- ○ velocity
- ○ acceleration

➤ understand and use the language of kinematics:
- ○ position
- ○ displacement
- ○ distance travelled
- ○ velocity
- ○ speed
- ○ acceleration

➤ understand, use and interpret graphs in kinematics for motion in a straight line:
- ○ displacement against time and interpretation of gradient
- ○ velocity against time and interpretation of gradient and area under the graph

➤ understand, use and derive the formulae for constant acceleration for motion in a straight line.

KEY POINTS

1 **Vectors** (with magnitude and direction)
Displacement
Position (displacement from a fixed origin)

Velocity (rate of change of position)
Acceleration (rate of change of velocity)

Scalars (magnitude only)
Distance
Speed (magnitude of velocity)
Time

Vertical is towards the centre of the Earth; **horizontal** is perpendicular to vertical.

2 Diagrams
- ■ Motion along a line can be illustrated vertically or horizontally (as shown).

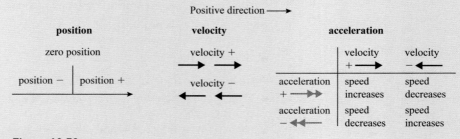

Figure 19.50

- Average speed = $\dfrac{\text{total distance travelled}}{\text{total time taken}}$

- Average velocity = $\dfrac{\text{displacement}}{\text{time taken}}$

- Average acceleration = $\dfrac{\text{change in velocity}}{\text{time taken}}$

3 Graphs
  ■ Position–time

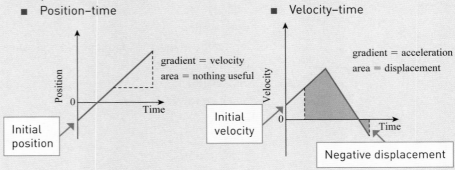

  ■ Velocity–time

  ■ Distance–time

  ■ Speed–time

Figure 19.51

4 The *suvat* equations
  ■ The equations for motion with constant acceleration are

  ① $v = u + at$

  ② $s = \dfrac{(u + v)}{2} \times t$

  ③ $s = ut + \dfrac{1}{2}at^2$

  ④ $v^2 = u^2 + 2as$

  ⑤ $s = vt - \dfrac{1}{2}at^2$

  ■ a is the constant acceleration
     s is the displacement from the starting position at time t
     v is the velocity at time t
     u is the velocity when $t = 0$.
  ■ If $s = s_0$ when $t = 0$, replace s in each formula with $(s - s_0)$.

5 Vertical motion under gravity
  ■ The acceleration due to gravity ($g\,\text{m s}^{-2}$) is $9.8\,\text{m s}^{-2}$ vertically downwards.
  ■ Always draw a diagram and decide in advance where your origin is and which way is positive.
  ■ Make sure that your units are compatible.

6 Using a mathematical model
- Make simplifying assumptions by deciding what is most relevant.
  For example: a car is a **particle** with no dimensions
  a road is a **straight line** with one dimension
  acceleration is constant.
- Define variables and set up equations.
- Solve the equations.
- Check that the answer is sensible. If not, think again.

FUTURE USES

- The work in this chapter is developed further in the next two chapters of this book, Chapter 20 and Chapter 21, where you meet Newton's laws of motion and variable acceleration.

20 Forces and Newton's laws of motion

The photo shows crates of supplies being dropped into a remote area by parachute. What forces are acting on a crate of supplies and the parachute?

One force which acts on every object near the Earth's surface is its own **weight**. This is the force of gravity pulling it towards the centre of the Earth. The weight of the crate acts on the crate and the weight of the parachute acts on the parachute.

1 Force diagrams

The parachute is designed to make use of **air resistance**. A resistance is present whenever a solid object moves through a liquid or gas. It acts in the opposite direction to the motion and it depends on the speed of the object. The crate also experiences air resistance, but to a lesser extent than the parachute.

Other forces are the **tensions** in the guy lines attaching the crate to the parachute. These pull upwards on the crate and downwards on the parachute.

All these forces can be shown most clearly if you draw **force diagrams** for the crate and the parachute (Figure 20.1).

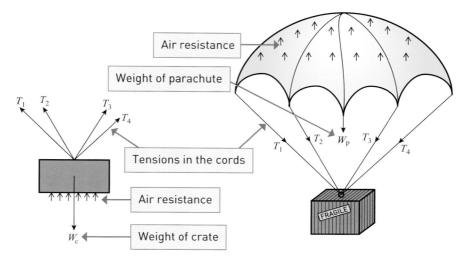

Figure 20.1 Forces acting on the crate and forces acting on the parachute

Force diagrams are essential for the understanding of most mechanical situations.

A force is a vector: it has a magnitude, or size and a direction. It also has a line of action. This line often passes through a point of particular interest. Any force diagram should show clearly

- the direction of the force

- the magnitude of the force

- the line of action.

In Figure 20.1 each force is shown by an arrow along its line of action. The air resistance has been depicted by a lot of separate arrows but this is not very satisfactory. It is much better if the combined effect can be shown by one arrow.

When you have learned more about vectors, you will see how the tensions in the guy lines can also be combined into one force if you wish. The forces on the crate and parachute can then be simplified.

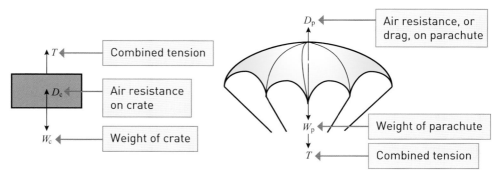

Figure 20.2 Forces acting on the crate and forces acting on the parachute

Centre of mass and the particle model

When you combine forces you are finding their **resultant**. The weights of the crate and parachute are also found by combining forces; they are the resultant of the weights of all their separate parts. Each weight acts through a point called the **centre of mass** or centre of gravity.

Think about balancing a pen on your finger. Figures 20.3 and 20.4 show the forces acting on the pen.

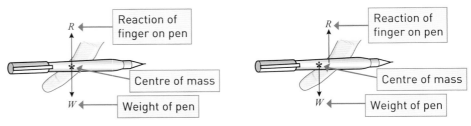

Figure 20.3   Figure 20.4

So long as you place your finger under the centre of mass of the pen, as in Figure 20.3, it will balance. There is a force called a **reaction** between your finger and the pen which balances the weight of the pen. The forces on the pen are then said to be in **equilibrium**. If you place your finger under another point, as in Figure 20.4, the pen will fall. The pen can only be in equilibrium if the two forces have the same line of action.

If you balance the pen on two fingers, there is a reaction between each finger and the pen at the point where it touches the pen. These reactions can be combined into one resultant vertical reaction acting through the centre of mass.

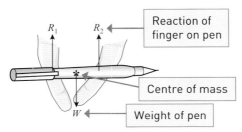

Figure 20.5

The behaviour of objects which are liable to rotate under the action of forces is covered in another chapter. Here you will only deal with situations where the resultant of the forces does not cause rotation. An object can then be modelled as a particle, that is a point mass, situated at its centre of mass.

Newton's third law of motion

Sir Isaac Newton (1642–1727) is famous for his work on gravity and the mechanics you learn in this course is often called Newtonian mechanics because it is based entirely on Newton's three laws of motion. These laws provide us with an extremely powerful model of how objects, ranging in size from specks of dust to planets and stars, behave when they are influenced by forces.

We start with Newton's **third law** which says that:

> When one object exerts a force on another there is always a reaction of the same kind which is equal, and opposite in direction, to the acting force.

You might have noticed that the combined tensions acting on the parachute and the crate in Figure 20.2 are both marked with the same letter, T. The crate applies a force on the parachute through the supporting guy lines and the parachute applies an equal and opposite force on the crate. When you apply a force to a chair by sitting on it, it responds with an equal and opposite force on you. Figure 20.6 shows the forces acting when someone sits on a chair.

Discussion point

→ Why is the weight of the person not shown on the force diagram of forces acting on the chair in Figure 20.6?

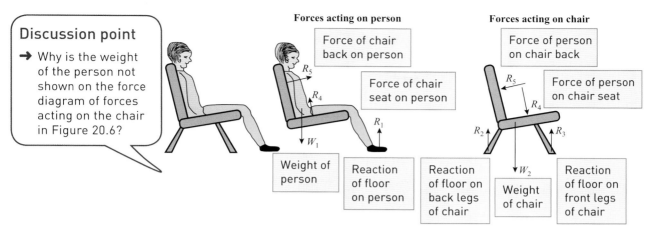

Forces acting on person

- Force of chair back on person
- Force of chair seat on person
- R_5
- R_4
- R_1
- W_1
- Weight of person
- Reaction of floor on person

Forces acting on chair

- Force of person on chair back
- Force of person on chair seat
- R_5
- R_4
- R_2
- R_3
- W_2
- Reaction of floor on back legs of chair
- Weight of chair
- Reaction of floor on front legs of chair

Figure 20.6

The reactions of the floor on the chair and on your feet act where there is contact with the floor. You can use R_1, R_2 and so on to show that they have different magnitudes. There are equal and opposite forces acting on the floor, but the forces on the floor are not being considered and so do not appear here.

Gravitational forces obey Newton's third law just as other forces between bodies. According to Newton's universal law of gravitation, the Earth pulls us towards its centre and we pull the Earth in the opposite direction. However, in this book we are only concerned with the gravitational force on us and not the force we exert on the earth.

All the forces you meet in Mechanics apart from the gravitational force are the result of physical contact. This might be between two solids or between a solid and a liquid or gas.

Friction and normal reaction

When you push your hand along a table, the table reacts in two ways.

- Firstly there are forces which stop your hand going through the table. Such forces are always present when there is any contact between your hand and the table. They are at right angles to the surface of the table and their resultant is called the **normal reaction** between your hand and the table.

- There is also another force which tends to prevent your hand from sliding. This is **friction** and it acts in a direction which opposes the sliding.

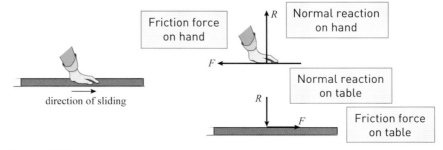

- Friction force on hand
- direction of sliding
- R
- Normal reaction on hand
- F
- Normal reaction on table
- R
- F
- Friction force on table

Figure 20.7

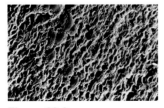

Figure 20.8

Figure 20.7 shows the reaction forces acting on your hand and on the table. By Newton's third law they are equal and opposite to each other. The frictional force is due to tiny bumps on the two surfaces (see Figure 20.8). When you hold your hands together you will feel the normal reaction between them. When you slide them against each other you will feel the friction.

When the friction between two surfaces is negligible, at least one of the surfaces is said to be **smooth**. This is a modelling assumption which you will meet frequently in this book. Oil can make surfaces smooth and ice is often modelled as a smooth surface.

When the contact between two surfaces is smooth, the only force between them is at right angles to any possible sliding and is just the normal reaction.

Discussion point

➜ What direction is the reaction between the sweeper's broom and the smooth ice?

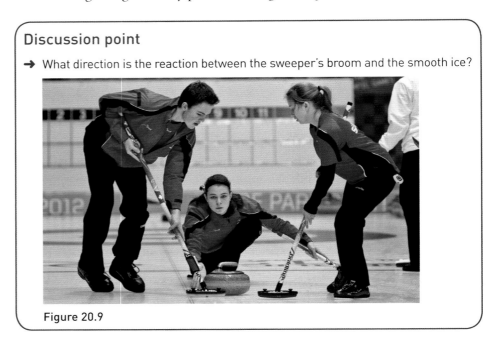

Figure 20.9

Example 20.1

A TV set is standing on a small table. Draw a diagram to show the forces acting on the TV and on the table as seen from the front.

Solution

Figure 20.10 shows the forces acting on the TV and on the table. They are all vertical because the weights are vertical and there are no horizontal forces acting.

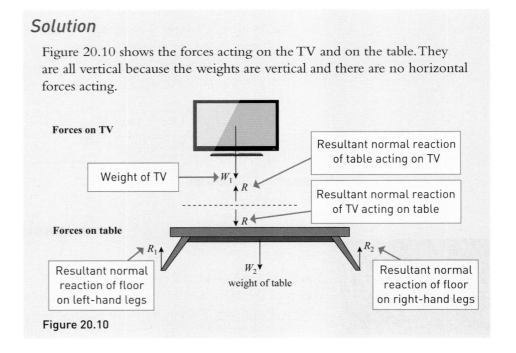

Forces on TV

Weight of TV → W_1

Resultant normal reaction of table acting on TV

R

Resultant normal reaction of TV acting on table

R

Forces on table

R_1

Resultant normal reaction of floor on left-hand legs

W_2
weight of table

R_2

Resultant normal reaction of floor on right-hand legs

Figure 20.10

Example 20.2

Draw diagrams to show the forces acting on a tennis ball which is hit horizontally across the court

(i)   at the instant it is hit by the racket

(ii)   as it crosses the net

(iii) at the instant it lands on the other side.

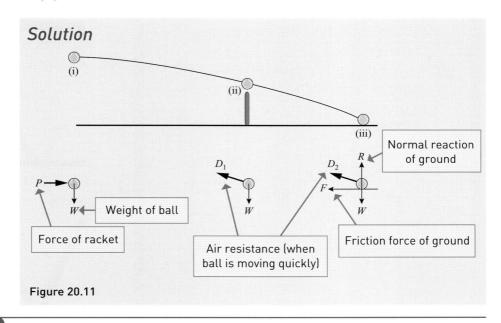

Solution

Figure 20.11

Exercise 20.1

In this exercise draw clear diagrams to show the forces acting on the objects named in italics. Clarity is more important than realism when drawing these diagrams.

Note

All questions involve modelling.

① Give three everyday examples of forces and their equal and opposite reactions.

② A *box* is placed at rest on a horizontal table. Which of the following diagrams shows the forces acting on the box?

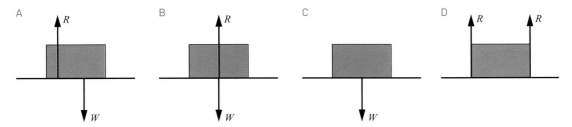

Figure 20.12

③ A *gymnast* hanging at rest on a bar.

④ A *light bulb* hanging from a ceiling.

⑤ A *book* lying at rest on a table.

⑥ A *book* at rest on a table but being pushed by a small horizontal force.

⑦ *Two books* lying on a table, one on top of the other.

⑧ A *horizontal plank* being used to bridge a stream (Figure 20.13).

Figure 20.13

⑨ A *snooker ball* on a table which can be assumed to be smooth
  (i) as it lies at rest on the table
  (ii) at the instant it is hit by the cue.

⑩ An *ice hockey puck*
  (i) at the instant it is hit when standing on smooth ice
  (ii) at the instant it is hit when standing on rough ice.

⑪ A *cricket ball* which follows the path shown in Figure 20.14.

Draw diagrams for each of the three positions A, B and C (include air resistance).

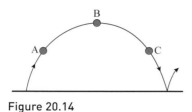

Figure 20.14

⑫ (i) *Two balls* colliding in mid-air.
  (ii) *Two balls* colliding on a snooker table.

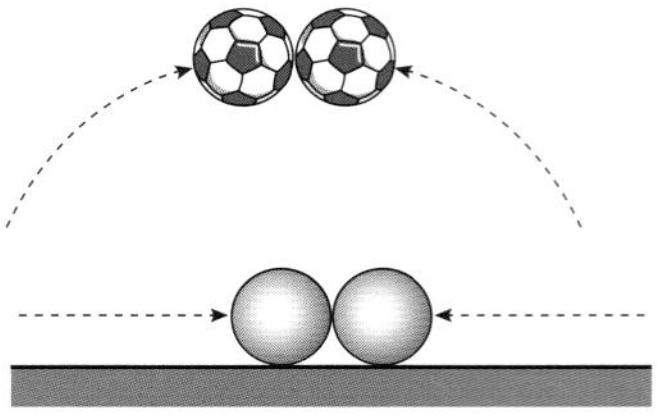

Figure 20.15

⑬ A *paving stone* leaning against a wall.

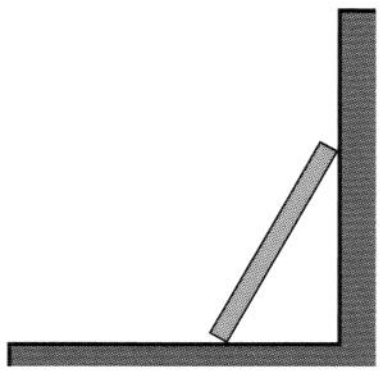

Figure 20.16

⑭ A *cylinder* at rest on smooth surfaces.

Figure 20.17

2 Force and motion
Newton's first law

Newton's **first law** can be stated as follows:

> Every particle continues in a state of rest or of uniform motion in a straight line unless acted on by a resultant external force.

Newton's first law provides a reason for the handles on trains and buses. When you are on a train which is stationary or moving at constant speed in a straight line you can easily stand without support. But when the velocity of the train changes, a force is required to change your velocity to match. This happens when the train slows down or speeds up. It also happens when the train goes round a bend even if the speed does not change. The velocity changes because the direction changes.

Discussion point

→ How are the rails and handles provided in buses and trains used by standing passengers?

Figure 20.18

Discussion point

→ Why is Josh's car in the pond?

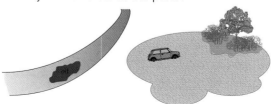

Figure 20.19

Example 20.3

A coin is balanced on your finger and then you move it upwards.

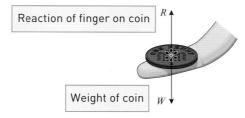

Reaction of finger on coin | R

Weight of coin | W

Figure 20.20

By considering Newton's first law, what can you say about W and R in each of these situations?

(i)   The coin is stationary.

(ii)   The coin is moving upwards with a constant velocity.

(iii) The speed of the coin is increasing as it moves upwards.

(iv) The speed of the coin is decreasing as it moves upwards.

Solution

(i)   When the coin is stationary the velocity does not change. The forces are in equilibrium and $R = W$.

(ii)   When the coin is moving upwards with a constant velocity, the velocity does not change. The forces are in equilibrium and $R = W$.

(iii) When the speed of the coin is increasing as it moves upwards there must be a net upward force to make the velocity increase in the upward direction so $R > W$. The net force is $R - W$.

(iv) When the speed of the coin is decreasing as it moves upwards there must be a net downward force to make the velocity decrease and slow the coin down as it moves upwards. In this case $W > R$ and the net force is $W - R$.

Figure 20.21 shows each of these situations.

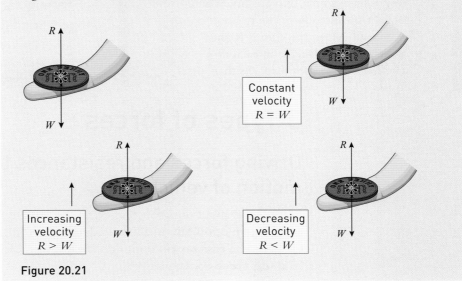

R

W

Constant
velocity
$R = W$

R

W

Increasing
velocity
$R > W$

R

W

Decreasing
velocity
$R < W$

R

W

Figure 20.21

> ### Note
> All questions involve modelling.

 ① An object always moves in the direction of the resultant force on it. True or false?

 ② The diagram shows a box. It is subject to an applied force T and its weight W.

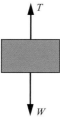

Figure 20.22

The box is accelerating upwards. Which of the following is true?

A $W = T$ B $T < W$

C $W < T$ D $W \leqslant T$

③ A book is resting on an otherwise empty table.

(i) Draw diagrams showing the forces acting on
  (a) the book
  (b) the table as seen from the side.

(ii) Write down equations connecting the forces acting on the book and on the table.

④ You balance a coin on your finger and move it up and down. The reaction of your finger on the coin is R and its weight is W. Decide in each case whether R is greater than, less than, or equal to W and describe the net force.

(i) The coin is moving downwards with a constant velocity.

(ii) The speed of the coin is increasing as it moves downwards.

(iii) The speed of the coin is decreasing as it moves downwards.

⑤ In each of the following situations say whether the forces acting on the object are in equilibrium by deciding whether its motion is changing.

(i) A car that has been stationary, as it moves away from a set of traffic lights.

(ii) A motorbike as it travels at a steady $60\,\mathrm{km\,h^{-1}}$ along a straight road.

(iii) A parachutist descending at a constant rate.

(iv) A box in the back of a lorry as the lorry picks up speed along a straight, level motorway.

(v) An ice hockey puck sliding across a smooth ice rink.

(vi) A book resting on a table.

(vii) An aeroplane flying at a constant speed in a straight line, but losing height at a constant rate.

(viii) A car going round a corner at constant speed.

⑥ Explain each of the following in the terms of Newton's laws.

(i) Seat belts should be worn in cars.

(ii) Head rests are necessary in a car to prevent neck injuries when there is a collision from the rear.

3 Types of forces

Driving forces and resistances to the motion of vehicles

In problems about things such as cycles, cars and trains, all the forces acting along the line of motion can usually be reduced to two or three: the **driving force** forwards, the **resistance** to motion (air resistance, friction, etc.) and possibly a **braking force** backwards.

Resistances due to air or water act in a direction opposite to the velocity of a vehicle or boat and are usually more significant for fast-moving objects.

Tension and thrust

The lines joining the crate of supplies to the parachute described at the beginning of this chapter are in tension. They pull upwards on the crate and downwards on the parachute.

You are familiar with tensions in ropes and strings, but rigid objects can also be in tension.

When you hold the ends of a pencil, one with each hand, and pull your hands apart, you are pulling on the pencil. What is the pencil doing to each of your hands? Draw the forces acting on your hands and on the pencil.

Now draw the forces acting on your hands and on the pencil when you push the pencil inwards.

Your first diagram might look like Figure 20.23(a). The pencil is in tension so there is an inward **tension** force on each hand.

Discussion point

➜ Which of the diagrams here is still possible if the pencil is replaced by a piece of string?

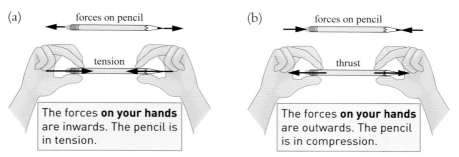

Figure 20.23

When you push the pencil inwards the forces on your hands are outwards as in Figure 20.23(b). The pencil is said to be in **compression** and the outward force on each hand is called a **thrust**.

If each hand applies a force of 2 units on the pencil, the tension or thrust acting on each hand is also 2 units because each hand is in equilibrium.

Resultant forces and equilibrium

You have already met the idea that a single force can have the same effect as several forces acting together. Imagine that several people are pushing a car. A single rope pulled by another car can have the same effect. The force of the rope is equivalent to the resultant of the forces of the people pushing the car. When there is no resultant force, the forces are in equilibrium and there is no change in motion.

Example 20.4

A car is using a towbar to pull a trailer along a straight, level road. There are resisting forces D acting on the car and S acting on the trailer. The driving force of the car is P and its braking force is B.

Draw diagrams showing the horizontal forces acting on the car and trailer

(i)   when the car is moving at constant speed

(ii) when the speed of the car is increasing

(iii) when the car brakes and slows down rapidly.

In each case write down the resultant force acting on the car and on the trailer.

Solution

(i) When the car moves at constant speed, the forces are as shown in Figure 20.24. The towbar is in tension and the effect is a forward force on the trailer and an equal and opposite backward force on the car.

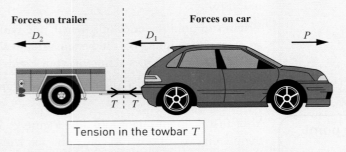

Forces on trailer   Forces on car

Tension in the towbar T

Figure 20.24 Car travelling at constant speed

There is no resultant force on either the car or the trailer when the speed is constant; the forces on each are in equilibrium.

For the trailer: $T - D_2 = 0$.

For the car: $P - D_1 - T = 0$.

(ii) When the car speeds up, the same diagram will do, but now the magnitudes of the forces are different. There is a resultant forward force on both the car and the trailer.

For the trailer: resultant $= T - D_2$ $\rightarrow$

For the car: resultant $= P - D_1 - T$ $\rightarrow$

(iii) When the car brakes a resultant **backward** force is required to slow down the trailer. When the resistance is not sufficiently large to do this, a thrust in the towbar comes into play as shown in Figure 20.25.

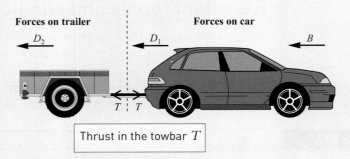

Forces on trailer   Forces on car

Thrust in the towbar T

Figure 20.25 Car braking

For the trailer: resultant $= T + D_2$ $\leftarrow$

For the car: resultant $= B + D_1 - T$ $\leftarrow$

Newton's second law

Newton's **second law** gives us more information about the relationship between the magnitude of the resultant force and the change in motion. Newton said that:

> The change in motion is proportional to the force.

For objects with constant mass, this can be interpreted as *the force is proportional to the acceleration*.

Resultant force = a constant × acceleration   ①

The constant in this equation is proportional to the mass of the object: a more massive object needs a larger force to produce the same acceleration. For example, you and your friends would be able to give a car a greater acceleration than you would be able to give a lorry.

Newton's second law is so important that a special unit of force, the **newton** (**N**), has been defined so that the constant in equation ① is actually equal to the mass. A force of 1 newton will give a mass of 1 kilogram an acceleration of $1\,\mathrm{m\,s^{-2}}$. The equation then becomes:

Resultant force = mass × acceleration   ②

This is written:

$$F = ma$$

The resultant force and the acceleration are always in the same direction.

Relating mass and weight

The **mass** of an object is related to the amount of matter in the object. It is a **scalar**.

The **weight** of an object is a force. It has magnitude and direction and so is a **vector**.

The mass of an astronaut on the Moon is the same as their mass on Earth but their weight is only about one-sixth of their weight on Earth. This is why astronauts can move around more easily on the Moon. The gravitational force on the Moon is less because the mass of the Moon is less than that of the Earth.

When Buzz Aldrin made the first landing on the Moon in 1969 with Neil Armstrong, one of the first things he did was to drop a feather and a hammer to demonstrate that they fell at the same rate. There is no air on the Moon so there is no air resistance, so that the accelerations on the two objects due to the gravitational force of the Moon were equal, even though they had very different masses. If other forces were negligible on Earth all objects would fall with an acceleration g.

When the weight is the only force acting on an object, Newton's second law means that

Weight in newtons = mass in kg × g in $\mathrm{m\,s^{-2}}$

Using standard letters:

$$W = mg$$

Even when there are other forces acting, the weight is still mg.

A good way to visualise a force of 1 N is to think of the weight of an apple. 1 kg of apples weighs approximately $(1 \times 10)\,\mathrm{N} = 10\,\mathrm{N}$. There are about 10 small to medium sized apples in 1 kg, so each apple weighs about 1 N.

Figure 20.26

Notice that the S.I. unit of mass is the kilogram and the S.I. unit of weight is the newton. Anyone who says 1 kg of apples weighs 1 kg is not strictly correct. The terms weight and mass are often confused in everyday language but it is very important for your study of mechanics that you should understand the difference.

Example 20.5

What is the weight of

(i)   a baby of mass 3 kg

(ii) a golf ball of mass 46 g?

Solution

(i)   The baby's weight is $3 \times 9.8 = 29.4$ N.

(ii) Mass of golf ball = 46 g = 0.046 kg.

Weight = $0.046 \times 9.8 = 0.45$ N.

Exercise 20.3

Data: on Earth $g = 9.8\,\text{m s}^{-2}$. On the Moon $g = 1.6\,\text{m s}^{-2}$.

1000 newtons (N) = 1 kilonewton (kN)

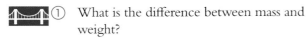 ① What is the difference between mass and weight?

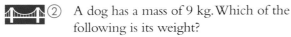 ② A dog has a mass of 9 kg. Which of the following is its weight?

   A 9 N   B 0.9 N

   C 88.2 N   D 1 N

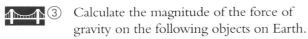 ③ Calculate the magnitude of the force of gravity on the following objects on Earth.

  (i)   A suitcase of mass 15 kg.

  (ii) A car of mass 1.2 tonnes.
       (1 tonne = 1000 kg)

  (iii) A letter of mass 50 g.

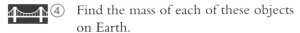 ④ Find the mass of each of these objects on Earth.

  (i)   A girl of weight 600 N.

  (ii) A lorry of weight 11 kN.

 ⑤ A person has mass 65 kg. Calculate the force of gravity

  (i)   of the Earth on the person

  (ii) of the person on the Earth.

⑥ What reaction force would an astronaut of mass 70 kg experience while standing on the moon?

⑦ Two balls of the same shape and size but with masses 1 kg and 3 kg are dropped from the same height.

  (i)   Which hits the ground first?

  (ii) If they were dropped on the Moon what difference would there be?

⑧ (i)   Estimate your mass in kilograms.

  (ii) Calculate your weight when you are on the Earth's surface.

  (iii) What would your weight be if you were on the Moon?

  (iv) When people say that a baby weighs 4 kg, what do they mean?

Discussion points

Most weighing machines have springs or some other means to measure force even though they are calibrated to show mass.

→ Would something appear to weigh the same on the Moon if you used one of these machines?

→ What could you use to find the mass of an object irrespective of where you measure it?

4 Pulleys

A pulley can be used to change the direction of a force; for example it is much easier to pull down on a rope than to lift a heavy weight. When a pulley is well designed it takes a relatively small force to make it turn and such a pulley is modelled as being **smooth and light**. Whatever the direction of the string passing over this pulley, its tension is the same on both sides.

Figure 20.27 shows the forces acting when a pulley is used to lift a heavy parcel.

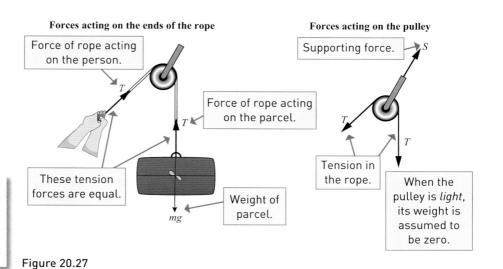

> ### Note
> The rope is in tension. It is not possible for a rope to exert a thrust force.

Figure 20.27

Example 20.6

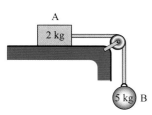

Figure 20.28

In Figure 20.28 the pulley is smooth and light and the 2 kg block, A, is on a rough surface.

(i)   Draw diagrams to show the forces acting on each of A and B.

(ii) If the block does not slip, find the tension in the string and calculate the magnitude of the friction force on the block.

(iii) Write down the resultant force acting on each of A and B if the block slips and accelerates.

Solution

(i)

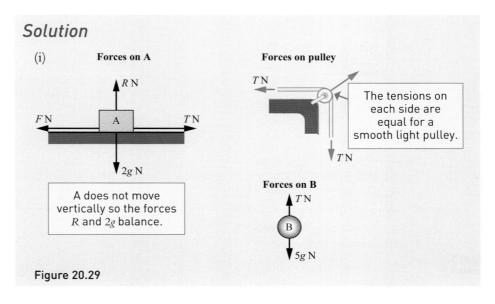

Figure 20.29

> **Note**
>
> The masses of 2 kg and 5 kg are not shown in the force diagram. The weights $2g$ N and $5g$ N are more appropriate.

(ii) When the block does not slip, the forces on B are in equilibrium so

$$5g - T = 0$$
$$T = 5g$$

The tension throughout the string is $5g$ N.

For A, the resultant horizontal force is zero so

$$T - F = 0$$
$$F = T = 5g$$

The friction force is $5g$ N towards the left.

(iii) When the block slips, the forces are not in equilibrium and T and F have different magnitudes.

The resultant horizontal force on A is $(T - F)$ N towards the right.

The resultant force on B is $(5g - T)$ N vertically downwards.

Exercise 20.4

In this exercise you are asked to draw force diagrams using the various types of force you have met in this chapter. Remember that all forces you need, other than weight, occur when objects are in contact or are joined together in some way. Where motion is involved, indicate its direction clearly.

> **Note**
>
> All questions involve modelling.

 ① A train, pulling a truck, is accelerating. Is the coupling between the train and the truck in tension or thrust? Justify your answer.

 ② A box is placed on a smooth horizontal table. It is subject to the forces shown in the diagram in Figure 20.30.

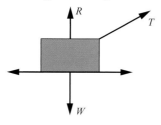

Figure 20.30

Which of the following is a correct description of the motion of the box?

A at rest

B moving with constant velocity

C accelerating to the right

D accelerating to the left.

 ③ Draw labelled diagrams showing the forces acting on the objects in *italics*.

(i) A *car* towing a caravan.

(ii) A *caravan* being towed by a car.

(iii) A *person* pushing a supermarket trolley.

(iv) A *suitcase* on a horizontal moving pavement (as at an airport)

(a) immediately after it has been put down

(b) when it is moving at the same speed as the pavement.

(v) A *sledge* being pulled uphill.

④ Ten boxes each of mass 5 kg are stacked on top of each other on the floor.

(i) What forces act on the top box?

(ii) What forces act on the bottom box?

⑤ The diagrams in Figure 20.31 show a box of mass m under different systems of forces.

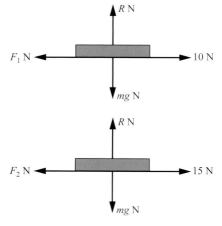

Figure 20.31

(i) In the first case above the box is at rest. State the value of F_1.

(ii) In the second case the box is slipping. Write down the resultant horizontal force acting on it.

⑥ In Figure 20.32, the pulleys are smooth and light, the strings are light, and the table is rough. The tension in the string attached to A is T_1 N and the tension in the string attached to C is T_2 N.

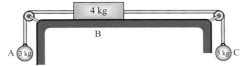

Figure 20.32

(i) What is the direction of the friction force on the block B?

(ii) Draw clear diagrams to show the forces on each of A, B and C.

(iii) By considering the equilibrium of A and C, calculate the tensions in the strings when there is no slipping.

(iv) Calculate the magnitude of the friction when there is no slipping.

Now suppose that there is insufficient friction to stop the block from slipping.

(v) Write down the resultant forces acting on A, B and C.

⑦ A man who weighs 720 N is doing some repairs to a shed. In each of these situations draw diagrams showing

(a) the forces the man exerts on the shed

(b) all the forces acting on the man (ignore any tools he might be using).

720 N

Figure 20.33

In each case, compare the reaction between the man and the floor with his weight of 720 N.

(i) He is pushing upwards on the ceiling with force U N.

(ii) He is pulling downwards on the ceiling with force D N.

(iii) He is pulling upwards on a nail in the floor with force F N.

(iv) He is pushing downwards on the floor with force T N.

⑧ Figure 20.34 shows a train, consisting of an engine of mass 50 000 kg pulling two trucks, A and B, each of mass 10 000 kg. The force of resistance on the engine is 2000 N and that on each of the trucks is 200 N. The train is travelling at constant speed.

Figure 20.34

(i) Draw a diagram showing the horizontal forces on the train as a whole. Hence, by considering the equilibrium of the train as a whole, find the driving force provided by the engine.

The coupling connecting truck A to the engine exerts a force T_1 N on the engine and the coupling connecting truck B to truck A exerts a force T_2 N on truck B.

(ii) Draw diagrams showing the horizontal forces on the engine and on truck B.

(iii) By considering the equilibrium of the engine alone, find T_1.

(iv) By considering the equilibrium of truck B alone, find T_2.

(v) Show that the forces on truck A are also in equilibrium.

5 Applying Newton's second law

Equation of motion along a straight line

Suppose you make the book accelerate upwards at $a\,\mathrm{m\,s^{-2}}$. Figure 20.35 shows the forces acting on the book and the acceleration.

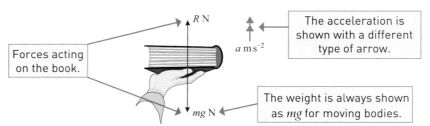

Forces acting on the book.

R N

$a\,\mathrm{m\,s^{-2}}$

The acceleration is shown with a different type of arrow.

The weight is always shown as mg for moving bodies.

mg N

Figure 20.35

By Newton's first law, a resultant force is required to produce an acceleration. In this case the resultant upward force is $R - mg$ newtons.

When the forces are in newtons, the mass in kilograms and the acceleration in metres per second squared, Newton's second law says:

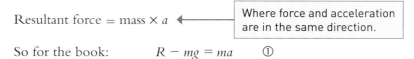

Resultant force = mass × a

Where force and acceleration are in the same direction.

So for the book: $\qquad R - mg = ma \qquad$ ①

When Newton's second law is applied, the resulting equation is called **the equation of motion**.

When you give a book of mass $0.8\,\mathrm{kg}$ an acceleration of $0.5\,\mathrm{m\,s^{-2}}$, equation ① becomes

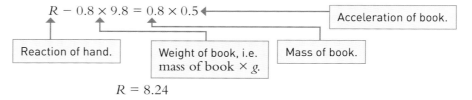

$$R - 0.8 \times 9.8 = 0.8 \times 0.5$$

Reaction of hand.

Weight of book, i.e. mass of book × g.

Mass of book.

Acceleration of book.

$$R = 8.24$$

When the book is accelerating upwards the reaction force of your hand on the book is $8.24\,\mathrm{N}$. This is equal and opposite to the force experienced by you so the book feels heavier than its actual weight, mg, which is $0.8 \times 9.8 = 7.84\,\mathrm{N}$.

Discussion points

Attach a weight to a spring balance and move it up and down.

→ What happens to the pointer on the balance?

→ What would you observe if you stood on some bathroom scales in a moving lift?

Hold a heavy book on your hand and move it up and down.

→ What force do you feel on your hand?

Figure 20.36

| Example 20.7 | A lift and its passengers have a total mass of 400 kg. Find the tension in the cable supporting the lift when |

(i)   the lift is at rest

(ii) the lift is moving at constant speed

(iii) the lift is accelerating upwards at $0.8 \, \text{m s}^{-2}$

(iv) the lift is accelerating downwards at $0.6 \, \text{m s}^{-2}$.

Solution

Before starting the calculations you must define a direction as positive. In this example the upward direction is chosen to be positive.

(i)   At rest

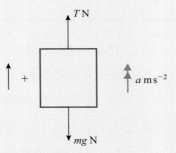

Figure 20.37

As the lift is at rest the forces must be in equilibrium. The equation of motion is

$$T - mg = 0$$
$$T - 400 \times 9.8 = 0$$
$$T = 3920$$

The tension in the cable is 3920 N.

(ii) Moving at constant speed

Again the forces on the lift must be in equilibrium because it is moving at a constant speed, so the tension is 3920 N.

(iii) Accelerating upwards

The resultant upward force on the lift is $T - mg$ so the equation of motion is

$$T - mg = ma \quad \text{which in this case gives}$$
$$T - 400 \times 9.8 = 400 \times 0.8$$
$$T - 3920 = 320$$
$$T = 4240$$

The tension in the cable is 4240 N.

(iv) Accelerating downwards

The equation of motion is

$$T - mg = ma$$

In this case, a is negative so

$$T - 400 \times 9.8 = 400 \times (-0.6)$$

$$T - 3920 = -240$$

$$T = 3680$$

A downward acceleration of $0.6\,\text{m s}^{-2}$ is an upward acceleration of $-0.6\,\text{m s}^{-2}$.

The tension in the cable is $3680\,\text{N}$.

Discussion point

→ How is it possible for the tension to be 3680 N upwards but the lift to accelerate downwards?

Prior knowledge

You will use the constant acceleration formulae from Chapter 19.

The following example shows how the **suvat** equations for motion with constant acceleration, which you met in Chapter 19, can be used with Newton's second law.

Example 20.8

Figure 20.37

A supertanker of mass 500 000 tonnes is travelling at a speed of $10\,\text{m s}^{-1}$ when its engines fail. It takes half an hour for the supertanker to stop.

(i) Find the force of resistance, assuming it to be constant, acting on the supertanker.

When the engines have been repaired it takes the supertanker 10 minutes to return to its full speed of $10\,\text{m s}^{-1}$.

(ii) Find the driving force produced by the engines, assuming this also to be constant.

Solution

Use the direction of motion as positive.

(i) First find the acceleration of the supertanker, which is constant for constant forces. Figure 20.38 shows the velocities and acceleration.

Since the supertanker is slowing down, you expect a to be negative.

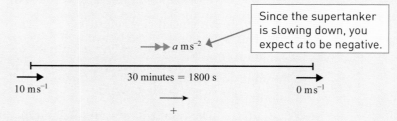

Figure 20.38

You know $u = 10$, $v = 0$, $t = 1800$ and you want to find a, so use

$$v = u + at$$

$$0 = 10 + 1800a$$

$$a = -\frac{1}{180}$$

The acceleration is negative because the supertanker is slowing down.

Now Newton's second law (N2L) can be used to write down the equation of motion.

Figure 20.39 shows the horizontal forces and the acceleration.

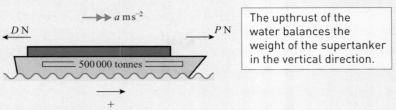

> The upthrust of the water balances the weight of the supertanker in the vertical direction.

Figure 20.39

The resultant force is $P - D$ newtons. When there is no driving force $P = 0$, so N2L gives

> The mass must be in kg.

$$0 - D = 500\ 000\ 000 \times a$$

so when $a = -\frac{1}{180}$,

$$-D = 500\ 000\ 000 \times -\left(\frac{1}{180}\right)$$

The resistance to motion is 2.78×10^6 N or 2780 kN (correct to 3 s.f.).

! You have to be very careful with signs here: the resultant force and the acceleration are both positive towards the right.

(ii) Now $u = 0$, $v = 10$ and $t = 600$, and you want a, so use

$$v = u + at$$
$$10 = 0 + a \times 600$$
$$a = \frac{1}{60}$$

Using N2L again

$$P - D = 500\ 000\ 000 \times a$$
$$P - 2.78 \times 10^6 = 500\ 000\ 000 \times \frac{1}{60}$$
$$P = 2.78 \times 10^6 + 8.33 \times 10^6$$

The driving force is 11.11×10^6 N or 11 100 kN (correct to 3 s.f.).

Equation of motion using vectors

Since force and acceleration are vector quantities they can be represented using the notation you met in Chapter 12. The following example shows how vectors allow you to work in two dimensions.

Example 20.9

Forces $\mathbf{F_1} = 8\mathbf{i} - 5\mathbf{j}$ newtons and $\mathbf{F_2} = -2\mathbf{i} + 3\mathbf{j}$ newtons act on a particle. The vectors $\mathbf{i}$ and $\mathbf{j}$ are perpendicular unit vectors in the x and y directions respectively. The particle has mass 0.2 kg.

Find

(i) the resultant force on the particle

(i) the acceleration of the particle

(i) the magnitude of the acceleration.

Solution

(i) The resultant force, $\mathbf{F} = \mathbf{F_1} + \mathbf{F_2}$
$$= 8\mathbf{i} - 5\mathbf{j} + -2\mathbf{i} + 3\mathbf{j}$$
$$= 6\mathbf{i} - 2\mathbf{j}$$

The resultant force is $6\mathbf{i} - 2\mathbf{j}$ newtons.

(ii) Use Newton's second law to find the acceleration

$F = ma$

$6\mathbf{i} - 2\mathbf{j} = 0.2\mathbf{a}$

$\mathbf{a} = 30\mathbf{i} - 10\mathbf{j}$

The acceleration of the particle is $30\mathbf{i} - 10\mathbf{j}$ m s^{-2}.

(iii) The magnitude of the acceleration $= \sqrt{30^2 + 10^2}$

$= \sqrt{1000}$

$= 31.6$ m s^{-2}

The magnitude of the acceleration is 31.6 m s^{-2}.

Notice that the vectors take care of the direction of the forces and the acceleration.

Tackling mechanics problems

When you tackle mechanics problems such as these you will find them easier if you:

- always draw a clear diagram
- clearly indicate the positive direction
- label each object (A, B, etc. or whatever is appropriate)
- show all the forces acting on each object
- make it clear which object you are referring to when writing an equation of motion.

Exercise 20.5

Note

All questions involve modelling.

 ① A gymnast of mass 60 kg balances on a beam. What force does she exert on the beam?

 ② Estimate the force, in newtons, needed to give a *Tyrannosaurus rex* of mass 8 tonnes an acceleration of 2 m s^{-2}.

  A 16 N   B 160 N

  C 1600 N   D 16 000 N

 ③ Calculate the resultant force in newtons required to produce the following accelerations.

(i) A car of mass 400 kg has acceleration 2 m s^{-2}.

(ii) A blue whale of mass 177 tonnes has acceleration $\frac{1}{2}$ m s^{-2}.

(iii) A pygmy mouse of mass 7.5 g has acceleration 3 m s^{-2}.

(iv) A freight train of mass 42 000 tonnes brakes with deceleration of 0.02 m s^{-2}.

(v) A bacterium of mass 2×10^{-16} g has acceleration 0.4 m s^{-2}.

(vi) A woman of mass 56 kg falling off a high building has acceleration 9.8 m s^{-2}.

(vii) A jumping flea of mass 0.05 mg accelerates at 1750 m s^{-2} during take-off.

(viii) A galaxy of mass 10^{42} kg has acceleration 10^{-12} m s^{-2}.

④ A resultant force of 100 N is applied to a body. Calculate the mass of the body when its acceleration is

(i) 0.5 m s^{-2} (ii) 2 m s^{-2}

(iii) 0.01 m s^{-2} (iv) 10g.

⑤ What is the reaction between a book of mass 0.8 kg and your hand when it is

(i) accelerating downwards at 0.3 m s^{-2}

(ii) moving upwards at constant speed?

⑥ A man pushes a car of mass 400 kg on level ground with a force of 200 N. The car is initially at rest and the man maintains this force until the car reaches a speed of 5 m s^{-1}. Ignoring any resistance forces, find

(i) the acceleration of the car

(ii) the distance the car travels while the man is pushing.

⑦ The engine of a car of mass 1.2 tonnes can produce a driving force of 2000 N. Ignoring any resistance forces, find

(i) the car's resulting acceleration

(ii) the time taken for the car to go from rest to 27 m s^{-1} (about 60 mph).

⑧ A top sprinter of mass 65 kg starting from rest reaches a speed of 10 m s^{-1} in 2 s.

(i) Calculate the force required to produce this acceleration, assuming it is uniform.

(ii) Compare this to the force exerted by a weight lifter holding a mass of 180 kg above the ground.

⑨ Forces $\mathbf{F}_1 = -4\mathbf{i} + \mathbf{j}$ newtons, $\mathbf{F}_2 = 5\mathbf{i} + 2\mathbf{j}$ newtons and $\mathbf{F}_3 = \mathbf{i} - \mathbf{j}$ newtons act on a particle. The vectors $\mathbf{i}$ and $\mathbf{j}$ are perpendicular unit vectors in the x and y directions respectively. The particle has mass 2 kg.

Find

(i) the resultant force on the particle

(ii) the magnitude of the acceleration of the particle.

⑩ An ice skater of mass 65 kg is initially moving with speed 2 m s^{-1} and glides to a halt over a distance of 10 m. Assuming that the force of resistance is constant, find

(i) the size of the resistance force

(ii) the distance he would travel gliding to rest from an initial speed of 6 m s^{-1}

(iii) the force he would need to apply to maintain a steady speed of 10 m s^{-1}.

⑪ A helicopter of mass 1000 kg is taking off vertically.

(i) Draw a labelled diagram showing the forces on the helicopter as it lifts off and the direction of its acceleration.

(ii) Its initial upward acceleration is 1.5 m s^{-2}. Calculate the upward force its rotors exert. Ignore the effects of air resistance.

⑫ Forces $\mathbf{F}_1 = -\mathbf{i} + 7\mathbf{j}$ newtons, $\mathbf{F}_2 = -3\mathbf{i} + 2\mathbf{j}$ newtons and $\mathbf{F}_3 = 8\mathbf{i} - 3\mathbf{j}$ newtons act on a particle. The vectors $\mathbf{i}$ and $\mathbf{j}$ are perpendicular unit vectors in the x and y directions respectively. The particle has mass 5 kg.
Find

(i) the acceleration of the particle

(ii) the angle of the resultant force with the vector $\mathbf{i}$.

⑬ Pat and Nicholas are controlling the movement of a canal barge by means of long ropes attached to each end. The tension in the ropes may be assumed to be horizontal and parallel to the line and direction of motion of the barge, as shown in Figure 20.40.

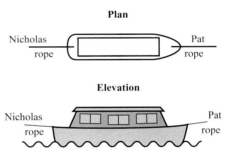

Plan

Elevation

Figure 20.40

The mass of the barge is 12 tonnes and the total resistance to forward motion may be taken to be 250 N at all times. Initially Pat pulls the barge forwards from rest with a force of 400 N and Nicholas leaves his rope slack.

(i) Write down the equation of motion for the barge and hence calculate its acceleration.

Pat continues to pull with the same force until the barge has moved 10 m.

(ii) What is the speed of the barge at this time and for what length of time did Pat pull?

Pat now lets her rope go slack and Nicholas brings the barge to rest by pulling with a constant force of 150 N.

(iii) Calculate

    (a) how long it takes the barge to come to rest

    (b) the total distance travelled by the barge from when it first moved

    (c) the total time taken for the motion.

⑭ A spaceship of mass 5000 kg is stationary in deep space. It fires its engines producing a forward thrust of 2000 N for 2.5 minutes, and then turns them off.

  (i) What is the speed of the spaceship at the end of the 2.5 minute period?

  (ii) Describe the subsequent motion of the spaceship.

The spaceship then enters a cloud of interstellar dust which brings it to a halt after a further distance of 7200 km.

  (iii) What is the force of resistance (assumed constant) on the spaceship from the interstellar dust cloud?

The spaceship is travelling in convoy with a second spaceship which is the same in all respects except that it carries an extra 500 kg of equipment. The second spaceship carries out exactly the same procedure as the first one.

  (iv) Which spaceship travels further into the dust cloud?

⑮ A crane is used to lift a hopper full of cement to a height of 20 m on a building site. The hopper has mass 200 kg and the mass of the cement is 500 kg. Initially the hopper accelerates upwards at $0.05\,\mathrm{m\,s^{-2}}$, then it travels at constant speed for some time before decelerating at $0.1\,\mathrm{m\,s^{-2}}$ until it is at rest. The hopper is then emptied.

  (i) Find the tension in the crane's cable during each of the three phases of the motion and after emptying.

The cable's maximum safe load is 10 000 N.

  (ii) What is the greatest mass of cement that can safely be transported in the same manner?

The cable is in fact faulty and on a later occasion breaks without the hopper leaving the ground. On that occasion the hopper is loaded with 720 kg of cement.

  (iii) What can you say about the strength of the cable?

⑯ The police estimate that for good road conditions the frictional force, F, on a skidding vehicle of mass m is given by $F = 0.8mg$. A car of mass 450 kg skids to a halt narrowly missing a child. The police measure the skid marks and find they are 12.0 m long.

  (i) Calculate the deceleration of the car when it was skidding to a halt.

The child's mother says the car was travelling well over the speed limit but the driver of the car says she was travelling at 30 mph and the child ran out in front of her. (There are approximately 1600 m in a mile.)

  (ii) Calculate the speed of the car when it started to skid. Who was telling the truth?

⑰ An object is hung from a spring balance suspended from the roof of a lift. When the lift is descending with uniform acceleration $0.8\,\mathrm{m\,s^{-2}}$ the balance indicates a weight of 245 N.

When the lift is ascending with uniform acceleration $a\,\mathrm{m\,s^{-2}}$ the reading is 294 N.

Find the value of a.

⑱ A train weighs 250 tonnes. The resistance to motion is $\frac{1}{100}$ of the weight of the train and its braking force is $\frac{1}{10}$ of the weight. The train starts from rest and accelerates uniformly until it reaches a speed of $45\,\mathrm{km\,h^{-1}}$. At this point the brakes are applied until the train stops. Find the time taken for the train to stop to the nearest second.

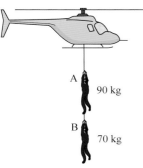

Figure 20.41

6 Newton's second law applied to connected objects

This section is about using Newton's second law for more than one object. It is important to be very clear which forces act on which object in these cases.

A stationary helicopter is raising two people of masses 90 kg and 70 kg as shown in Figure 20.41.

Discussion points

→ Imagine that you are each person in turn. Your eyes are shut so you cannot see the helicopter or the other person. What forces act on you?
   Remember that all the forces acting, apart from your weight, are due to contact between you and something else.

→ Which forces acting on A and B are equal in magnitude? What can you say about their accelerations?

Example 20.10

(i) Draw a diagram to show the forces acting on the two people being raised by the helicopter in Figure 20.42 and their acceleration.

(ii) Write down the equation of motion for each person.

(iii) When the force applied to the first person, A, by the helicopter is $180g$ N, calculate

    (a) the acceleration of the two people being raised

    (b) the tension in the ropes.

Solution

(i) Figure 20.42 shows the acceleration and forces acting on the two people.

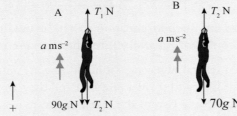

Figure 20.42

(ii) When the helicopter applies a force T_1 N to A, the resultant upward forces are

    A $(T_1 - 90g - T_2)$ N

    B $(T_2 - 70g)$ N

Their equations of motion are

    A($\uparrow$) $T_1 - 90g - T_2 = 90a$ ①

    B ($\uparrow$) $T_2 - 70g = 70a$ ②

(iii) You can eliminate T_2 from equations ① and ② by adding:

$$T_1 - 90g - T_2 + T_2 - 70g = 160a$$

$$T_1 - 160g = 160a \quad ③$$

When the force applied by the helicopter is $T_1 = 180g = 180 \times 9.8 = 1764$

$$20g = 160a$$

$$a = \frac{20 \times 9.8}{160} = 1.225$$

Substituting for a in equation ② gives $T_2 = 70 \times 1.225 + 70 \times 9.8$

$$T_2 = 771.75$$

The acceleration is $1.225\,\text{m s}^{-2}$ and the tensions in the ropes are $1764\,\text{N}$ and $772\,\text{N}$.

Discussion point

→ The force pulling downwards on A is 771.75 N. This is not equal to B's weight (686 N). Why are they different?

Treating the system as a whole

When two objects are moving in the same direction with the same velocity at all times they can be treated as one. In Example 20.9 the two people can be treated as one object and then the equal and opposite forces T_2 cancel out. They are **internal forces** similar to the forces between your head and your body.

The resultant upward force on both people is $T_1 - 90g - 70g$ and the total mass is $160\,\text{kg}$, so the equation of motion is:

$$T_1 - 90g - 70g = 160a$$

So you can find a directly

| As equation ③ in Example 20.9 above. |

when $T_1 = 180g$

$$20g = 160a$$

$$a = \frac{20 \times 9.8}{160} = 1.225$$

Treating the system as a whole finds a, but not the internal force T_2.

You need to consider the motion of B separately to obtain equation ②.

$$T_2 - 70g = 70a \qquad ②$$

$$T_2 = 771.75 \longleftarrow \boxed{\text{as before}}$$

Using this method, equation ① can be used to check your answers. Alternatively, you could use equation ① to find T_2 and equation ② to check your answers.

When several objects are joined there are always more equations possible than are necessary to solve a problem and they are not all independent. In the above example, only two of the equations were necessary to solve the problem. The trick is to choose the most relevant ones.

A note on mathematical modelling

Several modelling assumptions have been made in the solution to Example 20.9. It is assumed that:

■ the only forces acting on the people are the weights and the tensions in the ropes. Forces due to the wind or air turbulence are ignored

■ the motion is vertical and nobody swings from side to side

■ the ropes do not stretch (i.e. they are inextensible) so the accelerations of the two people are equal

■ the people are rigid bodies, i.e. they do not change shape and can be treated as particles.

All these modelling assumptions make the problem simpler. In reality, if you were trying to solve such a problem you might work through it first using these assumptions. You would then go back and decide which ones needed to be modified to produce a more realistic solution.

In the next example one person is moving vertically and the other horizontally. You might find it easier to decide on which forces are acting if you imagine you are Alvin or Bernard and you can't see the other person.

Example 20.11

Alvin is using a snowmobile to pull Bernard out of a crevasse. His rope passes over a smooth block of ice at the top of the crevasse as shown in Figure 20.43 and Bernard hangs freely away from the side. Alvin and his snowmobile together have a mass of 300 kg and Bernard's mass is 75 kg. Ignore any resistance to motion.

Figure 20.43

(i) Draw diagrams showing the forces on the snowmobile (including Alvin) and on Bernard.

(ii) Bernard's acceleration is $0.5 \, \text{m s}^{-2}$ upwards. Find the tension in the rope and the snowmobile's driving force.

(iii) How long will it take for Bernard's speed to reach $5 \, \text{m s}^{-1}$ starting from rest and how far will he have been raised in this time?

Solution

(i) Figure 20.44 shows the essential features of the problem.

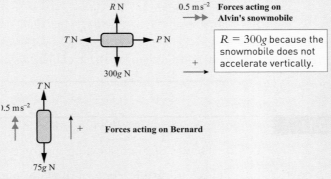

Figure 20.44

(ii) Alvin and Bernard have the same acceleration providing the rope does not stretch. The tension in the rope is T newtons and Alvin's driving force is P newtons.

The equations of motion are:

Alvin ($\rightarrow$) $P - T = 300 \times 0.5$

The force towards the right = mass × acceleration towards the right.

$P - T = 150$ ①

Bernard ($\uparrow$) $T - 75g = 75 \times 0.5$

The upwards force = mass × upwards acceleration.

$T - 75g = 37.5$ ②

$T = 37.5 + 75 \times 9.8$

$T = 772.5$

Substituting in equation ①

$P - 772.5 = 150$

$P = 922.5$

The driving force required is 922.5 N and the tension in the rope is 772.5 N.

(iii) When $u = 0$, $v = 5$, $a = 0.5$ and t is required

$v = u + at$

$5 = 0 + 0.5 \times t$

$t = 10$

The time taken is 10 seconds.

Then using $s = ut + \frac{1}{2}at^2$

$v^2 = u^2 + 2as$ would also give s.

$s = 0 + \frac{1}{2}at^2$

$s = \frac{1}{2} \times 0.5 \times 100$

$s = 25$

Bernard has been raised 25 m in 10 s.

Discussion points

Alvin thinks the rope will not stand a tension of more than 1.2 kN.

→ What is the maximum safe acceleration in this case? Under the circumstances, is Alvin likely to use this acceleration?

→ Make a list of the modelling assumptions made in this example and suggest what effect a change in each of these assumptions might have on the solution.

Example 20.12

A woman of mass 60 kg is standing in a lift.

(i) Draw a diagram showing the forces acting on the woman.

Find the normal reaction of the floor of the lift on the woman in the following cases.

(ii) The lift is moving upwards at a constant speed of $3\,\text{m}\,\text{s}^{-1}$.

(iii) The lift is moving upwards with an acceleration of $2\,\text{m}\,\text{s}^{-2}$ upwards.

(iv) The lift is moving downwards with an acceleration of $2\,\text{m}\,\text{s}^{-2}$ downwards.

(v) The lift is moving downwards and slowing down with a deceleration of $2\,\text{m}\,\text{s}^{-2}$.

In order to calculate the maximum number of occupants that can safely be carried in the lift, the following assumptions are made: the lift has mass 300 kg, all resistances to motion may be neglected, the mass of each occupant is 75 kg and the tension in the supporting cable should not exceed 12 000 N.

(vi) What is the greatest number of occupants that can be carried safely if the magnitude of the acceleration does not exceed $3\,\text{m}\,\text{s}^{-2}$?

Solution

(i) Figure 20.45 shows the forces acting on the woman and her acceleration.

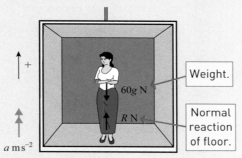

Figure 20.45

In general, when positive is upwards, her equation of motion is

$$(\uparrow) \qquad R - 60g = 60a$$

This equation contains all the mathematics in the situation. It can be used to solve parts (ii) to (iv).

(ii) When the speed is constant $a = 0$, so $R = 60g = 588$.

The normal reaction is 588 N.

(iii) When $a = 2$

$$R - 60g = 60 \times 2$$
$$R = 120 + 588$$
$$= 708$$

The normal reaction is 708 N.

(iv) When the acceleration is downwards $a = -2$ so

$$R - 60g = 60 \times (-2)$$
$$R = 468$$

The normal reaction is 468 N.

(v) When the lift is moving downwards and slowing down, the acceleration is negative downwards, so it is positive upwards, and $a = +2$. Then $R = 708$ as in part (iii).

(vi) When there are n passengers in the lift, the combined mass of these and the lift is $(300 + 75n)$ kg and their weight is $(300 + 75n)g$ N.

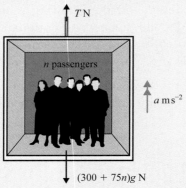

Figure 20.46

The equation of motion for the lift and passengers together is

$$T - (300 + 75n)g = (300 + 75n)a$$

So when $a = 3$ and $g = 9.8$,

$$T = (300 + 75n) \times 3 + (300 + 75n) \times 9.8$$

$$= 12.8(300 + 75n)$$

For a maximum tension of $12\,000$ N

$$12\,000 = 12.8(300 + 75n)$$

$$937.5 = 300 + 75n$$

$$637.5 = 75n$$

$$n = 8.5$$

The lift cannot carry more than 8 passengers.

Exercise 20.6

Remember to always make it clear which object each equation of motion refers to.

Note

All questions involve modelling.

 ① Two blocks are attached to the ends of a light inextensible string which hangs over a smooth pulley. One block is heavier and when the blocks are released it soon hits the ground. What difference does it make to the motion of the lighter block when that happens?

 ② A lift of weight W has one passenger of weight mg. The tension in the lift cable is T. The lift is accelerating upwards. Which of the following is true?

A $T < W + mg$ B $T > W + mg$

C $T + mg > W$ D $T + mg < W$

 ③ Blocks A and B with masses 100 g and 200 g are attached to the ends of a light, inextensible string which hangs over a smooth pulley as shown in Figure 20.47.

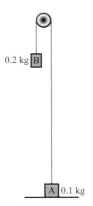

Figure 20.47

Initially B is held at rest 2 m above the ground and A rests on the ground with the string taut. Then B is released.

(i) At a later time A and B are moving with acceleration $a\,\text{m s}^{-2}$, B has not yet hit the ground. Draw a diagram for each block showing the forces acting on it and the direction of its acceleration at this time.

(ii) Write down the equation of motion of each block in the direction it moves using Newton's second law.

(iii) Use your equations to find a and the tension in the string.

(iv) Find the time taken for B to hit the ground.

④ A particle of mass 5 kg rests on a smooth horizontal table. It is connected by a light inextensible string passing over a smooth pulley at the edge of the table to a particle of mass 2 kg, which is hanging freely (Figure 20.48). Find the acceleration of the system and the tension in the string.

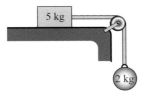

Figure 20.48

⑤ Figure 20.49 shows a lift containing a single passenger.

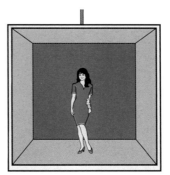

Figure 20.49

(i) Make clear diagrams to show the forces acting on the passenger and the forces acting on the lift using the following letters:
- the tension in the cable, $T\,\text{N}$
- the reaction of the lift on the passenger, $R_\text{P}\,\text{N}$
- the reaction of the passenger on the lift $R_\text{L}\,\text{N}$
- the weight of the passenger $mg\,\text{N}$
- the weight of the lift $Mg\,\text{N}$.

The masses of the lift and the passenger are 450 kg and 50 kg respectively.

(ii) Calculate T, R_P and R_L when the lift is stationary.

The lift then accelerates upwards at $0.8\,\text{m s}^{-2}$.

(iii) Find the new values of T, R_P and R_L.

⑥ In this question you should take g to be $10\,\text{m s}^{-2}$. The diagram shows a block of mass 5 kg lying on a smooth table. It is attached to blocks of mass 2 kg and 3 kg by strings which pass over smooth pulleys. The tensions in the strings are T_1 and T_2, as shown, and the blocks have acceleration $a\,\text{m s}^{-2}$.

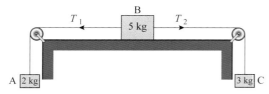

Figure 20.50

(i) Draw a diagram for each block showing all the forces acting on it and its acceleration.

(ii) Write down the equation of motion for each of the blocks.

(iii) Use your equations to find the values of a, T_1 and T_2.

In practice, the table is not truly smooth and a is found to be $0.5\,\text{m s}^{-2}$.

(iv) Repeat parts (i) and (ii) including a frictional force on B and use your new equations to find the frictional force that would produce this result.

(7) A car of mass 800 kg is pulling a caravan of mass 1000 kg along a straight, horizontal road. The caravan is connected to the car by means of a light, rigid towbar. The car is exerting a driving force of 1270 N. The resistances to the forward motion of the car and caravan are 400 N and 600 N respectively; you may assume that these resistances remain constant.

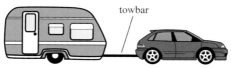

Figure 20.51

(i) Show that the acceleration of the car and caravan is $0.15\,\text{m s}^{-2}$.

(ii) Draw a diagram showing all the forces acting on the caravan along the line of its motion. Calculate the tension in the towbar.

The driving force is removed but the car's brakes are not applied.

(iii) Determine whether the towbar is now in tension or compression.

The car's brakes are then applied gradually. The brakes of the caravan come on automatically when the towbar is subjected to a compression force of at least 50 N.

(iv) Show that the acceleration of the caravan just before its brakes come on automatically is $-0.65\,\text{m s}^{-2}$ in the direction of its motion. Hence, calculate the braking force on the car necessary to make the caravan's brakes come on.

(8) Figure 20.52 shows a goods train consisting of an engine of mass 40 tonnes and two trucks of 20 tonnes each. The engine is producing a driving force of $5 \times 10^4\,\text{N}$, causing the train to accelerate. The ground is level and resistance forces may be neglected.

Figure 20.52

(i) By considering the motion of the whole train, find its acceleration.

(ii) Draw a diagram to show the forces acting on the engine and use this to help you to find the tension in the first coupling.

(iii) Find the tension in the second coupling.

The brakes on the first truck are faulty and suddenly engage, causing a resistance of $10^4\,\text{N}$.

(iv) What effect does this have on the tension in the coupling to the last truck?

9. A short train consists of two locomotives, each of mass 20 tonnes, with a truck of mass 10 tonnes coupled between them, as shown in Figure 20.53. The resistances to forward motion are 0.5 kN on the truck and 1 kN on each of the locomotives. The train is travelling along a straight, horizontal section of track.

Figure 20.53

Initially there is a driving force of 15 kN from the front locomotive only.

(i) Calculate the acceleration of the train.

(ii) Draw a diagram indicating the horizontal forces acting on each part of the train, including the forces in each of the couplings. Calculate the forces acting on the truck due to each coupling.

On another occasion each of the locomotives produce a driving force of 7.5 kN in the same direction and the resistances remain as before.

(iii) Find the acceleration of the train and the forces now acting on the truck due to each of the couplings. Compare your answer to this part with your answer to part (ii) and comment briefly.

10. A man of mass 70 kg is standing in a lift which has an upward acceleration $a\,\text{m}\,\text{s}^{-2}$.

(i) Draw a diagram showing the man's weight, the force, $R\,\text{N}$, that the lift floor exerts on him and the direction of his acceleration.

(ii) Taking g to be $10\,\text{m}\,\text{s}^{-2}$ find the value of a when $R = 770\,\text{N}$.

Figure 20.54 shows the value of R from the time ($t = 0$) when the man steps into the lift to the time ($t = 12$) when he steps out.

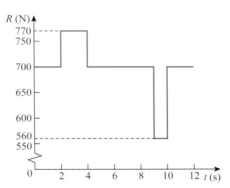

Figure 20.54

(iii) Explain what is happening in each section of the journey.

(iv) Draw the corresponding speed–time graph.

(v) To what height does the man ascend?

11. A lift in a mine shaft takes exactly one minute to descend 500 m. It starts from rest, accelerates uniformly for 12.5 seconds to a constant speed which it maintains for some time and then decelerates uniformly to stop at the bottom of the shaft.

The mass of the lift is 5 tonnes and on the day in question it is carrying 12 miners whose average mass is 80 kg.

(i) Sketch the speed–time graph of the lift.

During the first stage of the motion the tension in the cable is 53 640 N.

(ii) Find the acceleration of the lift during this stage.

(iii) Find the length of time for which the lift is travelling at constant speed and find the deceleration.

(iv) What is the maximum value of the tension in the cable?

(v) Just before the lift stops one miner experiences an upthrust of $1002\,N$ from the floor of the lift. What is the mass of the miner?

⑫ Two particles, A of mass $5\,kg$ and B of mass $8\,kg$, are connected by a light inextensible string which passes over a smooth fixed pulley.

The system is released from rest with the string taut and both particles at a height of $0.25\,m$ above the ground.

(i) Find the acceleration of the system and the tension in the string.

(ii) Find the greatest height reached by A, assuming that the pulley is of such a height that A never reaches it, and that B does not rebound when it hits the ground.

⑬ Particles A and B, of masses $2\,kg$ and $M\,kg$ respectively, are attached to the ends of a light inextensible string which passes over a smooth fixed pulley. B is held at rest on the horizontal floor and A hangs in equilibrium (Figure 20.55).

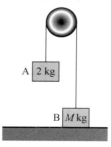

Figure 20.55

B is released and each particle starts to move vertically. A hits the floor $1.5\,s$ after B is released.

The speed of each particle when A hits the floor is $3.5\,m\,s^{-1}$.

(i) For the motion while A is moving downwards, find the acceleration of A and the tension in the string.

(ii) Find the value of M.

⑭ A train consists of a locomotive of mass 50 tonnes and two trucks each of mass 10 tonnes. The resistance to motion of the engine is $600\,N$ and that to each truck is $200\,N$. The driving force produced by the engine is $36\,000\,N$. All the couplings are light, rigid and horizontal.

(i) Show that the acceleration of the train is $0.5\,m\,s^{-2}$.

(ii) Draw a diagram showing all the forces acting on the end truck in the line of its motion. Calculate the force in the coupling between the two trucks.

With the driving force removed, brakes are applied, so adding a further resistance of $12\,000\,N$ to the total of resistance forces.

(iii) Find the new acceleration of the train.

(iv) Calculate the new force in the coupling between the trucks if the brakes are applied to

(a) the engine

(b) the end truck.

In each case state whether the force is a tension or a thrust.

⑮ A train consists of a locomotive of mass 50 tonnes and 25 trucks each of mass 10 tonnes. The force of resistance on the engine is $1500\,N$ and that on each truck is $100\,N$. The engine is producing a driving force of $34\,kN$.

(i) Find the acceleration of the train.

(ii) Find the force in the coupling between the engine and first truck.

(iii) Show that the force in the coupling between the nth and $(n + 1)$th truck is equal to $(27.5 - 1.1n)\,\text{kN}$.

(iv) Hence or otherwise, find the force in the coupling between the last two trucks.

⑯ A particle of mass M, which is subject to a constant resisting force, is observed to cover distances of s_1 and s_n respectively in the first and nth seconds of its motion. Find an expression in terms of s_1 and s_n for the magnitude of the resistance force.

⑰ A particle A of mass 8 kg rests on a rough horizontal table. Attached to A is a light inextensible string which runs parallel to the table, over a smooth peg at the edge of the table and down to a particle B of mass 2 kg hanging freely, 60 cm above the floor. The system is released from rest with A 1.2 m from the peg and subject to a uniform resistive force of 1.5 N.

(i) Find the time that elapses before B hits the ground.

(ii) Find the further time that elapses before A reaches the peg.

Reviewing models for air resistance

1 Read this short article about modelling air resistance. It ends with an experiment for you to do.

In mechanics it is often suggested that you should neglect air resistance.

In this article this is called **Model 1**. It can be written as air resistance = 0.

According to this model, all objects have the same acceleration of g when they fall to the ground. However, if you think about what will happen if you drop a feather and a hammer together, you will realise that Model 1 is not always realistic.

Sometimes you must take air resistance into account.

In **Model 2** it is assumed that air resistance is constant and the same for all objects. This is written

$$\text{air resistance} = D$$

Think of an object of mass m falling vertically through the air.

The equation of motion is

$$mg - D = ma$$

and so $\quad a = g - \dfrac{D}{m}$

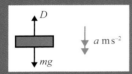

Figure 20.56

The model predicts that a heavy object will always have a greater acceleration than a lighter one because $\dfrac{D}{m}$ is smaller for larger m.

This seems to agree with our experience of dropping a feather and a hammer. The hammer is heavier and has a greater acceleration. However, if you think again you will realise this isn't always the case.

Try dropping two identical sheets of paper from a horizontal position, but fold one of them. You will find that the folded one lands first even though both sheets have the same mass. This contradicts the prediction of Model 2. A large surface area at right angles to the direction of motion seems to increase the resistance.

In **Model 3** it is assumed that air resistance is proportional to the area, A, perpendicular to the motion. This is written as

$$\text{air resistance} = kA \text{ where } k \text{ is a constant.}$$

In this case, for a falling object, the equation of motion is

$$mg - kA = ma$$

and so $\quad a = g - \dfrac{kA}{m}$

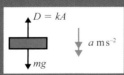

Figure 20.57

According to this model, the acceleration depends on the ratio of the area to the mass.

Here is an experiment you can do to test Model 3.

Experiment

For this experiment you will need some rigid corrugated card such as that used for packing or in grocery boxes (cereal box card is too thin), scissors and tape.

Cut out ten equal squares of side 8 cm. Stick two together by binding the edges with tape to make them smooth. Then stick three or four together in the same way so that you have four blocks A to D of different thickness as shown in Figure 20.58.

Figure 20.58

Cut out ten larger squares with 12 cm sides. Stick them together in the same way to make four blocks E to H.

Observe what happens when you hold one or two blocks horizontally at a height of about 2 m and let them fall. You do not need to measure anything in this experiment, unless you want to record the area and mass of each block, but write down your observations in an orderly fashion.

1   Drop each one separately. Could its acceleration be constant?

2   Compare A with B and C with D. Make sure you drop each pair from the same height and at the same instant of time. Do they take the same time to fall? Predict what will happen with other combinations and test your predictions.

3   Experiment in a similar way with E to H.

4   Now compare A with E; B with F; C with G; and D with H. Also compare the two blocks which have dimensions that are all in the same ratio, i.e. B and G.

5   Carry out the experiment in the article. Do your results suggest that **Model 3** might be better than **Model 2**?

6   The article and the experiment have taken you through the modelling cycle three times.

7   State what work was done each time in the various boxes in this diagram of the cycle (Figure 20.59).

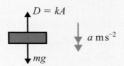

Figure 20.59

8   Models 1 to 3 all ignore one important aspect of air resistance. What is it?

TECHNOLOGY

You could use a spreadsheet to record your data. You can also use the spreadsheet to calculate the predictions using the model for comparison.

LEARNING OUTCOMES

When you have completed this chapter you should be able to:

➤ understand and use fundamental quantities and units in the S.I. system:
- ○ length
- ○ time
- ○ mass

➤ understand and use derived quantities and units:
- ○ velocity
- ○ acceleration
- ○ force
- ○ weight

➤ understand the concept of a force

➤ understand and use Newton's first law

➤ understand and use Newton's second law for motion in a straight line (restricted to forces in two perpendicular directions or simple cases of forces given as two-dimensional vectors)

➤ understand and use weight and motion in a straight line under gravity

➤ use gravitational acceleration, g, and its value in S.I. units to varying degrees of accuracy (knowledge of the inverse square law for gravitation is not required and g may be assumed to be constant, but students should be aware that g is not a universal constant but depends on location)

➤ understand and use Newton's third law

➤ understand equilibrium of forces on a particle and motion in a straight line (restricted to forces in two perpendicular directions or simple cases of forces given as two-dimensional vectors)

➤ understand application to problems involving smooth pulleys and connected particles.

KEY POINTS

1 **Newton's laws of motion**

1 Every object continues in a state of rest or uniform motion in a straight line unless it is acted on by a resultant external force.

2 Resultant force = mass × acceleration or $F = ma$.

3 When one object exerts a force on another there is always a reaction which is equal, and opposite in direction, to the acting force.

■ **Force** is a vector; **mass** is a scalar.

■ The **weight** of an object is the force of gravity pulling it towards the centre of the Earth. Weight $= mg$ vertically downwards.

2 **S.I. units**

- ■ length: metre (m)
- ■ time: second (s)
- ■ velocity: $\mathrm{m\,s^{-1}}$
- ■ acceleration: $\mathrm{m\,s^{-2}}$
- ■ mass: kilogram (kg)
- ■ force: newton (N)

3 **Force**

1 newton (N) is the force required to give a mass of 1 kg an acceleration of $1\,\mathrm{m\,s^{-2}}$.

A force of 1000 newtons (N) = 1 kilonewton (kN).

4 **Types of force**

- Forces due to contact between surfaces
- Forces in a joining rod or string

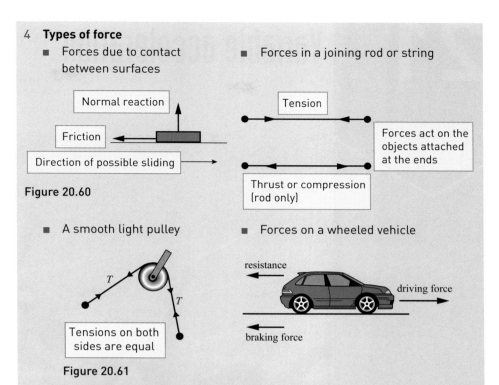

Figure 20.60

- A smooth light pulley
- Forces on a wheeled vehicle

Figure 20.61

5 **Commonly used modelling terms**

- inextensible   does not vary in length
- light   negligible mass
- negligible   small enough to ignore
- particle   negligible dimensions
- smooth   negligible friction
- uniform   the same throughout

6 **The equation of motion**

Newton's second law gives **the equation of motion** for an object.

The resultant force = mass × acceleration or $F = ma$.

The acceleration is always in the same direction as the resultant force.

7 **Connected objects**

- Draw separate force diagrams for each object in the system.
- Write down the equation of motion for each object.
- Solve the equations to find the acceleration of the system and the forces (tensions or thrusts) in the strings or rods connecting them.

8 **Modelling**

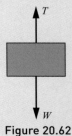

Figure 20.62

FUTURE USES

In this chapter you used Newton's second law in one dimension. You will go on to use it in two or three dimensions. You will also use Newton's third law to investigate equilibrium in more complicated situations.

21 Variable acceleration

So far you have studied motion with constant acceleration in a straight line, but the motion of a car round the Brand's Hatch racing circuit shown in Figure 21.1 is much more complex. In this chapter you will see how to deal with variable acceleration.

The equations you have used for constant acceleration do not apply when the acceleration varies. You need to go back to first principles.

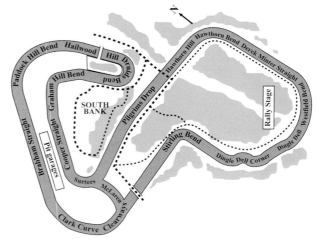

Figure 21.1

Consider how displacement, velocity and acceleration are related to each other. The velocity of an object is the rate at which its position changes with time. When the velocity is not constant the position–time graph is a curve.

The rate of change of the position is the gradient of the tangent to the curve. You can find this by differentiating.

$$v = \frac{ds}{dt} \qquad ①$$

Similarly, the acceleration is the rate at which the velocity changes, so

$$a = \frac{dv}{dt} = \frac{d^2s}{dt^2} \qquad ②$$

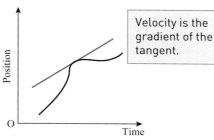

Velocity is the gradient of the tangent.

Figure 21.2

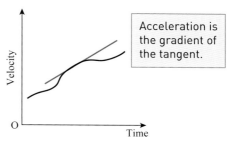

Acceleration is the gradient of the tangent.

Figure 21.3

Prior knowledge

In this chapter you will use the differentiation you met in Chapter 10 and the integration you met in Chapter 11.

Prior knowledge

You will be sketching polynomial curves as in Chapters 7 and 8.

1 Using differentiation

When you are given the position of a moving object in terms of time, you can use equations ① and ② to solve problems even when the acceleration is not constant.

Example 21.1

An object moves along a straight line so that its position at time t in seconds is given by

$$r = 2t^3 - 6t \text{ (in metres) } (t \geqslant 0).$$

(i) Find expressions for the velocity and acceleration of the object at time t.

(ii) Find the values of r, v and a when $t = 0, 1, 2$ and 3.

(iii) Sketch the graphs of r, v and a against time.

(iv) Describe the motion of the object.

Solution

(i) Position $r = 2t^3 - 6t$ ①

Velocity $v = \dfrac{dr}{dt} = 6t^2 - 6$ ②

Acceleration $a = \dfrac{dv}{dt} = 12t$ ③

You can now use these three equations to solve problems about the motion of the object.

(ii) When

$t =$	0	1	2	3
From ① $r =$	0	-4	4	36
From ② $v =$	-6	0	18	48
From ③ $a =$	0	12	24	36

(iii) The graphs are drawn under each other in Figure 21.4 so that you can see how they relate.

(iv) The object starts at the origin and moves towards the negative direction, gradually slowing down. At $t = 1$ it stops instantaneously and changes direction, returning to its initial position at approximately $t = 1.7$.

It then continues moving in the positive direction with increasing speed.

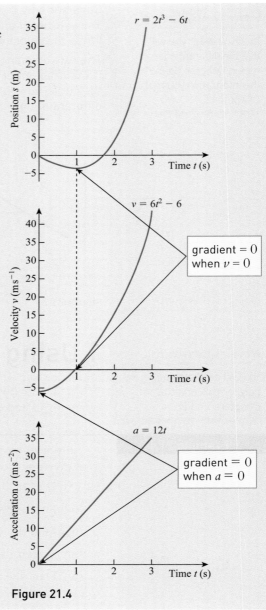

gradient $= 0$ when $v = 0$

gradient $= 0$ when $a = 0$

The acceleration is increasing at a constant rate. This cannot go on for much longer or the speed will become excessive.

Figure 21.4

Exercise 21.1

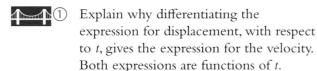 ① Explain why differentiating the expression for displacement, with respect to t, gives the expression for the velocity. Both expressions are functions of t.

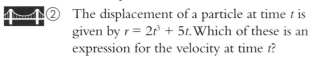 ② The displacement of a particle at time t is given by $r = 2t^3 + 5t$. Which of these is an expression for the velocity at time t?

A $2t^2 + 5$ B $3t^2 + 5$

C $6t^2 + 5$ D $6t + 5$

 ③ In each of the following cases

(a) find an expression for the velocity

(b) use your equations to write down the initial position and initial velocity

(c) find the time and position when the velocity is zero.

(i) $r = 10 + 2t - t^2$

(ii) $r = -4t + t^2$

(iii) $x = t^3 - 5t^2 + 4$

④ In each of the following cases

    (a) find an expression for the acceleration

    (b) use your equations to write down the initial velocity and acceleration.

    (i) $v = 4t + 3$

    (ii) $v = 6t^2 - 2t + 1$

    (iii) $v = 7t - 5$

⑤ The distance travelled by a cyclist is modelled by

$$r = 4t + 0.5t^2 \text{ in S.I. units.}$$

Find expressions for the velocity and the acceleration of the cyclist at time t.

⑥ A particle moves along the x axis, its position at time t is given by

$$x = 15t - 5t^2.$$

    (i) Find expressions for the velocity and the acceleration.

    (ii) Draw the acceleration–time graph and, below it, the velocity–time graph with the same scale for time and the origins in line.

    (iii) Describe how the two graphs for each object relate to each other.

    (iv) Describe how the velocity and acceleration change during the motion of each object.

⑦ A particle moves along the x axis, its position is given by

$$x = 6t^3 - 18t^2 - 6t + 3.$$

    (i) Find expressions for the velocity and the acceleration.

    (ii) Draw the acceleration–time graph and, below it, the velocity–time graph with the same scale for time and the origins in line.

    (iii) Describe how the two graphs for each object relate to each other.

    (iv) Describe how the velocity and acceleration change during the motion of each object.

⑧ A particle is moving in a straight line. Its distance, rm, from a fixed point in the line after ts is given by the equation $r = 24t - 22t^2 + 4t^3$.

Find

    (i) the velocity and acceleration after 2s

    (ii) the distance travelled between the two times when the velocity is instantaneously zero.

⑨ A particle moves along the x axis, its position at time t is given by

$$x = 21 - 15t + 9t^2 - t^3.$$

Find the acceleration when the particle is at rest.

2 Finding displacement from velocity

How can you find an expression for the position of an object when you know its velocity in terms of time?

One way of thinking about this is to remember that $v = \dfrac{dr}{dt}$, so you need to do the opposite of differentiation, that is to integrate, to find r.

$$r = \int v \, dt$$ ◄—————— The dt indicates that you must write v in terms of t before integrating.

Example 21.2

The velocity (in $m\,s^{-1}$) of a model train which is moving along straight rails is

$$v = 0.3t^2 - 0.5.$$

The initial position of the train is $r_0 = 2$. Find the position r

(i)   after time t

(ii) after 3 seconds.

Solution

(i)   The displacement at any time is $r = \int v\,dt$.

$$r = \int (0.3t^2 - 0.5)\,dt$$
$$r = 0.1t^3 - 0.5t + c$$

To find the train's position, put $r = 2$ when $t = 0$.

This gives $c = 2$ and so $r = 0.1t^3 - 0.5t + 2$.

You can use this equation to find the displacement at any time before the motion changes.

(ii) After 3 seconds, $t = 3$ and $r = 2.7 - 1.5 + 2 = 3.2$.

The position of the train after 3 s is 3.2 m.

> ! When using integration don't forget the constant. This is very important in mechanics problems and you are usually given some extra information to help you find the value of the constant.

The area under a velocity–time graph

In Chapter 19 you saw that the area under a velocity–time graph represents a displacement. Both the area under the graph and the displacement are found by integrating. To find a particular displacement you calculate the area under the velocity–time graph by definitive integration using suitable limits.

The distance travelled between the times T_1 and T_2 is shown by the shaded area on the graph in Figure 21.5.

$$r = \text{area} = \int_{T_1}^{T_2} v\,dt$$

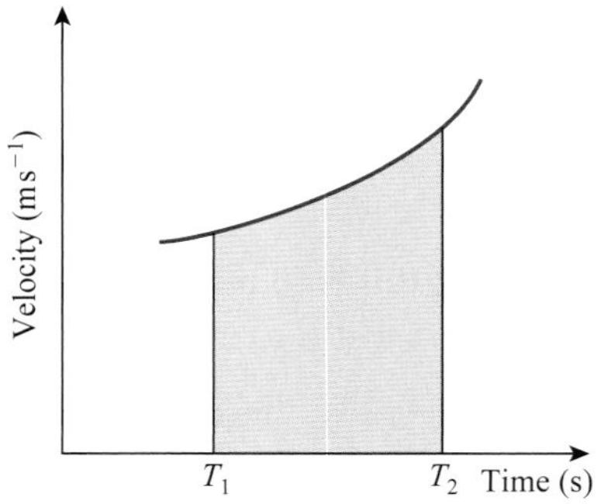

Figure 21.5

Example 21.3

21

Chapter 21 Variable acceleration

TECHNOLOGY

You could use a
spreadsheet to record
your data. You can also
use the spreadsheet
to calculate the
predictions using the
model for comparison.

A car moves between two sets of traffic lights, stopping at both. Its speed $v\,\text{m\,s}^{-1}$ at time $t\,\text{s}$ is modelled by

$$v = \frac{1}{20}t(40 - t),\ 0 \le t \le 40.$$

Find the times at which the car is stationary and the distance between the two sets of traffic lights.

Solution

The car is stationary when $v = 0$. Substituting this into the expression for the speed gives

$$0 = \frac{1}{20}t(40 - t)$$

$$\Rightarrow t = 0 \text{ or } t = 40.$$

These are the times when the car starts to move away from the first set of traffic lights and stops at the second set.

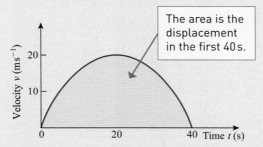

> The area is the displacement in the first 40 s.

Figure 21.6

The distance between the two sets of traffic lights is given by

$$\text{distance} = \int_0^{40} = \frac{1}{20}t(40 - t)\,\mathrm{d}t$$

$$= \frac{1}{20}\int_0^{40}(40t - t^2)\,\mathrm{d}t$$

> Notice that you write $t(40 - t) = 40t - t^2$. Otherwise you couldn't do the integration.

$$= \frac{1}{20}\left[20t^2 - \frac{t^3}{3}\right]_0^{40}$$

$$= 533.\dot{3}\,\text{m}$$

Finding velocity from acceleration

You can also find the velocity from the acceleration by using integration.

$$a = \frac{\mathrm{d}v}{\mathrm{d}t}$$

> You must write a in terms of t before integration.

$$\Rightarrow v = \int a\,\mathrm{d}t$$

The next example shows how you can obtain equations for motion using integration.

Example 21.4

The acceleration of a particle (in $m\,s^{-2}$) at time t seconds is given by
$$a = 6 - t.$$
The particle is initially at the origin with velocity $-2\,m\,s^{-1}$.

Find an expression for
(i) the velocity of the particle after $t\,s$

(ii) the position of the particle after $t\,s$.

Hence find the velocity and position $6\,s$ later.

Solution

The information given may be summarised as follows:

at $t = 0$, $r = 0$ and $v = -2$;

at time t, $a = 6 - t$ ①

(i) $\dfrac{dv}{dt} = a = 6 - t$

Integrating gives
$$v = 6t - \frac{1}{2}t^2 + c$$
When $t = 0$, $v = -2$

So $-2 = 0 - 0 + c$

     $c = -2$

At time t
$$v = 6t - \frac{1}{2}t^2 - 2 \qquad ②$$

(ii) $\dfrac{ds}{dt} = v = 6t - \frac{1}{2}t^2 - 2$

Integrating gives
$$r = 3t^2 - \frac{1}{6}t^3 - 2t + k$$
When $t = 0$, $r = 0$

So $0 = 0 - 0 - 0 + k$

     $k = 0$

At time t
$$r = 3t^2 - \frac{1}{6}t^3 - 2t \qquad ③$$

Equations ①, ② and ③ can now be used to give more information about the motion in a similar way to the **suvat** formulae.

> Remember that the **suvat** formulae only apply when the acceleration is constant.

When $t = 6$ $v = 36 - 18 - 2 = 16$ from ②

When $t = 6$ $r = 108 - 36 - 12 = 60$ from ③

The particle has a velocity of $+16\,m\,s^{-1}$ and it is at $+60\,m$ after $6\,s$.

! Notice that two different arbitrary constants (c and k) are necessary when you integrate twice. You could call them c_1 and c_2 if you wish.

Exercise 21.2

 ① Which two of the following expressions are expressions for velocity?

$$\frac{da}{dt} \qquad \frac{dr}{dt} \qquad \int a\,dt \qquad \int r\,dt$$

 ② Which of the following is an expression for the displacement of a particle at time t given that its velocity is $v = 3t^2 - 5$?

A $r = t^3 - 5t$ B $r = 3t^3 - 5t$

C $r = t^3 - 5t + r$ D $r = 6t$

 ③ Find expressions for the position in each of these cases.

(i) $v = 4t + 3$; initial position 0.

(ii) $v = 6t^3 - 2t^2 + 1$; when $t = 0$, $r = 1$.

(iii) $v = 7t^2 - 5$; when $t = 0, r = 2$.

④ The speed of a ball rolling down a hill is modelled by $v = 1.7t$ (in ms^{-1}).

(i) Draw the speed–time graph of the ball.

(ii) How far does the ball travel in 10 s?

⑤ Until it stops moving, the speed of a bullet t s after entering water is modelled by $v = 216 - t^3$ (in ms^{-1}).

(i) When does the bullet stop moving?

(ii) How far has it travelled by this time?

⑥ During braking the speed of a car is modelled by $v = 40 - 2t^2$ (in ms^{-1}) until it stops moving.

(i) How long does the car take to stop?

(ii) How far does it move before it stops?

⑦ In each case below, the object moves along a straight line with acceleration a in ms^{-2}. Find an expression for the velocity v (in ms^{-1}) and position x (m) of each object at time t s.

(i) $a = 10 + 3t - t^2$; the object is initially at the origin and at rest.

(ii) $a = 4t - 2t^2$; at $t = 0, x = 1$ and $v = 2$.

(iii) $a = 10 - 6t$; at $t = 1, x = 0$ and $v = -5$.

3 The constant acceleration formulae revisited

Discussion points

→ In which of the cases in question 3 in Exercise 21.2 is the acceleration constant?

→ Which constant acceleration formulae give the same results for s, v and a in this case?

→ Why would the constant acceleration formulae not apply in the other two cases?

ACTIVITY 21.1

How can you use these to derive the other equations for constant acceleration?

$s = \frac{1}{2}(u + v) + s_0$ ③

$v^2 - u^2 = 2a(s - s_0)$ ④

$s = vt - \frac{1}{2}at^2 + s_0$ ⑤

You can use integration to derive the equations for constant acceleration.

When a is constant (and only then)

$$v = \int a\,dt = at + c_1$$

When $t = 0, v = u$ $u = 0 + c_1$

$$\Rightarrow v = u + at$$

You can integrate this again to find

If $s = s_0$ when $t = 0, c_2 = s_0$ and

$s = ut + \frac{1}{2}at^2 + c_2$ ①

$s = ut + \frac{1}{2}at^2 + s_0$ ②

① Divide the following into groups of three which describe the same measure:

area under a velocity–time graph

gradient of a velocity–time graph

gradient of a displacement–time graph

displacement

velocity

acceleration

$$\int v \, dt \quad \frac{dv}{dr} \quad \frac{dr}{dt}$$

 ② Differentiate $s = ut + \frac{1}{2}at^2$.

A $s = ut + at$ B $s = u + at$

C $v = u + at$ D $v^2 = u^2 + 2as$

> **Note**
>
> All subsequent questions involve aspects of modelling.

③ A boy throws a ball up in the air from a height of 1.5 m and catches it at the same height. Its height in metres at time t seconds is

$$y = 1.5 + 15t - 5t^2.$$

(i) What is the vertical velocity $v\,\text{m s}^{-1}$ of the ball at time t?

(ii) Find the position, velocity and speed of the ball at $t = 1$ and $t = 2$.

(iii) Sketch the position–time, velocity–time and speed–time graphs for $0 \leqslant t \leqslant 3$.

(iv) When does the boy catch the ball?

(v) Explain why the distance travelled by the ball is not equal to $\int_0^3 v \, dt$ and state what information this expression does give.

④ An object moves along a straight line so that its position in metres at time t seconds is given by

$$r = t^3 - 3t^2 - t + 3 \quad (t \geqslant 0).$$

(i) Find the position, velocity and speed of the object at $t = 2$.

(ii) Find the smallest time when

(a) the position is zero

(b) the velocity is zero.

(iii) Sketch position–time, velocity–time and speed–time graphs for $0 \leqslant t \leqslant 3$.

(iv) Describe the motion of the object.

⑤ An object moves along a straight line so that its acceleration (in m s^{-2}) is given by $a = 4 - 2t$. It starts its motion at the origin with speed $4\,\text{m s}^{-1}$ in the direction of increasing x.

(i) Find as functions of t the velocity and position of the object.

(ii) Sketch the position–time, velocity–time and acceleration–time graphs for $0 \leqslant t \leqslant 2$.

(iii) Describe the motion of the object.

⑥ Nick watches a golfer putting her ball 24 m from the edge of the green and into the hole. He decides to model the motion of the ball. He assumes that the ball is a particle travelling along a straight line. In his model its distance, r metres, from the golfer at time t seconds is

$$r = -\frac{3}{2}t^2 + 12t \quad 0 \leqslant t \leqslant 4.$$

(i) Find the value of s when $t = 0, 1, 2, 3$ and 4.

(ii) Explain the restriction $0 \leqslant t \leqslant 4$.

(iii) Find the velocity of the ball at time t seconds.

(iv) With what speed does the ball enter the hole?

(v) Find the acceleration of the ball at time t seconds.

⑦

Figure 21.7

An object moves along a straight line AB so that its displacement s metres from O at time t seconds is given by

$$r = t^3 - 3t \text{ for } 0 \leqslant t \leqslant 2.$$

The displacement is positive in the direction OB.

(i)   Find the velocity and acceleration of the object at time t.

(ii)   When is the velocity zero?

(iii) Sketch the velocity–time and acceleration–time graphs for the motion for $0 \leqslant t \leqslant 2$.

(iv)   Describe the motion of the object for $0 \leqslant t \leqslant 2$.

(v)   Calculate the total distance travelled by the object from $t = 0$ to $t = 2$.

⑧ Andrew and Elizabeth are having a race over $100\,\text{m}$. Their accelerations (in m s^{-2}) are as follows:

Andrew   Elizabeth
$a = 4 - 0.8t$ $0 \leqslant t \leqslant 5$ $a = 4$ $0 \leqslant t \leqslant 2.4$

$a = 0$ $\quad\quad\quad t > 5$ $a = 0$ $t > 2.4$

(i)   Find the greatest speed of each runner.

(ii)   Sketch the speed–time graph for each runner.

(iii) Find the distance Elizabeth runs while reaching her greatest speed.

(iv)   How long does Elizabeth take to complete the race?

(v)   Who wins the race, by what time margin and by what distance?

On another day they race over $120\,\text{m}$, both running in exactly the same manner.

(vi)   What is the result now?

⑨ Cara is a parachutist. On one of her descents her vertical speed, $v\,\text{m s}^{-1}$, $t\,\text{s}$ after leaving an aircraft is modelled by

$v = 8.5t$ $\quad\quad\quad\quad 0 \leqslant t \leqslant 10$

$v = 5 + 0.8(t - 20)^2$ $\quad 10 < t \leqslant 20$

$v = 5$ $\quad\quad\quad\quad\quad 20 < t \leqslant 90$

$v = 0$ $\quad\quad\quad\quad\quad t > 90$

(i)   Sketch the speed–time graph for Cara's descent and explain the shape of each section.

(ii)   How high is the aircraft when Cara jumps out?

(iii) Write down expressions for the acceleration during the various phases of Cara's descent. What is the greatest magnitude of her acceleration?

⑩ Two objects move along the same straight line. The velocities of the objects (in m s^{-1}) are given by $v_1 = 16t - 6t^2$ and $v_2 = 2t - 10$ for $t \geqslant 0$.

Initially the objects are $32\,\text{m}$ apart. At what time do they collide?

⑪ A bird leaves its nest for a short horizontal flight along a straight line and then returns. Michelle models its distance, r metres, from the nest at time t seconds by

$$r = 25t - 2.5\,t^2;\, 0 \leqslant t \leqslant 10.$$

(i)   Find the value of r when $t = 2$.

(ii)   Explain the restriction $0 \leqslant t \leqslant 10$.

(iii) Find the velocity of the bird at time t seconds.

(iv)   What is the greatest distance of the bird from the nest?

Michelle's teacher tells her that a better model would be

$$r = 10t^2 - 2t^3 + \frac{1}{10}t^4.$$

(v)   Show that the two models agree about the time of the journey and the greatest distance travelled. Compare their predictions about velocity and suggest why the teacher's model is better.

⑫ A battery-operated toy dog starts at a point O and moves in a straight line. Its motion is modelled by the velocity–time graph in Figure 21.8.

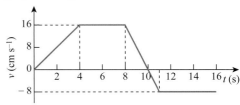

Figure 21.8

(i) Calculate the displacement from O of the toy

    (a) after 10 seconds

    (b) after 16 seconds.

(ii) Write down expressions for the velocity of the toy at time t seconds in the intervals $0 \leqslant t \leqslant 4$ and $4 \leqslant t \leqslant 8$.

(iii) Obtain expressions for the displacement from O of the toy at time t seconds in the intervals $0 \leqslant t \leqslant 4$ and $4 \leqslant t \leqslant 8$.

An alternative model for the motion of the toy in the interval $0 \leqslant t \leqslant 10$ is $v = \frac{2}{3}(10t - t^2)$, where v is the velocity in $\mathrm{cm\,s^{-1}}$.

(iv) Calculate the difference in the displacement from O after 10 seconds as predicted by the two models.

⑬ A train starts from rest at a station. Its acceleration is shown on the acceleration–time graph in Figure 21.9.

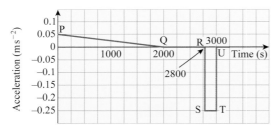

Figure 21.9

(i) Describe what is happening during the phases of the train's journey represented by the lines PQ, QR and ST.

(ii) The equation of the line PQ is of the form $a = mt + c$. Find the values of the constants m and c.

(iii) Find the maximum speed of the train.

(iv) What is the speed of the train at the time $t = 3000$?

(v) How far does the train travel during the first 3000 s?

⑭ A particle is moving in a straight line. Its position r at time t is given by

$r = 18 - 24t + 9t^2 - t^3 \quad 0 \leqslant t \leqslant 5$

(i) Find the velocity v at time t and the values of t for which $v = 0$.

(ii) Find the position of the particle at those times.

(iii) Find the total distance travelled in the interval $0 \leqslant t \leqslant 5$.

⑮ A particle starts from rest at a point O and moves in a straight line. After t s its velocity is $v = 4t^2 - t^4 \mathrm{\,m\,s^{-1}}$. Show that the particle is at rest after 2 s and find its distance from O at this time.

Find also the maximum velocity of the particle in the interval $0 \leqslant t \leqslant 2$.

⑯ A sprinter starts from rest at time $t = 0$ and runs in a straight line.

For $0 \leqslant t \leqslant 3$ the velocity of the sprinter is given by $v = 2.7t^2 - 0.6t^3$.

For $3 < t \leqslant 23$, the sprinter runs at a steady speed of $8.1 \mathrm{\,m\,s^{-1}}$.

For $t > 23$ the sprinter decelerates at a constant rate of $0.2 \mathrm{\,m\,s^{-2}}$.

(i) Find the distance travelled by the sprinter in the first three seconds.

(ii) Find the time taken by the sprinter to run 100 m.

(iii) Find the time taken by the sprinter to run 200 m.

⑰ The acceleration of a particle t seconds after starting from rest is $a = 3t - 1 \mathrm{\,m\,s^{-2}}$.

Prove that the particle returns to its starting point after one second, and find the distance of the particle from the starting point after a further second.

Find the times for which the velocity is zero and the distance of the particle from the starting point at those times. Hence find the total distance travelled in the interval $0 \leqslant t \leqslant 2$.

⑱ A particle P starts from rest at A at time $t = 0$, where t is in seconds, and moves in a straight line with constant acceleration $a\,\mathrm{m\,s^{-2}}$ for $10\,\mathrm{s}$. For $10 \leqslant t \leqslant 20$, P continues to move along the line with velocity $v\,\mathrm{m\,s^{-1}}$, where

$$v = \frac{(t - 20)^2}{20}.$$

Find

(i) the speed of P when $t = 10$ and the value of a

(ii) the value of t for which the acceleration of P is $-0.5\,\mathrm{m\,s^{-2}}$

(iii) the displacement of P from A when $t = 20$.

Human acceleration

You know that if you step off a high place your initial acceleration will be g or $9.8\,\text{m s}^{-2}$ in the vertically downwards direction.

What is the greatest acceleration a human being can generate in the horizontal direction?

Figure 21.10

No artificial help (like an engine, a spring or a rocket) is allowed. Acceleration is often given as a factor of g, like $4g$; express your answer in this way.

💻 TECHNOLOGY

You could use a spreadsheet to record your data. You can also use the spreadsheet to calculate the predictions using the model for comparison.

Problem specification and analysis

The problem is clearly specified and it is natural to think of sprinters when deciding how to go about it. A helpful start is to find 10 metre split times for a top 100-metre runner. These are the times at which the runner passes marks at $10\,\text{m}$, $20\,\text{m}$, . . . and so on up to the finish at $100\,\text{m}$. The data also give the runner's reaction time.

At this initial stage, you need to plan how to use those times to work out the runner's greatest acceleration. You can also, for your own interest, use them to work out the greatest speed.

Information collection

Collect figures for one or more suitable runners from the internet.

Processing and representation

Now that you have collected the figures, you need to put your plans from Stage 1 into practice.

Interpretation

Now that you have got an answer, here are some further things for you to think about.

- How can you check whether your answer is reasonable?

- How accurate do you think your answer is likely to be?

- Is there much variation between top runners? What about between males and females?

LEARNING OUTCOMES

When you have completed this chapter you should be able to:

➤ use calculus in kinematics for motion in a straight line:

$$v = \frac{dr}{dt}, \ a = \frac{dv}{dt} = \frac{d^2r}{dt^2}, \ r = \int v \, dt, \ v = \int a \, dt$$

KEY POINTS

1 Relationships between the variables describing motion

| Position | ⇒ | Velocity | ⇒ | Acceleration |

——————————————— *differentiate* ———————————————→

s $\qquad\qquad v = \dfrac{ds}{dt}$ $\qquad\qquad a = \dfrac{dv}{dt} = \dfrac{d^2s}{dt^2}$

| Acceleration ⇒ | Velocity | ⇒ | Position |

←——————————— *integrate* ———————————

a $\qquad\qquad v = \displaystyle\int a \, dt$ $\qquad\qquad s = \displaystyle\int v \, dt$

2 Acceleration may be due to change in direction or change in speed or both.

FUTURE USES

- Motion with a variable acceleration is a consequence of non-constant forces.
- This arises in situations involving elastic strings and springs and often results in simple harmonic motion.
- This also arises in situations where air resistance is involved and found to be dependent on velocity.
- Situations involving variable acceleration are often modelled using differential equations.

First name	Last name	Age	Distance from home	Cause	Injuries	Wearing a helmet?	Nights in hospital	Time of accident	Day of accident	Month of accident	Officer reporting
Jeremy	Marlow	55	6.5 km	lorry at roundabout	multiple fractures	y	2		Monday	June	97655
Michael	Marston	23	3 miles	car braked suddenly	head injuries	y	1	8.20am	Friday	November	97655
Joanne	Mason	26	2 miles	lorry turning left	concussion and broken arm		1	10am	Saturday	May	39014
Jennifer	Massey	10	100 m	fell off	suspected concussion		0	8 am	Saturday	Sept	39014
Justin	Matthews	6.5	1 mile	hit wall	broken arm	y	0	noon	Tuesday	August	97655
Sam	Maynard	28	1.5 km	pothole	unreadable	n	0	5.20 pm	Thursday	July	97655
Richard	McLennan	22	1 mile	car	bruising	y	0	8.10am	Wednesday	April	78264
Owen	Mitchell	61	1 mile	pothole	suspected concussion	n	1	9.30am	Tuesday	June	39813
Lisa	Montgomery	16	300 m	fell over	abrasions	n	0	3pm	Saturday	June	39813
James	Moore	20	2 km	trying to fix chain	partially severed finger	n	0	7am	Monday	June	97655
Glyn	Morgan	36	6 km	lorry	concussion	y	0	8.50am	Wednesday	September	39813
Luke	Murphy	20	3 miles	messing around	broken arm	y	0	10pm	Saturday	April	78264
Lewis	Ofan	36	4 miles	hit car	concussion, sprained wrists	y	0	10.30	Sunday	August	45211
Matthew	Ogunwe	25	10 km	car turning	broken leg	y	0	8.30am	Thursday	May	97655
Dylan	Omerod	42	8 km	bus pulled out	concussion, multiple fractures	y	2	8.45am		October	78264
Patrick	O'Toole	21	5 miles	slipped on oil patch	broken arm	y	0	8.20am	Monday	June	39813
Zoe	Painter	21	2 miles	skidded	abrasions		0	10am	Wednesday	July	78264
Eric	Passant	67	3 miles	slipped on leaves	abrasions, shock	n	1	9.50am	Friday	November	97655
Anthony	Patrick	17	3 miles	dragged under lorry	severe head injuries	y	5	4.30pm	Monday	April	97655
Cath	Pickin	16	half mile	puncture by nail	abrasions on leg	y	0	7.30am	Tuesday		79655
Kobi	Pitts	34	3 miles	bus	broken arm and sprained wrist	y	0	11am	Tuesday	August	97655
Simon	Porter	66	5 miles	hit cyclist	broken arm	y	0	3.30pm	Tuesday	June	39813
Kate	Price	17	5 miles	car collision	concussion	y	0	8am	Friday	June	39813
Dee	Pugh	11	about 1 mile	car hit	dislocated shoulder	y	1	3.30pm	Monday	August	97655
Simon	Rice	11	100m	hit friend	suspected concussion	y	0	4pm	Sunday	June	78264
Bob	Roberts	138	3 miles	lorry collision	broken arm	n	0	8.25am	Monday	January	79264
Benjamin	Ronan	60	1 mile	collision with cyclist	abrasions	n	0	8am	Monday	March	39813
Michael	Root	13	30 m	wing mirror	deep cuts	y	0	5pm	Monday	August	39813
Jonathan	Sanders	9	200 m	hit car	concussion	y	1	2.30pm	Tuesday	December	39813
Simon	Sefton	15	3 km	car hit	concussion	n	1		Tuesday	May	97655
Arvinder	Sethi	12	1200 m	lorry	multiple fractures	n	40	4.40pm	Thursday	May	39813
Dave	Smith	37	23 km	van pulled out	concussion, multiple fractures		2	3.30pm	Saturday	June	39014
John	Smith	45	3 km	skid	concussion	y	0	7.45	Tuesday	May	78264
Millie	Smith	88	2 miles	hit by car	bruising, shock	n	1	2pm	Tuesday	April	78264
Jacob	Squires	46	7 km	skidded on wet road	severe abrasions	y	0	8.50am	Sunday	September	45211
Jodie	Stanton	18	5 km	hit brick in road	abrasions to left leg	y	0	5.15pm	Thursday	January	97655
Sam	Thomas	9		hit tree	broken wrist	y	0	15.45	Wednesday	December	39813
Manny	Umberton	9	2 m	hit kerb	dislocated thumb	y	0	9.30am	Sunday	April	78264
Ian	Wade	12	2 km	car	concussion with complications	y	3	10 am	Tuesday	March	39813
Natalie	Walken	50	50 m	brakes failed, hit kerb	broken wrist, bruising	n	0	1300	Wednesday	March	39813
Agatha	Walker	64	1 mile	knocked by car	fractured wrist	n	0	10am	Wednesday	October	78264
Kerry	Wilde	52	3 km	car didn't look	abrasions to left arm	n	0	10	Saturday	May	97655
Jordan	Williams	35	4 km	hit fence	broken fingers	y	0	9.15am	Tuesday	May	97655
Marion	Wren	8.5	300 m	cyclist collision	bruising		0	10	Saturday	May	39014

Answers

Chapter 1

Discussion points (page 5)

The value of r would be 5.57 cm (to 3 s.f.)

Exercise 1.1 (page 6)

1. | | |
   |---|---|
   | multiple of 3 | $3n$ |
   | odd number | $2n + 1$ |
   | 7 more than x | $x + 7$ |
   | twice as big as y | $2y$ |

2. $x(x - 3)$
3. 10, 10 and 12
4. 8890 m^2 (3 s.f.)
5. 6
6. 6 kg
7. THE MEGGAN SPACE-FLEET IS APPROACHING PLEASE SEND HELP WE NEED MORE GALACTIC RUNABOUTS WITH HIGH SPEED LASER GUNS
8. 17
9. 18
10. (i) 110 minutes
    (ii) The model does not allow for speeding up at the start of the journey and slowing down at the end.
    (iii) 7 stops. Cannot be very confident as we don't know if there were delays on the journey. 7 stops suggests that it was 2 minutes late.
11. (i) (a) £30
       (b) £75
    (ii) More than 1250
    (iii) For example, $C = 70 + 0.02n$
12. It is a factor of $p - 1$
13. Many responses are possible.

Discussion point (page 10, top)

$\Leftrightarrow$

Discussion point (page 10, bottom)

Yes.
For example, 'If p is a prime > 2 then p is an odd number'.

Exercise 1.2 (page 11)

1. False, $n = 6$, 36 is a multiple of 4 but 6 is not.
2. $A \Rightarrow B$
3. (i) $A \Rightarrow B$
   (ii) $A \Leftarrow B$
   (iii) $A \Leftrightarrow B$
   (iv) $A \Leftrightarrow B$
   (v) $A \Rightarrow B$
   (vi) $A \Leftrightarrow B$
   (vii) $A \Rightarrow B$
   (viii) $A \Leftrightarrow B$
4. (i) If a triangle has two angles equal, then it has two sides equal. True
   (ii) If Alf is dead, then Fred murdered Alf. False
   (iii) Each of the angles of ABCD is 90° $\Rightarrow$ ABCD is a square. False
   (iv) A triangle with three equal angles has three equal sides. True
   (v) If Struan goes swimming, it is sunny. False
5. (i) $P \Leftarrow Q$
   (ii) $P \Leftarrow Q$
   (iii) $P \Leftarrow Q$
   (iv) $P \Rightarrow Q$
   (v) $P \Leftarrow Q$
   (vi) $P \Leftrightarrow Q$
   (vii) $P \Rightarrow Q$
   (viii) $P \Leftrightarrow Q$
6. (i) If x^2 is an integer, then x is an integer. False
   (ii) If $\angle PQR + \angle PSR = 180°$, the angles P, Q, R and S all lie on a circle. True
   (iii) If $x^2 = y^2$, then $x = y$. False
   (iv) $\angle x = \angle y \Rightarrow$ lines l and m are parallel. True
   (v) $n > 2 \Rightarrow n$ is an odd prime number. False
7. (i) No
   (ii) Yes
8. (i) (a) True
       (b) The six internal angles are all equal $\Rightarrow$ ABCDEF is a regular hexagon.
       (c) True
   (ii) (a) True
        (b) All the six sides are the same length $\Rightarrow$ ABCDEF is a regular hexagon.
        (c) False. The shape could look like this:

9. (i) True
   (ii) Triangles ABC and XYZ are congruent $\Rightarrow$ Together AB = XY, BC = YZ and angle BAC = angle XYZ.
   (iii) False – for example it could be that AB = YZ, BC = XZ and angle BAC = angle XZY.
10. 0, 1 or 4

Discussion point (page 13)

Only the prime numbers were necessary.

Exercise 1.3 (page 14)

1. Let the integers be $n, n + 1, n + 2, n + 3, n + 4, n + 5, n + 6$. Sum $= n + n + 1 + n + 2 + n + 3 + n + 4 + n + 5 + n + 6$

$= 7n + 21$

$= 7(n + 3)$ which is a multiple of 7

2   $n = 5$

3   (i)   True

   (ii)   True

   (iii)   False, e.g. $x = -1$, or any negative number

   (iv)   True

   (v)   False, e.g. $n = 11$

   (vi)   True

4   (i)   False, e.g. $n = 40$

   (ii)   False, e.g. $n = 41$

   (iii)   True

   (iv)   True

   (v)   True

   (vi)   True

5   (i) $w = 50\sqrt{2}$

6   Converse is: all numbers of the form $6n \pm 1$ are prime. The converse is false.

7   (i)   True

   (ii)   (a) True

      (b) False

   (iii)   True

8   (i)   These are 3, 7, 31 and 211 so demonstrate that none of them are divisible by the prime numbers less than their square root.

   (ii) $p_1\, p_2\, p_3 \cdots p_{n+1}$ is not divisible by any of p_1 $p_2\, p_3 \cdots p_n$ and so must either be a new prime or divisible by another prime which must be greater than p_n.

   (iii)   You can always find another prime by adding 1 to the product of those already known.

Chapter 2

Discussion points (page 23)

You cannot find a value for the answer when you substitute $x = 1$, $y = 1$. This is because in this case $x - \sqrt{y} = 0$ so you have multiplied the top and bottom by zero.

Exercise 2.1 (page 23)

1   (i)
$$\sqrt{72} = \sqrt{36 \times 2}$$
$$= \sqrt{36} \times \sqrt{2}$$
$$= 6\sqrt{2},$$
the wrong number has been extracted from the square root sign.

   (ii)
$$\sqrt{75} = \sqrt{25 \times 3}$$
$$= \sqrt{25} \times \sqrt{3}$$
$$= 5\sqrt{3},$$
25 was not square rooted when it was extracted from the square root sign.

   (iii) $\sqrt{32} = \sqrt{16 \times 2} = 4\sqrt{2}$, the largest square factor was not extracted from the square root sign.

2   (i) $2\sqrt{7}$

   (ii) $5\sqrt{3}$

   (iii) $8\sqrt{2}$

3   (i) $\sqrt{54}$

   (ii) $\sqrt{125}$

   (iii) $\sqrt{432}$

4   (i) $2\sqrt{2}$

   (ii) $\dfrac{\sqrt{3}}{6}$

   (iii) $\dfrac{3\sqrt{2}}{2}$

5   (i) $\dfrac{5}{7}$

   (ii) $\dfrac{4\sqrt{2}}{9}$

   (iii) $\dfrac{5}{7}$

6   (i) $8 + 3\sqrt{2}$

   (ii) $4\sqrt{2}$

   (iii) $8\sqrt{2} + 2\sqrt{5}$

   (iv) $8\sqrt{5}$

7   (i) $2\sqrt{y}$

   (ii) $2\sqrt{a} + 11\sqrt{b}$

8   (i) $11 + 6\sqrt{2}$

   (ii) $11 - 6\sqrt{2}$

   (iii)   7

9   (i) $3 - \sqrt{14}$

   (ii) $3 + \sqrt{14}$

   (iii) $p - 2q - \sqrt{pq}$

   (iv) $p - 2q + \sqrt{pq}$

10   (i) $\dfrac{\sqrt{3} - 1}{2}$

   (ii) $-\dfrac{(1 + \sqrt{5})}{4}$

   (iii) $\dfrac{3(4 + \sqrt{2})}{14}$

   (iv) $3(\sqrt{5} + 2)$

11   (i) $1 + \frac{1}{2}\sqrt{2}$

   (ii) $\frac{3}{2}\sqrt{3} - \frac{5}{2}$

   (iii) $1 - \frac{2}{5}\sqrt{5}$

   (iv) $\frac{9}{7} + \frac{4}{7}\sqrt{2}$

12   (i) $4\sqrt{2}$ cm; 32 cm^2

   (ii) $12\sqrt{2}$ cm

13   $4\sqrt{2}$ m

14   $1 : 64 : 729$

15   (i) $3 + \sqrt{6}$

   (ii) $\dfrac{14 + 5\sqrt{10}}{18}$

16   (i)   e.g. $a = 27, b = 12$

   (ii) a and b are of the form nx^2 and ny_2

Exercise 2.2 (page 30)

1   $PQ = 2 \times 10^2$, $QP = 2 \times 10^2$, $P \div Q = 8 \times 10^{-6}$, $Q \div P = 1.25 \times 10^5$

2   10^9

3   (i) 3^4

   (ii) 3^0

   (iii) 3^{-4}

   (iv) 3^{-2}

4   (i) $\dfrac{1}{32}$

   (ii) 3

   (iii) $\dfrac{1}{8}$

   (iv) 1

5   (i) $x^{\frac{3}{2}}$

   (ii) x^{-3}

   (iii) x^3

   (iv) $x^{\frac{1}{2}}$

6 (i) 4×10^9

(ii) 5×10^{-5}

7 (i) 3.78×10^{11} and 7.8×10^{10} metres

(ii) Greatest time 21 minutes, least time 4 minutes 20 seconds.

8 (i) $4x^3$

(ii) $4x$

(iii) $64x^{-10}$

9 (i) $x^2 - x$

(ii) $a - a^{-2}$

(iii) $p - 1$

10 (i) $-x^6$

(ii) $32x^9$

11 (i) $3\sqrt{2}$

(ii) $-3\sqrt[3]{4}$

(iii) $43\sqrt{7}$

12 $x = \dfrac{1}{2}$

13 (i) $(1 + x)^{\frac{1}{2}}(2 + x)$

(ii) $(x + y)^{\frac{1}{2}}(6 - 5x - 5y)$

(iii) $(x^2 - 2x + 3)^{\frac{1}{2}}(-x^2 + 2x - 2)$

14 $x = 12, y = 8$

15 $x = 0.5, y = -0.5; \dfrac{1}{5}, \sqrt{2}$

16 (i) $0.5, 0.25, 0.125, 0.0625,$ value approaches zero

(ii) $0, 0, 0, 0,$ value $= 0$

(iii) $1, 1, 1, 1,$ value $= 1$ (by definition)

Chapter 3

Activity 3.1 (page 35)

(i) $(x + 3)(x + 2)$

(ii) $(x + 2)(x + 3)$

(iii) $(x + 2)(x - 3)$

(iv) $(x - 3)(x - 2)$

(v) $(x - 3)(x + 2)$

(vi) $(x - 6)(x - 1)$

(vii) $(x - 6)(x + 1)$

(viii) not possible

(ix) $(x - 6)(x - 1)$

(x) not possible

(i) and (ii) give the same answer,

(iii) and (v) give the same answer,

(vi) and (ix) give the same answer

Exercise 3.1 (page 40)

1 (i) 2 and 3

(ii) 6 and -1

(iii) -15 and 1

(iv) 4 and -4

2 (i) $(x + 4)(x + 2)$

(ii) $(x - 4)(x - 2)$

(iii) $(y + 4)(y - 2)$

(iv) $(y - 4)(y + 2)$

(v) $(r + 5)(r - 3)$

(vi) $(r - 5)(r + 3)$

3 (i) $(s - 2)^2$

(ii) $(s + 2)^2$

(iii) $(p - 2)(p + 2)$

(iv) cannot be factorised

(v) $(a + 3)(a - 1)$

(vi) cannot be factorised

4 (i) $x = 1, 6$

(ii) $x = -1, -6$

(iii) $x = 2, 3$

(iv) $x = -2, -3$

(v) $x = 7, -1$

(vi) $x = -7, 1$

5 (i)

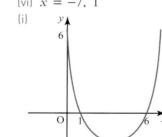

(ii)

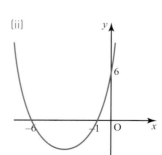

(iii)

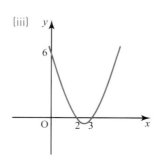

(iv)

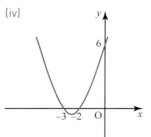

(v)

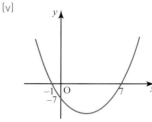

(vi)

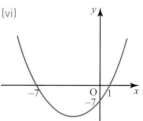

6 (i) $(4 - x)(1 + x)$

(ii) $(4 + x)(1 - x)$

(iii) $(4 - x)(3 + x)$

(iv) $(4 + x)(3 - x)$

(v) $(7 + x)(5 - x)$

(vi) $(7 - x)(5 + x)$

7 (i) $(x + 2)(2x + 1)$

(ii) $(x - 2)(2x - 1)$

(iii) $(5x + 1)(x + 2)$

(iv) $(5x - 1)(x - 2)$

(v) $2(x + 3)(x + 4)$

(vi) $2(x - 3)(x - 4)$

8 (i) $(1 + 3x)(1 - 2x)$

(ii) $(1 - 3x)(1 + 2x)$

(iii) $(1 + x)(5 - 2x)$

(iv) $(1 - x)(5 + 2x)$

9 (i) $x = \dfrac{1}{2}$ or $x = \dfrac{2}{3}$

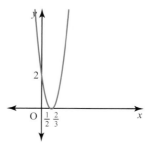

(ii) $x = \frac{2}{3}$ repeated

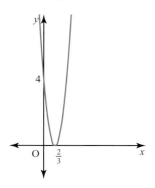

(iii) $x = \pm 1\frac{1}{2}$

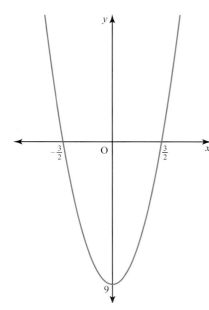

10 (i) $x = 4$ or $x = -1\frac{1}{2}$

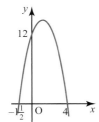

(ii) $x = -6$ or $x = 1\frac{2}{3}$

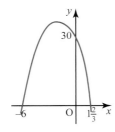

(iii) $x = \pm\frac{4}{5}$

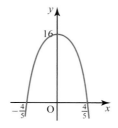

11 (i) $x = \pm 2, \pm 3$
(ii) $x = \pm\sqrt{2}$
(iii) $x = 1, 9$
(iv) $x = 1, -8$

12 (i) $A = w(30 + w)$
(ii) Width $= 80\,\text{m}$, perimeter $= 380\,\text{m}$

13 (i) $A = 2\pi rh + 2\pi r^2$
(ii) $54\pi\,\text{cm}^3$
(iii) $250\pi\,\text{cm}^3$

14 19.1 cm

15 (i) $x = 2y$
(ii) $4, 3, 5; 8, 15, 17; 12, 35, 37$
(iii) (a) $y = 10$
(b) The triangle is right-angled but doesn't fit that pattern.

16 (i) Volume $= 20x^2 - 600x + 4000\,\text{cm}^3$
(ii) 36 cm by 72 cm

Discussion point (page 44)

If the coefficient of x^2 is negative, then the quadratic function has the greatest value (i.e. the graph has a maximum point).

Activity 3.3 (page 45)

(i) Two numbers which multiply to make 2 must be either 1 and 2, or -1 and -2. Neither of these add to give -6.
(ii) 5.65 and 0.35
(iii) Answers should be close to zero
(iv) Substituting into the equation does not give exactly zero.

Exercise 3.2 (page 46)

1 $x^2 - 4x + 4 \quad (x + 1)^2 \quad 49$

2 $(3, 5)$

3 (i) $x^2 + 4x + 1$
(ii) $x^2 + 8x + 12$
(iii) $x^2 - 2x + 3$
(iv) $x^2 - 20x + 112$
(v) $x^2 - x + 1$
(vi) $x^2 + 0.2x + 1$

4 (i) $x = -1 \pm \sqrt{10}$
(ii) $x = 2 \pm \sqrt{5}$
(iii) $x = 3 \pm \sqrt{5}$
(iv) $x = \dfrac{-1 \pm \sqrt{3}}{2}$
(v) $x = \dfrac{3 \pm \sqrt{12}}{2}$
(vi) $x = \dfrac{1 \pm \sqrt{18}}{3}$

5 (i) $(x + 2)^2 + 1$
(ii) $(x - 3)^2 - 6$
(iii) $(x + 1)^2 - 6$
(iv) $(x - 4)^2 - 20$

6 (i) (a) $y = (x + 2)^2 + 5$
(b) $x = -2, (-2, 5)$
(c)

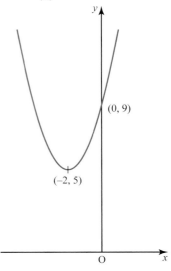

(ii) (a) $y = (x - 2)^2 + 5$
(b) $x = 2, (2, 5)$

(c)

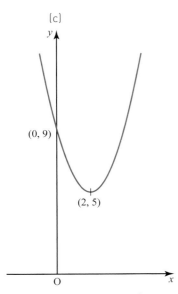

(0, 9)

(2, 5)

O

(iii) (a) $y = (x + 2)^2 - 1$
(b) $x = -2, (-2, -1)$
(c)

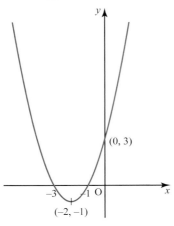

(0, 3)

−3   −1 O

(−2, −1)

(iv) (a) $y = (x - 2)^2 - 1$
(b) $x = 2, (2, -1)$
(c)

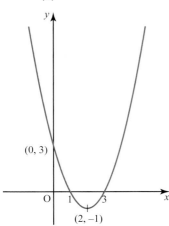

(0, 3)

O   1   3

(2, −1)

(v) (a) $y = (x + 3)^2 - 10$
(b) $x = -3, (-3, -10)$
(c)

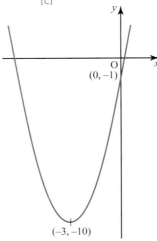

O
(0, −1)

(−3, −10)

(vi) (a) $y = (x - 3)^2 - 10$
(b) $x = 3, (3, -10)$
(c)

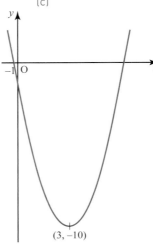

−1 O

(3, −10)

7   (i) (a) $y = 2(x + 1)^2 + 4$
(b) $(-1, 4), x = -1$

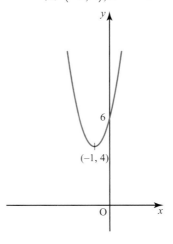

6

(−1, 4)

O

(ii) (a) $y = 2(x - 1)^2 + 4$
(b) $(1, 4), x = 1$

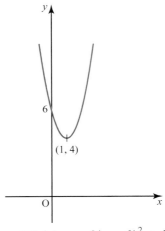

6

(1, 4)

O

(iii) (a) $y = 3(x - 3)^2 + 3$
(b) $(3, 3), x = 3$

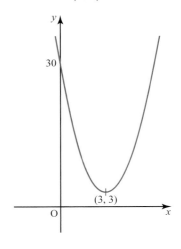

30

(3, 3)

O

(iv) (a) $y = 3(x - 3)^2 + 3$
(b) $(-3, 3), x = -3$

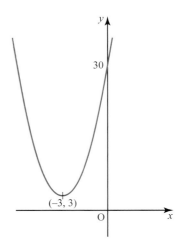

30

(−3, 3)

O

8 (i) (a) $y = -(x + 1)^2 + 6$
(b) $(-1, 6), x = -1$

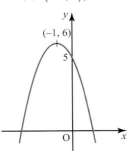

(ii) (a) $y = -(x - 1)^2 + 6$
(b) $(1, 6), x = 1$

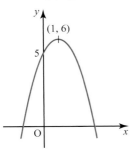

(iii) (a) $y = -3(x + 2)^2 - 3$
(b) $(-2, -3), x = -2$

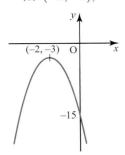

(iv) (a) $y = -3(x - 2)^2 - 3$
(b) $(2, -3), x = 2$

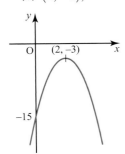

9 (i) (a) $y = (x - 2)^2 + 3$
(b) $(2, 3), x = 2$

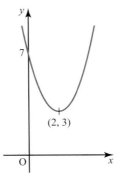

(ii) (a) $y = 2(x - 2)^2 - 3$
(b) $(2, -3), x = 2$

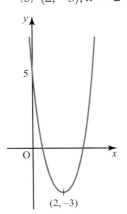

(iii) (a) $y = -(x + 4)^2 + 9$
(b) $(-4, 9), x = -4$

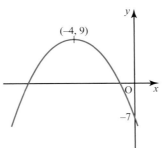

(iv) (a) $y = -3(x - \frac{2}{3})^2 - 3\frac{2}{3}$
(b) $(\frac{2}{3}, -3\frac{2}{3}), x = \frac{2}{3}$

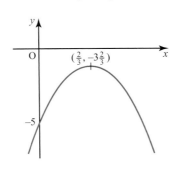

10 (i) $b = -6, c = 10$
(ii) $b = 2, c = 0$
(iii) $b = -8, c = 16$
(iv) $b = 6, c = 11$

11 (i) $(x - 4.5)^2 - 12.25$
(ii) $(4.5, -12.25)$
(iii) $(1, 0), (8, 0), (0, 8)$
(iv)

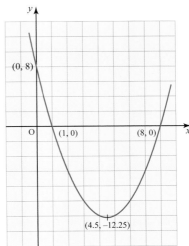

12 (i) $(x - 2)^2 + 2, (2, 2)$
(ii) $c > 4$

13 (i) (a) $(x + 11)^2 - 36$
(b) $x = -5$ or $x = -17$
(ii) (a) $(x - 12)^2 - 81$
(b) $x = 3$ or $x = 21$

14 (i) $3(x - 1)^2 + 11$
(ii) $x = 1, (1, 11)$

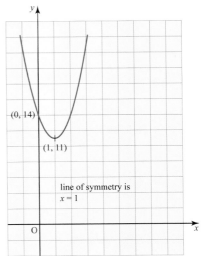

15 (i) $a = \pm 6$

(ii) Curves are reflections of each other in the y axis.

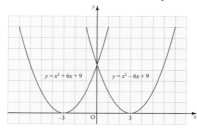

Exercise 3.3 (page 50)

1 (i) $\dfrac{-3 \pm \sqrt{(-5)^2 - 4 \times 1 \times 5}}{2 \times 3}$

has -3 substituted for b instead of -5, and 5 for a instead of 3

(ii) $\dfrac{-5 \pm \sqrt{-5^2 - 4 \times 1 \times 3}}{2 \times 3}$

has 5 rather than -5 for b then the -5^2 is without brackets and so is incorrect

(iii) $\dfrac{5 \pm \sqrt{5^2 - 4 \times 1 \times 3}}{2}$

has only 2 in the denominator rather than 2×3

(iv) $\dfrac{5 \pm \sqrt{5^2 - 4 \times 1 \times 3}}{2 \times 3}$

is correct

(v) $\dfrac{5 \pm \sqrt{(-5)^2 - 4 \times 1 \times 3}}{2 \times 3}$

is correct

(vi) $5 \pm \dfrac{\sqrt{(-5)^2 - 4 \times 1 \times 3}}{2 \times 3}$

has not included $5 \pm$ in the numerator

2 (i) $-0.683, -7.32$

(ii) $1.24, -3.24$

(iii) $7.52, -2.52$

(iv) $0.303, -3.30$

(v) $1.37, -0.366$

(vi) $1.57, -1.91$

3 (i) $-0.23, -1.43$

(ii) $1.43, 0.23$

(iii) $0.34, -5.84$

(iv) $-0.39, -5.11$

(v) $1.89, 0.11$

(vi) $2.10, -0.10$

4 (i) (a) 20

(b) 2

(c) No

(ii) (a) 0

(b) 1

(c) Yes

(iii) (a) -12

(b) 0

(c) No

(iv) (a) 49

(b) 2

(c) Yes

(v) (a) 0

(b) 1

(c) Yes

(vi) (a) -275

(b) 0

(c) No

5 (i) $2 \pm \sqrt{3}$

(ii) $\dfrac{-1 \pm \sqrt{13}}{6}$

(iii) $\dfrac{-1 \pm \sqrt{13}}{-6}$

(iv) $\dfrac{5 \pm \sqrt{29}}{2}$

(v) $4 \pm \sqrt{15}$

(vi) $\dfrac{-3 \pm \sqrt{89}}{8}$

6 (i) (a) $16 - 4c$

(b) 4

(c) $c \leqslant 4$

(ii) (a) $36 - 8c$

(b) 4.5

(c) $c \leqslant 4.5$

(iii) (a) $16 - 12c$

(b) $1\frac{1}{3}$

(c) $c \leqslant 1\frac{1}{3}$

(iv) (a) $4 + 20c$

(b) $-\frac{1}{5}$

(c) $c \geqslant -\frac{1}{5}$

7 (i) $A \Rightarrow B$

(ii) $A \Leftrightarrow B$

(iii) $A \Leftrightarrow B$

(iv) $A \Leftarrow B$

(v) $A \Leftrightarrow B$

8 (i) (a) $b^2 - 4ac = 0$

$\Rightarrow$ the graph of $y = ax^2 + bx + c$ touches the x axis

(b) True

(ii) (a) $ax^2 + bx + c$ cannot be factorised

$\Rightarrow b^2 - 4ac < 0$

(b) False

(iii) (a) $x^2 - 9 = 0$

$\Rightarrow x = 3$

(b) False

9 (i) $t = 1$ and 2

(ii) $t = 3.065$

(iii) 12.25 m

10 $k = 1$ or $k = 2$

12 (i) $k \leqslant \frac{13}{9}$ or $k \geqslant 3$

(ii) $k = 3$, repeated root $= -4$; $k = -\frac{13}{9}$, repeated root $= \frac{8}{3}$

Chapter 4

Opening activity (page 53)

A packet of nuts costs £1.20 and a packet of crisps costs 80p.

In the second question, both statements give the same information, so there is not enough information to solve the problem.

Discussion points (page 54)

There are an infinite number of pairs of values for x and y that satisfy the equation. You need two equations.

Three equations

Activity 4.1 (page 55)

$(1, 3)$

Approximately $(1.7, 1.2)$

Discussion points (page 57)

There may be no points where the line meets the curve. If you try to solve the equations algebraically, you will obtain a quadratic equation with negative discriminant (so it has no real roots).

There may be just one point where the line meets the curve, in which case the line will touch the curve without crossing it. If you try to solve the equations algebraically, you will obtain a quadratic equation with discriminant zero (so there is a repeated root).

Exercise 4.1 (page 57)

1 $y = -2x - 1$ and $4x + 3y = 1$
2 $x = -1, y = 3$
3 (i) $x = 7, y = 2$
    (ii) $x = 8, y = 2$
    (iii) $a = 5, b = -2$
4 (i) $x = 3, y = 7$
    (ii) $x = 1, y = 7$
    (iii) $x = 5.5, y = 1.5$
    (iv) $a = 1, b = -2$
5 (i) $x = 0.5, y = 1$
    (ii) $x = 3, y = -1$
    (iii) $l = -1, m = -2$
    (iv) $r = 5, s = -1$
6 (i) $5p + 8h = 20$
       $10p + 6h = 20$
    (ii) Paperback costs 80p,
       Hardback costs £2
7 (i) Adult costs £30, Child
       costs £15
    (ii) £135
8 (i) $t_1 + t_2 = 4$
       $110t_1 + 70t_2 = 380$
    (ii) 275 km motorway,
       105 km country roads
9 £6.80
10 (i) $x = 1, y = 1$
       $x = 4, y = 16$
    (ii) $x = 0.5, y = 2.5$
       $x = -2, y = -5$
11 (i) $x = 3, y = 1$ or
       $x = 1, y = 3$
    (ii) $x = 4, y = 2$ or
       $x = -4, y = -2$
    (iii) $k = -1, m = -7$ or
       $k = 4, m = -2$
    (iv) $x = 1, y = -2$ or
       $x = -2\frac{3}{7}, y = -\frac{2}{7}$

12 (i)
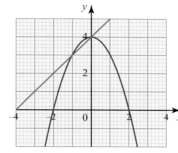

    (ii) $(0, 4)$ and $(-1, 3)$
13 (i)
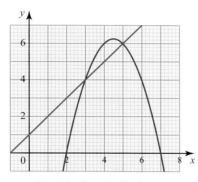

    (ii) $(3, 4)$ and $(5, 6)$
14 (i) $(3x + 2y)(2x + y)$ m^2
    (iii) $x = \frac{1}{2}, y = \frac{1}{4}$
15 (i) $h + 4r = 100$,
       $2\pi rh + 2\pi^2 = 1400\pi$
    (ii) 6000π or $\frac{98000}{27}\pi$ cm^2
16 (i) $x = 1, y = 3$
    (ii) Line is a tangent to the
       curve.
17 (i) $x = \frac{5 \pm \sqrt{-19}}{2}$
    (ii) No solution since there
       is a square root of a
       negative number.
    (iii) The line and the curve
       don't meet.

Activity 4.2 (page 59)

$60m + 40d \leq 2500$

$m + d \geq 50$

She could buy 25 of each, or she could buy fewer muffins and more doughnuts, e.g. if she bought 23 muffins she could either buy 27 or 28 doughnuts.

Exercise 4.2 (page 62)

1
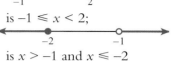

is $x \leq 2$;

is $x > -2$;

is $-1 \leq x < 2$;

is $x > -1$ and $x \leq -2$
2 $x \leq -3, x > -1$
3 (i) $x \geq 4$
    (ii) $x < 3$
    (iii) $-2 \leq x \leq 5$
    (iv) $-4 < x < -1$
    (v) $5 \leq x < 7$
    (vi) $-6 < x \leq 0$
4 (i)

    (ii)

    (iii)

    (iv)

    (v)

    (vi)
5 (i) $-2 \leq x \leq 3$
    (ii) $4 < x < 7$
    (iii) $x \leq -1$ or $x \geq 2$
6 (i) $x < 7$
    (ii) $x > 3$
    (iii) $x \leq 4$
    (iv) $x \geq -2$
    (v) $p < 7$
    (vi) $s \geq 6$
7 (i) $2 \leq x \leq 4$

    (ii) $-1 < x < 3$

    (iii) $0.5 \geq x \geq -1.5$

    (iv) $1 > x > -1$

    (v) $2 \leq x \leq 5$

(vi) $-0.5 > x > -1.5$

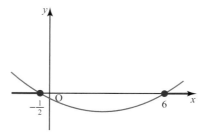
-1.5 -0.5

8 (i) $c > -2$

(ii) $d \leqslant -\frac{4}{3}$

(iii) $e > 7$

(iv) $f > -1$

(v) $g \leqslant 1.4$

(vi) $h < 0$

9 (i) $1 < p < 4$

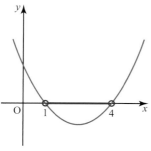

(ii) $-2 \leqslant x \leqslant -1$

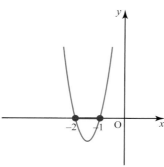

(iii) $x < -2$ or $x > -1$

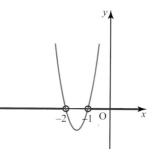

(iv) $-2 < x < \frac{1}{3}$

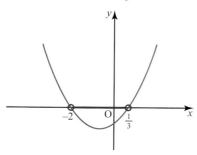

(v) $x \leqslant -\frac{1}{2}$ or $x \geqslant 6$

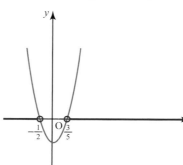

(vi) $x < -\frac{1}{2}$ or $x > \frac{3}{5}$

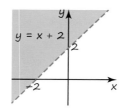

10 (i) $\{y : y < -1 \cup y > 3\}$

(ii) $\{z : z \geqslant -4 \cap z \leqslant 5\}$

(iii) $\{x : x \leqslant 1 \cup x \geqslant 3\}$

(iv) $\{a : a < -3 \cup a > 2\}$

(v) $\{a : a \geqslant -4 \cap a \leqslant 2\}$

(vi) $\{s : s > -1 \cap s < \frac{1}{3}\}$

11 (i)

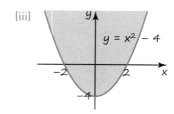

(ii)

(iii)

(vi)

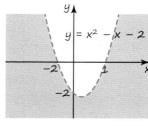

12 $0 < k < 4$

13 (i) Single point solution
$a = 2$

(ii) No real value of b

(iii) Doesn't factorise –
quadratic formula
needed

14 36 to 57 years

Chapter 5

Opening activity (page 65)

There are 9 pieces 1 m long.
There are 8 pieces 2 m long.
Assuming the ant takes the
shortest route, it travels 7 m.
Assuming the mouse takes
the shortest route, it travels
$1 + 2\sqrt{5} = 5.47$ m.

The bee travels 5 m.

Activity 5.1 (page 66)

(i) $M(5, 3)$

(ii) $\sqrt{52}$

Discussion point (page 67)

No, the top and bottom lines
of the fraction in the gradient
formula would have the same
magnitude but the opposite sign,
so m would be unchanged.

Activity 5.2 (page 67)

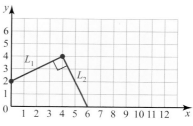

$m_1 = \frac{1}{2}$; $m_2 = -2$

The gradients are negative reciprocals of each other; $m_1 m_2 = -1$.

This is true for any pair of perpendicular lines.

Exercise 5.1 (page 70)

1

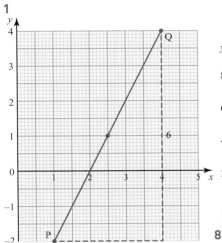

$$PQ = \sqrt{6^2 + 3^2} = \sqrt{45} = 6.71$$

$$\text{Gradient} = \frac{6}{3} = 2$$

2 (ii) $-\frac{3}{4}$, the change in y is -3 for a change in x of 4, or subtracting the other way the change in y is 3 for a change in x of -4.

3 (i) (a) $(4, 6.5)$
(b) 5
(c) $\frac{3}{4}$
(d) $-\frac{4}{3}$

(ii) (a) $(-4, -6.5)$
(b) 5
(c) $\frac{3}{4}$
(d) $-\frac{4}{3}$

(iii) (a) $(2, 1.5)$
(b) $\sqrt{233}$
(c) $-\frac{13}{8}$
(d) $-\frac{8}{13}$

(iv) (a) $(2, -1.5)$
(b) $\sqrt{233}$
(c) $\frac{13}{8}$
(d) $\frac{8}{13}$

4 $q = 5$

5 $y = 1$

6 (i) $\dfrac{2y}{x}$

(ii) $(2x, 3y)$

(iii) $\sqrt{4x^2 + 16y^2}$

7 (i) AB: $\frac{1}{2}$, BC: $\frac{3}{2}$, CD: $\frac{1}{2}$, DA: $\frac{3}{2}$

(ii) Parallelogram

(iii)

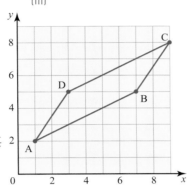

8 (i)

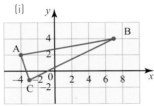

(ii) $AB = BC = \sqrt{125}$

(iii) $(-3\frac{1}{2}, \frac{1}{2})$

(iv) 17.5 square units

9 (i) 6

(ii) $AB = \sqrt{20}$, $BC = \sqrt{5}$

(iii) 5 square units

10 (i) 1 or 5

(ii) 7

(iii) 9

(iv) 1

11 (i) 18

(ii) -2

(iii) 0 or 8

(iv) 8

12 (i)

(ii) Gradient BC = Gradient AD = $\frac{1}{2}$

(iii) $(6, 3)$

13 Diagonals have gradients $\frac{2}{3}$ and $-\frac{3}{2}$ so are perpendicular. Mid-points of both diagonals are $(4, 4)$ so they bisect each other.

52 square units.

14 (i) $\angle ABE = 90° - \theta$ so $\angle DBC = \theta$ which is the same as $\angle BAE$. Both triangles have a right angle.
Since $AB = BC$ the triangles ABE and BCD are congruent (*AAS*: two angles and a side are equal).

(ii) Hence $AE = BD$ and $BE = DC$
So $m_1 = \dfrac{BE}{AE}$ and $m_2 = -\dfrac{BD}{DC} = -\dfrac{AE}{BE}$
So $m_1 m_2 = \dfrac{BE}{AE} \times \left(-\dfrac{AE}{BE}\right)$
$= -1$ as required.

Activity 5.3 (page 73)

A $\dfrac{y - y_1}{y_2 - y_1} = \dfrac{x - x_1}{x_2 - x_1}$
Multiply both sides by $\dfrac{y_2 - y_1}{x - x_1}$ gives
$\dfrac{y - y_1}{x - x_1} = \dfrac{y_2 - y_1}{x_2 - x_1}$

B $m = \dfrac{3 - 4}{5 - 2} = -\dfrac{1}{3}$
Now substitute into
$y - y_1 = m(x - x_1)$
$y - 4 = -\dfrac{1}{3}(x - 2)$
$3y - 12 = 2 - x$
$x + 3y - 14 = 0$
Using $\dfrac{y - y_1}{y_2 - y_1} = \dfrac{x - x_1}{x_2 - x_1}$
gives:
$\dfrac{y - 3}{4 - 3} = \dfrac{x - 5}{2 - 5}$
$\Rightarrow \dfrac{y - 3}{1} = \dfrac{x - 5}{-3}$
$\Rightarrow -3(y - 3) = x - 5$
$\Rightarrow -3y + 9 = x - 5$
$x + 3y - 14 = 0$

Using $\dfrac{y - y_1}{x - x_1} = \dfrac{y_2 - y_1}{x_2 - x_1}$
gives:

$$\dfrac{y - 3}{x - 5} = \dfrac{4 - 3}{2 - 5}$$

$$\dfrac{y - 3}{x - 5} = \dfrac{1}{-3}$$

$$-3(y - 3) = x - 5$$

$$-3y + 9 = x - 5$$

$$x + 3y - 14 = 0$$

Discussion points (page 74)

Equivalent forms include:

$3y = x + 14$

$y = \frac{1}{3}x + \frac{14}{3}$

It is tidier to write the equation of the line in a form that doesn't involve fractions.

Discussion points (page 75)

Interest earned on savings in a bank account (gradient gives the multiplier for the interest rate; simplifying assumption: money is not credited or debited from account, interest is simple; these assumptions are not realistic in real life, i.e. a savings account would normally have compound interest which does not fit a linear model).

Tax paid versus earnings (gradient gives tax rate; simplifying assumption: tax paid only at the lower rate).

Mass of candle versus length of time it is burning (gradient gives rate of change of mass; simplifying assumption: candle is uniform thickness).

Cost of apples versus mass of apples (gradient gives cost per unit mass of apples; simplifying assumption: no discount offered for a bulk buy).

Distance travelled by car against time (gradient gives speed of car; simplifying assumption: car is travelling at constant speed; these assumptions are not realistic in

real life, i.e. a car is unlikely to travel at constant speed).

Mobile phone bill against number of texts sent (gradient gives cost of one text; simplifying assumption: each text costs same amount, no calls or data used; these assumptions would only apply to a very simple pay-as-you-go tariff with no 'bonuses' or 'rewards' for topping up).

Profit of ice-cream seller against number of sales (gradient gives profit per ice-cream sold; simplifying assumption: profit on each sale is the same).

Mass of gold bars against volume of gold bars (gradient gives density of gold bars; simplifying assumption: all gold bars have same purity).

Length of spring versus mass of weights attached (gradient gives the extension of the spring per unit mass added).

Exercise 5.2 (page 76)

1. P is $y + 2x - 1 = 0$;
   Q is $y + 3x = 1$;
   R is $y = x + 6$;
   S is $y = 2x - 3$

2. $2y + x - 6 = 0$

3.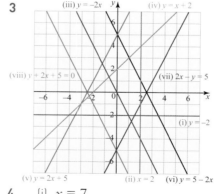

4. (i) $x = 7$
   (ii) $y = 5$
   (iii) $y = 2x$
   (iv) $x + y = 2$
   (v) $x + 4y + 12 = 0$

5. (i) $y = 3x$
   (ii) $y = 3x - 1$
   (iii) $y = 1 - 2x$
   (iv) $y = 3x - 17$
   (v) $x + 2y + 12 = 0$

6. (i) $3y + x = 0$
   (ii) $y = -\frac{1}{2}x + 5$
   (iii) $y = \frac{1}{2}x - 5$
   (iv) $y = -2x - 5$
   (v) $y = \frac{3}{2}x + 3$

7. (i) $y = 3x - 8$
   (ii) $y = 3x + 8$
   (iii) $y = -3x - 8$
   (iv) $y = 8 - 3x$
   (v) $3y = x + 8$

8. The gradient of the lines
   $y = \frac{2}{3}x + 1$ and
   $3y - 2x + 1 = 0$ is $\frac{2}{3}$
   so these lines are parallel.
   The gradient of the lines
   $y = 1 - \dfrac{3x}{2}$ and
   $2y + 3x + 5 = 0$ is $-\frac{3}{2}$
   so these lines are parallel.
   $\frac{2}{3} \times \left(-\frac{3}{2}\right) = -1$ so the lines form a quadrilateral with 2 pairs of parallel lines and four right angles which is a rectangle.

9. (i) $y = 7 - x$
   (ii) $y = 7 - x$
   (iii) $y = -x - 7$
   (iv) $y = x + 7$
   (v) $y = x - 7$

10. (i)

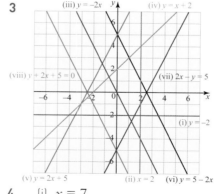

(ii) $y = x$; $x + 2y - 6 = 0$;
$2x + y - 6 = 0$

11 (i)

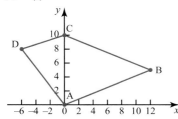

(ii) AB: $\frac{5}{12}$; BC: $-\frac{5}{12}$; CD: $\frac{1}{3}$; AD: $-\frac{4}{3}$

(iii) AB $=13$; BC $=13$; CD $= \sqrt{40}$; AD $= 10$

(iv) AB: $5x - 12y = 0$; BC: $5x + 12y - 120 = 0$; CD: $x - 3y + 30 = 0$; AD: $4x + 3y = 0$

(v) 90 square units

12 (i)

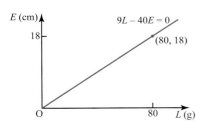

(ii) $10P + N = 300$

(iii) £21.20

(iv) 63

13 (i) $2x + y - 5 = 0$

(ii) 5 m

(iii) 5 m 59 cm

14 (i)

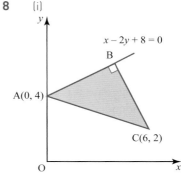

(ii) 10.8 cm

(iii) $44\frac{4}{1}$ g

(iv) $133\frac{1}{3}$ g

15 Take (x_1, y_1) to be $(0, b)$ and (x_2, y_2) to be $(a, 0)$.

The formula gives
$$\frac{y - b}{0 - b} = \frac{x - 0}{a - 0}$$
$$\Rightarrow \frac{y - b}{-b} = \frac{x}{a}$$
$$\Rightarrow \frac{y - b}{b} = -\frac{x}{a}$$
$$\Rightarrow \frac{y}{b} - \frac{b}{b} = -\frac{x}{a}$$
$$\Rightarrow \frac{y}{b} - 1 = -\frac{x}{a}$$
$$\Rightarrow \frac{x}{a} + \frac{y}{b} = 1$$

Discussion point (page 78)

The lines are parallel so there is no point of intersection.

Exercise 5.3 (page 78)

1 $y = x + 4$ and $y = -\frac{1}{2}x - \frac{1}{2}$ intersect at $(-3, 1)$; $y = -\frac{1}{2}x - \frac{1}{2}$ and $y = -3x + 12$ intersect at $(5, -3)$; $y = x + 4$ and $y = -3x + 12$ intersect at $(2, 6)$

2 $(2, -3)$

3 (i) $(\frac{1}{2}, 4)$

(ii) $(-2, 8)$

(iii) $(1, \frac{1}{2})$

4 (i) (a) $(3, 2)$

(b) $(1, 3)$

(ii) $y = -\frac{1}{2}x + 1$ or $-\frac{1}{2}x + 6$

(iii) 5 square units

5 (i) A $(1, 1)$; B $(5, 3)$; C $(-1, 10)$

(ii) BC = AC = $\sqrt{85}$

6 (i) $y = \frac{1}{2}x + 1, y = -2x + 6$

(ii) Gradients $= \frac{1}{2}$ and -2 $\Rightarrow$ AC and BD are perpendicular. Intersection $= (2, 2) =$ midpoint of both AC and BD.

(iii) AC = BD = $\sqrt{20}$

(iv) Square

7 (i) A$(\frac{5}{2}, 0)$, B$(-2, 0)$

(ii) P$(2, 8)$

(iii) $\frac{48}{5}$ square units

8 (i)

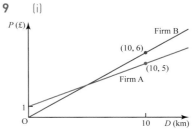

(ii) AC: $x + 3y - 12 = 0$, BC: $2x + y - 14 = 0$

(iii) AB $= \sqrt{20}$, BC $= \sqrt{20}$, area $= 10$ square units

(iv) $\sqrt{10}$

9 (i)

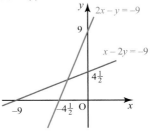

(ii) A: $2D - 5P + 5 = 0$, B: $3D - 5P = 0$

(iii) 5 km

(iv) A

10 (i)

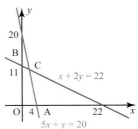

(ii) $(-3, 3)$

(iii) $2x - y = 3$; $x - 2y = 0$

(iv) $(-6, -3), (5, 7)$

11 (i)

(ii) A$(4, 0)$, B$(0, 11)$, C$(2, 10)$

(iii) 11

(iv) $(-2, 21)$

12 (i) Supply:
$L - 500W + 500 = 0$;
Demand:
$L + 750W - 4750 = 0$

(ii) $L = 1600$; $W = 4.2$

(iii) Wage rate is the independent variable.

13 (i)

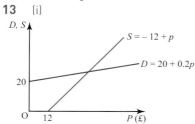

(ii) £40; 28 articles

14 (i) $(2, 4)$

(ii) $(0, 3)$

15 7.5 square units

Activity 5.4 (page 81)

(i) For the graph to look like a circle, the same scale needs to be used for both axes.

(ii) Sophie has used $y = \sqrt{(9 - x^2)}$ and so only has the part of the circle which lies above the x axis, i.e. where y is positive.

Activity 5.5 (page 82)

Multiplying out gives
$x^2 - 2ax + a^2 + y^2 - 2by + b^2 = r^2$.

Rearranging gives $x^2 + y^2 - 2ax - 2by + (a^2 + b^2 - r^2) = 0$

Discussion points (page 83)

For the angle in a semicircle, join O to the vertex and use the fact that the triangles are isosceles, together with the angle sum of a triangle and the angle on a straight line.

Converse: When triangle ABC has a right angle at B, then AC forms the diameter of a circle. For the perpendicular bisecting the chord, join O to both ends of the chord and prove that the triangles are congruent.

Converse: The centre of a circle lies on the perpendicular bisector of a chord.

For the tangent and radius, use a symmetry argument.

Converse: The radius of a circle through a point is perpendicular to the tangent at that point.

Exercise 5.4 (page 85)

1 (i) $(x - 2)^2 + (y - 3)^2 = 1$

(ii) $(x - 2)^2 + (y + 3)^2 = 4$

(iii) $(x + 2)^2 + (y - 3)^2 = 9$

(iv) $(x + 2)^2 + (y + 3)^2 = 16$

2 (i) (a) $(0, 0)$
(b) radius: 1

(ii) (a) centre: $(0, 2)$
(b) radius: $\sqrt{2}$

(iii) (a) centre: $(2, 0)$
(b) radius: $\sqrt{3}$

(iv) (a) centre: $(-2, -2)$
(b) radius: 2

(v) (a) centre: $(2, -2)$
(b) radius: $\sqrt{5}$

3

Point	Inside	Outside	On
$(3, -2)$	✓		
$(-2, -5)$		✓	
$(6, -6)$	✓		
$(4, 3)$			✓
$(0, 2)$	✓		
$(-2, -3)$			✓

4

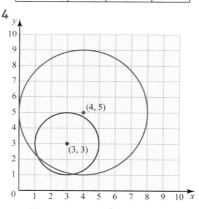

Two points of intersection.

5

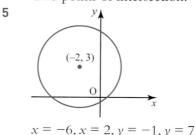

$x = -6, x = 2, y = -1, y = 7$

6 (i) $(\pm 5, 0), (0, \pm 5)$

(ii) $(0, -2), (0, -8), (4, 0)$

(iii) $(0, 0), (0, 16), (-12, 0)$

7 $(x - 1)^2 + (y - 7)^2 = 169$

8 $r = 2$; $(-1, 2)$; 2

9

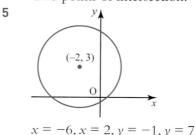

$(x - 4)^2 + (y - 4)^2 = 16$

10 Find the perpendicular bisector of AB.

11 (i) (a) $(3, 1)$
(b) $r = 4$

(ii) (a) $(-1, -3)$
(b) $r = 4$

(iii) (a) $(1, -4)$
(b) $r = 3$

12 (i) $AB = \sqrt{20}$, $BC = \sqrt{45}$, $AC = \sqrt{65}$

(ii) $AB^2 + BC^2 = 65 = AC^2$
$\Rightarrow$ triangle ABC is right-angled, with the right angle at B (converse of Pythagoras' theorem)
$\Rightarrow$ B is angle in a semi-circle (converse of the angle in a semi-circle is a right angle) i.e. AC is a diameter of the circle, as required.

(iii) 15 square units

13 (i) $(2, 11)$; $\sqrt{10}$

(ii) $(x - 2)^2 + (y - 11)^2 = 10$

14 (i) Centre: $(3, -1)$; B: $(5, 0)$

15 $\frac{169}{24}$ square units

16 $(x - 5)^2 + (y - 4)^2 = 25$ or $(x - 5)^2 + (y + 4)^2 = 25$

17 (i) $\sqrt{45}$

(ii) $(x - 11)^2 + (y - 8)^2 = 50$

18 $(x - 3)^2 + (y - 2)^2 = 25$

Discussion points (page 89)

Look for any like terms. You can see that x^2 appears in both equations, so you can eliminate x^2 by rewriting $y = x^2 - 4$ as $x^2 = y + 4$ and then substituting into the equation of the circle. This method is more efficient.

Exercise 5.5 (page 90)

1 $(3.6, 0.2)$
2 $(2, 7)$
3 $(1, 1); (-\frac{1}{5}, -\frac{7}{5})$
4 $(1, -2)$; the line forms a tangent to the circle
5 (ii)

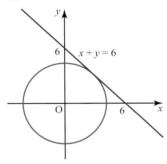

   (iii) $(-3, -3); x + y = -6$
6 (i) $(1, 2); (-5, -10)$
   (ii) no real roots
7 $(0, 5); (-3, 4). (0, 1); (1, 2); (2, 3)$
8 (i) $14\,\text{cm}$
   (ii) $x^2 + (y-8)^2 = 100$
   (iii) $264\,\text{cm}^2$ to 3 s.f., or 84π
9 (i) $(-3, 4)$ and $(4, -3); \sqrt{98}$
   (ii) No; the diameter of the circle is 10 and AB is $\sqrt{98} < 10$, hence AB is a chord not a diameter.
10 (i) $k = 2$
    (ii) $(2, 2)$
11 $k = -6$ or $k = 2$
12 5 square units

Practice questions (page 96)

1 Outline answer

   (i) $\sqrt{2\frac{2}{3}} = \sqrt{\frac{8}{3}}$

       $= \sqrt{\frac{4 \times 2}{3}}$ [1]

       $= 2\sqrt{\frac{2}{3}}$ [1]

   (ii)
   $$\frac{\sqrt{3}+1}{\sqrt{3}-1} = \frac{(\sqrt{3}+1)(\sqrt{3}+1)}{(\sqrt{3}-1)(\sqrt{3}+1)}$$ [1]

       $\dfrac{4 + 2\sqrt{3}}{2} = 2 + \sqrt{3}$ [1]

2 Outline answer

   (i) $2^{3x} = 2^{2(x+4)}$ [1]
       $3x = 2x + 8$ [1]
       $x = 8$ [1]
   (ii) $x = 0$ [1]

3 Outline answer

   (i) $x^2 - 4x + 1 = 7 - x^2$ [1]

       $2x^2 - 4x - 6 = 0$ [1]
       $x^2 - 2x - 3 = 0$

       $(x - 3)(x + 1) = 0$ [1]
       $(3, -2)$ and $(-1, 6)$ [1,1]

   (ii) $x^2 - 4x + 1 = -2x$ [1]

       $x^2 - 2x + 1 = 0$

       $(x - 1)^2 = 0$ [1]

       Exactly one solution so the line is a tangent to the curve. [1]
       $(1, -2)$ [1]

4 Outline answer

   (i) $(x^2 + 6x) + 7$

       $((x + 3)^2 - 9) + 7$ [1]

       $(x + 3)^2 - 2$ [1,1]

   (ii) $(-3, -2)$ [1, 1]
       Minimum [1]

   (iii)

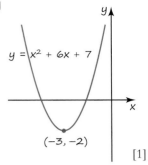

   $y = x^2 + 6x + 7$
   $(-3, -2)$ [1]

   $(x + 3)^2 - 2 > 0$ [1]
   $(x + 3)^2 > 2$
   $x + 3 > \sqrt{2}$ or
   $x + 3 < -\sqrt{2}$
   $x > -3 + \sqrt{2}$ or
   $x < -3 - \sqrt{2}$ [1,1]

5 Outline answer

   (i) Centre lies on perpendicular bisector of AB [1]
       Midpoint of AB is $(0.5, 2.5)$ [1]
       Gradient of AB is $\frac{3}{3} = 1$ [1,1]
       Gradient of perpendicular bisector is -1 [1]
       Equation of perpendicular bisector is $(y - 2.5) = -(x - 0.5)$ [1]
       Centre on $x + y = 3$ [1]
   (ii) Centre $(3, 0)$ [1,1]
       Radius^2 = $1^2 + 4^2 = 17$ [1,1]
       Equation $(x - 3)^2 + y^2 = 17$ [1]

6 Outline answer

   (i) $A(-2, 0), B(2, 0)$ [1,1]
       $OC^2 = BC^2 - OB^2$
       $OC^2 = 16 - 4 = 12$ [1]
       $C(0, \sqrt{12})$ or $(0, 2\sqrt{3})$ [1]
   (ii) Gradient $-\dfrac{\sqrt{12}}{2} = -\sqrt{3}$ [1]

$$y = \sqrt{12} - x\sqrt{3}$$
$$= \sqrt{3}(2 - x) \qquad [1]$$

(iii) Method 1

GFC is an equilateral triangle (hence similar to ABC) [1]

Suppose GF = 4d

The height of triangle GFC is $d\sqrt{12}$ [1]

FE = $\sqrt{12} - d\sqrt{12}$

$= \sqrt{12}(1 - d)$ [1]

Area of rectangle

$= 4d\sqrt{12}(1 - d)$ [1]

$d(1 - d)$ is a quadratic

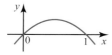

Maximum at line of symmetry, i.e. when $d = 0.5$ [1,1]

Maximum area

$= \sqrt{12}$ [1]

Method 2

F has coordinates $(x, \sqrt{3}(2 - x))$ [1,1]

Area of rectangle

$= 2x\sqrt{3}(2 - x)$ [1,1]

$x(2 - x)$ is a quadratic

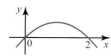

Maximum at line of symmetry, i.e. when $x = 1$ [1,1]

Maximum area = $\sqrt{12}$ [1]

7 Outline answer

(i) (a) The points for thinking distance lie in a straight line. [1]

(b) Distance travelled is proportional to

speed at constant time so this means the thinking time is the same for all speeds. [1]

Reasonable as it is to do with reaction time. [1]

(c) $d = 0.3x$ [1]

(ii) (a) -9.609 (m) [1]

(b) It is not possible to have negative stopping distance. [1]

(iii) (a) 12.138 (m) [1]

(b) Possible reason such as:
- the model does not include all factors
- distances in the Highway Code have been rounded. [1]

Chapter 6

Opening activity (page 99)

→ Angle between train and mountain at each point, the angles of elevation of the point on the mountain as viewed from the train and the speed of the train, hence the distance between the points.

→ Angle of elevation and height of lighthouse.

Exercise 6.1 (page 103)

1 Sometimes true, when $\theta = 90°$, $\sin \theta = 1$; never true for values > 1 as $\sin \theta$ has a maximum value of 1.

2 $\dfrac{24}{25}$

3 (i) Converse of Pythagoras' theorem

(ii) $\dfrac{8}{17}, \dfrac{15}{17}, \dfrac{8}{15}$

4 (i) $\dfrac{3}{2}$ (ii) $\dfrac{1}{3}$

(iii) $\dfrac{1}{2}$ (iv) $\sqrt{3}$

(v) 3 (vi) $\dfrac{3\sqrt{3}}{\sqrt{2}}$

6 (i) 5 cm

7 (i) $\dfrac{8}{9}\sqrt{3}$ cm

Activity 6.1 (page 105)

Discussion point (page 108)

The oscillations continue to the left.

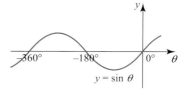

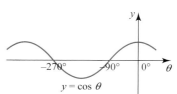

Activity 6.2 (page 109)

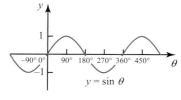

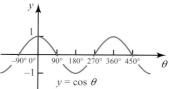

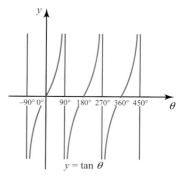

Discussion point (page 109)

$y = \sin\theta$: the graph has rotational symmetry order 2 about the origin.

$y = \cos\theta$: the graph is symmetrical about the y axis.

$y = \tan\theta$: the graph has rotational symmetry order 2 about the origin.

Exercise 6.2 (page 111)

1   $\sin 170° = \sin 10°$;
    $\cos 350° = \cos 10°$;
    $\sin 260° = \cos 170°$;
    $\sin 190° = \cos 100°$

2   $\tan 40°$

3   (i) (a) $\dfrac{\sqrt{3}}{2}$

         (b) $\dfrac{1}{2}$

         (c) $\sqrt{3}$

    (ii) (a) $-\dfrac{\sqrt{3}}{2}$

         (b) $\dfrac{1}{2}$

         (c) $-\sqrt{3}$

4   (i) (a) $\dfrac{\sqrt{3}}{2}$

         (b) $-\dfrac{1}{2}$

         (c) $-\sqrt{3}$

    (ii) (a) $\dfrac{1}{2}$

         (b) $-\dfrac{\sqrt{3}}{2}$

         (c) $\dfrac{1}{\sqrt{3}}$

    (iii) (a) $\dfrac{1}{\sqrt{2}}$

         (b) $\dfrac{1}{\sqrt{2}}$

         (c) 1

5   $\dfrac{25\sqrt{3}}{2}$

6   (i) α between 0° and 90°, 360° and 450°, 720° and 810°, etc. (and corresponding negative values)

    (ii) No: since $\tan\alpha = \dfrac{\sin\alpha}{\cos\alpha}$, all must be positive or one positive and two negative

    (iii) No: $\sin\alpha = \cos\alpha$ $\Rightarrow \alpha = 45°, 225°$, etc.

---

but $\tan\alpha = \pm 1$ for these values of α, and $\sin\alpha = \cos\alpha = \dfrac{1}{\sqrt{2}}$

7   (i)

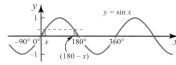

shaded areas are congruent

    (ii) (a) False
         (b) True
         (c) False
         (d) True

8   (i) $\dfrac{4}{5}$

    (ii) $\dfrac{4}{3}$

9   (i) $-\dfrac{1}{2}$

    (ii) $-\dfrac{1}{\sqrt{3}}$

10 (i) $-\dfrac{\sqrt{7}}{4}$

    (ii) $-\dfrac{3}{\sqrt{7}}$

11 (i) $\sin\theta$

    (ii) $\cos\theta$

    (iii) 2

13 (i) $\sqrt{1-k^2}$

    (ii) $\dfrac{\sqrt{1-k^2}}{k}$

14 n is an integer
    (i) $(90°n - 45°)$
    (ii) $(180°n - 90°)$ and $180°n$
    (iii) $180°n - 90°$ and $180°n$
    (iv) $180°n - 90°$

Discussion point (page 113)

Add or subtract multiples of 180°.

Discussion point (page 116)

The graph crosses any horizontal line (in this case $y = \sqrt{3}$) four times.

Exercise 6.3 (page 117)

1   $\sin 60° = -\cos 210° = -\sin 300°$;
    $\tan 60° = -\tan 120° = \tan 240°$;
    $\cos 120° = -\sin 150° = -\cos 300°$;

$\tan 150° = -\tan 210° = \tan 330°$

2   126.9°

3   (i) 120°, 300°
    (ii) 30°, 210°
    (iii) 60°, 300°
    (iv) 120°, 240°
    (v) 45°, 135°
    (vi) 225°, 315°

4   (i) 36.9°, 143.1°
    (ii) 216.9°, 323.1°
    (iii) 53.1°, 306.9°
    (iv) 126.9°, 233.1°
    (v) 31.0°, 211.0°
    (vi) 149.0°, 329.0°

5   (i) $-45.6°$
    (ii) $-158.2°$
    (iii) 53.1°

6   (i) 60°, 300°
    (ii) 199.5°, 340.5°
    (iii) 60°, 120°, 240°, 300°
    (iv) 0°, 180°, 360°
    (v) 18.4°, 71.6°, 198.4°, 251.6°
    (vi) 180°

7   (i) 60°, 180°, 300°
    (ii) 0°, 90°, 270°, 360°
    (iii) 0°, 180°, 360°
    (iv) 54.7°, 125.3°, 234.7°, 305.3°
    (v) 60°, 300°
    (vi) 120°, 240°

8   (i) 70°, 310°
    (ii) 10°, 190°
    (iii) 30°, 120°
    (iv) 10°, 70°, 130°, 190°, 250°, 310°
    (v) 120°, 240°
    (vi) 60°, 120°, 240°, 300°
    (vii) 20°, 100°, 140°, 220°, 260°, 340°
    (viii) 120°, 150°, 300°, 330°
    (ix) 0°, 60°, 120°, 180°, 240°, 300°, 360°

9   9°, 45°, 81°

10 A = (38.2°, 0.786), B = (141.8°, −0.786)

11 (i) 26.6°, 206.6°
    (ii) 45°, 225°
    (iii) 54.7°, 125.3°, 234.7°, 305.3°

12 (i) 45°, 75°, 225°, 255°
    (ii) 13.3°, 103.3°, 193.3°, 283.3°
    (iii) 15°, 75°, 135°, 195°, 255°, 315°
    (iv) 75°, 165°, 255°, 345°
    (v) 60°, 300°
    (vi) 90°, 270°
13 90°, 270°

Discussion point (page 121)

$a^2 = b^2 + c^2 - 2bc \cos A$

$\Rightarrow 2bc \cos A = b^2 + c^2 - a^2$

$\Rightarrow \cos A = \dfrac{b^2 + c^2 - a^2}{2bc}$

Exercise 6.4 (page 123)

1 It must be wrong as the triangle is impossible with an answer < 2 (7 − 5). The error was working out $5^2 + 7^2 - 2 \times 5 \times 7 \; (= 4)$ and then multiplying all of that by cos 82°, rather than just the $2 \times 5 \times 7$.

2 (i) 8.0 cm
    (ii) 7.4 cm
3 (i) 10.14 cm
    (ii) 5.57 cm
4 (i) 42.8°
    (ii) 47.9°
    The diagram shows that θ is acute, so 132.1° is relevant.
5 (i) 57.1°
    (ii) 97.4°
6 (i) 5 cm
    (ii) 90.7°
7 8.8 km
8 10.7 km
9 3.28 km
10 (i) 18.6°
    (ii) 76.9 m
    (iii) 35.6 m
11 (i) 1011 m
    (ii) 1082 m
    (iii) 065°

12 (i)

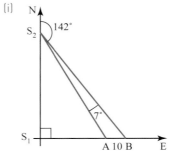

    (ii) 64.7 km
    (iii) 27.7 km h^{-1}
13 14.6 km h^{-1}
14 (i) 3.72 km
    (ii) 3.32 km
    (iii) 94.8 km h^{-1}
15 (i)

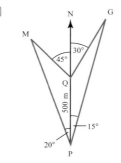

    (ii) 837 m
    (iii) 556 m
16 (i) QR = QT = a, RT = $a\sqrt{2 - \sqrt{3}}$

Activity 6.3 (page 126)

1 (i) $\frac{1}{2} bc \sin A$
    (ii) $\frac{1}{2} ac \sin B$
    (iii) $\frac{1}{2} ab \sin C$
2 Each expression gives the area of the triangle which is the same however it is worked out.
3 $\frac{1}{2} bc \sin A = \frac{1}{2} ac \sin B = \frac{1}{2} ab \sin C$;

$\dfrac{\sin A}{a} = \dfrac{\sin B}{b} = \dfrac{\sin C}{c}$;

the sine rule

Exercise 6.5 (page 127)

1 $\frac{1}{2} ab \sin A$
2 $\frac{1}{2} \times 7.8 \times 8.4 \times \sin 51°$

3 (i) 7.21 cm^2
    (ii) 8.45 cm^2
    (iii) 6.77 cm^2
    (iv) 6.13 cm^2
4 (i) 2.25 m^2
    (ii) 0.3375 m^3
5 27.4°, 152.6°
6 77.94 cm^2
7 (i) 4.35 m
    (ii) 7.38 m
    (iii) 47.38 m^2
    (iv) 7.29 m
8 11 011 m^2
9 5412 m^2

Chapter 7

Discussion points (page 133)

Order 3.

Order $m + n$.

Line A: $x^3 + 3x - 2$ has been multiplied by x^2.

Line B: $x^3 + 3x - 2$ has been multiplied by $-2x$.

Line C: $x^3 + 3x - 2$ has been multiplied by -4.

Line D: lines A, B and C have been added together.

The multiplication has been laid out in columns so that each column contains a different power of x.

Discussion point (page 136)

$x = 0$ on the y axis. Substituting $x = 0$ in the polynomial makes all terms zero except the constant term.

Discussion point (page 137)

If there is a factor $(x - a)^3$, the curve is horizontal at the x axis but crosses it. If there is a factor $(x - a)^4$, the curve touches the x axis, but is flatter than if there were a factor $(x - a)^2$.

Exercise 7.1 (page 138)

1

	Order		
	3	4	5
$x \to \infty$ $f(x) \to \infty$	R	QT	
$x \to \infty$ $f(x) \to -\infty$	S	U	P

2   −2

3   (i) 3
     (ii) 4
     (iii) 2

4   (i) $2x^3$
     (ii) $2x^3 + 6x + 4$
     (iii) $-2x^2 - 6x - 4$

5   (i) $2x^3 + 7x^2 + 9x + 10$
     (ii) $9x^2 + 9x + 10$
     (iii) $2x^3 + 7x^2 + 12x + 17$
     (iv) $6x^2 + 7x + 9$

6   (i) $x^4 + 4x^3 + 6x^2 + 4x + 1$
     (ii) $x^3 - 7x + 6$

7   (i) D
     (ii) A
     (iii) C
     (iv) B

8   (i)
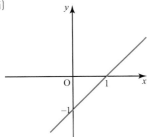
     (ii)
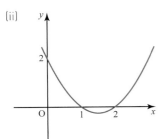
     (iii)
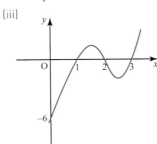

(iv)
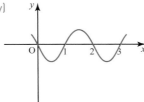

9   (i) $-2x^2 + 2x$
     (ii) $10x^2$

10   (i) $x^3 - 7x - 6$
     (ii) $-2x^3 + 5x^2 + 4x - 3$

11   (i)
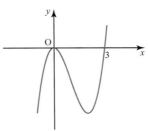
     (ii)
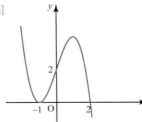
     (iii)
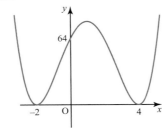

12   (i) For example, $x = 7$
          gives $y = 21$, which
          does not fit the graph.
     (ii) k is approximately $\frac{1}{3}$.
     (iii) The graph is fairly flat
          near the origin, and
          Fatima's equation gives
          a repeated root at $x = 0$.
     (iv) p is approximately 0.05.

13   (i)
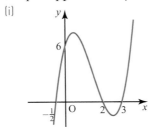

(ii) $f(x) = 2x^3 - 9x^2 + 7x + 6$

14   (i) Possibly
          $y = -(x - 2)^2(x - 4)$.
          Could be any multiple
          of this since no y axis
          intercept is given.

     (ii) Possibly
          $y = \frac{1}{2}(x + 1)^2(x - 2)^2$.
          Other equations like
          $y = \frac{1}{2}(x + 1)^4(x - 2)^2$
          would fit the intersections
          with the axes, but
          substituting $x = 1$ shows
          that this does not fit the
          rest of the graph.

Discussion points (page 140)

Order 3
Order $m - n$

Exercise 7.2 (page 141)

1   (i) $x^2 - 2x - 3 =$
          $(x - 3)(x + 1)$
     (ii) $x^3 - 3x^2 - 10x + 24 =$
          $(x^2 - x - 12)(1x - 2)$
     (iii) $2x^3 + x^2 - 7x - 6 =$
          $(x + 1)(2x^2 - 1x - 6)$

2   (i) $x + 3$
     (ii) $x - 1$
     (iii) $x^2 + 3x$

3   (i) $3x + 1$
     (ii) $5x - 4$
     (iii) $2x + 1$

4   (i) $2x - 1$
     (ii) $4x + 3$
     (iii) $2x - 3$

5   (i) $x^2 + x + 2$
     (ii) Solve $13 = x + 3$ to get
          $x = 10$ and substitute
          into the quadratic.
     (iii) Solve $8 = x + 3$ to get
          $x = 5$ and substitute
          into the quadratic.

6   (i) $x^2 + 3x + 2$
     (ii) $2x^2 - 5x + 5$

7   (i) $x^2 + x + 2$
     (ii) $2x^2 + 3$

8   (i) $x^2 + 3x$
     (ii) $x^3 + x^2 + 2x + 2$
     (iii) $2x^2 + 3$

9 (i) $x^3 - x^2 + 2x - 2$
　　(ii) $3x^3 + 2x^2 + 2x + 4$

Activity 7.2 (page 142)

1.689

Discussion point (page 143)

Three roots have already been found and a cubic only has 3 roots.

Discussion point (page 144)

Any integer root must be a factor (positive or negative) of the constant term.

Exercise 7.3 (page 146)

1 (i) $x + 1$ is a factor of f(x), f(-1) = 0, when $x = -1$, f(x) = 0
　　　　$x - 1$ is a factor of f(x), f(1) = 0, when $x = 1$, f(x) = 0
　　(ii) $x + 2$ is a factor of f(x), f(-2) = 0, when $x = -2$, f(x) = 0

2 $x + 1$

3 Show substituting $x = 1$ gives a zero answer.

4 Show substituting $x = -2$ gives a non-zero answer.

5 $x = 0$

6 $k = -4$

7 Show substituting gives a non-zero answer in each case.

8 (i) $f(0) = 30; f(3) = 0;$
　　　　$(x - 3)$
　　(ii) $p = 2; q = -15$
　　(iii) $x = 2, 3, -5$
　　(iv)

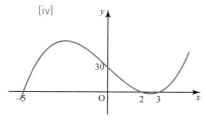

9 $a = 2, 3$

10 $a = 0, b = -7, (x + 3)$

11 $x = -\frac{1}{2}, 1, 3$

12 $4p + q = -26; 9p + q = 39;$
　　$p = 13, q = -78$

13 $(x + 2)(x + 3)(x - 4)$

14 (i) $f(x) = (x - 1)^2(x + 4)$
　　(ii) A($-4, 0$); B($1, 0$); C($0, 4$)
　　(iii) $(2\frac{1}{4}, 9\frac{49}{64})$

15 (i) $\dfrac{18}{x^2}$
　　(iv) $x = 3$ or 3.62 (3 s.f.) or -6.52 (3 s.f.)
　　　　Dimensions $3 \times 3 \times 2$ or $3.62 \times 3.62 \times 1.37$ (3 s.f.)

Chapter 8

Discussion point (page 148)

For the solid curve, the first turning point is $(0.5, 0.08)$. Between 0 and 1 it has roots $x = 0$ and $x = 1$, so is $x(1 - x)$ scaled by factor $\frac{0.08}{0.25} = 0.32$

You can therefore consider it to be a two part function consisting of 2 quadratic functions.

　　$f(x) = 0.32x(1 - x)$ for $0 \le x \le 1$
and　$f(x) = 0.32(x - 1)(x - 2)$ for $1 < x \le 2$.

Likewise for the dashed curve:

　　$g(x) = -0.32x(1 - x)$ for $0 \le x \le 1$
and　$g(x) = -0.32(x - 1)(x - 2)$ for $1 < x \le 2$.

Activity 8.1 (page 151)

(i) A: $y = x^3$; B: $y = x^2$;
　　C: $y = x^4$ and D: $y = x^5$
(ii) (a) When n is even the curve is U shaped. As n increases the curve is flatter at the origin and increases more steeply.
　　(b) When n is odd the curve is ⌐-shaped. As n increases the curve is flatter at the origin and increases more steeply.
(iii) They are reflections in the x axis of the curves of $y = x^n$.

Exercise 8.1 (page 153)

1

	No negative values		Some negative values	
Continuous	$y = kx^2$	$y = k\sqrt{x}$	$y = kx$	$y = kx^3$
Discontinuous	$y = \dfrac{k}{x^2}$		$y = \dfrac{k}{x}$	$y = \dfrac{k}{x^3}$

2 $y = x^2 - 3$

3 (i) E, (ii) D, (iii) C, (iv) A, (v) B

4 (i) Asymptotes at $x = 0$ and $y = 0$

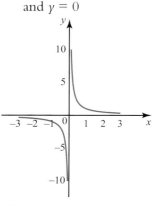

(ii) Asymptotes at $x = 0$ and $y = 0$

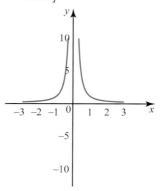

(iii) Asymptotes at $x = 0$ and $y = 0$

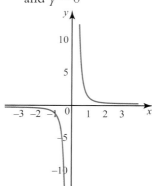

(iv) Asymptotes at $x = 0$
and $y = 0$

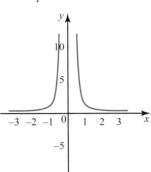

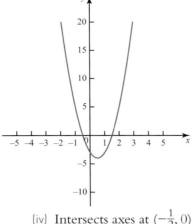

(ii) Intersects axes at $(\frac{3}{4}, 0)$,
$(\frac{4}{3}, 0)$ and $(0, -36)$

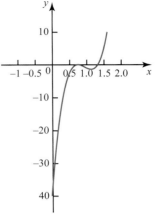

5 (i) Intersects axes at $(1, 0)$,
$(-3, 0)$ and $(0, -3)$

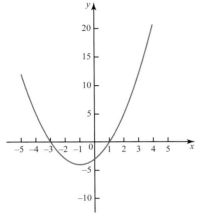

(iv) Intersects axes at $(-\frac{1}{2}, 0)$,
$(0, 0)$ and $(\frac{3}{2}, 0)$

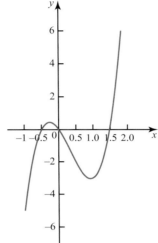

(iii) Intersects axes at $(-2, 0)$,
$(4, 0)$ and $(0, 64)$

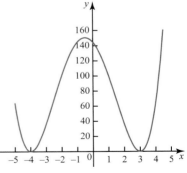

(ii) Intersects axes at $(-1, 0)$,
$(-2, 0)$, $(-3, 0)$ and $(0, 6)$

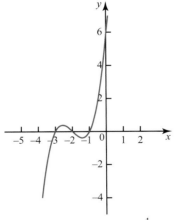

6 (i) Intersects axes at $(-1, 0)$,
$(3, 0)$ and $(0, 3)$

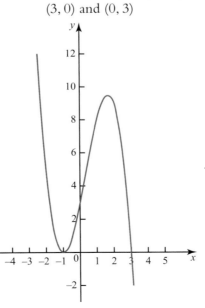

(iv) Intersects axes at $(-4, 0)$,
$(3, 0)$ and $(0, 144)$

(iii) Intersects axes at $(-\frac{1}{2}, 0)$,
$(\frac{3}{2}, 0)$ and $(0, -3)$

7 (i)

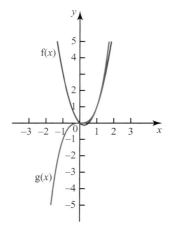

(ii) 2

(iii) (a) $(0, 0)$

(b) $(1, 1)$

8 (i) $y = \sqrt{2x}$

(ii)

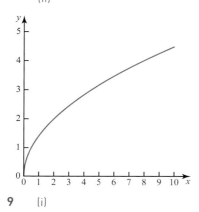

9 (i)

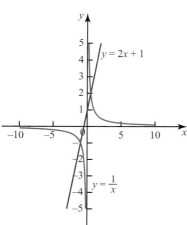

(ii) $(-1, -1), (\frac{1}{2}, 2)$

(iii) $y = \frac{1}{x}$ shows a proportional relationship as y is proportional to $\frac{1}{x}$.

$y = 2x + 1$ does not show direct proportion as it doesn't pass through the origin.

10 (i) $y = \dfrac{3}{x^2}$

(ii) $y = 0.03$

(iii) $x = \pm 2$

11 (i) $y = \sqrt{\dfrac{2}{x}}$

(ii)

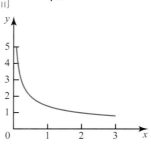

12 $2, \left(\dfrac{\sqrt{2}}{2}, 2\sqrt{2}\right), \left(-\dfrac{\sqrt{2}}{2}, -2\sqrt{2}\right)$

Activity 8.2 (page 155)

A All the graphs are translations of each other in the y direction.

B All the graphs are translations of each other in the x direction.

Activity 8.3 (page 158)

A $y = -f(x)$ is a reflection of $y = f(x)$ in the x axis

B $y = f(-x)$ is a reflection of $y = f(x)$ in the y axis

Activity 8.4 (page 159)

A The graphs are stretched in the y direction.

B The graphs are stretched in the x direction.

Discussion point (page 160)

A stretch in the negative direction, so this means that the graph will be reflected in the x axis as well as stretched.

Exercise 8.2 (page 161)

1 Translations: $y = \sin(x - 2)$, $y = 2 + \sin x$, $y = \sin x - 2$; reflections: $y = -\sin x$;

stretches: $y = 2\sin x$, $y = \frac{1}{2}\sin x$, $y = \sin 2x$

2 $y = (x - 2)^2$

3

x	-3	-1	0	2
$f(x)$	3	5	2	-1
$f(x) + 2$	5	7	4	1
$f(x) - 2$	1	3	0	-3
$2f(x)$	6	10	4	-2
$-2f(x)$	-6	-10	-4	2

4 (i) translation by $\begin{pmatrix} -5 \\ 0 \end{pmatrix}$

(ii) one way stretch, scale factor $\frac{1}{5}$, parallel to the x axis

(iii) one way stretch, scale factor 5, parallel to the y axis

(iv) reflection in the y axis

(v) translation by $\begin{pmatrix} 0 \\ 5 \end{pmatrix}$.

5 (i)

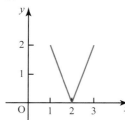

(ii)

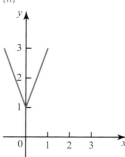

(iii)

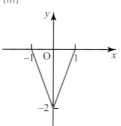

(iv)

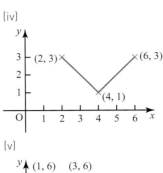

(v)

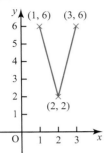

(vi)

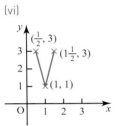

6 (i) 9
(ii) 5
(iii) 1
(iv) 13
(v) 15
(vi) 3.5

7 (i)

(ii)

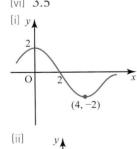

(iii)

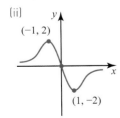

(iv)

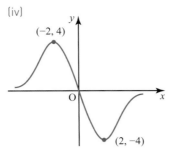

(v)

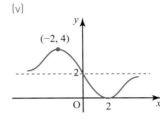

(vi)

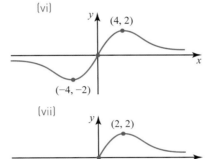

(vii)

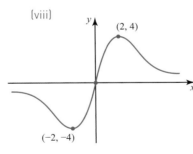

(viii)

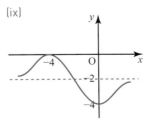

(ix)

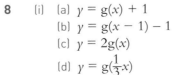

8 (i) (a) $y = g(x) + 1$
(b) $y = g(x - 1) - 1$
(c) $y = 2g(x)$
(d) $y = g(\frac{1}{3}x)$

(ii) (a)

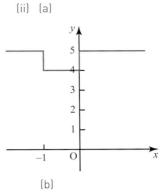

(b)

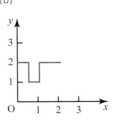

(c)

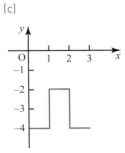

9 (i) (a) $(1, 0); (-3, 0), (0, 3),$
$(-1, 4)$
(b) $(-1, 0); (3, 0), (0, -3),$
$(1, -4)$
(c) $(-\frac{1}{3}, 0); (1, 0), (0, 3),$
$(\frac{1}{3}, 4)$
(d) $(-1, 0); (3, 0),$
$(0, -\frac{3}{2}), (1, -2)$

(ii) $a = 1$

10 (i) $y = f(x + 2)$
(ii) $y = -f(x)$
(iii) $y = f\left(\frac{x}{2}\right)$
(iv) $y = f(x) - 3$
(v) $y = f(-x)$ (or $y = 2 - f(x)$)
(vi) $y = \frac{3}{2}f(x)$

Discussion points (page 164)

Completing the square uses
an approach starting from the
algebra of the equation in the
form $y = ax^2 + bx + c$. Using
transformations starts best
from the graph, identifying the
equation as $y = (x - p)^2 + q$.

Yes. An identity is true for all values of the variable so any particular values can be substituted to give an equation.

Exercise 8.3 (page 166)

1 $y = (x + 5)^2 - 1$ is a translation of $\begin{pmatrix} -5 \\ -1 \end{pmatrix}$ with minimum point $(-5, -1)$
$y = (x - 1)^2 - 5$ is a translation of $\begin{pmatrix} 1 \\ -5 \end{pmatrix}$ with minimum point $(1, -5)$
$y = (x + 1)^2 + 5$ is a translation of $\begin{pmatrix} -1 \\ 5 \end{pmatrix}$ with minimum point $(-1, 5)$

2 $(-2, -3)$

3 (i) (a) $y = x^3 + 4$
    (b) $y = (x + 3)^3$
    (c) $y = (x - 3)^3 - 4$
  (ii) (a) $y = 2x^3$
    (b) $y = \frac{1}{8}x^3$
    (c) $y = 3x^3$
    (d) $y = 8x^3$

4 (i)
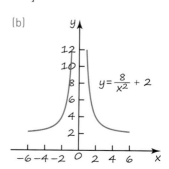

  (ii) $(y - 1) = 2(x - 4) + 3$;
    $y = 2x - 4$

5 (i)

(ii) -3
(iii) 1

6 (i) $y = 2x - 10$
  (ii) $y = 2x + 10$
  (iii) $y = 2x$

7 (i) One way stretch scale factor -2 parallel to the y axis.
  (ii) Translation by the vector $\begin{pmatrix} 0 \\ -2 \end{pmatrix}$
  (iii) Translation by the vector $\begin{pmatrix} 2 \\ 0 \end{pmatrix}$
  (iv) Translation by the vector $\begin{pmatrix} 2 \\ -2 \end{pmatrix}$

8 (i) One way stretch, stretch factor 4 parallel to the y axis, or one way stretch, stretch factor $\frac{1}{2}$ parallel to the x axis.
  (ii) One way stretch, stretch factor $\frac{1}{3}$ parallel to the y axis, or one way stretch, stretch factor $\sqrt{3}$ parallel to the x axis.

9 (i) $\begin{pmatrix} 0 \\ 4 \end{pmatrix}$; $x = 0$
  (ii) $\begin{pmatrix} -4 \\ 0 \end{pmatrix}$; $x = -4$
  (iii) $\begin{pmatrix} 0 \\ -3 \end{pmatrix}$; $x = 0$
  (iv) $\begin{pmatrix} 3 \\ 0 \end{pmatrix}$; $x = 3$
  (v) $\begin{pmatrix} 4 \\ 3 \end{pmatrix}$; $x = 4$
  (vi) $\begin{pmatrix} -3 \\ 4 \end{pmatrix}$; $x = -3$
  (vii) $\begin{pmatrix} 2 \\ -4 \end{pmatrix}$; $x = 2$
  (viii) $\begin{pmatrix} 2 \\ -1 \end{pmatrix}$; $x = 2$

10 (i) $a = 3$ and $b = -4$
  (ii)
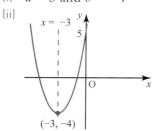

11 $y = 3(x - 4)(x - 1)(x + 2)$

12 (i) $y = \frac{8}{x^2}$
  (ii)
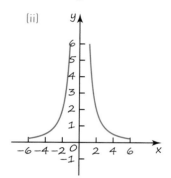

Asymptotes at $x = 0$ and $y = 0$

  (iii) (a)
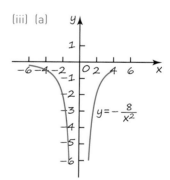

Asymptotes at $x = 0$ and $y = 0$

    (b)
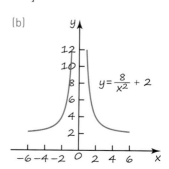

Asymptotes at $x = 0$ and $y = 2$

(c)

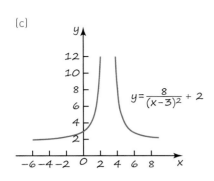

$$y = \frac{8}{(x-3)^2} + 2$$

Crosses y axis at $(0, 2\frac{8}{9})$
Asymptotes at $x = 3$ and
$y = 2$

(iv) (a) $x = 5$
   (b) $(5, 4)$

13 $y = x^2 + 20x + 91$

14 (i) $p = 3$ and $q = 2$
   (ii)

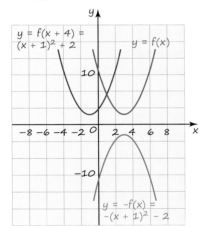

15 (i)

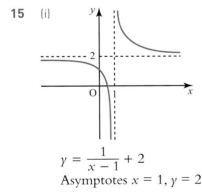

$$y = \frac{1}{x - 1} + 2$$
Asymptotes $x = 1$, $y = 2$

(ii)

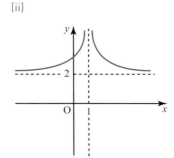

$$y = \frac{1}{(x - 1)^2} + 2$$
Asymptotes $x = 1$, $y = 2$

Exercise 8.4 (page 169)

1

	Maximum value of the function		
	1	2	3
Maximum at $x = 90°$	P, R		Q
Maximum not at $x = 90°$	U	T	S

2 $y = \sin x + 3$

3 (i) $-1 \leq f(x) \leq 1$
   (ii) $-3 \leq f(x) \leq 3$
   (iii) $-2 \leq f(x) \leq 0$
   (iv) $2 \leq f(x) \leq 4$

4 (i) $360°$
   (ii) $180°$
   (iii) $720°$
   (iv) $120°$

5 (i) (a)

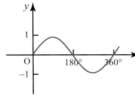

   (b) $y = \sin x$
   (ii) (a)

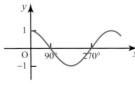

   (b) $y = \cos x$

(iii) (a)

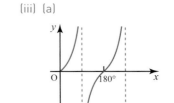

   (b) $y = \tan x$

(iv) (a)

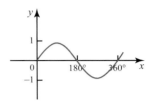

   (b) $y = \sin x$

(v) (a)

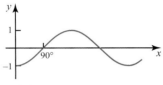

   (b) $y = -\cos x$

6 These can be checked using graphing software.
   (i) Translation $\begin{pmatrix} 90° \\ 0 \end{pmatrix}$
   (ii) One way stretch parallel to x axis of scale factor $\frac{1}{3}$
   (iii) One way stretch parallel to x axis of scale factor $\frac{1}{2}$
   (iv) One way stretch parallel to x axis of scale factor 2
   (v) Translation $\begin{pmatrix} 0 \\ 2 \end{pmatrix}$
   (vi) Reflection in the x axis

7 (i) (a) $y = \tan x + 4$
   (b)

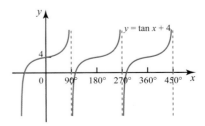

(ii) (a) $y = \tan(x + 30°)$

(b)

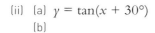

(iii) (a) $y = \tan 0.5x$

(b)

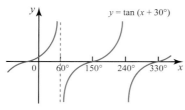

8 (i) $(60°, 0), (0, \frac{\sqrt{3}}{2})$

(ii) $(60°, 0), (0, -\sqrt{3})$

(iii) $(90°, 0), (0, \frac{1}{2})$

(iv) $(360°, 0), (0, 0)$

(v) $(180°, 0), (0, 2)$

9 (i) $y = \cos(x - 45°);$
$y = \sin(x + 45°)$

(ii) $P(0, \frac{\sqrt{2}}{2}), Q(45, 1)$

10 (i) $y = 4\sin x$

(ii) $-2\sqrt{3}$

11 (i) False

(ii) True

(iii) True

(iv) False

(v) True

Chapter 9

Opening activity (page 172)

6 routes from A to B
4 routes from A to C

Discussion points (page 178)

Yes
No

A Pascal puzzle (page 178)

1.61051. This is $1 + 5 \times (0.1) + 10 \times (0.1)^2 + 10 \times (0.1)^3 + 5 \times (0.1)^4 + 1 \times (0.1)^5$ and
1, 5, 10, 10, 5, 1 are the binomial coefficients for $n = 5$.

Exercise 9.1 (page 179)

1 The '$-2x$' term is squared so the term should be positive and there should either be brackets around '$-2x$' or it should be simplified to $4x^2$.

2 $10 \times 1^2 \times (-2x)^3$

3 (i) $1 + 3x + 3x^2 + x^3$

(ii) $x^5 + 5x^4 + 10x^3 + 10x^2 + 5x + 1$

(iii) $x^4 + 8x^3 + 24x^2 + 32x + 16$

4 (i) $27 + 27x + 9x^2 + x^3$

(ii) $16x^4 - 96x^3 + 216x^2 - 216x + 81$

(iii) $x^5 + 10x^4y + 40x^3y^2 + 80x^2y^3 + 80xy^4 + 32y^5$

5 (i) 15

(ii) 1

(iii) 220

(iv) 1

(v) 36

6 (i) 35

(ii) -120

(iii) 180

(iv) 135

(v) 70

7 $12x + 18x^2 + 36x^3 + 15x^4$

8 $1 - 9x + 27x^2 - 27x^3;$
$2 - 17x + 45x^2 - 27x^3 - 27x^4$

9 (i) $x^3 - 6x^2 + 12x - 8$

(ii) $x = 0, 2$

10 (i) $1 - 6x + 12x^2 - 8x^3$

(ii) $1 - 3x - 6x^2 + 28x^3 - 24x^4$

11 (i) $1 + 3x + 3x^2 + x^3$

(ii) $1 + 3y - 5y^3 + 3y^5 - y^6$

12 (i)
```
1  2
1  3  2
1  4  5  2
1  5  9  7  2
1  6  14  16  9  2
1  7  20  30  25  11  2
```

(ii) Row sums are 3, 6, 12, 24, 48, 96, i.e. start with 3 and double, whereas the 'normal' Pascal's triangle starts with 2 and doubles.

(iii)
```
1  3
1  4  3
1  5  7  3
1  6  12  10  3
1  7  18  22  13  3
1  8  25  40  35  16  3
```
Row sums are 4, 8, 16,..., i.e. start with 4 and double.

(iv) Starting with 1, n has row sums $(1 + n), 2(1 + n), 4(1 + n), 8(1 + n)$, etc.

Discussion point (page 181, top)

Gary could have put the bricks in order by chance. A probability of $\frac{1}{120}$ is small but not very small. What would be really convincing is if he could repeat the task whenever he was given the bricks.

Discussion point (page 181, bottom)

No, it does not matter.

Discussion point (page 182, top)

Multiply top and bottom by 53!

$$\frac{59 \times 58 \times 57 \times 56 \times 55 \times 54}{6!} \times \frac{53!}{53!}$$

$$= \frac{59!}{6!53!}$$

This works out to be 45 057 474.

Discussion point (page 182, bottom)

By following the same argument as for the National Lottery example but using n instead of 59 and r instead of 6.

Exercise 9.2 (page 183)

1 P no, Q no, R yes, S yes, T yes

2 56

3 24

4 $\frac{1}{593775}$

5 40 320

6 715

7 (i) 120

  (ii) $\dfrac{1}{120}$

8 280

9 (i) 14!

  (ii) $\dfrac{1}{14!}$

10 (i) 715

  (ii) 5

  (iii) $\dfrac{1}{143}$

  (iv) The applicants are all equally suitable and so the jobs are given at random.

11 (i) 126

  (ii) (a) $\dfrac{1}{126}$

   (b) $\dfrac{45}{126}$

12 (i) $\dfrac{1}{120}$

  (ii) $\dfrac{1}{7893600}$

13 (i) $\dfrac{1}{10!}$

  (ii) Lose 7.2p

  (iii) £362 000 (to the nearest £1000)

14 (i) 495

  (ii) 45

  (iii) $\dfrac{1}{11}$

Practice questions (page 186)

1 (i) Even, because $y > 0$ for negative and positive large x [1]

  (ii) x as factor [1]
   $(x + 1)$ as factor [1]
   $(x - 2)$ as factor [1]
   $y = x(x + 1)(x - 2)^2$ [1]

2 (i) $y(1) = 1 - a + a - 1 = 0$
   …so goes through $(1, 0)$ [1]
   $y(0) = 0 - 0 + 0 - 1 = -1$
   … so goes through $(0, -1)$ [1]

  (ii) Goes through $(1, 0)$ so try translation by 1 to the right $y = (x - 1)^3$ [1]
   $= x^3 - 3x^2 + 3x - 1$
   which is the equation with $a = 3$ [1]

  (iii) $y = x^3 + x^2 - x - 1$
   Root at $(1, 0)$ so factor $(x - 1)$ [1]
   $x^3 + x^2 - x - 1$
   $= (x - 1)(x^2 + 2x + 1)$ [1]
   $= (x - 1)(x + 1)^2$ so repeated root [at $x = -1$] [1]

3 (i) E.g. spacecraft travels at maximum speed for the whole journey Acceleration and deceleration can be ignored. Distance travelled is 2.25×10^9 metres whatever the speed [1]

  (ii) Graph showing inverse proportionality [1]

  (iii) Journey time =
   $\dfrac{1.65 \times 10^{11}}{11000} =$
   1.5×10^7 seconds [1]
   $= \dfrac{1.5 \times 10}{24 \times 60 \times 60}$
   $= 173.6$ days [1]
   which is less than 175 so supplies appear sufficient [1]

  (iv) 173.6 is close to 175, so probably need a better model to be sure that supplies are sufficient. Would not trust model. [1]

4 (i) $y = 1 - \cos x$: period 360° [1]

  (ii) $2\sin^2 x = 1 - \cos x$
   $2 - 2\cos^2 x = 1 - \cos x$ [1]
   $2\cos^2 x - \cos x - 1 = 0$
   $(2\cos x + 1)(\cos x - 1) = 0$ [1]
   $\cos x = -\frac{1}{2}$ or 1 [1]
   $x = 0°, 120°, 240°, 360°$ [1, 1]

5 (i) $\dfrac{\sin\theta}{8} = \dfrac{\sin 20°}{5}$ [1]

$\theta = \arcsin 0.547$
  $= 33.2°$ [1]
  or 146.8° [1]

  (ii) Sketch(es) showing both triangles [1]
   A $= 180° - (33.2° + 20°)$
   $= 126.8°$
   Or
   A $= 180° - (146.8° + 20°)$
   $= 13.2°$ [1]
   Area $=$
   $\frac{1}{2} \times 5 \times 8 \times \sin A$ [1]
   Correct choice of triangle [1]
   Area $= 16.0$ units^2 [1]

6 (i) A $y = (2x)^3$;
   B $y = 2x^3$;
   C $y = x^3$;
   D $y = \dfrac{x^3}{2}$ [1, 1, 1]

  (ii) A $y = \cos(x + 30)$;
   B $y = \cos x$;
   C $y = \sin(x + 30)$;
   D $y = \sin x$ [1, 1, 1]

  (iii) A $y = x^4 + 1$;
   B $y = x^3 + 1$;
   C $y = -x^3 + 1$;
   D $y = -x^3$ [1, 1, 1]

7 (i) $x^2 = 1^2 + 2^2 -$
   $2 \times 1 \times 2\cos 20°$ [1]
   $x = 1.114$ [1]

  (ii) $x^2 = 1^2 + 2^2 -$
   $2 \times 1 \times 2 \times \cos\theta$ [1]
   $R = x^2 = 5 - 4\cos\theta$ [1]

  (iii) $5 - 4\cos\theta = 4$
   (or inequality) [1]
   $\cos\theta = \frac{1}{4}$ (or inequality) [1]
   So $R < 4$ for
   $[0 <] \theta < 75.5°$ [1]

8 (i) $(2 + x)^3 = 2^3 + 3.2^2 x +$
   $3.2.x^2 + x^3$ [1]
   $= 8 + 12x + 6x^2 + x^3$ [1]
   $(1 - x)^3 = 1 - 3x +$
   $3x^2 - x^3$ [1, 1]
   (a) $y = 9 + 9x + 9x^2$
    $= 9(x^2 + x + 1)$
    $= 9((x + \frac{1}{2})^2 + \frac{3}{4})$ [1,1]
    which is quadratic with line of symmetry
    $x = -\frac{1}{2}$ [1]

(b) $y = 2x^3 + 3x^2 + 15x + 7$

$y(-\frac{1}{2}) = -\frac{1}{4} + \frac{3}{4} - \frac{15}{2} + 7 = 0$

so root at $x = -\frac{1}{2}$ [1]

$y = (2x + 1)(x^2 + x + 7)$
              [1, 1]

Discriminant of quadratic factor = $1 - 28 < 0$ [1]

So quadratic factor has no zero

So cubic has only one root [at $x = -\frac{1}{2}$]   [1]

Chapter 10

Activity 10.1 (page 191)

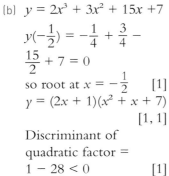

(i) For the curve $y = x^2$, the gradient is positive for $x > 0$, and negative for $x < 0$. The gradient is zero at $x = 0$.

(ii) For positive values of x, as x gets bigger the gradient also increases.

(iii) For negative values of x, as x gets more negative the gradient also becomes more negative.

(iv) Because the curve is symmetrical about the y axis, the gradient at $x = -2$ has the same magnitude as the gradient at $x = 2$ but the opposite sign. In general for the curve $y = x^2$

(gradient at $x = a$)
$= -$(gradient at $x = -a$)

(v)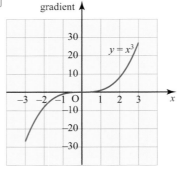

For the curve $y = x^3$, the gradient is positive for both $x > 0$ and $x < 0$. The gradient is zero at $x = 0$. For positive values of x, as x gets bigger the gradient also increases. For negative values of x, as x gets more negative the gradient increases. The gradient at $x = -2$ is the same as the gradient at $x = 2$, and in general for $y = x^3$

(gradient at $x = a$)
$= $ (gradient at $x = -a$)

Activity 10.2 (page 192)

(x_1, y_1)	(x_2, y_2)	Gradient $= \frac{y_2 - y_1}{x_2 - x_1}$
(1,1)	(3,9)	4
(1,1)	(2,4)	3
(1,1)	(1.5,2.25)	2.5
(1,1)	(1.25,1.5625)	2.25
(1,1)	(1.1,1.21)	2.1
(1,1)	(1.01,1.0201)	2.01
(1,1)	(1.001, 1.002001)	2.001

1. The gradients are getting closer and closer to 2.

2. The gradient of $y = x^2$ at the point $(1,1)$ appears to be 2.

Exercise 10.1 (page 193)

1. $\dfrac{9 - 8.41}{3 - 2.9}$ $\dfrac{16 - 4}{-4 - -2}$ $\dfrac{3.4225 - 2.89}{1.85 - 1.7}$

2. (i)

(x_1, y_1)	Gradient of $y = x^3$ at x_1
(−3,−27)	27
(−2,−8)	12
(−1,−1)	3
(0,0)	0
(1,1)	3
(2,8)	12
(3,27)	27

(ii)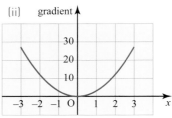

(iii) $f'(x) = 3x^2$

3. (i)

(x_1, y_1)	Gradient of $y = x^4$ at x_1
(−3,81)	−108
(−2,16)	−32
(−1,1)	−4
(0,0)	0
(1,1)	4
(2,16)	32
(3,81)	108

(ii)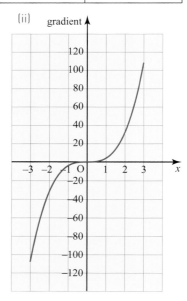

(iii) $f'(x) = 4x^3$

4 (i) $f'(x) = 5x^4$

  (ii) 80

  (iii) Yes

5 (i) (a) $f'(x) = 6x^5$

    (b) $f'(x) = 7x^6$

  (ii) $f'(x) = nx^{n-1}$

Discussion point (page 195, top)

The transformation stretches the graph by a scale factor of a parallel to the y axis. This multiplies the gradient by a.

Discussion points (page 195, bottom)

(i) This follows from the fact that the graph of $y = c$ is a horizontal line with gradient zero.

(ii) This follows from the fact that the graph of $y = kx$ is a straight line with gradient k.

Exercise 10.2 (page 197)

1 $12x$ and 12, $3x^2$ and $6x$, $4x^2$ and $8x$, $2x^3$ and $6x^2$, $4x^3$ and $12x^2$

2 $3x^2 - 6x$

3 (i) $\dfrac{dy}{dx} = 7x^6$

  (ii) $\dfrac{dy}{dx} = 11x^{10}$

  (iii) $\dfrac{dy}{dx} = 14x^6 - 33x^{10}$

4 (i) $\dfrac{dV}{dx} = 3x^2$

  (ii) $\dfrac{dx}{dt} = 4t - 5$

  (iii) $\dfrac{dz}{dl} = 15l^4 - 2l + 5$

5 (i) -1

  (ii) 9

  (iii) 0

6 (i) -8

  (ii) 4

  (iii) $\dfrac{35}{3}$

7 $(-1, 8)$ and $(4, -57)$

8 (i) $5x^3 - x^2 + 15x - 3$

  (ii) $\dfrac{dy}{dx} = 15x^2 - 2x + 15$

  (iii) He has differentiated each bracket separately and then multiplied the results.

9 (i) $2x + 3$

  (ii) $\dfrac{dy}{dx} = 2$

10 (i) $\dfrac{dy}{dx} = 4x^3 + 6x^2 - 4$

  (ii) $\dfrac{dy}{dx} = 6x - \dfrac{3}{2}$

11 (i)

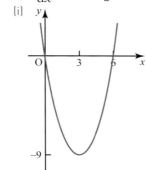

  (ii) $\dfrac{dy}{dx} = 2x - 6$

  (iii) Gradient is 0

  (iv) Turning point of curve

12 (i)

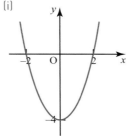

  (ii) $(2, 0)$ and $(-2, 0)$

  (iii) $\dfrac{dy}{dx} = 2x$

  (iv) At $(2, 0)$ gradient is 4, at $(-2, 0)$ gradient is -4

  (v) $(0, -8)$

13 (i) $x^3 - 6x^2 + 11x - 6 = (x - 1)(x - 2)(x - 3)$.

    Curve crosses x axis at $(1, 0), (2, 0), (3, 0)$.

  (ii)

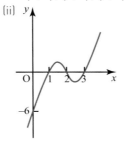

  (iii) $\dfrac{dy}{dx} = 3x^2 - 12x + 11$

  (iv) Gradients are $2, -1$ and 2. Tangents are parallel at $(1, 0)$ and $(3, 0)$.

14 (i)

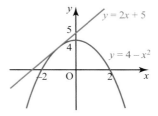

  (iii) $\dfrac{dy}{dx} = -2x$, gradient is 2

  (iv) Yes

  (v) Yes

15 (i)

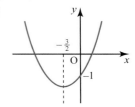

  (ii) $\dfrac{dy}{dx} = 2x + 3$

  (iii) $(1, 3)$

  (iv) No

16 $a = 2, b = -3$

17 (i)

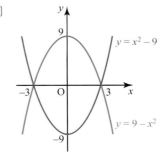

  (ii)

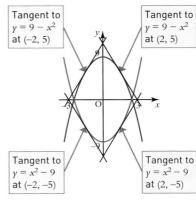

  Rhombus – opposite sides parallel, all sides of equal length.

Exercise 10.3 (page 200)

1

f(x)	f'(x)	Gradient of tangent when $x = 2$	Gradient of normal when $x = 2$
$f(x) = x^2 - 3x$	$2x - 3$	1	-1
$f(x) = 2x^3$	$6x^2$	24	$-\dfrac{1}{24}$
$f(x) = x^4 - 3x^2$	$4x^3 - 6x$	20	$\dfrac{1}{20}$

2   (i) and (v)

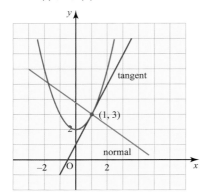

(ii) $\dfrac{\mathrm{d}y}{\mathrm{d}x} = 2x$

(iii) 2

(iv) $-\dfrac{1}{2}$

3   (i), (iii), (iv)

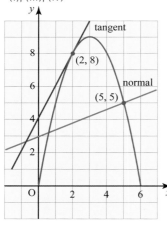

(ii) $\dfrac{\mathrm{d}y}{\mathrm{d}x} = 6 - 2x$

(iii) 2

(iv) $\dfrac{1}{4}$

4   (i), (iii), (iv)

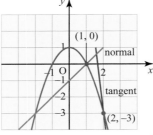

Draw the curve $y = 1 - x^2$, using the same scale on both axes.

(ii) $\dfrac{\mathrm{d}y}{\mathrm{d}x} = -2x$

(iii) $y + 4x - 5 = 0$

(iv) $2y - x + 1 = 0$

(v) $\left(\dfrac{11}{9}, \dfrac{1}{9}\right)$

5   (i), (iii), (iv)

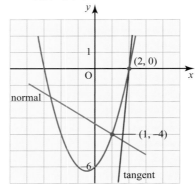

(ii) $\dfrac{\mathrm{d}y}{\mathrm{d}x} = 2x + 1$

(iii) $y - 5x + 10 = 0$

(iv) $3y + x + 11 = 0$

(v) $\dfrac{845}{32}$ square units

6   (i) and (iii)

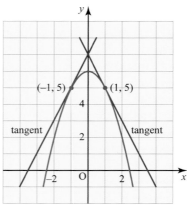

(ii) 2, -2

(iii) $y - 2x - 7 = 0$,
$y + 2x - 7 = 0$

(iv) $(0, 7)$

7   (i) and (iv)

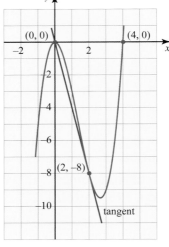

(ii) $\dfrac{\mathrm{d}y}{\mathrm{d}x} = 3x^2 - 8x$

(iii) -4

(v) $(0, 0)$

8   (i) $y - 6x - 28 = 0$

(ii) $(3, 45)$

(iii) $6y + x - 273 = 0$

9   64 square units

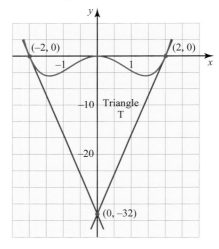

Activity 10.3 (page 202)

(i) and (ii)

(a) Increasing for no values of x

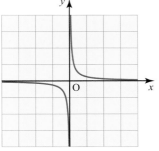

(b) Increasing for all values of x

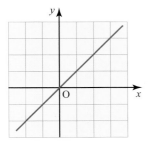

(c) Increasing for some values of x (all except $x = 0$)

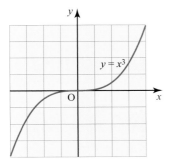

(d) Increasing for all values of x

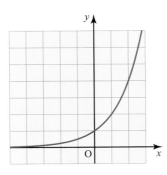

Activity 10.4 (page 203)

	What are the coordinates of the turning point?	Is the gradient just to the left of the turning point positive or negative?	What is the gradient at the turning point?	Is the gradient just to the right of the turning point positive or negative?
Maximum	$(-1, 2)$	Positive	0	Negative
Minimum	$(1, -2)$	Negative	0	Positive

Exercise 10.4 (page 205)

1 False because when $x = 0$, it is stationary, not decreasing.
2 All x except $x = 0$
3 (i) $-1 < x < 3$
   (ii) $x < -1,\ x > 3$
4 (i) $x > 0$
   (ii) $x > 3$
   (iii) $x > -5$
5 (i) $x > -1$
   (ii) $x = -3,\ x = 1$
   (iii)

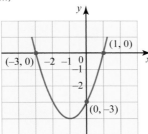

   (iv) By symmetry, turning point is at $x = -1$. Curve increasing after turning point.
6 Maximum at $(-2, 18)$ and minimum at $(2, -14)$.
7 (i) $x < 2$
   (ii) $y = (x - 2)^2 - 11,\ (2, -11)$
   (iii)

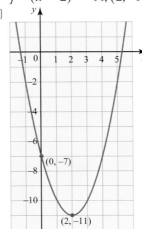

(iv) Function decreasing before turning point at $x = 2$.
8 (i) $\dfrac{dy}{dx} = 3x^2 + 6x - 24$
   (ii) $-4 < x < 2$
9 (i) $y = x^4 - 4x^3 + 4x^2$,
        $\dfrac{dy}{dx} = 4x^3 - 12x^2 + 8x$
   (ii) $(0,0)$ minimum,
        $(1,1)$ maximum,
        $(2,0)$ minimum
   (iii)

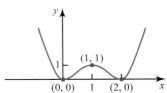

10 $x < 3,\ x > 7$
12 (i) $\dfrac{dy}{dx} = 3x^2 - 12x + 15$
    (ii) $\dfrac{dy}{dx} = 3(x - 2)^2 + 3 > 0$

Discussion points (page 207)

(i) $m = \dfrac{5}{2}$. Gradient graph is horizontal line $y = \dfrac{5}{2}$
(ii) Horizontal line $y = m$
(iii) The line $y = 3$ is horizontal and therefore has a constant gradient of zero. The gradient curve would lie along the x axis. The line $x = 3$ is vertical and has an infinite gradient. It is not possible to draw a gradient curve for this straight line.

Exercise 10.5 (page 208)

1 The function has turning points at −6 and 1, corresponding to the gradient function intersecting the x-axis. The function has a negative gradient until $x = -6$, then a positive gradient until $x = 1$, then negative again. This corresponds to the gradient function being below the axis until $x = -6$, then above it until $x = 1$, then below it again.

2 B

3

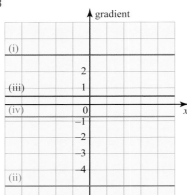

4

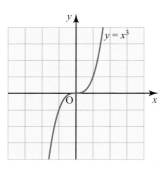

(iv) The gradient is 0 when $x = 0$; everywhere else it is positive.

5 C

6 (i) $(-2, -8)$, $(2, 8)$
(ii) $-2 < x < 2$
(iii) $x < -2$, $x > 2$
(iv)

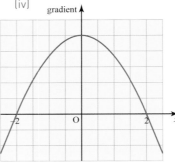

7 (i) (a) $\Leftarrow$
(b) $\Rightarrow$
(c) $\Leftarrow$
(d) None
(ii)

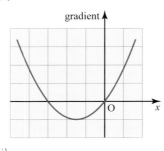

8 (i)

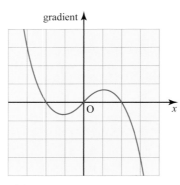

(ii)

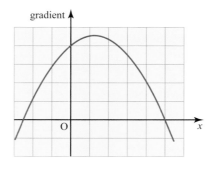

9 (i)

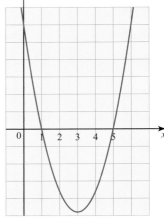

(ii) $\dfrac{dy}{dx} = 3x^2 - 18x + 15$

10

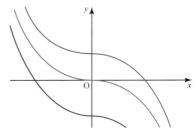

The graphs are the same curve translated up or down – each graph is a translation by $\begin{pmatrix} 0 \\ a \end{pmatrix}$ for some value a of every other graph.

Activity 10.5 (page 210)

(i) (a) Using graphing software, the gradient of the tangent at $x = 4$ is 0.25.
(b) Using the formula $\dfrac{dy}{dx} = \dfrac{1}{2\sqrt{x}} = \dfrac{1}{4}$. The results are the same.
(c) At $x = 1$, using graphing software, the gradient of the tangent is 0.5. Using the formula $\dfrac{dy}{dx} = \dfrac{1}{2}$

(ii) The result is true for other values of n.

Exercise 10.6 (page 212)

1 (i) $-\dfrac{5}{2}, -\dfrac{3}{2}, -\dfrac{1}{2}, \dfrac{1}{2}, \dfrac{3}{2}, \dfrac{5}{2}$

  (ii) $-\dfrac{8}{3}, -\dfrac{5}{3}, -\dfrac{2}{3}, \dfrac{1}{3}, \dfrac{4}{3}, \dfrac{7}{3}$

2 $-\dfrac{1}{2\sqrt{x}}$

3 (i) $\dfrac{dy}{dx} = -20x^{-6}$

  (ii) $\dfrac{dy}{dx} = 3x^{-\frac{1}{2}}$

  (iii) $\dfrac{dy}{dx} = -\dfrac{2}{3}x^{-\frac{5}{3}}$

4 (i) $\dfrac{dz}{dt} = -21t^{-4}$

  (ii) $\dfrac{dx}{dt} = -2t^{-\frac{3}{2}}$

  (iii) $\dfrac{dp}{dr} = -12r^{-3} + 3$

6 (i) $-\dfrac{1}{8}$

  (ii) 2

7 (i) $\dfrac{dy}{dx} = \dfrac{7x^2\sqrt{x}}{2} + \dfrac{1}{x^2}$

  (ii) $\dfrac{dy}{dx} = -\dfrac{1}{2x\sqrt{x}} + \dfrac{5}{x^6}$

8 (i) $\left(-\dfrac{1}{2}, 0\right)$

  (ii)

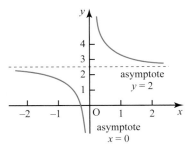

  (iii) $\dfrac{dy}{dx} = -\dfrac{1}{x^2}$

  (iv) -4

9 (i) $\dfrac{dy}{dx} = -\dfrac{8}{x^3} + 1$

  (iii) 2

  (iv) $(2, 3)$

10 (i) and (iv)

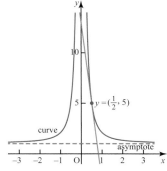

  (iii) $\dfrac{dy}{dx} = \dfrac{-2}{x^3}, -16$

  (iv) $y = -16x + 13$

11 (i) $\dfrac{dy}{dx} = \dfrac{1}{2\sqrt{x}}$

  (ii) $\left(\dfrac{1}{16}, -\dfrac{3}{4}\right)$

  (iii) No

12 (i) $y = -2$

  (ii) $\dfrac{dy}{dx} = 3 + \dfrac{2}{x^3}$

  (iii) 1

  (iv) $y - x + 1 = 0$

13 (i) $\dfrac{dy}{dx} = 2x - \dfrac{1}{x^2}$

  (ii) 1

  (iv) $\left(-1 + \sqrt{2}, 4 - \sqrt{2}\right)$,

   $\left(-1 - \sqrt{2}, 4 + \sqrt{2}\right)$

14 $\dfrac{3}{4}$

Discussion points (page 213)

(i) No

(ii) By differentiating for a third time; third derivative; rate of change of $\dfrac{d^2y}{dx^2}$ with respect to x.

Discussion points (page 216)

(i) $\dfrac{dy}{dx} = 0$

Exercise 10.7 (page 217)

1 $\dfrac{1}{2}x^4 - x^2 + 3, 2x^3 - 2x, 6x^2 - 2;$
 $2x^3 - 3x, 6x^2 - 3, 12x$

2 $24x - 2$

3 (i) $\dfrac{dy}{dx} = 4x^3, \dfrac{d^2y}{dx^2} = 12x^2$

  (ii) $\dfrac{dy}{dx} = 15x^4 - 6x^2 + 2x,$

   $\dfrac{d^2y}{dx^2} = 60x^3 - 12x + 2$

4 (i) $\dfrac{dy}{dx} = 2x + \dfrac{1}{x^2},$

   $\dfrac{d^2y}{dx^2} = 2 - \dfrac{2}{x^3}$

  (ii) $\dfrac{dy}{dx} = 3\sqrt{x}, \dfrac{d^2y}{dx^2} = \dfrac{3}{2\sqrt{x}}$

5 (i) $\dfrac{dy}{dx} = 2x + 2, (-1, -9)$

  (ii) $y = (x + 1)^2 - 9$

  (iii) $\dfrac{d^2y}{dx^2} = 2$, minimum, yes

  (iv)

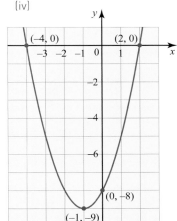

6 (i) $\dfrac{dy}{dx} = 4x^3$

  (ii) Minimum at $(0, -16)$

  (iii)

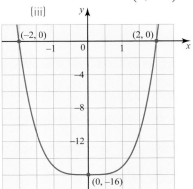

7 (i) $y = x^3 - 5x^2 + 7x - 3$,

$\dfrac{dy}{dx} = 3x^2 - 10x + 7$,

$\dfrac{d^2y}{dx^2} = 6x - 10$

(ii) $(1, 0)$ maximum,

$\left(\dfrac{7}{3}, -\dfrac{32}{27}\right)$ minimum

(iii)

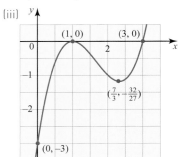

8 (i) $\dfrac{dy}{dx} = 1 - \dfrac{2}{\sqrt{x}}$,

$\dfrac{d^2y}{dx^2} = \dfrac{1}{x\sqrt{x}}$

(ii) $(4, -4)$ minimum

9 $x = 2$

10 (i) $p = 2, q = -3$

(ii) Minimum

(iii) $(0, 0)$ maximum

(iv)

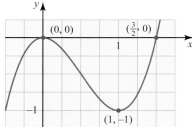

Exercise 10.8 (page 219)

1 A x is 10 less than y;

$y - 10 = x$

B the sum of x and y is 10; $x + y = 10$

C the mean of x and y is 10; $x + y = 20$

D y is 10 less than x;

$x = y + 10$

2 $x = 10y$

3 (i) $y = 20 - x$

(ii) $P = x(20 - x)$

(iii) $\dfrac{dP}{dx} = 20 - 2x$,

$\dfrac{d^2P}{dx^2} = -2$

(iv) $P = 100$

4 (i) $y = \dfrac{400}{x}$

(ii) $S = x + \dfrac{400}{x}$

(iii) $\dfrac{dS}{dx} = 1 - \dfrac{400}{x^2}$,

$\dfrac{d^2S}{dx^2} = \dfrac{800}{x^3}$

(iv) $S = 40$

5 (i) $y = 8 - x$

(ii) $S = 2x^2 - 16x + 64$

(iii) $\dfrac{dS}{dx} = 4x - 16$,

$\dfrac{d^2S}{dx^2} = 4$

6 (i) $y = 60 - x$

(ii) $y = x(60 - x)$

(iii)

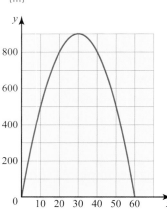

(iv) $\dfrac{dA}{dx} = 60 - 2x$,

$\dfrac{d^2A}{dx^2} = -2$

(v) Maximum value $x = 30, y = 30, A = 900$

7 (i) $A = xy$

(ii) $T = 2x + y$

(iv) $\dfrac{dT}{dx} = 2 - \dfrac{18}{x^2}$,

$\dfrac{d^2T}{dx^2} = \dfrac{36}{x^3}$

(v) $x = 3, y = 6, T = 12$

---

8 (i) $4 - 2x$, $(4 - 2x)^2$

(iii) $x = \dfrac{8}{7}$

(iv) $A = \dfrac{48}{7}$

10 $144\,cm^3$

11 (i) $y = \dfrac{16}{x}$

(ii) $S = x^2 + \dfrac{256}{x^2}$

(iii) $S = 32$

(iv) $4\sqrt{2}$

12 (i) $r = \dfrac{15 - 2x}{\pi}$

(iii) $\dfrac{120}{4 + \pi}, \dfrac{30\pi}{4 + \pi}$

13 $\dfrac{40}{3}\sqrt{\dfrac{40}{3\pi}}$

14 £2052

Exercise 10.9 (page 224)

1 (i) 5, (ii) 3, (iii) 6, (iv) 0

2 $h^2 + 7h + 9$

3 $h + 6$

4 (i) 1

(ii) $(1 + h)^3$

(iii) $h^2 + 3h + 3$

5 (i) -2

(ii) $2(-1 + h)^2 - 4$

(iv) -4

6 (i) 18

(ii) $(3 + h)^2 + 3(3 + h)$

(iii) $h + 9$

(iv) 9

7 (i) $y_1 = x_1^3$

$y_2 = \left(x_1^3 + h\right)^3 =$

$x_1^3 + 3x_1^2h + 3x_1h^2 + h^3$

(iii) $3x^2$

8 (i) $y_1 = 2x_1^3 + 1$,

$y_2 = 2\left(x_1 + h\right)^3 + 1 =$

$2x_1^3 + 6x_1^2h + 6x_1h^2 + 2h^3 + 1$

(iii) $6x^2$

9 (i) $y_1 = x_1^2 + 5x_1$

(iii) $2x_1 + h + 5$

(iv) $2x + 5$

10 (i) $y_1^2 - x_1 - 6$,

$y_2 (x_1 + h)^2 - (x_1 + h) - 6 =$
$x_1^2 + 2x_1 h + h^2 - x_1 - h - 6$

(ii) $2x_1 + h - 1$

(iii) $2x - 1$

Chapter 11

Opening activity 11.1 (page 229)

They all have gradient function $\dfrac{dy}{dx} = 3x^2$

$y = x^3 + k$

The gradient is the same for all four curves for every value of x.

e.g. $y = x^3 + 100$,

$y = x^3 - \pi, y = x^3 + \sqrt{2}$

$y = x^3 + c$, where c is any constant.

Activity 11.1 (page 230)

(i) e.g. $y = x^2, y = x^2 + 1$,

$y = x^2 - 1$

(ii) The curves are parallel in the sense that they have the same gradient for every value of x.

(iii) $y = x^2 + c$, where c is any constant

(iv) (a) $y = x^3 + c$

(b) $y = \dfrac{x^3}{3} + c$

(c) $y = x^4 + c$

(d) $y = \dfrac{x^4}{4} + c$

(v) $y = \dfrac{x^{n+1}}{n + 1} + c$. It does not work for $n = -1$, because you cannot divide by zero.

Discussion point (page 230)

Write 1 as x^0, and then apply the rule with $n = 0$.

Discussion point (page 231)

c represents an arbitrary constant and does not have an assigned value. So it is not correct to say that the constant in part (iv) is three times the constant in the other three parts.

Activity 11.2 (page 232)

(i) $y = x^3 + c$

(ii) (a) $y = x^3 + 3$

(b) $y = x^3 + 12$

(c) $y = x^3 - 7$

(iii) (a) $y = x^3 + 2$

(b) $y = x^3 + 2$

(c) $y = x^3 + 2$

Exercise 11.1 (page 233)

1 $y = x^2 + 3x - 5$

2 $\dfrac{x^4}{4} + c$

3 (i) $\dfrac{x^7}{7} + c$

(ii) $\dfrac{x^8}{4} + c$

(iii) $\dfrac{x^7}{7} + \dfrac{x^8}{4} + c$

(iv) $\dfrac{5x^7}{7} + c$

$\int (x^6 + 2x^7)\, dx =$
$\int x^6\, dx + \int 2x^7\, dx$
$\int 5x^6\, dx = 5\int x^6\, dx$

4 (i) $y = 2x^4 + c$

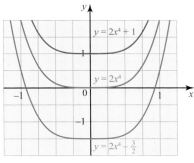

(ii) $y = x^2 - x + c$

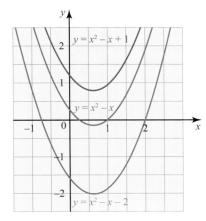

5 $f(x) = \dfrac{3x^2}{2} - \dfrac{x^3}{6} + c$

6 (i) $y = 5x + c$

(ii) $y = 5x + 3$

(iii)

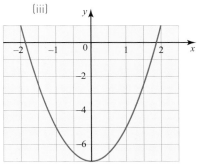

7 (i) $y = 2x^2 - 7$

(ii) $y = -5$

(iii)

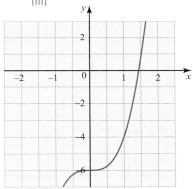

8 (i) $y = 2x^3 - 6$

(iii)

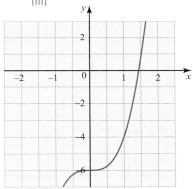

9 (i) $y = 4x^3 + 2x^2 + c$

(ii) $y = 4x^3 + 2x^2 + 3$

10 (i) $\dfrac{x^3}{3} + x^2 - 3x + c$

(ii) $\dfrac{x^2}{2} + 7x - \dfrac{2x^5}{5} + c$

11 (i) $2z^3 - \dfrac{11z}{2} + 3z + c$

(ii) $\dfrac{2t^5}{5} - \dfrac{4t^3}{3} + \dfrac{t^2}{2} + c$

12 (i) $y = x^2 - 6x + 9$

(ii) Through the point

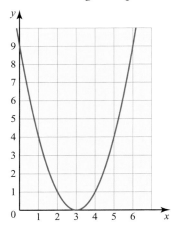

13 (i) Minimum at $x = 1$, maximum at $x = -1$.

(ii) $y = x^3 - 3x + 3$

(iii)

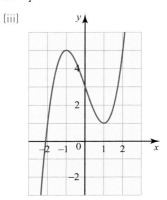

14 $y = -x^3 + 3x^2 - x + 2$

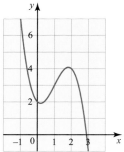

15 $y = x^4 + 4x^3 - 2x^2 - 12x + 1$

Exercise 11.2 (page 237)

1 $= \dfrac{256}{3} - 12 - \dfrac{32}{3} + 6$; sign error

2 (i) 16

(ii) 52

(iii) 7

3 (i) $\dfrac{7}{3}$

(ii) $\dfrac{19}{3}$

(iii) $\dfrac{26}{3}$

The answer to (iii) is the sum of the answers to (i) and (ii).

4 (i) $\dfrac{32}{5}$

(ii) 32

The answer to (ii) is 5 times the answer to (i).

5 (i) (a) $\dfrac{19}{3}$

(b) $\dfrac{19}{3}$

(ii) Answers the same because curve is symmetrical in the y axis.

6 (i) 15

(ii) 40

(iii) -8

7 (i) and (iii)

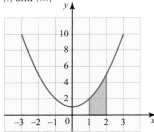

(ii) $3\dfrac{1}{3}$

8 (i) and (ii)

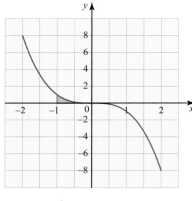

(iii) $\dfrac{1}{4}$

9 (i)

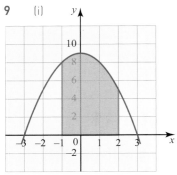

(ii) 24 square units

10 (i)

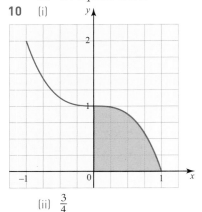

(ii) $\dfrac{3}{4}$

11 (i) and (ii)

$x^3 - 6x^2 + 11x - 6$
$= (x - 1)(x - 2)(x - 3)$

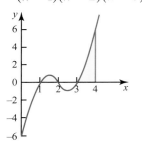

(iii) $\dfrac{1}{4}, \dfrac{9}{4}$

(iv) $\dfrac{9}{64}$. Maximum is between 1 and 1.5

12

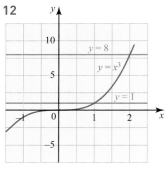

$\dfrac{45}{4}$

13 (i) 45, 135

(ii) No, in general
$$\int_a^b k\mathrm{f}(x)\,\mathrm{d}x \neq \int_{ka}^{kb} \mathrm{f}(x)\,\mathrm{d}x$$

Activity 11.3 (page 238)

(i) 0

(ii)

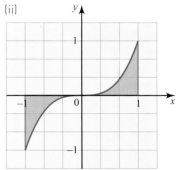

(iii) Integral between −1 and 0 is negative, integral between 0 and 1 is positive, and they cancel each other out.

(iv) $2 \times \int_0^1 x^3\,\mathrm{d}x = 2 \times \frac{1}{4} = \frac{1}{2}$

Discussion point (page 238)

An area cannot be negative. The integral coming out as a negative number tells you that the area is below the x axis; the area is the positive value.

Discussion point (page 240)

$\frac{15}{2}$

Negative answer for region below x axis is cancelling out positive answer for region above x axis.

Exercise 11.3 (page 240)

1 (i) and (ii)

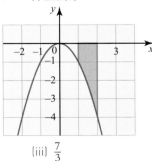

(iii) $\frac{7}{3}$

2 (i) and (ii)

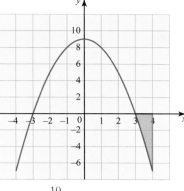

(iii) $\frac{10}{3}$

3 (i) and (ii)

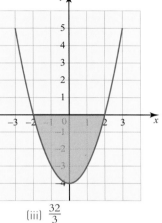

(iii) $\frac{32}{3}$

4 (i) and (ii)

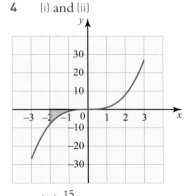

(iii) $\frac{15}{4}$

5

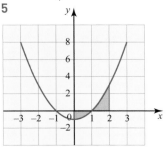

Area = 2 square units

6

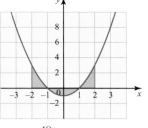

Area = $\frac{49}{6}$ square units

7

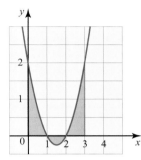

Area = $\frac{11}{6}$ square units

8 (i) $-\frac{20}{3}$

(ii) $\frac{28}{3}$

(iii) Below the x axis, y is negative, resulting in a negative integral. Answer to (i) is Area B − Area A − Area C, not Area A + Area B + Area C.

9 (i) $\frac{\mathrm{d}y}{\mathrm{d}x} = 20x^3 - 5x^4$, $(0,0),(4,256)$

(ii) $\frac{3125}{6}$

(iii) 0, region between $x = 5$ and $x = 6$ is below the x axis, and the same area as the shaded region.

10 $x^3 + x^2 - 2x = x(x + 2)(x - 1)$

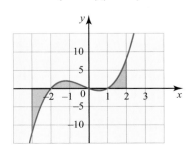

Area = $\frac{133}{12}$ square units

Exercise 11.4 (page 243)

1 $30x$ and $\dfrac{6x}{\frac{1}{5}}$; $\dfrac{15x}{2}$ and $\dfrac{5x}{\frac{2}{3}}$;

   $\dfrac{10x}{3}$ and $\dfrac{5x}{\frac{3}{2}}$; $\dfrac{3x}{\frac{5}{2}}$ and $\dfrac{6x}{5}$

2 $x^{-\frac{3}{2}}$

3 $y = \sqrt{x} + c$

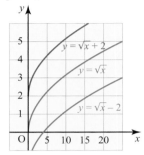

4 (i) $f(x) = -\dfrac{1}{2}x^{-2} + c$

   (ii) $f(x) = -\dfrac{1}{2}x^{-4} + c$

   (iii) $f(x) = -\dfrac{1}{3x^3} + c$

   (iv) $f(x) = -\dfrac{2}{3x^6} + c$

5 (i) $\dfrac{3}{4}x^{\frac{4}{3}} + c$

   (ii) $\dfrac{8}{3}x^{\frac{3}{4}} + c$

   (iii) $\dfrac{4}{5}x^{\frac{5}{4}} + c$

   (iv) $-\dfrac{4}{\sqrt{x}} + c$

6 (i) 4

   (ii) $\dfrac{64}{5}$

   (iii) $-\dfrac{3}{2}$

7 (i) $\dfrac{1}{3}$

   (ii) $\dfrac{2}{3}$

   (iii) $\dfrac{1}{3}$

8 (i) $\dfrac{5}{3}x^3 + \dfrac{1}{2}x^{\frac{2}{3}} + c$

   (ii) $y = 6x - \dfrac{2}{3}x^{\frac{3}{2}}$

      $- \dfrac{1}{4x^2} + c$

   (iii) $\dfrac{5x^{\frac{8}{3}}}{8} + c$

   (iv) $y = -\dfrac{3}{2x^2} + \dfrac{1}{3x^{\frac{3}{2}}} + c$

9 (i) $y = \dfrac{2x^{\frac{3}{2}}}{3} + c$

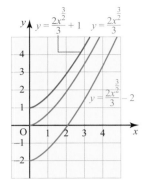

   (ii) x cannot be negative because you cannot take the square root of a negative number.

   (iii) $y = \dfrac{2x^{\frac{3}{2}}}{3} + 2$

10 $x = \dfrac{4}{3}$

11 $v = 4t^{\frac{1}{2}} + 1$

12 (i) $\dfrac{707}{192}$

   (ii) $\dfrac{335}{12}$

   (iii) $\dfrac{6237}{10}$

13 (i) $P(-4,3), Q(-2,0),$
       $R(2,0), S(4,3)$

   (ii) 8 square units

14 (i)

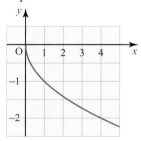

   (ii) $y - 2x + 3 = 0$

   (iii) $\left(\dfrac{3}{2}, 0\right)$

   (iv) $\dfrac{11}{12}$ square units

15 $\dfrac{5}{4}$ square units

Chapter 12

Exercise 12.1 (page 251)

1 Component form: $6\mathbf{i} - \mathbf{j}$,
   $\begin{pmatrix} -2 \\ 4 \end{pmatrix}, (0,0), \begin{pmatrix} 0 \\ 0 \end{pmatrix}$
   Magnitude-direction form:
   $(5, 25°), (0.5, 124°)$

2 $-45°$

3 (i) (a) $3\mathbf{i} + 2\mathbf{j}$
        (b) $\sqrt{13}$

   (ii) (a) $5\mathbf{i} - 4\mathbf{j}$
        (b) $\sqrt{41}$

   (iii) (a) $3\mathbf{i}$
        (b) 3

   (iv) (a) $-3\mathbf{i} - \mathbf{j}$
        (b) $\sqrt{10}$

   (v) (a) $2\mathbf{j}$
        (b) 2

4 Three − first he needs to say where to start from (for example, one corner of the field) and then he needs to give a distance and a direction.

5 (i)

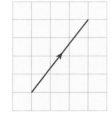

   $(5, 53.1°)$

   (ii)

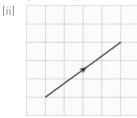

   $(5, 36.9°)$

   (iii)

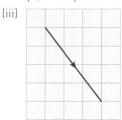

   $(5, 306.9°)$

(iv)

(5, 143.1°)

(v)

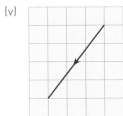

(5, 233.1°)

6 (i)

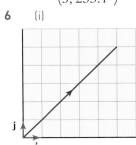

$7.07\mathbf{i} + 7.07\mathbf{j}$

(ii)

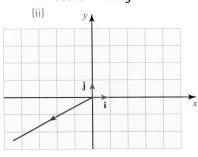

$-4.33\mathbf{i} - 2.5\mathbf{j}$

(iii)

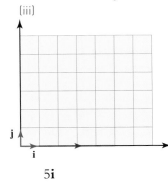

$5\mathbf{i}$

(iv)

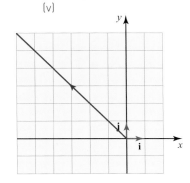

$-10\mathbf{j}$

(v)

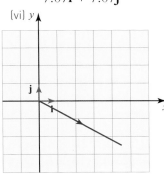

$-7.07\mathbf{i} + 7.07\mathbf{j}$

(vi)

$4.33\mathbf{i} - 2.5\mathbf{j}$

7 (i) $2\mathbf{i} - 2\mathbf{j}$
(ii) $-2\mathbf{i} + 2\mathbf{j}$
(iii) $6\mathbf{i} + 4\mathbf{j}$
(iv) $-2\mathbf{i} - 2\mathbf{j}$
(v) $-2\mathbf{i} + 2\mathbf{j}$
(vi) $2\mathbf{i} - 2\mathbf{j}$

8 (i) (a) Median at A: $\begin{pmatrix} 1.5 \\ 3 \end{pmatrix}$;

median at B: $\begin{pmatrix} -3 \\ -1.5 \end{pmatrix}$;

median at C: $\begin{pmatrix} 1.5 \\ -1.5 \end{pmatrix}$

(b) Magnitude of
median at A is $\frac{3}{2}\sqrt{5}$;
magnitude of median
at B is $\frac{3}{2}\sqrt{5}$;
magnitude of
median at C is $\frac{3}{2}\sqrt{2}$

(ii) $|\overrightarrow{AX}| = 2\sqrt{2}$

9 (i) Possible – R is on the
perpendicular bisector
of PQ.

(ii) Possible – PQR is a
straight line and R is on
the same side of both P
and Q.

(iii) Not possible – if (i) is
true then R is between
P and Q, so (ii) cannot
be true.

Activity 12.1 (page 257)

The diagram shows the vector
$a\mathbf{i} + b\mathbf{j}$.

$|a\mathbf{i} + b\mathbf{j}| = \sqrt{a^2 + b^2}$

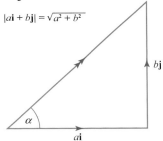

Magnitude of
$\frac{a}{\sqrt{a^2+b^2}}\mathbf{i} + \frac{b}{\sqrt{a^2+b^2}}\mathbf{j}$ is

$$\sqrt{\left(\frac{a}{\sqrt{a^2+b^2}}\right)^2 + \left(\frac{b}{\sqrt{a^2+b^2}}\right)^2}$$

$$= \frac{a^2}{a^2+b^2} + \frac{b^2}{a^2+b^2}$$

$$= \frac{a^2+b^2}{a^2+b^2}$$

$$= 1$$

Direction of $\frac{a}{\sqrt{a^2+b^2}}\mathbf{i} +$

$\frac{b}{\sqrt{a^2+b^2}}\mathbf{j}$ is given by

$$\tan\alpha = \frac{b/\sqrt{a^2+b^2}}{a/\sqrt{a^2+b^2}} = \frac{b}{a}$$

Direction of $a\mathbf{i} + b\mathbf{j}$ is given by $\tan\alpha = \dfrac{b}{a}$

Hence the two vectors have the same direction.

Exercise 12.2 (page 258)

1 $\begin{pmatrix} \frac{1}{\sqrt{2}} \\ \frac{1}{\sqrt{2}} \end{pmatrix}$ $\begin{pmatrix} -1 \\ 0 \end{pmatrix}$

2 $\begin{pmatrix} -1 \\ 2 \end{pmatrix}$

3 (i) $\begin{pmatrix} 6 \\ 8 \end{pmatrix}$ (ii) $\begin{pmatrix} -1 \\ -1 \end{pmatrix}$

(iii) $\begin{pmatrix} 0 \\ 0 \end{pmatrix}$ (iv) $\begin{pmatrix} -7 \\ -2 \end{pmatrix}$

(v) $2\mathbf{i} + \mathbf{j}$
(vi) $2\mathbf{i} + 7\mathbf{j}$

5 (i) $\begin{pmatrix} 2 \\ 3 \end{pmatrix}$ (ii) $\begin{pmatrix} 1 \\ 0 \end{pmatrix}$ (iii) $\begin{pmatrix} 0 \\ 1 \end{pmatrix}$

(iv) $\begin{pmatrix} 0 \\ 0 \end{pmatrix}$ (v) $\begin{pmatrix} 1 \\ 4 \end{pmatrix}$ (vi) $\begin{pmatrix} 3 \\ 2 \end{pmatrix}$

6 (i) $a = 4, b = 6$
(ii) $p = -1, q = 2$
7 (i) $x = 4$
(ii) $x = -16$
(iii) $x = -3$
8 (i) $\dfrac{3}{5}\mathbf{i} + \dfrac{4}{5}\mathbf{j}$

(ii) $\begin{pmatrix} -\frac{3}{5} \\ \frac{4}{5} \end{pmatrix}$

(iii) $\dfrac{4}{5}\mathbf{i} - \dfrac{3}{5}\mathbf{j}$

(iv) $\dfrac{2}{\sqrt{13}}\mathbf{i} + \dfrac{3}{\sqrt{13}}\mathbf{j}$

(v) $\begin{pmatrix} \frac{3}{\sqrt{13}} \\ \frac{2}{\sqrt{13}} \end{pmatrix}$

(vi) $\begin{pmatrix} -\frac{3}{\sqrt{13}} \\ -\frac{2}{\sqrt{13}} \end{pmatrix}$

(vii) $\mathbf{i}$
(viii) $-\mathbf{j}$

9 (i) $\begin{pmatrix} \mathbf{F_2} \\ \mathbf{F_1} \end{pmatrix}$

(ii) $\begin{pmatrix} \mathbf{F_6} - \mathbf{F_5} \\ \mathbf{F_4} \end{pmatrix}$

(iii) $\begin{pmatrix} \mathbf{F_8} - \mathbf{F_{10}} \\ \mathbf{F_7} - \mathbf{F_9} \end{pmatrix}$

10 (i) $2\mathbf{i}$
(ii)

11 $\begin{pmatrix} 16 \\ 9 \end{pmatrix}$

12 (i) $\begin{pmatrix} \cos\alpha \\ \sin\alpha \end{pmatrix}$

(ii) $\begin{pmatrix} \cos\beta \\ \sin\beta \end{pmatrix}$

13 (i) O, A and B are not in a straight line.
(ii) O, A and B all lie on a straight line.

Exercise 12.3 (page 261)

1 Always true
2 $-4\mathbf{i} + 6\mathbf{j}, 2\mathbf{i} - 3\mathbf{j},$
$\dfrac{2}{\sqrt{13}}\mathbf{i} - \dfrac{3}{\sqrt{13}}\mathbf{j};$
$6\mathbf{i} + 9\mathbf{j}, 2\mathbf{i} + 3\mathbf{j};$
$3\mathbf{i} + 2\mathbf{j}, -\dfrac{3}{\sqrt{13}}\mathbf{i} - \dfrac{2}{\sqrt{13}}\mathbf{j},$
$-12\mathbf{i} - 8\mathbf{j}$

3 (i) (a) $\begin{pmatrix} 3 \\ 7 \end{pmatrix}$

(b) $\begin{pmatrix} 2 \\ 4 \end{pmatrix}$

(c) $\begin{pmatrix} -1 \\ -3 \end{pmatrix}$

(d) $\sqrt{10}$

(ii) (a) $\begin{pmatrix} -3 \\ -7 \end{pmatrix}$

(b) $\begin{pmatrix} -2 \\ -4 \end{pmatrix}$

(c) $\begin{pmatrix} 1 \\ 3 \end{pmatrix}$

(d) $\sqrt{10}$

(iii) (a) $\begin{pmatrix} -3 \\ -7 \end{pmatrix}$

(b) $\begin{pmatrix} 2 \\ 4 \end{pmatrix}$

(c) $\begin{pmatrix} 5 \\ 11 \end{pmatrix}$

(d) $\sqrt{146}$

(iv) (a) $\begin{pmatrix} 3 \\ 7 \end{pmatrix}$

(b) $\begin{pmatrix} -2 \\ -4 \end{pmatrix}$

(c) $\begin{pmatrix} -5 \\ -11 \end{pmatrix}$

(d) $\sqrt{146}$

(v) (a) $\begin{pmatrix} 3 \\ -7 \end{pmatrix}$

(b) $\begin{pmatrix} -2 \\ 4 \end{pmatrix}$

(c) $\begin{pmatrix} -5 \\ 11 \end{pmatrix}$

(d) $\sqrt{146}$

4 (i) A: $2\mathbf{i} + 3\mathbf{j}$, C: $-2\mathbf{i} + \mathbf{j}$
(ii) $\overrightarrow{AB} = -2\mathbf{i} + \mathbf{j}$,
$\overrightarrow{CB} = 2\mathbf{i} + 3\mathbf{j}$
(iii) (a) $\overrightarrow{AB} = \overrightarrow{OC}$
(b) $\overrightarrow{CB} = \overrightarrow{OA}$
(iv) A parallelogram

5 (i) $\overrightarrow{OA} = \begin{pmatrix} 2 \\ 5 \end{pmatrix}, \overrightarrow{OB} = \begin{pmatrix} 4 \\ 9 \end{pmatrix},$

$$\overrightarrow{OC} = \begin{pmatrix} -3 \\ -5 \end{pmatrix}$$

(ii) $\overrightarrow{AB} = \begin{pmatrix} 2 \\ 4 \end{pmatrix}$,

$\overrightarrow{BC} = \begin{pmatrix} -7 \\ -14 \end{pmatrix}$

(iii) $\overrightarrow{BC} = \dfrac{-7}{2} \times \overrightarrow{AB}$
therefore the vectors
have the same direction.
Both vectors pass
through B, hence points
are collinear.

6 (i) $\overrightarrow{PR} = \mathbf{a} + \mathbf{b}$,

$\overrightarrow{QS} = -\mathbf{a} + \mathbf{b}$

(ii) $\overrightarrow{PM} = \frac{1}{2}(\mathbf{a} + \mathbf{b})$,

$\overrightarrow{QM} = \frac{1}{2}(-\mathbf{a} + \mathbf{b})$

(iii) For a parallelogram
PQRS, $\overrightarrow{PM} = \frac{1}{2}\overrightarrow{PR}$,
and $\overrightarrow{QM} = \frac{1}{2}\overrightarrow{QS}$

7 (i) (a) $\mathbf{i}$
     (b) $2\mathbf{i}$
     (c) $\mathbf{i} - \mathbf{j}$
     (d) $-\mathbf{i} - 2\mathbf{j}$

(ii) $|\overrightarrow{AB}| = |\overrightarrow{BC}| = \sqrt{2}$,
$|\overrightarrow{AD}| = |\overrightarrow{CD}| = \sqrt{5}$

8 (i) $6\mathbf{q} - 9\mathbf{p}$
   (ii) $2\mathbf{q} - 3\mathbf{p}$
   (iii) $6\mathbf{p} - 4\mathbf{q}$
   (iv) $6\mathbf{p} + 2\mathbf{q}$

9 (i) $\overrightarrow{BC} = -\mathbf{p} + \mathbf{q}$,
$\overrightarrow{NM} = -\frac{1}{2}\mathbf{p} + \frac{1}{2}\mathbf{q}$

(ii) $\overrightarrow{NM} = \frac{1}{2}\overrightarrow{BC}$, and
similarly $\overrightarrow{NL} = \frac{1}{2}\overrightarrow{AC}$
and $\overrightarrow{ML} = \frac{1}{2}\overrightarrow{AB}$

So the triangles ABC
and LMN are similar,
with the lengths of the
sides of LMN half the
corresponding lengths of
the sides of ABC.

10 (i) ABCD is a quadrilateral
and $\overrightarrow{AB}$ and $\overrightarrow{DC}$ are
parallel.

(ii) $\overrightarrow{AC} = 2\mathbf{a} + \mathbf{b}$,
$\overrightarrow{BC} = -\mathbf{a} + \mathbf{b}$

(iii) $\overrightarrow{AP} = \mu(2\mathbf{a} + \mathbf{b})$
$\overrightarrow{BP} = \mu(2\mathbf{a} + \mathbf{b}) - 3\mathbf{a}$

(iv) $\mu = \lambda = \dfrac{3}{5}$

Chapter 13

Opening activity (page 264)
(i) 32
(ii) 1024
About 20 days

Activity 13.1 (page 265)
(i) (a) A = (2, 4), B = (4, 16)
     (b) $0 < x < 2$ and $x > 4$
(iii) (a) Yes (b) Yes

Discussion points (page 266)
If the graphs are translated
vertically by k units, the
horizontal asymptote will be
$y = k$ and the graphs will go
through the point $(0, k)$.

If the graphs are translated
horizontally by k units, the
horizontal asymptote will be
$y = 0$ and the graphs will both go
through $(k, 1)$.

If the graphs are stretched
horizontally, the horizontal
asymptote will be $y = 0$ and the
graphs will both go through $(0, 1)$.

If the graphs are stretched vertically
by a scale factor k, the horizontal
asymptote will be $y = 0$ and the
graphs will both go through $(0, k)$.

Exercise 13.1 (page 267)
1 Increasing: $y = 8^x$, $P = 1.5^t$
Decreasing: $N = 0.7^t$, $y = 3^{-x}$,
$y = 0.3^x$, $N = 20^{-t}$, $y = 0.8^x$

2 $y = 0.6^x$

3 (i)
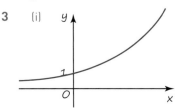

(ii)
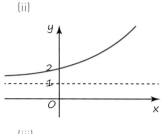

(iii)
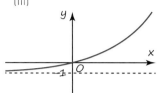

4 (i)
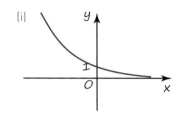

(ii)
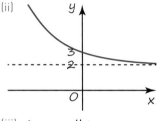

(iii)
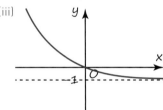

5 (i) Initial population =
10 000; population after
5 years = 31 623

(ii)
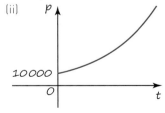

6 (i) 1 m
   (ii) 4.61 m
   (iii) Just over 6 years
   (iv) 20 m

7 (i) 520
   (ii) 210

(iii)

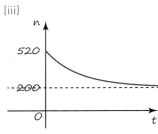

(iv) 200 birds

8 (ii) $x = 1.8$

9 (i) 1976 Hz

(ii) A, 3 octaves below the middle.

(iii) 48 (from E two octaves below the middle, up to C 5 octaves above the middle)

Activity 13.2 (page 271)

The graphs are reflections of each other in $y = x$.

Exercise 13.2 (page 272)

1 $5, 4.2, 7, 3.1, x, x$

2 4

3 (i) $\log_3 9 = x; x = 2$

(ii) $\log_4 64 = x; x = 3$

(iii) $\log_2 \frac{1}{4} = x; x = -2$

(iv) $\log_5 \frac{1}{5} = x; x = -1$

(v) $\log_7 1 = x; x = 0$

(vi) $\log_{16} 2 = x; x = \frac{1}{4}$

4 (i) $3^y = 81; y = 4$

(ii) $5^y = 125; y = 3$

(iii) $4^y = 2; y = \frac{1}{2}$

(iv) $6^y = 1; y = 0$

(v) $5^y = \frac{1}{125}; y = -3$

5 (i) $\log 10$

(ii) $\log 2$

(iii) $\log 36$

(iv) $\log \frac{1}{7}$

(v) $\log 3$

6 (i) $2 \log x$

(ii) $3 \log x$

(iii) $\frac{1}{2} \log x$

(iv) $6 \log x$

(v) $\frac{5}{2} \log x$

7 (i)

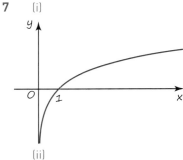

(ii)

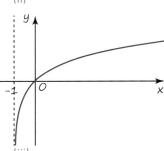

(iii)

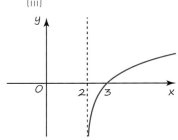

8 (i) $\log 4$

(ii) $\log 4$

(iii) $\log \frac{1}{3}$

(iv) $\log \frac{1}{2}$

(v) $\log 12$

9 (i) $\log \frac{x^2}{7}$

(ii) $x = 21$

10 (i) $x = 19.93$

(ii) $x = -9.97$

(iii) $x = 9.01$

(iv) $x = 48.32$

(v) $x = 1375$

11 (i) 6.6

(ii) 1.58×10^{12}

(iii) 794 times more energy

12 (i) 1.259 (3 d.p.)

(iii) 3 decibels

(iv) It should be
$$\frac{10^{3.7} - 10^{3.5}}{10^{3.5}} \times 100 = 58.5\%$$

13 (i) (a) 7

(b) 5.30

(c) 7.80

(ii) 3.98×10^{-5} and 3.16×10^{-6}

Discussion point (page 274)

£21.05

Exercise 13.3 (page 276)

1 A

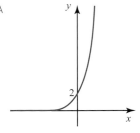

is stretch of 2 in the y direction

B

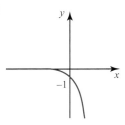

is reflection in the x axis

C

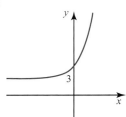

is translation of $\begin{pmatrix} 0 \\ 2 \end{pmatrix}$

2 $y = e^{\frac{x}{5}}$

3 (i)

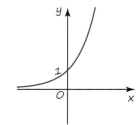

(ii)

(iii)
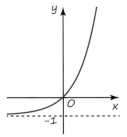

4 (i)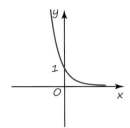

(ii)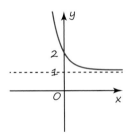

(iii)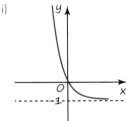

5 (i) $\dfrac{dy}{dx} = 2e^{2x}$

(ii) $\dfrac{dy}{dx} = -3e^{-3x}$

(iii) $\dfrac{dy}{dx} = 0.5e^{0.5x}$

(iv) $\dfrac{dy}{dx} = -6e^{-6x}$

6 (i)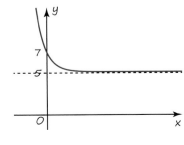

(ii)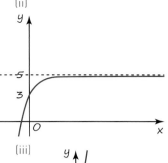

(iii)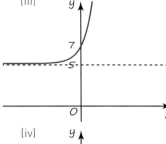

(iv)

7 (i) £20 000
(ii) £14 816
(iii) (a) £2000 per year
(b) £1637 per year
(iv)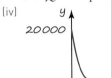

(v) The value of the car would probably fall more quickly in practice. The model implies that the value of the car would eventually reach zero, which is not the case as it would have some value as scrap metal.

8 (i) £2210
(ii) £7788
9 (i) 100
(ii)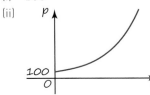

(iii) 8.24 people per year
(iv) 1218
(v) 165
10 0.231
11 (i) £12 649
(ii) 9.12 years
12 (i) Gradient of both graphs is e^e.
(ii)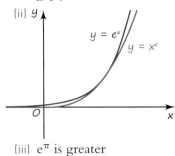

(iii) e^π is greater

Exercise 13.4 (page 279)
1 $2, x, 2^x, 5^t, x, 3.8, 4t$

2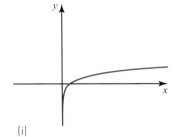

3 (i)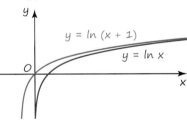

(ii)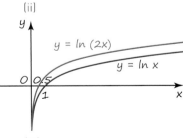

(iii)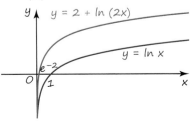

4 (i) $x = x_0 e^{kt}$

(ii) $t = \frac{1}{k} \ln\left(\frac{s_0}{s}\right)$

(iii) $p = 25e^{-0.02t}$

(iv) $x = \ln\left(\frac{\gamma - 5}{\gamma - \gamma_0}\right)$

5 (i) $x = \ln 6$

(ii) $x = \frac{1}{2}\ln 6$

(iii) $x = 1 + \ln 6$

(iv) $x = e^5$

(v) $x = 2e^5$

(vi) $x = \frac{1}{2}e^5$

6 $(0.34, 0.5)$

7 (i)

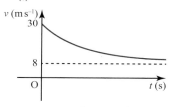

(ii) $621.5\,\text{m}$

(iii) 8.07 a.m.

(iv) No

(v) Never, according to this model.

8 (i)

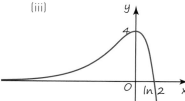

(ii) $30\,\text{m s}^{-1}, 8\,\text{m s}^{-1}$

(iii) $8.33\,\text{m s}^{-1}$

(iv) 8.7 seconds

9 (i) $(0, 4)$ and $(\ln 2, 0)$

(iii)

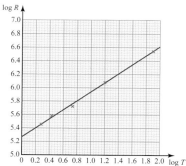

Exercise 13.5 (page 283)

1 $y = ab^x$ and $\ln y = \ln a + x \ln b$;

$y = ax^b$ and $\ln y = \ln a + b \ln x$;

$y = ba^x$ and $\ln y = \ln b + x \ln a$;

$y = bx^a$ and $\ln y = \ln b + a \ln x$

2 $\log_{10} P = a\log_{10} t + \log_{10} k$

3 Students could also use ln.

(i) $\log A = \log k + b \log t$, $m = b, c = \log k$

(ii) $\log N = \log k + a \log t$, $m = a, c = \log k$

(iii) $\log P = \log k + n \log V$, $m = n, c = \log k$

4 (ii) $b = 1.4, k = 0.9$

(iii) (a) 2.4 days

(b) $2.9\,\text{cm}^2$

5 (ii) $n = 1.5, k = 1200$

(iii) $2600\,\text{m}$

(iv) The train won't continue to accelerate for that length of time.

6 (ii)

t	0	20	40	50	60	80
$\ln N$	14.98	14.71	14.40	14.11	13.82	13.51

(iii)

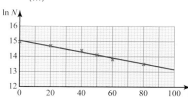

(iv) $a = 0.98, k = 3\,400\,000$

(v) In the year 2005

7 (ii)

Moon	Tethys	Dione	Rhea	Titan	Iapetus
$\log R$	5.46	5.58	5.72	6.09	6.55
$\log T$	0.278	0.431	0.732	1.201	1.899

(iii)

(iv) $n = 0.68, k = 180\,000$

(v) 0.7 days

8 (ii) $y = 400 \times 0.63^t$

(iii) $y = 0.4$. The infection is under control.

9 (i) $P = 4000 \times t^{0.2}$

(ii) 4000

(iii) The model predicts growth in local user numbers without limit, which is impossible. However, the growth becomes so slow that this may well not be a problem.

Practice questions (page 288)

1 $\overrightarrow{AB} = -2\mathbf{i} + 3\mathbf{j}$,

$\overrightarrow{DC} = -1\mathbf{i} + \frac{3}{2}\mathbf{j}$ [1]

so $\overrightarrow{DC} = \frac{1}{2}\overrightarrow{AB}$, [1]

and DC is therefore parallel to AB, and ABCD is a trapezium.

[1]

2 (i) $52\,300 = ae^0 = a$ [1]

$58\,500 = 52\,100\,e^{5k}$

[1]

$5k = \ln\left(\frac{58\,500}{52\,300}\right)$ [1]

$k = 0.0224$ [1]

(ii) $62\,600$ [1]

3 (i) $\int_{-1}^{1}(3x - x^3)\,\mathrm{d}x$

$= \left[\frac{3}{2}x^2 - \frac{1}{4}x^4\right]_{-1}^{1}$ [1]

$= \left(\frac{3}{2}1^2 - \frac{1}{4}1^4\right)$

$- \left(\frac{3}{2}(-1)^2 - \frac{1}{4}(-1)^4\right)$ [1]

$= 0$ [1]

(ii) $\frac{\mathrm{d}y}{\mathrm{d}x} = 3 - 3x^2$ [1]

when $x = 1$,

$\frac{\mathrm{d}y}{\mathrm{d}x} = 3 - 3 = 0$;

when $x = 1$,

$y = 3 \times 1 - 1^3 = 2$; [2]

solve $3 - 3x^2 = 0$ to get

$x = -1$ or 1

when $x = -1$,

$y = 3 \times (-1) - (-1)^3$

$= -2.$ [2]

So $(-1, -2)$ and $(1, 2)$ are stationary points.

(iii)

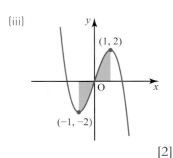

[2]

(iv) There are equal areas above and below the x axis, which cancel each other out when working out the integral. [1]

4 (i) $\dfrac{(13.23 - 12)}{0.1}$ [2]

(ii) 2.001, 12.012, 12.003 [3]

(iii) 12 [1]

(iv) $y = 3x^2$, so $\dfrac{dy}{dx} = 6x$ [1]

When $x = 2$,

$\dfrac{dy}{dx} = 6 \times 2 = 12$ [1]

5 (i) Box area
$= xy + xy + 5y + 5y + 5x$
$= 2xy + 5x + 10y$ [1]

Sleeve area
$= 5y + 5x + 5y + 5x$
$= 10x + 10y$ [1]

so $A = 2xy + 15x + 20y$ [1]

(ii) $V = 5xy = 60$ [1]

so $y = \dfrac{12}{x}$, and

$A = 24 + 15x + \dfrac{240}{x}$

as required. [1]

(iii) $\dfrac{dA}{dx} = 15 - \dfrac{240}{x^2} = 0$ [1]

so $\dfrac{240}{x^2} = 15$,

$x^2 = \dfrac{240}{15} = 16$,

$x = 4$; [2]

$y = \dfrac{12}{4} = 3$, so

dimensions of the box are 3 cm by 4 cm by 5 cm.

$\dfrac{d^2 A}{dx^2} = \dfrac{480}{x^3} > 0$

for $x = 4$. so the volume is a minimum. [1]

6 (i) $\overrightarrow{OD} = \mathbf{v} + \dfrac{1}{2}\mathbf{u}$, so

$\overrightarrow{OE} = \dfrac{2}{3}(\dfrac{1}{2}\mathbf{u} + \mathbf{v})$

$= \dfrac{1}{3}\mathbf{u} + \dfrac{2}{3}\mathbf{v}$ [2]

(ii) $\overrightarrow{OF} = \overrightarrow{OA} + \overrightarrow{AF}$

$= \overrightarrow{OA} + \dfrac{2}{3}\overrightarrow{AC}$ [1]

$= \mathbf{u} + \dfrac{2}{3}(\mathbf{v} - \mathbf{u})$

$= \dfrac{1}{3}\mathbf{u} + \dfrac{2}{3}\mathbf{v}$ [2]

So $\overrightarrow{OE} = \overrightarrow{OF}$, and E and F are the same point. [1]

7 (i) P is (6.685, 4.614),
Q is (9.210, 7.364) [2]

(ii) $b = \dfrac{(7.364 - 4.614)}{(9.210 - 6.685)}$

$= 1.089 = 1.09$ (2 d.p.) [2]

$4.614 = a + 1.089 \times 6.685$ so $a = -2.67$ (2 d.p.) [2]

(iii) $t = e^{-2.67} \times d^{1.09}$
$= 0.069 \times d^{1.09}$ [2]

(iv) When $d = 100$,
$t = 10.44$, [1]
so the record is below what is predicted by the model. [1]
The model is for middle and long distances but seems not to be accurate for sprints. [1]

8 (i) $100\left(1 + \dfrac{2}{100}\right)^{12}$ [1]

$= 126.82$, so the APR is 26.82% [1]

(ii) $100\left(1 + \dfrac{p}{100}\right)^{12} = 150$ [1]

$\Rightarrow \left(1 + \dfrac{p}{100}\right)^{12} = 1.5$ so

$\ln\left[\left(1 + \dfrac{p}{100}\right)^{12}\right]$

$= \text{in } 1.5$ [1]

$\Rightarrow 12\ln\left[\left(1 + \dfrac{p}{100}\right)\right]$

$= \text{in } 1.5$ [1]

$\Rightarrow \ln\left[\left(1 + \dfrac{p}{100}\right)\right]$

$= \dfrac{\ln 1.5}{12}$

$= 0.033788\ldots$ [1]

$\Rightarrow = \left(1 + \dfrac{p}{100}\right) = e^{0.033788\ldots}$

$= 1.034$

$\Rightarrow p = 3.4\%$ [1]

The same calculation could be done using logs rather than lns, in which case

$\left(1 + \dfrac{p}{100}\right) = 10^{0.01467\ldots}$

$= 1.034$ etc.

Chapter 14

Discussion point (page 295)

The fairest answer is that there is not enough information. Ignoring the journalistic prose, '… our town council rates somewhere between savages and barbarians…', the facts given are broadly correct. However, to say whether or not the council is negligent one would need to compare accident statistics with other *similar* communities. Also, one would need to ask who is responsible for a cyclist's risk of having or not having an accident. Perhaps parents should ensure there is adequate training given to their children, and so on.

Exercise 14.1 (page 296)

1 An outlier may be a genuine item of data but an error is not.

2 $8\dfrac{5}{7}$

4 (i) 30

(ii) 34 under the age of 18

(iii) Approx. 3.4 km

(depending on how the data is cleaned)

(iv) Approx. 33% (depending on how the data is cleaned)

(v) Approx. 47%

(vi) Approx. 37%

(vii) (a) 74%

(b) Approx. 27%

(viii) From this data 39014 and 54211 never reported whether the cyclist was wearing a helmet. On one occasion 78264 did not record this information.

5 While this is the correct average of the police officers' numbers it is quite meaningless. The numbers are a way of identifying the officers. There is no more sense in taking their average than there would be in trying to find an average name for them. So, she is not right.

6 and 7 There are various correct answers to these questions. Students could compare their diagrams with other students and discuss which graphs show the information in the best way.

8 He must establish well defined classes so that each accident can be assigned to one of them. These may include 'None of these' and 'Multiple'.

9 Calculations and displays showing the distribution of those having accidents among the age groups.

10 Robin should address the question of how likely it is that a cyclist, particularly a child, is involved in a preventable accident. His article should be based on data and not mere rhetoric. A key question is how many of the

accidents were caused by cars, lorries and other heavy traffic.

11 Robin is focusing on two aspects of the investigation: he is looking at cycling accidents in the area over a period of time and he is considering the distribution of ages of accident victims. He should also try to estimate the overall number of cyclists, not just those who have accidents, and their age distribution. Another thing he might consider is to investigate accidents in a similar community in order to be able to make comparisons. He should also look at the traffic management in the area and the availability (or otherwise) of cycle paths. He could also look at whether adequate safe cycling training is available in Avonford or nearby.

Discussion points (page 297)

There is no information on how the sample was selected but with a size of only 30 it cannot be representative of all the different groups: male and female; different ages; different income levels; etc.

A sample of 30 is nowhere near large enough. Samples for polls about voting intentions are typically over 1000.

It may well matter that the right question was not asked; it certainly would not be safe to imagine otherwise.

Discussion points (page 298)

The population is made up of the MP's constituents. The sample is part of that population of constituents. Without information relating

to how the constituents' views were elicited, the views obtained seem to be biased towards those constituents who bother to write to their MP. The population is made up of Manchester households. We are not told how the sample is chosen. Even if a random sample of households were chosen the views obtained are still likely to be biased as the interview timing excludes the possibility of obtaining views of most of the residents in employment.

Discussion points (page 300)

Each student is equally likely to be chosen but samples including two or more students from the same class are not permissible so not all samples are equally likely. Yes

Exercise 14.2 (page 302)

1 It is more efficient as it is cheaper and quicker than a simple random sample.

2 A sampling frame.

3 Systematic sampling

4 (i) Simple random sampling

(ii) $\frac{1}{25}$

5 (i) Cluster sampling

(ii) No. The streets are chosen at random and then 15 houses are chosen at random. However not every sample of 15 (throughout the town) can be chosen.

6 (i) Quota sampling

(ii) No

(iii) The sample size is 160 which should be large enough to give some useful information, providing they are reasonably representative of people who might possibly shop at Teegan's boutique.

7 In most of the situations there are several possible sampling procedures and so no single right answer.

(i) Cluster sampling. Choose representative streets or areas and sample from these streets or areas.

(ii) Stratified sample. Identify routes of interest and randomly sample trains from each route.

(iii) Stratified sample. Choose representative areas in the town and randomly sample from each area as appropriate.

(iv) If the candidate has a list of the electorate (which they should do), simple random sampling, or systematic sampling. Otherwise the candidate might take a number of locations and use cluster sampling.

(v) Depends on method of data collection. If survey is, say, via a postal enquiry, then a random sample may be selected from a register of addresses.

(vi) Cluster sampling. Routes and times are chosen and a traffic sampling station is established to randomly stop vehicles to test tyres.

(vii) Cluster sampling. Areas are chosen and households are then randomly chosen.

(viii) Cluster sampling. A period (or periods) is chosen to sample and speeds are surveyed.

(ix) Cluster sampling. Meeting places for 18-year-olds are identified: night clubs, pubs, etc. and samples of 18-year-olds are surveyed, probably via a method to maintain privacy. This might be a questionnaire to ascertain required information.

(x) Simple random sampling. The school student register is used as a sampling frame to establish a simple random sample within the school.

8 (i) $\frac{1}{10}$

(ii) Years 1 and 2: 14 students each; Years 3 and 4: 10 students each; Year 5: 12 students.

9 (i) 28 light vans, 2 company cars and 1 large-load vehicle.

(ii) Randomly choose the appropriate number of vehicles from each type. This is stratified sampling.

10 (i) $\frac{1}{8}$

(ii) 0–5: 5; 6–12: 10; 13–21: 13 or 14; 22–35: 25 or 26; 36–50: 22 or 23; 51+: 3 or 4

11 (i) Systematic sampling. Easy to set up but may be difficult to track down the student once they have been identified.

(ii) Self-selected sample. The responses may be from those with strong views that are not representative of the majority of the students.

(iii) The sample will be biased. Easy to survey. Those using the canteen will be surveyed.

(iv) Cluster sampling. Assumes first and second year students are representative of the whole college. (If there are only first and second year students this will be true. The sampling procedure is then stratified.) Similar to (i), that is, once students have been chosen from the lists they have to be located to seek their views.

12 Justification: $\frac{1153}{1235} = 93\%$ – and this is from people who chose to buy the product so it must be the brand they prefer!
Issues: only a quarter of the 5000 responded, which means that three quarters either didn't feel it was better, or didn't feel it was significantly better and worth replying. Only 80% who replied actually think the product is better than the current moisturiser. The group was self-selecting which means that they already had a preference (in this case to the company's products) and would be expected to be favourably inclined towards the new product.

13 (i) $\frac{5 \times 24 \times 7}{100\,000} = 0.0084$

(ii) It is a systematic sample. It is easy to apply. It allows the output of each machine to be checked on a regular schedule.

14 All production lines are identified. If it is judged they are equivalent then one (or more) can be chosen to

produce a sample. This is cluster sampling. From this (or these) production line(s) a day (or days) is chosen to be the time when a sample is taken. A reasonable number of strip lights is chosen and then tested to destruction, that is, tested until they are exhausted. An estimate is found from the mean life of the sample chosen.

15 The map of the forest is covered with a grid. Each grid square is numbered. A sample is chosen by randomly selecting the squares. The tree (or trees) in each of the chosen squares is sampled.

16 Depending on the number of staff, one could carry out a census of all staff or, if more appropriate, a stratified sample based on part-time staff, full-time staff, academic staff, support staff, etc.

17 Identify different courses in your school/college. Access students from each of these courses, choosing them at random in order to elicit their views. This is a stratified sample.

Chapter 15

Exercise 15.1 (page 309)
1 (i) They are both the most frequently occurring.
   (ii) The mode refers to a single data value and the modal class refers to a group of data values.
2 Discrete.
3 (i) Heart and coronary disease.
   (ii) They refer to the percentages of deaths in the various categories which can be attributed to smoking.

(iii) (a) Heart and coronary disease.
   (b) Heart and coronary disease and lung cancer.
4 (i) Product A (it has approx. doubled in size and increased by over £750 000).
   (ii) Sales decreased in 2013 but have continued to increase since.
5 (i) Tertiary
   (ii) Secondary
   (iii) Primary education has probably remained universal. The increase in expenditure is due to inflation and an increased population. Secondary education is now universal or nearly so. The expenditure on it has increased more than that on primary education and is now slightly more than on primary education, possibly because it is more expensive to run. Tertiary education is now much more widespread but the lower expenditure on it suggests it is not universal.
6 (Typical answers)
   (i) Comment about overall number of NEETs – general falling trend, only increases in Oct–Dec 2014. Comment about gender differences, e.g. female NEETs always higher than male NEETs, biggest difference in Oct–Dec 2014.
   (ii) Figure 15.8 shows both time periods and gender differences so patterns over time can be identified. Figure 15.9 shows the total NEETs

and allows comparision of gender differences.
   (iii) Pie charts can be used to show how a total quantity is divided into categories. This is not suitable for the male NEET data as individuals may apprear in more than one of the time periods and so would be double-counted in the sectors of the pie chart. The total quantity is not meaningful.
7 (i) Compound bar chart
   (ii) PE & Sport has the largest change – a decrease of 40%. Classics show the smallest change – an increase of 16.7%.
   (iii) The numbers are much larger for the new subjects – a new scale would be needed, which may mean that small changes are difficult to detect.
8 A TRUE. If you add together the sections covering those with two adults, it is more than half.
   B TRUE. Although there are slightly more females, the split is roughly equal.
   C FALSE. Although most people own at least part of their home, about half of those who do have a mortgage so the lending body owns part of their home until this is paid off.
   D UNCERTAIN. There are about 4m households in the category 'Single 65 and over' and half of the population is male. However there is not enough evidence to say that these two are independent and

therefore 2m men over 65 live on their own. A longer life expectancy for women would lead to the assumption that fewer than half the 4m people aged over 65 living alone were men.

E FALSE. Although the housing status is roughly equally shared, there is not enough evidence to say that this applies equally across all household types. For instance young adults are more likely to live in rented accommodation, while older couples are more likely to have finished paying off their mortgage.

9 A good chart for proportions is a pie chart when viewed from above; for comparing numbers a bar chart is better. 3-D charts often mislead people as they give prominence to those groups which are 'closer' to the viewer as you see more of the side of these sectors. Colour can also be used to mislead people as darker colours tend to dominate.

Discussion points (page 313)

Range = 182 − 74 = 108 minutes

IQR = 119 − 77 = 42 minutes

The IQR is the better measure. The range was unduly influenced by Sally who took much longer than all the other athletes.

Exercise 15.2 (page 314)

1 Measures of spread are range and interquartile range; measures of central tendency are median, mode and mean.

2 Mean

3 (i)

	Life expectancy for people born in different London boroughs		Ranking	
	Female	Male	Female	Male
Barking and Dagenham	79.7	74.7	9	7
Barnet	81.8	77.6	3	2.5
Camden	80.5	74.3	6.5	8
Croydon	80.5	76.7	6.5	5.5
Greenwich	80.1	74.2	8	9
Islington	79.1	73.5	10	10
Kensington & Chelsea	84.0	78.8	1	1
Kingston upon Thames	81.3	77.6	5	2.5
Redbridge	81.5	76.7	4	5.5
Tower Hamlets	78.9	72.7	11	11
Westminster	82.7	77.0	2	4

(ii)

	Female	Male
Minimum	78.9	**72.7**
Q_1	79.7	**74.2**
Q_2	80.5	**76.7**
Q_3	**81.8**	**77.6**
Maximum	84.0	**78.8**

(iii)

Female

Male

72 74 76 78 80 82 84 86

(iv) The distribution of female life expectancies is higher than for males. All of the female values, minimum, Q_1, Q_2, Q_3 and maximum are higher than the male equivalent. Even the **maximum** value for the male life expectancy (Kensington and Chelsea, 78.8) is below the **minimum** value for females (Tower Hamlets, 78.9). On average (median) a female in London is expected to outlive a male by approximately 4 years.

(v) (a) As different boroughs have different costs linked to living within them (such as housing, travel) and different environmental conditions (such as being central with few green spaces or towards the outer edges with more access to green spaces) there are different standards of living which could contribute to life expectancy.

(b) Generally women are found to live longer than men across the UK and across the world.

4 (i) 0.74 g
(ii) 0.89 − 0.7 = 0.19 g
(iii) The majority of the cereals have a salt content which is 0.8 g or less. The mid-point of the amber range for salt (0.3 g − 1.5 g) is 0.90 g. Half the cereals are below 0.74 g or $\frac{3}{4}$

of the cereals are below 0.89 g related to the amber range of (0.3 g − 1.5 g) for salt. Very few of the cereals are coloured green and so deemed healthy.

5 (i)

	1st	2nd	3rd	4th
A	Scotland	Ireland	England	Wales
B	England	Wales	Ireland	Scotland
C	Scotland	Ireland	England	Wales
D	Scotland	Ireland	England	Wales

(ii) The median ranks are based upon the times.

(iii) A and B both take the full team into account. A does not penalise a disaster as much as B so is perhaps fairer.

6 (i)

Europe		A & C	
Q_1	56.75	Q_1	27.75
Q_2	70	Q_2	49
Q_3	87	Q_3	76.5
IQR	30.25	IQR	48.75

(ii) Europe has a higher median, indicating that on average a greater proportion of the population in Europe has access to the internet. The IQR for America and the Caribbean is greater which suggests that there is a greater spread in the data. The distribution for Europe is bimodal, suggesting that there may be two sorts of countries involved. These might, for example, be those in the west and the east of the continent; however, without the original data it is impossible to say.

The distribution for America and the Caribbean has three modes. Since some of these counties are small island states and others, like the USA and Canada, have large populations, the variability shown by the stem-and-leaf plot is unsurprising. However, as with the European countries, the original data are not available so no explanation of the distribution can be more than speculation.

7 A TRUE. The lowest wage in 2011 (£445) is above the highest wage in 1997 (£400).

B TRUE. These can be read from the boxplots.

C FALSE. The range in 1997 was approx. £115 while in 2011 it is approx. £205. Similarly the interquartile range has increased from about £15 to about £30.

D FALSE. In both years the median is closer to the lower quartile than the upper quartile, suggesting the higher earning regions have a greater spread. Similarly for the ranges.

E UNCERTAIN. It is true that the vertical scale has been truncated; it starts at £280 instead of zero. It is also true that this gives a false impression, making the increase from 1997 to 2011 look proportionately greater.

However there is no information to indicate that this was done deliberately.

8 For a list of n items of data, an *Excel* spreadsheet uses the 'method of hinges'. It places the median, Q_2, at position $\frac{n+1}{2}$, the lower quartile, Q_1, at position $\frac{1}{2}\left(1 + \frac{n+1}{2}\right) = \frac{1}{2} + \frac{n+1}{4}$ and the upper quartile, Q_3, at position $\frac{1}{2}\left(\frac{n+1}{2} + n\right) = \frac{3(n+1)}{4} - \frac{1}{2}$.

Whilst the quartiles Q_1 and Q_3 differ from those obtained with a graphical calculator, either method is acceptable.

With a large set of data, results are likely to be less affected than with a small data set, but different methods should be taken into account when analysing data and checking on the results of others.

9

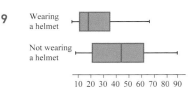

Possible responses include:

Younger people tend to wear helmets; the boxplot is skewed towards the younger ages.

The oldest people don't wear helmets.

We don't know the relative sizes of the groups: 55 were wearing helmets and 29 were not wearing helmets, but boxplots don't show the relative sizes of the groups.

Activity 15.1 (page 318)

	Numbers of goals scored by teams	Goals	Frequency
0	〣〣〣〣〣〣〣I	0	36
1	〣〣〣〣〣〣〣〣IIII	1	44
2	〣〣〣〣〣II	2	27
3	〣〣II	3	12
4	〣I	4	6
5	II	5	2
6		6	0
7	I	7	1

Discussion point (page 319)

(One among many possible answers.)

You know the mean GDP per person for each country in the world and want to find the mean GDP per person for the whole world. In this case you would find a weighted mean, multiplying the figure for each country by its population, adding all these figures together to get the total world income and then dividing by the world's population.

Discussion point (page 320)

It makes no difference mathematically but seems more natural to rank the best performance 1.

Discussion point (page 321)

You would treat it as an ordinary frequency table, with the range of the scores in the left hand column replaced by their mid-points.

Midpoint, x	Frequency, f	xf
4.5	1	4.5
14.5	2	29.0
24.5	4	98.0
34.5	7	241.5
44.5	14	623.0
54.5	15	817.5
64.5	17	1096.5
74.5	13	968.5
84.5	5	422.5
94.5	2	189.0
Σ	80	4490

Estimated mean =

$$\frac{\sum xf}{\sum f} = \frac{4490}{80} = 56$$

(to the nearest whole number).

Discussion points (page 322)

There may be too much data to for it to be practical to enter each item on a stem-and-leaf diagram and the variable may take too many values for a vertical line chart to be suitable.

For example, the marks obtained by candidates on a public examination with an entry of several thousand people.

The data may cover too large a number of possible values of the variable.

For example, the cost in £ of a sample of second-hand cars.

The data may be spread out over too wide a range.

For example, people's winnings in £ in the National Lottery one Saturday.

Exercise 15.3 (page 322)

1 You can still access the detail of the data.

2 Discrete data.

3 (i)

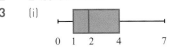

(ii) The vertical line graph as it retains more data for this small sample.

4 (i)

Wickets	0	1	2	3	4	5	6
Matches	2	5	0	5	7	5	2

(ii) Mode: 4; mean: 3.3.

(iii) She will use the mode as it is the largest.

5 (i) These 12 females produce 28 fledglings; they do so 3 times so produce 84 fledglings. 42 of these are females, 25% survive which is 10 to the nearest bird.

So at the next generation these 12 females will be replaced by just 10.

So the birds will reduce in numbers.

(ii) The birds eat insects and the farmer is using insecticides so this is a possible cause.

There are many other possible causes, e.g. predation. The study is quite small and the data cover only one year.

6 (i) The mode is the most useful average to use in this situation as this is the shoe size which is most often bought. The median might also be the mode, but this just tells you which size is the middle size, not whether it is popular, while the mean may not even be a recognised shoe size.

(ii) He could ask every student in the sixth form but this may take a long time and not everyone might accurately know their height.

He could measure a sample of 60 students as they enter sixth form one morning and record their heights in a table. He should ensure that the groups are not overlapping and he may want to consider using narrower groups for the central values which are likely to be more common.

Answers

541

7 (i) (a) US$8160 (b) US$6964

(ii) The weighted mean

8 The mean wages are £10.91 per hour, which would justify the manager's claim. However, less than one third of the workers are paid over this, so a fairer advertisement would be to use the median or mode, both of which are £7.50 per hour.

9 (i) 14

(ii) 196

(iii) 0.7

(iv) The mode and median are both zero so are not representative. So the mean is the most representative measure.

10 A UNCERTAIN. It is true that there was no day on which all 8 hens laid so it is possible that one hen did not lay any eggs at all, but it is also possible that they all laid some eggs but happened not to do so on the same day.

B FALSE. The totals of the four columns are 90, 92, 92 and 91 making the 365 days in the year. The February was in 2015 which was not a leap year.

C TRUE. The total number of eggs was 1085 and 1085 ÷ 365 = 2.972..., which is 2.97 when rounded to 2 d.p.

D FALSE. A pie chart could be used to illustrate and compare the total numbers of eggs in the four seasons but it would not be showing 'All the data in the table'; some of the detail would have been lost.

11 mean = 30 (rounded); median = 22; mode = 9

The best to use may depend on what you are trying to state. In Chapter 14, Robin was asked to report on the growing local concern about accidents involving children. In this case, the mode supports this argument. However, it can be argued that the mean is better as it uses all of the data or that the median is better as the extreme ages reported do not affect this measure.

Discussion point (page 329)

No. You worked out the frequency densities to draw the histogram. The modal class is that with the highest value.

Exercise 15.4 (page 332)

1 The estimate for the greatest value will be at the upper end of the class which is the first one to reach the total. Ignore further classes that do not contain data. The estimate for the least value is the lower end of the class which has the first non-zero frequency.

2 Mode

3 (i)

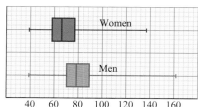

(ii) The median mass of the men is greater than the upper quartile of the women, so 50% of the men are heavier than 75% of the women. The central 50% of the women's masses is less symmetrical than the men's with the median towards the lower end. Both sets of data have long upper tails, suggesting both men and women's data have a few individuals who are heavier.

(iii) It is easier to see what is going on with the boxplots. They show you the actual distributions of the data whereas the cumulative frequency curves do not.

4 (i) 133 g

(ii) Median 135, lower quartile 118, upper quartile 156, IQR 38.

(iii) The point for 160 should be plotted at 247.

(iv) David's error affects the upper quartile and the IQR. They should be 144 and 26.

(v) This graph on its own is not very useful; a boxplot (which can be generated from the graph) would be a better way of showing where the median lies and how spread out the data are.

5 (i) It is not clear in which interval a weight of, say, 2.5 kg should be recorded. The table does not allow for very small or very large babies. The intervals are wide and so there is a risk of losing important information.

(ii) The majority of cases are within the usual weight range. There are some babies which are large and may cause issues during the birth process. There are many babies born with a low (below 2.5 kg) birth weight.

(iii) The doctors may need to know whether this trend is similar to others in similar areas. There may be other medical factors such as multiple births (twins and triplets), whether there are links to other diseases, whether these are first-born babies.

6 A FALSE. It is true that the data have been grouped, and you could describe that process as rounding. At first sight the vertical line for, say, 100 could mean 'At least 95 and less than 105' which would correspond to rounding to the nearest 10. However, if you look at the very first vertical line this is marked 0 and that would correspond to an interval of −5 to 5. Clearly you cannot have a negative number of mobile phone accounts, so this interpretation must be wrong. The alternative that, for example, 100 means 'At least 100 and less than 110' must be the case.

B FALSE. The modal class is 110 to 120. The mode is a single value. If the statement had been 'The mode of the numbers displayed' is 110, it would have been true.

C TRUE. The total number of dots on the lines to the left of 100 is 89 and this is less than half of the total of 189.

D UNCERTAIN. It may well be that the statement is true but the diagram does not provide you with the information to know it.

E FALSE. Each dot represents one country and different countries have different populations.

F TRUE. The raw data are means per 100 people, so they are numbers to 2 decimal places. Such data can be regarded as continuous so could indeed be represented by a frequency chart or a histogram.

7 On weekdays a histogram will show two maxima, one for each 'rush hour' when most people are travelling. This is more marked if school-age children who have accidents during the holidays are omitted from the sample.
If the other data are grouped into weekends and children during school holidays the times are more evenly distributed.

Discussion point (page 335)
Place is ranked, Team is categorical and the other 7 variables are all discrete numerical.

Discussion point (page 336)
Manchester United and Liverpool are quite a long way from the rest of the teams so could be considered outliers. To be safe you should check that they are correctly shown on the scatter diagram.

Exercise 15.5 (page 339)
1 Any opinion should be supported with an argument.
2 Positive
3 A positive correlation
  B strong negative correlation

C no correlation
D positive correlation
4 (i) There is a weak positive correlation between wingspan and mass.
  (ii) One blackbird has wingspan of 122 mm and has mass of 128 g; this blackbird is much heavier than other birds with similar wingspans. Another blackbird has wingspan 134 mm but a mass of only 81 g. This blackbird seems underweight compared with the rest of the sample.
  (iii) This bird is clearly underweight. Maybe it is sick, or hungry, or both.
  (iv) (a) 98 g. Yes it is consistent with the data on the scatter diagram.
      (b) The points on the scatter diagram for a wingspan of 125 mm correspond to masses between 88 and 108 g so these are ± 10 g on a value of 98, so about $\pm 10\%$. It might be safer to give wider margins, say $\pm 15\%$.
5 (i) The data are bivariate; they allow correlation/association between the variables to be identified.
  (ii) Positive correlation: as the maximum monthly temperature increases the total monthly rainfall increases.
  (iii) (14.4, 215). The total rainfall was unusually high for the maximum monthly temperature.

(iv) $C = 15.6$ (Note: the actual value that year was 15.0°C.)

(v) 269 mm
1975 is quite a long time before the data used to calculate the regression lines were collected. So there is an assumption that the climate has not changed in the intervening years.

6 (i) (a) The distinct sections are female elephant seals and male elephant seals.

(b) The outlier at the top right of the graph is a very large elephant bull seal, probably a dominant bull seal. The outlier near the middle of the graph (length 4.2 m) is possibly a young male.

(ii) A regression line would not be suitable due to there being two distinct groups within the sample. These two groups do not follow the same general pattern. Also, 15 is a small sample from which to produce a regression line.

(iii) $w = 775 \Rightarrow l = 2.48$. On the scatter diagram this point is clearly among the female elephant seals.

7 A TRUE. The probabilities have been calculated correctly, dividing the number of homicides by the population in the same year.

B FALSE. The number of homicides in given years is random variable but the years themselves are not.

C TRUE. The source of the data is not given but the National Census is carried out every 10 years, when the dates end in 1. So it was carried out in the years quoted, 1901, 1911, 1991 and 2001.

D FALSE. The estimate is based on extrapolation well beyond the period covered by the data.

E FALSE. The data form two islands on the scatter diagram and in that situation it is quite wrong to infer correlation from a line joining them (see page 337). Within each of the subsets the points do not fit a straight line particularly well, and there is a large gap in the data between 1911 and 1989. A different issue is that that the variables are not both random.

8 Using all of the data, you get a scatter graph like this which shows some areas or islands of data.

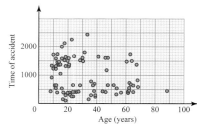

The effect becomes more noticeable if you look at the data from children who had accidents on weekdays not in school holidays.

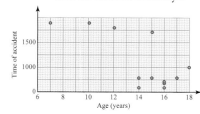

Discussion points (page 343)

$$\sum (x - \bar{x})^2 = \sum (x^2 - x\bar{x} + \bar{x}^2)$$
$$= \sum x^2 - x\bar{x}\sum x + \sum \bar{x}$$
$$= \sum x^2 - 2\bar{x}\left(n \times \sum \frac{x}{n}\right) + n\bar{x}^2$$

$$\boxed{\sum \frac{x}{n} = \bar{x}} \quad \boxed{\sum \bar{x}^2 = n\bar{x}^2}$$

$$= \sum x^2 - 2n\bar{x}^2 + n\bar{x}^2$$
$$= \sum x^2 - n\bar{x}^2$$

If you are working by hand and the mean, $\bar{x}$, has several decimal places, working out the individual deviations can be tedious. So in such cases it is easier to use the $\sum x^2 - n\bar{x}^2$ form. However, it is good practice to get into the habit of using the statistical functions on your calculator or spreadsheet.

Activity 15.2 (page 344)

$S_{xx} = 6$, $s^2 = \dfrac{6}{9} = 0.666...$,

$s = \sqrt{0.666...} = 0.816...$

Discussion point (page 345)

No, standard deviation cannot be negative. If you work out a standard deviation and get a negative value, you have definitely made a mistake.

Exercise 15.6 (page 346)

1 The standard deviation is larger as it is divided by a smaller number.

2 $\sum x^2 - \bar{x}^2$

3 (i) 0.14
(ii) 1.88

Number of (mm) away from 1 m	Mid point x	Fre-quency f	$f \times$ midpoint	fx^2
$-5 \leqslant$ (l) < -3	-4	6	-24	96
$-3 \leqslant$ (l) < -1	-2	12	-24	48
$-1 \leqslant$ (l) < 1	0	58	0	0
$1 \leqslant$ (l) < 3	2	17	34	68
$3 \leqslant$ (l) < 5	4	7	28	112
		100	14	324
		(i)	$\bar{x}$	0.14
		(ii)	σ	1.8

(iii) The estimates in (i) and (ii) assumed that all of the values in a group were the same value, the midpoint. The sample mean and standard deviation from the raw data was calculated using the actual measurements.

4 Both students have used the same data so the variation in plant height within the sample is the same. The different figures for the standard deviation are as a result of the different units of measurement chosen by the two students. In Alisa's case the standard deviation is 7.46 mm whilst Bjorn's standard deviation is 0.746 cm; these are equivalent distances.

5 (ii) Mean 58, standard deviation 14.9
(iii) (a) $58 + 2 \times 14.9$ $= 87.8 < 96$
(b) Outlier

6 (i) The box shows that the central part of the distribution is very nearly symmetrical about the median. The whiskers show that the distribution is skewed to the right.
(ii) 50–60 16×10 $= 160$
60–70 51×10 $= 510$
70–80 80×10 $= \underline{800}$
Total   1470

(iii) $81.7 \pm 2 \times 15.7$ so 50.3 to 113.1
(iv) $3152 - (16 \times 9.7 + 51 \times 10 + 80 \times 10 + 81 \times 10 + 48 \times 10 + 21 \times 10 + 3.1 \times 8) = 162$
(v) There are more very heavy people. The boxplot shows the distribution is skewed and this can also be seen on the histogram with more very high values than very low ones.

7 (i) Total weight is 1963.951 g
(ii) $\sum x^2 = 33819.03$
(iii) $n = 202$, $\sum x = 3196.285$, $\sum x^2 = 53748.02$
(iv) Mean = 15.823 g, standard deviation = 3.978

8 Estimated numbers of patients are as follows: within 1 sd of the mean 2170, so 68.8% - very close to 68%
within 2 sds of the mean 2990, so 94.9% - extremely close to 95%
within 3 sd of the mean 3126. so 99.2% - very close to 99.75%.

9 (These values are rounded and answers may vary according to how the data is cleaned.) The mean is 30 with a standard deviation of 20 (rounded); the median is 22 with a semi-IQR of 16. The mean takes extreme values into account, so the very young and very old can make a big difference to the mean – this is demonstrated by the large standard deviation. The median with the smaller semi-IQR may give a better

representative value for the data set in this case. In this case the standard deviation and semi-IQR are both relatively large suggesting a varied set of data.

Chapter 16

Discussion point (page 361)
Throwing a single, normal, die.

Discussion point (page 363)
You would work out the probability that it is not flooded in 5 years and then subtract this from 1.

So for 5 years

P(the street is flooded in at least one year)
$= 1 - $ P(it is not flooded in 5 years)
$= 1 - \left(\dfrac{29}{30}\right)^5 \approx \dfrac{1}{6.4}$

The risk is about once in every 6 or 7 years.

Exercise 16.1 (page 363)
1 $P(F \cap G) = 0.31$ and $P(G \cap F') = 0.27$
2 0.88
3 $\dfrac{66}{534}$, assuming each faulty torch has only one fault.
4 A TRUE $- \dfrac{90}{120} = \dfrac{3}{4}$
B UNCERTAIN $-$ not enough information but it seems unlikely.
C TRUE $-$ the expected number is $\dfrac{80}{30} = 2\dfrac{2}{3}$ fledglings.
D UNCERTAIN $-$ not enough information. You don't know how many of the existing adult females and the female fledglings will have died by next year.
E TRUE $- \dfrac{1}{2}$ of 80 is 40 and $\dfrac{40}{120} = \dfrac{1}{3}$

5 (i) $\dfrac{1}{13}$

(ii) $\dfrac{36}{52} = \dfrac{9}{13}$

(iii) $\dfrac{9}{13}$

(iv) $\dfrac{16}{52} = \dfrac{4}{13}$

6 (i) $\dfrac{12}{100} = \dfrac{3}{25}$

(ii) $\dfrac{53}{100}$

(iii) $\dfrac{45}{100} = \dfrac{9}{20}$

(iv) $\dfrac{42}{100} = \dfrac{21}{50}$

(v) $\dfrac{56}{100} = \dfrac{14}{25}$

(vi) $\dfrac{5}{100} = \dfrac{1}{20}$

7 (i) 0.4

(ii) 0.5

(iii) 2, 2.2

8 (i) 0.2

(ii) 0.6

(iii) 0.26. The readings are independent

9 (ii) Approximately $\dfrac{19}{60} = 0.32$

(iii) Approximately $\dfrac{19}{60} \times \dfrac{19}{60} = 0.1$

(iv) Consecutive journey times are independent of each other. However, the same roadworks could cause delays to both journeys, so the times would not be independent.

10 (i) $\dfrac{5}{2000}$

(ii) $\dfrac{1995}{2000}$

(iii) Lose £100, if all tickets are sold

(iv) 25p

(v) 2500

11 (i) 0.35

(ii) They might draw

(iii) 0.45

(iv) 0.45

12 (i) $k = 0.4$

r	2	4	6	8
P($X = r$)	0.1	0.2	0.3	0.4

(ii) (a) 0.3

(b) 0.35

13 (i)

		First die				
+	1	2	3	4	5	6
1	2	3	4	5	6	7
2	3	4	5	6	7	8
3	4	5	6	7	8	9
4	5	6	7	8	9	10
5	6	7	8	9	10	11
6	7	8	9	10	11	12

(Second die)

(ii) $\dfrac{3}{36} = \dfrac{1}{12}$

(iii) 7

(iv) The different outcomes are not all equally probable.

14

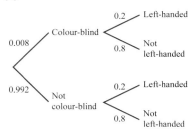

(i) 0.0016

(ii) 0.0064

(iii) 0.2064

(iv) 0.7936

15 1 in 2.6

16 (i) 0.020

(ii) 0.65

(iii) 0.99

17 (i) (a) $\dfrac{15}{365}$

(b) $\dfrac{20}{365}$

(c) $\dfrac{330}{365}$

(ii) $\dfrac{55}{730\,000} = 0.000\,075\ldots$

(iii) Heat presents the greater risk.

18 (i) $k = 0.08$

r	0	1	2	3	4
P($X = r$)	0.2	0.24	0.32	0.24	0

(ii) Let Y represent the number of chicks

r	0	1	2	3
P($Y = r$)	0.351 04	0.449 28	0.184 32	0.015 36

Chapter 17

Exercise 17.1 (page 376)

1

P($X = r$)	Formula	n	p	q
P($X = 2$)	$\dbinom{5}{2}\left(\dfrac{1}{3}\right)^2\left(\dfrac{2}{3}\right)^3$	5	$\dfrac{1}{3}$	$\dfrac{2}{3}$
P($X = 4$)	$\dbinom{6}{4}\left(\dfrac{1}{4}\right)^4\left(\dfrac{3}{4}\right)^2$	6	$\dfrac{1}{4}$	$\dfrac{3}{4}$
P($X = 3$)	$\dbinom{7}{3}\left(\dfrac{2}{3}\right)^3\left(\dfrac{1}{3}\right)^4$	7	$\dfrac{2}{3}$	$\dfrac{1}{3}$
P($X = 4$)	$\dbinom{10}{4}\left(\dfrac{1}{5}\right)^4\left(\dfrac{4}{5}\right)^6$	10	$\dfrac{1}{5}$	$\dfrac{4}{5}$

2 $\dbinom{20}{7}\left(\dfrac{1}{5}\right)^7\left(\dfrac{4}{5}\right)^{13}$

3 $\dfrac{15}{64}$

4 0.271

5 0.214

6 0.294

7 (i) 0.146

(ii) Poor visibility might depend on the time of day, or might vary with the time of year. If so, this simple binomial model would not be applicable.

8 (i) $\dfrac{1}{8}$

(ii) $\dfrac{3}{8}$

(iii) $\dfrac{3}{8}$

(iv) $\dfrac{1}{8}$

9 (i) 0.246

(ii) Exactly 7 heads

10 (i) (a) 0.058

(b) 0.198

(c) 0.296

(d) 0.448

(ii) 2

11 (i) (a) 0.264

(b) 0.368

(c) 0.239

(d) 0.129

(ii) Assumed the probability of being born in January $= \frac{31}{365}$. This ignores the possibility of leap years and seasonal variations in the pattern of births throughout the year.

12 The three possible outcomes are not equally likely: 'one head and one tail' can arise in two ways (HT or TH) and is therefore twice as probable as 'two heads' or 'two tails'.

Exercise 17.2 (page 380)

1 The score you expect to get is the one with the highest probability and is a value of the data. The expectation of the score is a statistic and may not take a value that is a possible data item.

2 5

3 (i) (a) 0.000129
   (b) 0.0322
   (c) 0.402
   (ii) 0 and 1 are equally likely

4 (i) 2
   (ii) 0.388
   (iii) 0.323

5 (i) (a) 0.240
   (b) 0.412
   (c) 0.265
   (d) 0.512
   (e) 0.384
   (f) 0.096
   (g) 0.317
   (ii) Assumption: the men and women in the office are randomly chosen from the population (as far as their weights are concerned).

6 (i) (a) $\frac{1}{81}$
   (b) $\frac{8}{81}$
   (c) $\frac{24}{81}$
   (d) $\frac{32}{81}$
   (ii) 2 min 40 s

7 (i) He must be an even number of steps

from the shop. (The numbers of steps he goes east or west are either both even or both odd, since their sum is 12, and in both cases the difference between them, which gives his distance from the shop, is even.)
   (ii) 0.00293
   (iii) At the shop
   (iv) 0.242
   (v) 0

8 (i) 0.0735
   (ii) 2.7
   (iii) 0.267

9 (i) (a) 0.349
   (b) 0.387
   (c) 0.194
   (ii) 0.070
   (iii) 0.678

Chapter 18

Discussion point (page 384)

Assuming both types of parents have the same fertility, boys born would outnumber girls in the ratio 3 : 1. In a generation's time there would be a marked shortage of women of child-bearing age.

Discussion point (page 388)

It does seem a bit harsh when the number 1 came up more often than the others. You cannot actually prove a result with statistics but even trying to show it beyond reasonable doubt, in this case using a 5% significance level, can be difficult. This is particularly so if the sample size is small; it would have been better is the die had been thrown a lot more than 20 times.

Exercise 18.1 (page 390)

1 The p-value is calculated in a hypothesis test

and compared with the significance level. The value of p is the default probability that is used to do the calculation of the p-value.

2 $H_0 : p = \frac{1}{6}$

   $H_1 : p > \frac{1}{6}$

3 (i) Null hypothesis: $p = 0.25$; alternative hypothesis: $p > 0.25$
   (ii) 0.0139
   (iii) 5%
   (iv) Yes

4 0.1275   Accept H_0

5 $0.0547 > 5\%$ Accept H_0

6 H_0: probability that toast lands butter-side down = 0.5
   H_1: probability that toast lands butter-side down > 0.5
   0.240   Accept H_0

7 0.048   Reject H_0. There is evidence that the complaints are justified at the 5% significance level, though Mr McTaggart might object that the candidates were not randomly chosen.

8 0.104   Accept H_0. Insufficient evidence at the 5% significance level that the machine needs servicing.

9 (i) 0.590
   (ii) 0.044
   (iii) 0.0000712
   (iv) 0.0292
   (v) H_0: P(long question right) = 0.5; H_1: P(long question right) > 0.5
   (vi) No

Discussion point (page 393)
$X \leqslant 4$

Exercise 18.2 (page 396)

1 A critical value is a single value at one end of the distribution that marks the start of the critical region

at that end. The critical region may consist of several values. For a 2-tail test there is a critical value and a critical region at both ends of the distribution, even if one is empty.

2 $H_0 : p = \frac{1}{2}$

   $H_1 : p \neq \frac{1}{2}$

3 (i) Henry is only interested in whether tails is more likely so he should carry out a 1-tail test. Mandy is interested in knowing about any bias so should carry out a two-tail test.

   (ii) Henry: for 29 tails $p = 0.8987$ and for 30 tails $p = 0.9405$ so 30 is the critical value. Mandy: for 18 tails $p = 0.0325$ and for 19 tails $p = 0.0595$. For 30 tails $p = 0.9405$ and for 31 tails $p = 0.9675$ so her critical values are 18 and 31.

   (ii) Henry will accept any number of tails up to and including 29; Mandy will accept any number of tails between and including 19 and 30.

   (iii) Henry will say the coin is biased if there are 30 or more tails.

   (iv) Mandy will say the coin is biased if there are 18 or fewer tails or if there are 31 or more tails.

   (v) Henry has only looked at whether tails is more likely; he has no way of knowing if heads is more likely. Mandy will be able to say whether either heads or tails is more likely and therefore whether the coin is biased.

4 $P(X \leqslant 7) = 0.1316 > 5\%$
   Accept H_0

5 $P(X \geqslant 13) = 0.0106 < 2\frac{1}{2}\%$
   Reject H_0

6 $P(X \geqslant 9) = 0.0730 > 2\frac{1}{2}\%$
   Accept H_0

7 $P(X \geqslant 10) = 0.0139 < 5\%$
   Reject H_0, but data not independent

8 $P(X \geqslant 6) = 0.1018 > 5\%$
   Accept H_0

9 (i) $0.0395 < 5\%$
        Reject H_0
   (ii) $0.0395 > 2\frac{1}{2}\%$
        Accept H_0

10 $\leqslant 1$ or > 9 males

11 $\leqslant 1$ or > 8 correct

12 Critical region is $\leqslant 3$ or $\geqslant 13$ letter Zs

13 (i) 20
   (ii) 0.0623
   (iii) Complaint justified

Practice questions (page 399)

1 (i) $0.75^3 = \frac{27}{64}$ (or 0.421 875) [2]
   (ii) $0.75^3 + 0.2^3 + 0.05^3 = 0.43$ [2]

2 (i) Pie charts (and variants such as this chart) are used to represent proportions of a total. In this case there is no such total as not all goals scored are shown. [1]
        This type of chart makes comparisons difficult, particularly when the quantities are similar to one another. [1]
   (ii) A bar chart would make for easier visual comparison. [1]
        In this case there is little to be gained from having any sort of chart. The list of goal scorers and their numbers of goals is sufficient. [1]

NB: other sensible answers are possible.

3 (i) Simple random sampling is sampling in which all possible samples of the required size have an equal chance of selection. [1]
        In this case simple random sampling would not be possible because there is no sampling frame. That is, there is no list of all customers from which a sample could be constructed. [1]

   (ii) In opportunity sampling the interviewer(s) would interview any convenient customers until a total of 200 had been reached. [1]
        In quota sampling the interviewer(s) would be given target numbers of interviewees in different groups, e.g. 100 male, 100 female; equal numbers under and over 40 years of age. [1]

        Quota sampling is preferable. [1]

        Because it attempts to make the sample representative of the target population. [1]

4 (i) Eruptions typically last between 1.5 and 5.5 minutes, with intervals between eruptions typically being between 40 and 95 minutes. [1]
        There appear to be two different types of eruption, short and long. Similarly there are short and long intervals between eruptions. [1]

        Short eruptions are typically followed by short intervals, long eruptions by long intervals. [1]

(ii) The data points separate out into two fairly distinct clusters. Within each cluster there is only a slight tendency for longer eruptions to be followed by longer times to the next eruption. [1]

(iii) The histogram of *Length of eruption* would be bimodal. [1]
The histogram of *Time to next eruption* would show little or no bimodality. [1]

(iv) The mean, 3.5 minutes, represents a length of eruption that is very unlikely to occur. So in that sense it could be misleading. [1]
However, it could be useful in other ways. For example, it is a figure that would be needed to estimate the proportion of the total time for which the geyser was erupting. [1]

NB: other sensible answers are possible.

5 (i) The description of the year involves two years. [1]
Some data may be included in both of 2005–06 and 2006. [1]

(ii) The data varies with no particular pattern over the 9 years. [1]
Removing the data point for 2008 gives a slight downward trend. [1]
The variation in weekly cheese consumption is only about 10 g overall. [1]

(iii) It is unlikely that this data point represents a real change in consumption; almost certainly this is just a random fluctuation in the data. (The actual change is of very small

magnitude; it looks bigger because of the scale.) [2]

(iv) There would be some merit in fitting a straight line model to the data set. [1]
There is a slight downward trend. [1]
Interpolation or extrapolation would be meaningless as the data varies considerably from year to year. [1]

6 (i) Mean 93.1, standard deviation 13.26. [3]

(ii) Mean 92.1, standard deviation 8.11. [3]

(iii) The households in the South East eat less sugar than households in the North West. [1]
The weekly sugar consumption of households in the North West varies more than for households in the South East. [1]

(iv) The difference in the mean mass of sugar eaten per person per week supports this view. However, the difference in the standard deviation suggests that there are other factors involved besides location. [2]

NB: other sensible answers are possible.

7 (i) $(1, 2), (2, 1), (2, 2)$. [1]
36 equally likely outcomes so the probability is $\frac{3}{36} = \frac{1}{12}$. [1]

(ii) $k = \frac{1}{36}$ [1]
$P(X = 6) = 11k = \frac{11}{36}$ [1]

(iii) $P(X = Y) = k^2 + (3k)^2 + (5k)^2 + (7k)^2 + (9k)^2 + (11k)^2 = 286k^2$ [1]

$P(X \neq Y) = 1 - 286k^2$
$(= 0.77932)$ [1]
A and B are equally likely to win. [1]
So P(B beats A)
$= 0.390$ to 3 d.p. [1]

8 (i) Expected number is $0.15 \times 25 = 3.75$ cars. [1]

(ii) Use $X \sim B(25, 0.15)$ to find $P(X \leq 3)$ [1]
$P(X \leq 3) = 0.471$ (to 3 d.p.) [1]

(iii) $H_0: p = 0.15$,
$H_1: p < 0.15$,
where p is the proportion of cars with unsafe tyres. [2]
Use $X \sim B(50, 0.15)$ to find $P(X \leq 5)$
$= 0.219$ [1]
Observe that $0.219 > 0.05$ so the observed result is not in the 5% critical region, hence insufficient evidence that p has reduced. [1]

(iv) Use $X \sim B(100, 0.15)$ to find
$P(X \leq 7) = 0.012 > 0.01$
and
$P(X \leq 6) = 0.005 < 0.01$. [2]
Hence the possible values of k are $0, 1 \ldots, 6$. [1]

Chapter 19

Discussion point (page 404)
$-4, 0, -5$
(i) $+4$
(ii) -5

Discussion point (page 405)
The marble is below the origin.

Exercise 19.1 (page 405)

1 Position is where you are relative to a fixed origin, displacement is where you are relative to where you

started from. Both can be described by vectors.

2 0 m
3 (i) +1 m
   (ii) +2.25 m
4 (i) 3.5 m, 6 m, 6.9 m, 6 m, 3.5 m, 0 m
   (ii) 0 m, 2.5 m, 3.4 m, 2.5 m, 0 m, −3.5 m
   (iii) (a) 3.4 m
         (b) 10.3 m
5 (i) 2 m, 0 m, −0.25 m, 0 m, 2 m, 6 m, 12 m
   (ii)

$t = 1, 2$ $t = 0, 3$ $t = 4$ $t = 5$
$t = 1.5$

−2 0 2 4 6 8 10 12 x (m)

   (iii) 0 m, −2 m, −2.25 m, −2 m, 0 m, 4 m, 10 m
   (iv) 14.5 m
6 (i) 0 m, −16 m, −20 m, 0 m, 56 m
   (ii)

−20   0   56
        −16   0

   (iii)

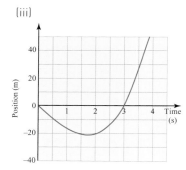

   (iv) $t = 0$ moving backwards, $t = 3$ moving forwards.
7 (i) height above river (m)
       40

       0 → time

       positive direction

   (ii) height above ground (m)
       2000

       0 → time

   (iii) height above bridge (m)
       0 → time

       positive direction

8 (i) The ride starts at $t = 0$. At A it changes direction

and returns to pass the starting point at B, continuing past to C, where it changes direction again, returning to its initial position at D.
   (ii) An oscillating ride such as a swing boat.

Discussion point (page 407)

10, 0, −10. The gradient represents the velocity.

Discussion point (page 408, top)

The graph would curve where the gradient changes. Little effect over this period.

Discussion point (page 408, bottom)

$+5\,\text{m s}^{-1}, 0\,\text{m s}^{-1}, -5\,\text{m s}^{-1}, -6\,\text{m s}^{-1}$.
The velocity decreases at a steady rate.

Exercise 19.2 (page 409)

1 A speed–time graph records the magnitude whereas a velocity–time graph shows the direction as well. A speed–time graph is always on or above the time axis whereas a velocity–time graph may go below it.
2 Graph A
3

Speed
(m s^{-1})
10

5

    100 200 300 400 Time (s)

4 (i) The person is waiting at the bus stop.
   (ii) It is faster.
   (iii)

Speed
(km h^{-1})
24

16

8

0   9.15   9.30   9.45 Time

   (iv) Constant speed, infinite acceleration
5 (i) (a) 2 m, 8 m
        (b) 6 m
        (c) 6 m
        (d) $2\,\text{m s}^{-1}, 2\,\text{m s}^{-1}$
        (e) $2\,\text{m s}^{-1}$
        (f) $2\,\text{m s}^{-1}$
   (ii) (a) 60 km, 0 km
        (b) −60 km
        (c) 60 km
        (d) $-90\,\text{km h}^{-1}, 90\,\text{km h}^{-1}$
        (e) $-90\,\text{km h}^{-1}$
        (f) $90\,\text{km h}^{-1}$
   (iii) (a) 0 m, −10 m
        (b) −10 m
        (c) 50 m
        (d) OA: $10\,\text{m s}^{-1}, 10\,\text{m s}^{-1}$;
             AB: $0\,\text{m s}^{-1}, 0\,\text{m s}^{-1}$;
             BC: $-15\,\text{m s}^{-1}, 15\,\text{m s}^{-1}$
        (e) $-1.67\,\text{m s}^{-1}$
        (f) $8.33\,\text{m s}^{-1}$
   (iv) (a) 0 km, 25 km
        (b) 25 km
        (c) 65 km
        (d) AB: $-10\,\text{km h}^{-1}$, $10\,\text{km h}^{-1}$;
             BC: $11.25\,\text{km h}^{-1}$, $11.25\,\text{km h}^{-1}$
        (e) $4.167\,\text{km h}^{-1}$
        (f) $10.83\,\text{km h}^{-1}$
6 $10.44\,\text{m s}^{-1}, 37.58\,\text{km h}^{-1}$
7 $20.59\,\text{km h}^{-1}$
8 $1238.71\,\text{km h}^{-1}$
9 $40\,\text{km h}^{-1}$
10 (i) $32\,\text{km h}^{-1}$;
    (ii) $35.7\,\text{km h}^{-1}$
11 (i) (a) $56.25\,\text{km h}^{-1}$
         (b) $97.02\,\text{km h}^{-1}$
         (c) $46.15\,\text{km h}^{-1}$
    (ii) The ratio of distances must be in the ratio 10:3.
12 (i) $4.48\,\text{km h}^{-1}$, $36.73\,\text{km h}^{-1}$, $18.32\,\text{km h}^{-1}$; $26.15\,\text{km h}^{-1}$
    (ii) B finishes first.

Discussion point (page 412)

(i) D
(ii) B, C, E
(iii) A

Exercise 19.3 (page 413)

1 Either an acceleration in the opposite direction, or a deceleration.

2 Graph B

3 (i) (a) $0.8\,\mathrm{m\,s^{-2}}$
   (b) $-1.4\,\mathrm{m\,s^{-2}}$
   (c) $0.67\,\mathrm{m\,s^{-2}}$
   (d) $0\,\mathrm{m\,s^{-2}}$
   (e) $0.5\,\mathrm{m\,s^{-2}}$

(ii)

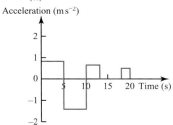

4 (i) $20\,\mathrm{m\,s^{-1}}$
   (ii) $7.5\,\mathrm{s}$

5 (i) $6\,\mathrm{m\,s^{-2}}$
   (ii) $6\,\mathrm{m\,s^{-1}}$
   (iii) $a = 3t$
   (iv) $t = 5$

6 (i)

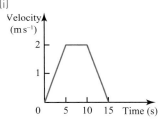

(ii)

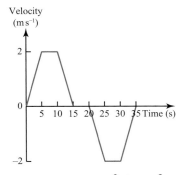

(iii) $+0.4\,\mathrm{m\,s^{-2}}, 0\,\mathrm{m\,s^{-2}},$
      $-0.4\,\mathrm{m\,s^{-2}}, 0\,\mathrm{m\,s^{-2}},$
      $-0.4\,\mathrm{m\,s^{-2}}, 0\,\mathrm{m\,s^{-2}},$
      $+0.4\,\mathrm{m\,s^{-2}}$

(iv)

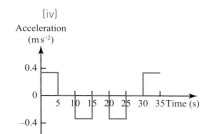

7 (i) $20\,\mathrm{m\,s^{-1}}, 40\,\mathrm{m\,s^{-1}},$
      $10\,\mathrm{m\,s^{-1}}$

(ii)

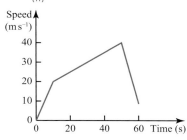

(iii) $1550\,\mathrm{m}$

8 (i)

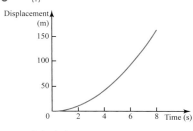

(ii) (a)

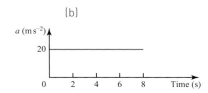

(b)

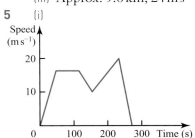

Discussion points (page 415)

(i) 5
(ii) 20
(iii) 45
They are the same.

Discussion point (page 416)

It represents the displacement.

Discussion point (page 417)

Approximately $460\,\mathrm{m}$.

Discussion point (page 418)

No, as long as the lengths of the parallel sides are unchanged, the trapezium has the same area.

Exercise 19.4 (page 418)

1 The area under a speed–time graph represents distance travelled whereas the area under a velocity–time graph represents the displacement from the starting point.

2 The acceleration is $0.8\,\mathrm{m\,s^{-2}}$.

3 (i) (A) $0.4\,\mathrm{m\,s^{-2}},$
         $0\,\mathrm{m\,s^{-2}}, 3\,\mathrm{m\,s^{-2}}$
      (B) $-1.375\,\mathrm{m\,s^{-2}},$
         $-0.5\,\mathrm{m\,s^{-2}}, 0\,\mathrm{m\,s^{-2}},$
         $2\,\mathrm{m\,s^{-2}}$

   (ii) (A) $62.5\,\mathrm{m}$
        (B) $108\,\mathrm{m}$

   (iii) (A) $4.17\,\mathrm{m\,s^{-1}}$
         (B) $3.6\,\mathrm{m\,s^{-1}}$

4 (i) Enters motorway at $10\,\mathrm{m\,s^{-1}}$, accelerates to $30\,\mathrm{m\,s^{-1}}$ and maintains this speed for about $150\,\mathrm{s}$. Slows down to a stop after a total of $400\,\mathrm{s}$.

   (ii) Approx. $0.4\,\mathrm{m\,s^{-2}},$
        $-0.4\,\mathrm{m\,s^{-2}}$

   (iii) Approx. $9.6\,\mathrm{km}, 24\,\mathrm{m\,s^{-1}}$

5 (i)

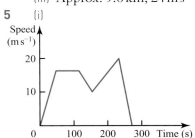

   (ii) $3562.5\,\mathrm{m}$

6 (i)

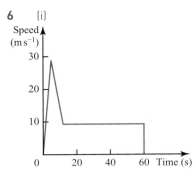

(ii) 558 m

7 (i)

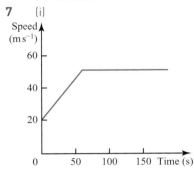

(ii) 60 s

(iii) 6600 m

(iv) $v = 20 + 0.5t$,
$0 \leqslant t \leqslant 60$;
$v = 50, t \geqslant 60$

8 (i)

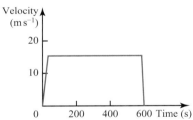

(ii) $15\,\mathrm{m\,s^{-1}}, -1\,\mathrm{m\,s^{-2}}$,
8.66 km

9 (i)

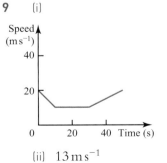

(ii) $13\,\mathrm{m\,s^{-1}}$

10 (i)

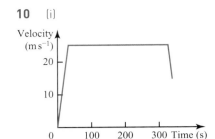

(ii) 337 s

11 (i) BC: decelerates uniformly,
CD: stopped, DE:
accelerates uniformly

(ii) $a = -0.5\,\mathrm{m\,s^{-2}}, 2500\,\mathrm{m}$

(iii) $0.2\,\mathrm{m\,s^{-2}}, 6250\,\mathrm{m}$

(iv) 325 s

(v)

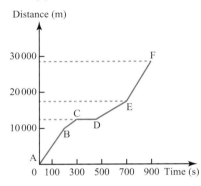

12 462.5 s

13 40 s

14 (i) $10\,\mathrm{m\,s^{-1}}, 0.7\,\mathrm{s}$

(ii)

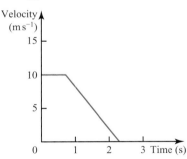

(iii) $6.25\,\mathrm{m\,s^{-2}}$

(iv) 33.9 m

15 (i) 7.5 m

(ii) $1.2\,\mathrm{m\,s^{-1}}$

(iii) 10 s

(iv) $a = -0.4\,\mathrm{m\,s^{-2}}$

16 (i) A: 17 min 30 seconds
after starting; B: 16 min
27 seconds after starting

(ii) 16 min 35 seconds after
A starts, 9450 m from
the start

17 (i) 11.125 s

(ii) 10 s

Discussion point (page 422)

$s = \frac{1}{2}(2u + at) \times t$

$s = \left(u + \frac{1}{2}at\right) \times t$

$s = ut + \frac{1}{2}at^2$

Discussion point (page 424)

$u = 13.\dot{3}\,\mathrm{m\,s^{-1}}, v = 26.\dot{6}\,\mathrm{m\,s^{-1}}$,
$t = 5, s = 100\,\mathrm{m}$

Exercise 19.5 (page 425)

1 $s = ut + \frac{1}{2}at^2$, because it is
the only one with those
four letters.

2 $s = \frac{1}{2}(u + v)t$

3 (i) $22\,\mathrm{m\,s^{-1}}$

(ii) 120 m

(iii) 0 m

(iv) $-10\,\mathrm{m\,s^{-2}}$

4 (i) $v^2 = u^2 + 2as$

(ii) $v = u + at$

(iii) $s = ut + \frac{1}{2}at^2$

(iv) $s = \frac{1}{2}(u + v)t$

(v) $v^2 = u^2 + 2as$

(vi) $s = ut + \frac{1}{2}at^2$

(vii) $v^2 = u^2 + 2as$

(viii) $s = vt - \frac{1}{2}at^2$

5 (i) $9.8\,\mathrm{m\,s^{-1}}, 98\,\mathrm{m\,s^{-1}}$

(ii) 4.9 m, 490 m

(iii) 2 s, speed and distance after
10 s, both over-estimates

6 $2.08\,\mathrm{m\,s^{-2}}, 150\,\mathrm{m}$, assume
constant acceleration

7 $4.5\,\mathrm{m\,s^{-2}}, 9\,\mathrm{m}$

8 $-8\,\mathrm{m\,s^{-2}}, 3\,\mathrm{s}$

9 $a = -0.85\,\mathrm{m\,s^{-2}}, 382\,\mathrm{m}$

10 3.55 s

11 (i) $s = 16t - 4t^2$
$v = 16 - 8t$

(ii) (a) 2 s

(b) 4 s

(iii)

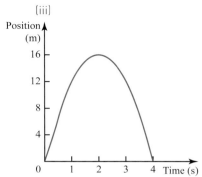

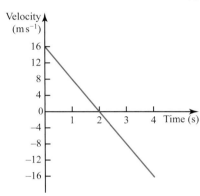

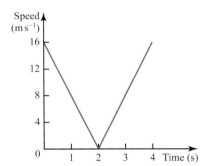

Discussion point (page 427)

$u = -15.0$. No.

Exercise 19.6 (page 429)

1. Acceleration = 4 m s^{-2} and time = 5 s
2. $15 = aT$
3. 604.9 s, 9.04 km
4. (i) $v = 2 + 0.4t$
   (ii) $s = 2t + 0.2t^2$
   (iii) 18 m s^{-1}
5. No, 10 m behind
6. (i) $h = 4 + 2t - 4.9t^2$
   (ii) 1.13 s
   (iii) 9.08 m s^{-1}
   (iv) t greater, v less
7. (i) 11.75 m s^{-1}
   (ii) 8.29 m
   (iii) 12.75 m s^{-1}
   (iv) 5.31 m

(v) Underestimate

8. (i)
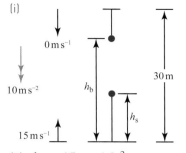
   (ii) $h_S = 15t - 4.9t^2$
   (iii) $h_b = 30 - 4.9t^2$
   (iv) $t = 2s$
   (v) 10.4 m
9. (i) 5.4 m s^{-1}
   (ii) -4.4 m s^{-1}
   (iii) 1 m s^{-1} gain
   (iv) 9 m s^{-1}
   (v) Too fast
10. 2.94 m
11. 43.75 m
12. (ii) $15.15 = u + 5a$
    (iii) 14.4 m s^{-1}
    (iv) No, distance at constant a is 166.5 m
13. 22.7 m s^{-1}
14. $-\frac{5}{12}$ m s^{-2}, $40\frac{5}{6}$ m
15. 2.5 m
16. $3\frac{1}{3}$ m s^{-1}, $\frac{1}{9}$ m s^{-2}, 9.5 m from the bus
17. 107.5 m

Chapter 20

Discussion point (page 437)

The reaction between the chair and the person acts on the chair. The person's weight acts on the person only.

Discussion point (page 438)

Vertically up

Exercise 20.1 (page 439)

In these diagrams, W represents a weight, R a normal reaction with another surface, T the tension, F a friction force, D drag and P another force.

1. Your weight and the normal reaction from the chair you are sitting on, weight of a cup and the normal reaction from the table, a ladder leaning against a wall and the reaction of the wall are some examples.

2. Graph B

3.

4.

5.

6.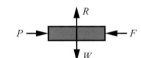

7.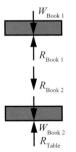

8.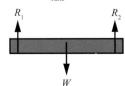

9. (i) 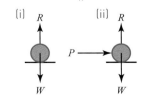 (ii)

10 (i)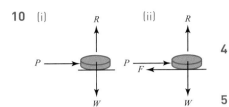
(ii)

11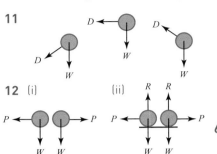

12 (i)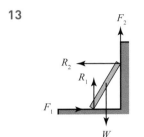
(ii)

13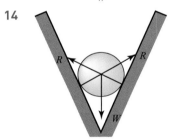

14

Discussion point (page 440, left)

To provide forces when the velocity changes.

Discussion point (page 440, right)

The friction force was insufficient to enable his car to change direction at the bend.

Exercise 20.2 (page 442)

1 False. A resultant force may decelerate an object moving in the opposite direction, such as a ball that is hit up into the air.

2 $W < T$

3 (i)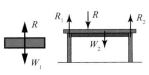

(ii) (a) $R = W_1$
(b) $R_1 + R_2 = W_2 + R$

4 (i) $R = W, 0$
(ii) $W > R, W - R$, down
(iii) $R > W, R - W$ up

5 (i) No
(ii) Yes
(iii) Yes
(iv) No
(v) Yes
(vi) Yes
(vii) Yes
(viii) No

6 Forces are required to give passengers the same acceleration as the car.
(i) A seat belt provides a backwards force.
(ii) The seat provides a forwards force on the body and the head rest is required to make the head move with the body.

Discussion point (page 443)

The first one, with the pencil in tension.

Exercise 20.3 (page 446)

1 Mass is a measure of the amount of matter in an object; it is a scalar quantity. Weight is the force that gravity exerts upon the object; it is a vector quantity.

2 88.2 N

3 (i) 147 N
(ii) 11 760 N = 11.76 kN
(iii) 0.49 N

4 (i) 61.2 kg
(ii) 1120 kg = 1.12 tonne

5 (i) 637 N
(ii) 637 N

6 112 N

7 (i) Both hit the ground together.
(ii) The balls take longer to hit the ground on the Moon, but still do so together.

8 *Answers for 60 kg:*
(ii) 588 N
(iii) 96 N
(iv) Its mass is 4 kg.

Discussion points (page 446)

No.
Scales which measure by balancing an object against fixed masses (weights).

Exercise 20.4 (page 448)

In these diagrams, mg represents a weight, R a normal reaction with another surface, F a friction force, D drag, T a tension or thrust, P a driving force and Q another force.

1 Tension; consider the force on the truck, it needs to accelerate so the force must be towards the train so in tension.

2 (c)

3 (i)
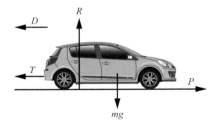

(ii)
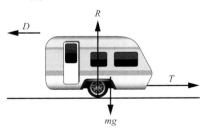

(iii)
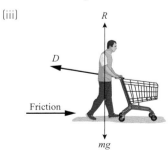

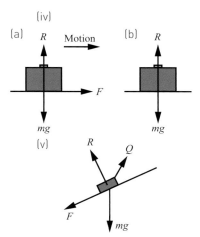

(iv)
(a) ... (b) ...

(v)

4 (i) Weight $5g$ down and reaction ($= 5g$) up.
  (ii) Weight $5g$ down, reaction with box above ($= 45g$ down) and reaction with ground ($50g$ up).
5 (i) $F_1 = 10$ N
  (ii) $15 - F_2$ N
6 (i) Towards the left.
  (ii)

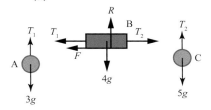

  (iii) $3g$ N, $5g$ N
  (iv) $2g$ N
  (v) $T_1 - 3g \uparrow$,
      $T_2 - T_1 - F \rightarrow, 5g - T_2 \downarrow$
7 All forces are in newtons.
  (i) greater   (ii) less

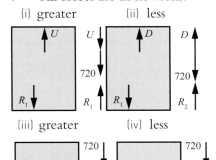

  (iii) greater   (iv) less

8 (i) 2400 N

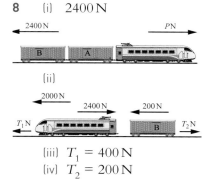

  (ii)

  (iii) $T_1 = 400$ N
  (iv) $T_2 = 200$ N

Discussion points (page 450)

The pointer moves up and down as the force on the spring varies. Your weight would seem to change as the speed of the lift changed. You feel the reaction force between your hand and the book which varies as you move the book up and down.

Discussion point (page 452)

There is a resultant downward force because the weight is greater than the tension.

Exercise 20.5 (page 454)

1 A force equal to her weight.
2 16 000 N
3 (i) 800 N
  (ii) 88 500 N
  (iii) 0.0225 N
  (iv) 840 000 N
  (v) 8×10^{-20} N
  (vi) 548.8 N
  (vii) 8.75×10^{-5} N
  (viii) 10^{30} N
4 (i) 200 kg
  (ii) 50 kg
  (iii) 10 000 kg
  (iv) 1.02 kg
5 (i) 7.6 N
  (ii) 7.84 N
6 (i) 0.5 m s^{-2}
  (ii) 25 m
7 (i) 1.67 m s^{-2}
  (ii) 16.2 s
8 (i) 325 N
  (ii) 1764 N
9 (i) $2\mathbf{i} + 2\mathbf{j}$ newtons
  (ii) 1.41 m s^{-2}
10 (i) 13 N

  (ii) 90 m
  (iii) 13 N
11 (i)

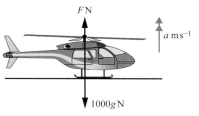

  (ii) 11 300 N
12 (i) $0.8\mathbf{i} + 1.2\mathbf{j}$ newtons
  (ii) 56.3°
13 (i) $400 - 250 = 12 000a$,
      $a = 0.0125$ m s^{-2}
  (ii) 0.5 m s^{-1}, 40 s
  (iii) (a) 15 s
       (b) 13.75 m
       (c) 55 s
14 (i) 60 m s^{-1}
  (ii) Continues at 60 m s^{-1}
  (iii) 1.25 N
  (iv) The first one by 655 km
15 (i) 6895 N, 6860 N, 6790 N, 1960 N
  (ii) 815 kg
  (iii) Max $T < 9016$ N
16 (i) 7.84 m s^{-2}
  (ii) 13.7 m s^{-1} which is just over 30 mph.
17 1 m s^{-2}
18 12 s

Discussion points (page 457)

Your own weight acts on you and the tensions in the ropes with which you have contact; the other person's weight acts on them. The tension forces acting at the ends of the rope AB are equal and opposite. The accelerations of A and B are equal because they must always travel the same distance in each interval of time, assuming the rope does not stretch.

Discussion point (page 458)

The tension in the rope joining A and B must be greater than B's weight because there must be a resultant force on B to produce an acceleration.

Discussion points (page 460)

Using $v = u + at$ with $u = 0$ and maximum $a = 6.2$, the speed after 1 second would be 6.2 m s^{-1} or about 14 mph. Under the circumstances, a careful driver is unlikely to accelerate at this rate.

Alvin and his snowmobile and Bernard are two particles each moving in a straight line, otherwise Bernard could swing from side to side; contact between the ice and the rope is smooth, otherwise the tensions acting on Alvin and Bernard are different; the rope is light, otherwise its tension could be affected by its weight; the rope is of constant length, otherwise the accelerations would not be equal; there is no air resistance, otherwise the equations of motion would involve a force to allow for it.

Exercise 20.6 (page 462)

1   The string becomes slack and so the only force acting on the lighter block is its own weight and its acceleration is in the downwards direction. It reaches a maximum height when its speed is zero and then starts moving towards the ground. However, the string will become taut again before the lighter block reaches the ground.

2   $T > W + mg$

3   (i)
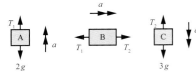

(ii)   A: $T - 0.98 = 0.1a$
B: $1.96 - T = 0.2a$
(iii) $3.27 \text{ m s}^{-2}, 1.31 \text{ N}$
(iv) 1.11 s

4   $2.8 \text{ m s}^{-2}, 14 \text{ N}$

5   (i)
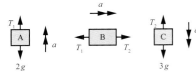

(ii) $R_p = R_L = 490 \text{ N}$,
$T = 4900 \text{ N}$
(iii) $R_p = R_L = 530 \text{ N}$,
$T = 5300 \text{ N}$

6   (i)

(ii)   A: $T_1 - 2g = 2a$
B: $T_2 - T_1 = 5a$
C: $3g - T_2 = 3a$
(iii) $a = 1 \text{ m s}^{-2}, T_1 = 22 \text{ N}$,
$T_2 = 27 \text{ N}$
(iv) 5 N

7
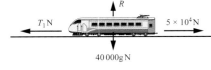

(ii) $T = 750 \text{ N}$
(iii)   Tension $= 44.4 \text{ N}$
(iv) 170 N

8   (i) 0.625 m s^{-2}
(ii) $25\,000 \text{ N}$

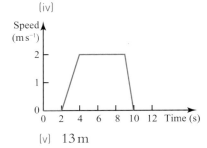

(iii) $12\,500 \text{ N}$
(iv)   Reduced to $10\,000 \text{ N}$

9   (i) 0.25 m s^{-2}
(ii) $T_1 = 9 \text{ kN}, T_2 = 6 \text{ kN}$

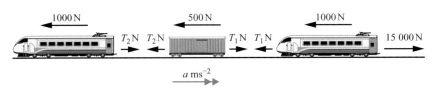

(iii) $0.25 \text{ m s}^{-2}, T_1 = 1.5 \text{ kN}$ in tension, $T_2 = -1.5 \text{ kN}$ in thrust. The second locomotive is now pushing rather than pulling back on the truck.

10   (i)
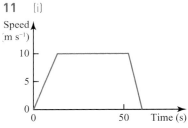

(ii) 1 m s^{-2}
(iii)   Stationary for 2 s, accelerating at 1 m s^{-2} for 2 s, at constant speed for 5 s, decelerating at 2 m s^{-2} for 1 s, stationary for 2 s.

(iv)
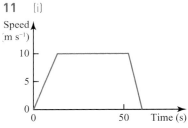

(v) 13 m

11   (i)

(ii) 0.8 m s^{-2}
(iii) $40 \text{ s}^{-1}, -1.33 \text{ m s}^{-2}$
(iv) 66.4 kN
(v) 90 kg

12   (i) $2.26 \text{ m s}^{-2}, 60.3 \text{ N}$
(ii) 0.56 m above ground

13 (i) $2\frac{1}{3}\,\mathrm{m\,s^{-2}}$, $14.9\,\mathrm{N}$

   (ii) $1.23\,\mathrm{kg}$

14 (ii) $T = 5200\,\mathrm{N}$

   (iii) $-0.186\,\mathrm{m\,s^{-2}}$

   (iv) (a) $T = -1660\,\mathrm{N}$

   　　　　 (compression)

   　　(b) $T = 10340\,\mathrm{N}$

   　　　　 (tension)

15 (i) $0.1\,\mathrm{m\,s^{-2}}$

   (ii) $27.5\,\mathrm{kN}$

   (iv) $1.1\,\mathrm{kN}$

16 $\dfrac{(S_1 - S_n)}{(n-1)}\,M$

17 (i) $0.81\,\mathrm{s}$

   (ii) $0.42\,\mathrm{s}$

Chapter 21

Exercise 21.1 (page 474)

1 Velocity is the rate of change of displacement with respect to time and the gradient of a displacement–time graph. Differentiating is finding the rate of change.

2 $6t^2 + 5$

3 (i) (a) $v = 2 - 2t$

   　　(b) $10, 2$

   　　(c) $1, 11$

   (ii) (a) $v = -4 + 2t$

   　　(b) $0, -4$

   　　(c) $2, -4$

   (iii) (a) $v = 3t^2 - 10t$

   　　(b) $4, 0$

   　　(c) $0, 4$ and $3\frac{1}{3}, -14.5$

4 (i) (a) $a = 4$

   　　(b) $3, 4$

   (ii) (a) $a = 12t - 2$

   　　(b) $1, -2$

   (iii) (a) $a = 7$

   　　(b) $-5, 7$

5 $v = 4 + t$　$a = 1$

6 (i) $v = 15 - 10t$　$a = -10$

   (ii)

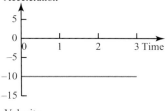

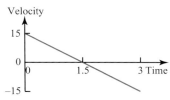

   (iii) The acceleration is the gradient of the velocity-time graph

   (iv) The acceleration is constant; the velocity decreases at a constant rate

7 (i) $v = 18t^2 - 36t - 6$

   　　$a = 36t - 36$

   (ii)

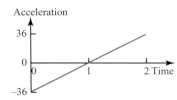

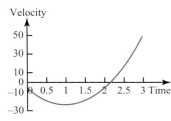

   (iii) The acceleration is the gradient of the velocity-time graph; the velocity is at a minimum when the acceleration is zero.

   (iv) It starts in the negative direction, v is initially -6 and decreases to -24 when $t = 1$, before increasing rapidly to 0, where the object turns round to move in the positive direction.

8 (i) $-16\,\mathrm{m\,s^{-1}}$, $4\,\mathrm{m\,s^{-2}}$

   (ii) $25.41\,\mathrm{m}$

9 $12\,\mathrm{m\,s^{-2}}$ when $t = 1$, $-12\,\mathrm{m\,s^{-2}}$ when $t = 5$

Exercise 21.2 (page 479)

1 $\dfrac{\mathrm{d}r}{\mathrm{d}t}$　$\displaystyle\int a\,\mathrm{d}t$

2 $r = t^3 - 5t + c$

3 (i) $r = 2t^2 + 3t$

   (ii) $r = \frac{3}{2}t^4 - \frac{2}{3}t^3 + t + 1$

   (iii) $r = \frac{7}{3}t^3 - 5t + 2$

4 (i)

   (ii) $85\,\mathrm{m}$

5 (i) When $t = 6$

   (ii) $972\,\mathrm{m}$

6 (i) $4.47\,\mathrm{s}$

   (ii) $119\,\mathrm{m}$

7 (i) $v = 10t + \frac{3}{2}t^2 - \frac{1}{3}t^3$,

   　　$x = 5t^2 + \frac{1}{2}t^3 - \frac{1}{12}t^4$

   (ii) $v = 2 + 2t^2 - \frac{2}{3}t^3$,

   　　$x = 1 + 2t + \frac{2}{3}t^3 - \frac{1}{6}t^4$

   (iii) $v = -12 + 10t - 3t^2$,

   　　$x = 8 - 12t + 5t^2 - t^3$

Discussion points (page 479)

Case (i)

$s = ut + \frac{1}{2}at^2$;

$v = u + at$;

$a = 4, u = 3$.

In the other two cases, the acceleration is not constant.

Activity 21.1 (page 479)

Substituting in $at = v - u$ ② gives

$s = ut + \frac{1}{2}(v - u)t + s_0$

$s = \frac{1}{2}(u + v)t + s_0$ ③

Substituting $v - u = at$ and
$v + u = \frac{2}{t}(s - s_0)$

$\Rightarrow (v - u)(v + u) = at \times \frac{2}{t}(s - s_0)$

$v^2 - u^2 = 2a(s - s_0)$ ④

Substituting $u = v - at$ in ② gives
$s = (v - at)t + \frac{1}{2}at^2 + s_0$

$s = vt - \frac{1}{2}at^2 + s_0$ ⑤

Exercise 21.3 (page 480)

1 Displacement, area under a velocity–time graph, $\int v \, dt$; velocity, gradient of a displacement–time graph, $\frac{dr}{dt}$; acceleration, gradient of a velocity–time graph, $\frac{dv}{dt}$

2 $v = u + at$

3 (i) $v = 15 - 10t$

(ii) $11.5\,\mathrm{m}, 5\,\mathrm{m\,s^{-1}}, 5\,\mathrm{m\,s^{-1}}$;
$11.5\,\mathrm{m}, -5\,\mathrm{m\,s^{-1}}, 5\,\mathrm{m\,s^{-1}}$

(iii)

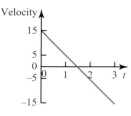

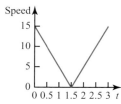

(iv) 3 s

(v) This gives the displacement of the ball, which is 0

4 (i) $-3\,\mathrm{m}, -1\,\mathrm{m\,s^{-1}}, 1\,\mathrm{m\,s^{-1}}$

(ii) (a) 1 s
(b) 2.15 s

(iii)

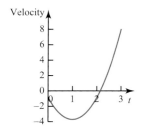

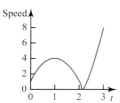

(iv) The object starts at 3 and moves in the negative direction from 3 to -3, which is reached when $t = 2.15$. It then moves in the positive direction with increasing speed.

5 (i) $v = 4 + 4t - t^2$
$x = 4t + 2t^2 - \frac{1}{3}t^3$

(ii)

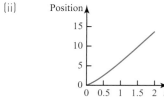

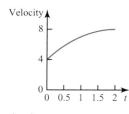

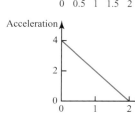

(iii) The object starts at the origin and moves in the positive direction with increasing speed reaching a maximum speed of $8\,\mathrm{m\,s^{-1}}$ after 2 s.

6 (i) 0, 10.5, 18, 22.5, 24

(ii) The ball reaches the hole at 4 s.

(iii) $v = 12 - 3t$

(iv) $0\,\mathrm{m\,s^{-1}}$

(v) $a = -3\,\mathrm{m\,s^{-2}}$

7 (i) $v = 3t^2 - 3, a = 6t$

(ii) $t = 1$

(iii)

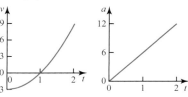

(iv) Object starts at O and moves towards A, slowing down to 0 when $t = 1$, then accelerates back towards B reaching a speed of $9\,\mathrm{m\,s^{-1}}$ when $t = 2$.

(v) 6 m

8 (i) Andrew: $10\,\mathrm{m\,s^{-1}}$, Elizabeth: $9.6\,\mathrm{m\,s^{-1}}$

(ii)

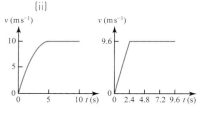

(iii) 11.52 m

(iv) 11.62 s

(v) Elizabeth by 0.05 s and 0.5 m

(vi) Andrew wins

9 (i)

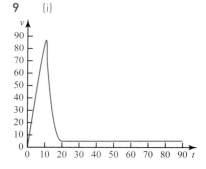

Cara is in free fall until $t = 10\,\text{s}$, then the parachute opens and she slows down to terminal velocity of $5\,\text{m}\,\text{s}^{-1}$.

(ii) $1092\,\text{m}$

(iii) $8.5\,\text{m}\,\text{s}^{-2}$, $1.6t - 32$, $0\,\text{m}\,\text{s}^{-2}$, $16\,\text{m}\,\text{s}^{-2}$

10    $2\,\text{s}$

11   (i) $40\,\text{m}$

  (ii) $r = 0$ when $t = 0$ and 10

  (iii) $25 - 5t$

  (iv) $62.5\,\text{m}$

  (v) $r = 0$ for $t = 0$ and $t = 10$ for both models. In Michelle's model, the velocity starts at $25\,\text{m}\,\text{s}^{-1}$ and decreases to $-25\,\text{m}\,\text{s}^{-1}$ at $t = 10$. The teacher's model is better as the velocity starts and ends at 0.

12   (i)   (a) $112\,\text{cm}$

       (b) $68\,\text{cm}$

  (ii) $4t$, 16

  (iii) $2t^2$, $16t - 32$

  (iv) $\frac{8}{9}\,\text{cm}$ less

13   (i)   PQ: train speeds up with gradually decreasing acceleration.
           QR: train travels at constant speed.
           ST: train slows down with constant deceleration.

  (ii) $a = -0.000025t + 0.05$

  (iii) $50\,\text{m}\,\text{s}^{-1}$

  (iv) $0\,\text{m}\,\text{s}^{-1}$

  (v) $111\frac{2}{3}\,\text{km}$

14   (i) $-24 + 18t - 3t^2$, 2, 4

  (ii) -2, 2

  (iii) 28

15    $4\frac{4}{15}\,\text{m}$, $4\,\text{m}\,\text{s}^{-1}$

16   (i) $12.15\,\text{m}$

  (ii) $13.85\,\text{s}$

  (iii) $26.33\,\text{s}$

17    $s(2) = 2$, 0 and $\frac{2}{3}\,s$, $s(0) = 0$ $s\left(\frac{2}{3}\right) = -\frac{2}{27}$,

   distance = $2\frac{4}{27}\,\text{m}$

18   (i) $5\,\text{m}\,\text{s}^{-1}$, $0.5\,\text{m}\,\text{s}^{-2}$

  (ii) $15\,\text{s}$

  (iii) $41\frac{2}{3}\,\text{m}$

Practice questions (page 487)

1    $s = 2800\,\text{m}$, $u = 0$,
    $v = 70\,\text{m}\,\text{s}^{-1}$, $a = $, $t = ?$
    $s = \frac{1}{2}(u + v)t$
    $2800 = \frac{1}{2}(0 + 70)t$ [1]
    $t = \frac{2800}{35} = 80\,\text{s}$ [1]

2    $x = 5 + 2.1t^2 - 0.07t^3$
    $v = \frac{dx}{dt} = 4.2t - 0.021t^2$ [1]
    When $t = 7$,
    $v = 4.2 \times 7 - 0.21 \times 7^2 =$
    $19.11\,\text{m}\,\text{s}^{-1}$ ($19.1\,\text{m}\,\text{s}^{-1}$ to 3 s.f.)   [1, 1]

3    $2\mathbf{F}_1 - 3\mathbf{F}_2 + \mathbf{F}_3 = 0$ [1]
    $2(7\mathbf{i} - 2\mathbf{j}) - 3(9\mathbf{i} - 3\mathbf{j}) + \mathbf{F}_3 = 0$ [1]
    $-13\mathbf{i} + 5\mathbf{j} + \mathbf{F}_3 = 0$
    $\mathbf{F}_3 = 13\mathbf{i} - 5\mathbf{j}$ [1]

4   (i)   Diagram showing weight for both objects   [1]
           Tension with arrows and labels   [1]
           Normal reaction and friction in correct directions   [1]

  (ii) $T = F$ and $T = 1.25g$ [1]
           $F = 1.25g$ ($= 12.3\,\text{N}$ to 3 s.f.)   [1]

  (iii) Inextensible string   [1]

  (iv)   N2L for block:
           $T - F = 5a$ [1]
           N2L for hanging mass:
           $2g - T = 2a$ [1]
           Solving simultaneous equations
           $2g - 1.25g = 7a$ [1]
           $a = 1.05\,\text{m}\,\text{s}^{-2}$ [1]

5   (i)   Displacement = area under graph   [1]
           $d = \frac{1}{2} \times 7 \times 9 = 31.5\,\text{m}$ [1]

  (ii)   Distance to go = $100 - 31.5 = 68.5\,\text{m}$ [1]

Time at maximum velocity
    $= \frac{68.5}{9} = 7.61\,\text{s}$ [1]
Total time = $7 + 7.61 =$
    $14.61\,\text{s}$ (14.6 to 3 s.f.)   [1]

  (iii) Displacement =
    $\int_0^6 0.9t^2 - 0.1t^3\,dt = \left[0.9\frac{t^3}{3} - 0.1\frac{t^4}{4}\right]_0^6$ [1, 1]
    $= 32.4\,\text{m}$ [1]

  (iv)   EITHER distance to cover = $100 - 32.4$
    $= 67.6\,\text{m}$ [1]
    Time = $\frac{67.6}{10.8} = 6.26\,\text{s}$ [1]

    Total time = $6 + 6.26$
    $= 12.26\,\text{s}$ (12.3 to 3 s.f.)
    which is less than
    $14.61\,\text{s}$ for Sunil,   [1]
    so Mo will win   [1]
    OR distance Mo covers
    by the time Sunil
    finishes   [1]
    $= 32.4 + 10.8 \times$
    $(14.6 - 6) = 125\,\text{m}$
    which is beyond the
    finish,   [1,1]
    so Mo will win   [1]

  (v)   Distance = $100\,\text{m}$ [1]
    $100 = \frac{1}{2}(12.26 +$
    $(12.26 - 7))v$ [1]
    $v = \frac{200}{17.52} = 11.4\,\text{m}\,\text{s}^{-1}$ [1]

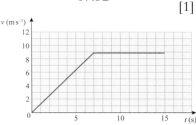

6   (i)   Returns when $s = 0$
           $u = 25$, $v = $,
           $a = -g$, $t = ?$
           $s = ut + \frac{1}{2}at^2$
           $s = 25t - 4.9t^2 = 0$ [1]
           $t(25 - 4.9t) = 0$ [1]
           $t = 0$, $5.1\,\text{s}$ [1]

Assume that air resistance is negligible, assume stone is a particle, assume gravity is constant, assume no horizontal motion [any two]. [1, 1]

(ii) Max. height when
$v = 0$ [1]
$s = h, u = u, v = 0,$
$a = -g, t =,$
$v^2 = u^2 + 2as$
$0 = u^2 + 2gh$ [1]
$h = \dfrac{u^2}{2g}$ [1]

(iii) Height above point of projection $= h -$ distance fallen in t seconds [1]
$s =, u = 0, v =,$
$a = -g, t = t$
height $= h - \left(ut + \frac{1}{2}at^2\right)$
$= h - \left(0 + \frac{1}{2}gt^2\right)$
$= h - \frac{1}{2}gt^2$
$= \dfrac{u^2}{2g} - \frac{1}{2}gt^2$ [1, 1]

(iv) Height of second stone t seconds after launch
$s = ut - \frac{1}{2}gt^2$
Stones cross when equal heights. [1]
$h - \frac{1}{2}gt^2 = ut - \frac{1}{2}gt^2$

$\dfrac{u^2}{2g} = ut$

$t = \dfrac{u}{2g}$ [1]

Height
$s = ut - \frac{1}{2}gt^2$
$= u\dfrac{u}{2g} - \frac{1}{2}g\left(\dfrac{u}{2g}\right)^2$ [1]
$\dfrac{u^2}{2g} - \dfrac{u^2}{8g} = \dfrac{3u^2}{8g}$
$= \frac{3}{4}\left(\dfrac{u^2}{2g}\right) = \frac{3}{4}h$ [1]

7 (i) In cell C3 , we need acceleration during the 200 m journey
$s = 200\,\text{m}, u = 22\,\text{ms}^{-1},$
$v = 18\,\text{ms}^{-1}, a = ?,$
$t =, v^2 = u^2 + 2as$ [1]
$18^2 = 22^2 + 2 \times 200a$
$a = \dfrac{18^2 - 22^2}{400} = -0.4$ [1]

In cell D3, we need magnitude of the resistance force [1]
$= ma = 800 \times 0.4$
$= 320\,\text{N}$ [1]

(ii) $s = 200\,\text{m}, u = 18\,\text{ms}^{-1},$
$v =, a = -0.4\,\text{ms}^{-2},$
$t =, v^2 = u^2 + 2as$ [1]
$v^2 = 18^2 - 2 \times 0.4 \times 200$
$v = \sqrt{164} = 12.8\,\text{ms}^{-1}$ [1]

(iii) $s =, u = 22\,\text{ms}^{-1}, v = 0,$
$a = -0.4\,\text{ms}^{-2}, t =,$
$v = u + at$ [1]
$0 = 22 - 0.4t$
$t = \dfrac{22}{0.4} = 55\,\text{s}$ [1]

(iv) Predicted time (55 s) is much shorter than the actual time (70 s) [1]
so model A is not a suitable model. [1]

(v) Calculation – for example
To come to rest in 74.2 s, constant acceleration model gives
$v = u + at$ [1]
$0 = 22 + 74.2a$
$a = -\dfrac{22}{74.2}$
$= -0.296$ [1]

Resistance $= 800 \times 0.296$
$= 237\,\text{N}$, which is less than the force in the first 200 m, so the force must decrease as the car slows down. [1]

Index

Index